Algebraic-Geometric Codes

Mathematics and Its Applications (*Soviet Series*)

Volume 58

Algebraic-Geometric Codes

by

M. A. Tsfasman
Institute of Information Transmission,
Academy of Sciences,
Moscow, U.S.S.R.

and

S. G. Vlăduţ
Central Economical Mathematical Institute,
Academy of Sciences,
Moscow, U.S.S.R.

SPRINGER-SCIENCE+BUSINESS MEDIA, B.V.

Library of Congress Cataloging-in-Publication Data

Tsfasman, M. A. (Michael A.), 1954-
Algebraic-geometric codes / by M.A. Tsfasman and S.G. Vlăduţ.
p. cm. -- (Mathematics and its applications. Soviet series ; 58)
Translated from Russian.
Includes bibliographical references and index.
ISBN 978-0-7923-0727-3 ISBN 978-94-011-3810-9 (eBook)
DOI 10.1007/978-94-011-3810-9
1. Coding theory. 2. Geometry, Algebraic. I. Vlăduţ, S. G. (Serge G.), 1954- . II. Title. III. Series: Mathematics and its applications (Kluwer Academic Publishers). Soviet series ; 58.
QA268.T75 1991
003.54--dc20 91-12192
CIP

ISBN 978-0-7923-0727-3

Printed on acid-free paper

SERIES EDITOR'S PREFACE

'Et moi, ..., si j'avait su comment en revenir, je n'y serais point allé.'
Jules Verne

The series is divergent; therefore we may be able to do something with it.
O. Heaviside

One service mathematics has rendered the human race. It has put common sense back where it belongs, on the topmost shelf next to the dusty canister labelled 'discarded non-sense'.
Eric T. Bell

Mathematics is a tool for thought. A highly necessary tool in a world where both feedback and non-linearities abound. Similarly, all kinds of parts of mathematics serve as tools for other parts and for other sciences.

Applying a simple rewriting rule to the quote on the right above one finds such statements as: 'One service topology has rendered mathematical physics ...'; 'One service logic has rendered computer science ...'; 'One service category theory has rendered mathematics ...'. All arguably true. And all statements obtainable this way form part of the raison d'être of this series.

This series, *Mathematics and Its Applications*, started in 1977. Now that over one hundred volumes have appeared it seems opportune to reexamine its scope. At the time I wrote

> "Growing specialization and diversification have brought a host of monographs and textbooks on increasingly specialized topics. However, the 'tree' of knowledge of mathematics and related fields does not grow only by putting forth new branches. It also happens, quite often in fact, that branches which were thought to be completely disparate are suddenly seen to be related. Further, the kind and level of sophistication of mathematics applied in various sciences has changed drastically in recent years: measure theory is used (non-trivially) in regional and theoretical economics; algebraic geometry interacts with physics; the Minkowsky lemma, coding theory and the structure of water meet one another in packing and covering theory; quantum fields, crystal defects and mathematical programming profit from homotopy theory; Lie algebras are relevant to filtering; and prediction and electrical engineering can use Stein spaces. And in addition to this there are such new emerging subdisciplines as 'experimental mathematics', 'CFD', 'completely integrable systems', 'chaos, synergetics and large-scale order', which are almost impossible to fit into the existing classification schemes. They draw upon widely different sections of mathematics."

By and large, all this still applies today. It is still true that at first sight mathematics seems rather fragmented and that to find, see, and exploit the deeper underlying interrelations more effort is needed and so are books that can help mathematicians and scientists do so. Accordingly MIA will continue to try to make such books available.

If anything, the description I gave in 1977 is now an understatement. To the examples of interaction areas one should add string theory where Riemann surfaces, algebraic geometry, modular functions, knots, quantum field theory, Kac-Moody algebras, monstrous moonshine (and more) all come together. And to the examples of things which can be usefully applied let me add the topic 'finite geometry'; a combination of words which sounds like it might not even exist, let alone be applicable. And yet it is being applied: to statistics via designs, to radar/sonar detection arrays (via finite projective planes), and to bus connections of VLSI chips (via difference sets). There seems to be no part of (so-called pure) mathematics that is not in immediate danger of being applied. And, accordingly, the applied mathematician needs to be aware of much more. Besides analysis and numerics, the traditional workhorses, he may need all kinds of combinatorics, algebra, probability, and so on.

In addition, the applied scientist needs to cope increasingly with the nonlinear world and the extra mathematical sophistication that this requires. For that is where the rewards are. Linear models are honest and a bit sad and depressing: proportional efforts and results. It is in the non-linear world that infinitesimal inputs may result in macroscopic outputs (or vice versa). To appreciate what I am hinting at: if electronics were linear we would have no fun with transistors and computers; we would have no TV; in fact you would not be reading these lines.

There is also no safety in ignoring such outlandish things as nonstandard analysis, superspace and anticommuting integration, p-adic and ultrametric space. All three have applications in both electrical engineering and physics. Once, complex numbers were equally outlandish, but they frequently proved the shortest path between 'real' results. Similarly, the first two topics named have already provided a number of 'wormhole' paths. There is no telling where all this is leading - fortunately.

Thus the original scope of the series, which for various (sound) reasons now comprises five sub-series: white (Japan), yellow (China), red (USSR), blue (Eastern Europe), and green (everything else), still applies. It has been enlarged a bit to include books treating of the tools from one subdiscipline which are used in others. Thus the series still aims at books dealing with:

- a central concept which plays an important role in several different mathematical and/or scientific specialization areas;
- new applications of the results and ideas from one area of scientific endeavour into another;
- influences which the results, problems and concepts of one field of enquiry have, and have had, on the development of another.

In designing an (error correcting) code one naturally, given certain desiderata such as the number of errors that can be automatically corrected, desires to code as efficiently as possible. For a long time there was a conjectural bound, the Gilbert-Varshanov bound. Then came Goppa with his algebraic-geometry based codes (now usually called Goppa codes) and using these ideas the bound was broken by Th. Zink and the authors of the present volume (1981).

There are now two books on these codes in this series and they complement and supplement each other. The first, by Goppa himself, stresses the intuitive basic ideas. But these codes are based on algebraic geometry, mostly the problem of the number of rational points of curves defined over finite fields. And that is a technically complicated subject. The present book is a complete full treatment of the subject by two authors who have been in the forefront of the research on these codes for the full 10 years of their history.

The shortest path between two truths in the real domain passes through the complex domain.

J. Hadamard

La physique ne nous donne pas seulement l'occasion de résoudre des problèmes ... elle nous fait pressentir la solution.

H. Poincaré

Never lend books, for no one ever returns them; the only books I have in my library are books that other folk have lent me.

Anatole France

The function of an expert is not to be more right than other people, but to be wrong for more sophisticated reasons.

David Butler

Bussum, 5 March 1991

Michiel Hazewinkel

CONTENTS

Preface

This book is devoted to algebraic-geometric codes - the recently emerged area which combines algebraic geometry and coding theory. Its interdisciplinary nature - which makes it so beautiful - also leads to some difficulties for those who want to master it. We try to help all those coming in touch with this field, from students to specialists. Starting from essentially first principles, the text leads the reader to the most recent published and unpublished results. We do not assume any initial knowledge either of coding theory or of algebraic geometry.

Algebraic-geometric codes open new rather exciting possibilities in coding theory and related topics (such as sphere packings). They also lead to a lot of deep and sometimes very difficult questions in algebraic geometry and number theory. We think that the next decade will witness many new interesting results, methods, and problems in this fruitful area. We hope this book to be of some use for those who choose this area for their own.

To make reading more interesting we have included a lot of exercises (with hints to those that do not look easy to us) and many open research problems.

The book is divided into five parts, first two being introductory. The first part contains basic notions of error-correcting codes; there is no algebraic geometry in it, but the style of exposition and some statements are motivated by AG-codes. In comparison with the classical book by MacWilliams and Sloane there are almost no new results. The main methodological difference is that we are always speaking about codes over an arbitrary finite field $\mathbb{F}_q$. The notion of $[n,k,d]_q$-systems makes the exposition more geometric.

The second part contains an introduction to the theory of algebraic curves. The choice of its material is determined by needs of AG-codes. The main topics are the Riemann-Roch theorem and properties of curves over finite fields. Relatively new results here include asymptotic bounds for the number of points on a curve and its Jacobian and the theorem on the structure of the group of rational points of an elliptic curve over a finite field.

In the third part we describe various constructions of AG-codes, their spectra, some examples, decoding algorithm, etc. We also analyze various bounds of algebraic-geometric origin, which demonstrate the possibilities of AG-codes.

The fourth part is devoted to modular curves (both classical and Drinfeld) and to codes obtained from them. Two main properties are proved: asymptotically good behaviour and polynomiality of their construction complexity. This part is the most difficult to read, since it involves some subtle algebraic-geometric techniques and rather cumbersome arguments.

In the fifth part we use AG-codes to construct dense sphere packings in $\mathbb{R}^n$. The most interesting algebraic-geometric and number-theoretic constructions of packings are direct analogues of AG-codes.

The appendix contains equations and tables of some asymptotic bounds, tables of parameters of certain families of codes, and tables of dense sphere packings.

Algebraic-geometric codes (AG-codes) were discovered by V.D.Goppa [Go 1]. This beautiful discovery came as a result of many years of thinking over the possible generalizations of Reed-Solomon codes, BCH-codes and "classical" Goppa codes. In the beginning of 1981 he presented the construction at several seminars. Yu.I.Manin was the first to understand and appreciate the sudden strange link between algebraic geometry and coding theory: he talked about it at his seminar, where the authors were also present. His question about the number of points on curves over a finite field quite soon (Spring 1981) led Th.Zink and the authors [Ts/Vl/Z] to ameliorating asymptotic Gilbert-Varshamov bound for q large enough. This paper started both our interest to coding theory, and those of coding theorists to algebraic-geometric constructions. Our book describes the results of the first ten years of fruitful development of algebraic-geometric codes, started at that point. More detailed information about the history is given in historical and bibliographic notes at the end of each part.

We are deeply grateful to our teacher Yu.I.Manin (who has drawn our attention to this theme), to V.G.Drinfeld (who has explained his theory to us), to G.L.Katsman (who has taught us coding theory), to S.N.Litsyn (who has drawn our attention to sphere packings), to A.M.Barg (who computed most of the numerical tables of asymptotic bounds), to A.N.Skorobogatov (conversations with whom were extremely useful), to M.Yu.Rosenbloom (whose ideas influenced our attitude to analogues of algebraic-geometric codes a lot), to G.Lachaud (for his interest in our work), and to many other mathematicians for their invaluable help. We are also grateful to N.Fliorova (who helped us to translate some part of this book into English), to L.Kadlubowich (who typed most of the English text), and to J.F.Kaashoek (who helped us with the final output). We would like to express our gratitude to the editor of this series M.Hazewinkel, to D.J.Larner, and to the staff of Kluwer Academic Publishers. The first named author wants to thank his parents for their care and his wife for her tender love. The second named author is also deeply grateful to his family.

* *

*

We are pure mathematicians, working mostly in the beautiful domain where algebraic geometry and number theory meet. This determines our viewpoint on coding theory, which may of course differ from an "insiders" point of view, we shall now express this impression of codes (to be precise, of block error-correcting codes only).

A finite-dimensional vector space over a normed field ($\mathbb{C}$, $\mathbb{R}$, $\mathbb{Q}_p$, etc.) has a natural metric. A vector space over $\mathbb{Q}$, or over another algebraic number field, or a free $\mathbb{Z}$-module of finite rank, has a number of metrics corresponding to different possible embeddings into normed fields or rings. The same happens to a global field of a finite characteristic, i.e. to $\mathbb{F}_q(T)$ or to its finite extension.

How is it possible to define a metric space structure on $\mathbb{F}_q^n$, a finite-dimensional space over a finite field? We know nothing more natural than the Hamming metric:

$$d(x,y) = |\{i \mid x_i \neq y_i\}| ,$$

i.e. the distance equals the number of positions where x and y differ; the corresponding norm is

$$\|x\| = |\{i \mid x_i \neq 0\}| \quad .$$

This norm has a significant disadvantage, it depends on the choice of basis in $\mathbb{F}_q^n$, but, alas, we know nothing better.

The space $\mathbb{F}_q^n$ being metricized, we arrive at a problem, which is quite parallel to the classical problem of finding dense sphere packings in $\mathbb{R}^n$; to find an arrangement of spheres of equal diameter d in $\mathbb{F}_q^n$ which has the largest possible density (the notion of volume in $\mathbb{F}_q^n$ is much more natural than the notion of metric; the volume of a subset is just its cardinality). The problem almost equivalent to this is to find (the largest possible) subset $C \subseteq \mathbb{F}_q^n$ such that the distance between each two elements of C is at least d. Each such subset C is called an $[n,k,d]_q$*-code*, where $k = \log_q|C|$. The question of finding the largest possible k for given q, n, and d looks quite natural.

Among all codes, those which are linear subspaces are of particular interest, and we call them *linear*. There are at least three sound reasons to study linear codes. First, the theory of linear codes is parallel to the theory of lattice packings of spheres in $\mathbb{R}^n$. Next, they are easier to construct, and they produce quite a number of nice examples. The third reason which is of particular importance to us deserves some explanation: linear $[n,k,d]_q$-codes correspond to systems $\mathcal{P}$ of n $\mathbb{F}_q$-points in the $(k-1)$-dimensional projective space $\mathbb{P}^{k-1}$ over $\mathbb{F}_q$, and $(n-d)$ equals the maximum number of points in $\mathcal{P}$ lying in a hyperplane (for a more detailed exposition see Section 1.1.2, we call such $\mathcal{P}$ a *projective* $[n,k,d]_q$*-system*). So we arrive at another problem: how to put n points in $\mathbb{P}^{k-1}$ so that there are many of them out of any given hyperplane (a condition of the "general position" type). Note that here we have dispensed with the choice of basis and with the metric which looked a bit artificial.

Having read the last paragraph, every algebraic geometer would ask: what if for $\mathcal{P}$ we take the whole or a part of the set of $\mathbb{F}_q$-points of an algebraic variety W ? For a curve the answer is rather simple. Let V be an algebraic curve defined over $\mathbb{F}_q$ together with a projective embedding $V \hookrightarrow \mathbb{P}^{k-1}$, then n is at most the number $|V(\mathbb{F}_q)|$ of $\mathbb{F}_q$-points of V and $(n - d)$ is just the degree of the curve $V \subset \mathbb{P}^{k-1}$ and the question about the possible values of n, k, d is that of algebraic geometry. The most part of the book is devoted to the properties of such codes.

The point we have just mentioned determines in a way our point of view on codes. Codes over all finite fields $\mathbb{F}_q$ are equally interesting to us (though for applications binary codes, i.e. those over $\mathbb{F}_2$, are most important), and linear codes are of especial interest. The main problem for us is the problem of finding out the possible parameters of a code (and also of actually finding a code with such parameters). This problem can be stated in several possible ways, which we are going to explain. Let $\mathbb{F}_q$ be fixed.

Let first the code length n be fixed. For a given d find the largest $k = K(n,d)$ such that there exists an $[n,k,d]_q$-code (or a linear $[n,k,d]_q$-code). An almost equivalent question is to find the largest $d = D(n,k)$ for a fixed k . Let us remark that if there is given an $[n,k,d]_q$-code it is possible to construct $[n + 1, k, d]_q$- , $[n, k - 1, d]_q$- , and $[n, k, d - 1]_q$- codes, therefore solving these problems we also describe *all* the possible parameters (not only the best). The third question of the same type is to find the least possible $n = N(k,d)$ for fixed k and d.

Since nowadays we are unable to solve these problems, the question is divided into parts (for simplicity we concentrate now on the $K(n,d)$ problem). On the one hand it is useful to look for some conditions satisfied by the parameters of any code. Such conditions bound the possible volume of a code from above $K(n,d) \leq K_{up}(n,d)$. Therefore they are called *upper bounds* or *possibility bounds*. On the other hand it would be good to show some concrete codes with more or less good parameters; for a given n and d each such code gives a *lower bound* or an *existence bound* $K_{low}(n,d) \leq K(n,d)$; of course normally a code appears not alone but as a member of a large class. If by chance for a given pair (n,d) these bound coincide, then the problem is solved.

The principal disadvantage of stating the problem this way is that we obtain not one but infinitely many problems, and right now we are quite far from being able to solve them. A partial answer for small n is given by tables of known $K_{up}(n,d)$ and $K_{low}(n,d)$. These tables help one to compare new methods of constructing codes with the known ones.

Another type of question arises when we are interested in the behaviour of the parameters when n grows. How can k and d behave for a family of codes with $n \longrightarrow \infty$? The correct statement of an asymptotic problem always depends on the actual asymptotic behaviour of parameters. In our case there are at least three natural asymptotic problems (for precise information see Chapter 1.3):

The first one is how k depends on n for a fixed d. Here is how the answer behaves: set

$$\kappa_q(d) = \liminf_{n \longrightarrow \infty} \left(\frac{n - K(n,d)}{\log_q n} \right) ,$$

then $0 < \kappa_q(d) < \infty$ and the question is that of its value. One is able to give lower and upper bounds for $\kappa_q(d)/(d-1)$ which do not depend on d. Roughly speaking, for any d we have

$$\frac{1}{2} \le \kappa_q(d)/(d-1) \le (q-1)/q \quad .$$

The second problem is how d depends on n for a fixed k. Here we have even more precise answer. Set

$$\delta_q(k) = \limsup_{n \longrightarrow \infty} \left(\frac{D(n,k)}{n} \right) \quad ,$$

then

$$\delta_q(k) = \frac{q^{k-1}(q-1)}{q^k - 1} \quad .$$

The third problem looks most interesting, and we call it the *principal asymptotic problem*. Let $n, k, d \longrightarrow \infty$, $k/n \longrightarrow R$, $d/n \longrightarrow \delta$, how do R and δ depend on each other? This problem can be stated precisely, and the dependence is non-trivial. A significant part of this book is devoted to this problem. There exist some other possible statements of asymptotic problems, but they seem somehow more artificial.

This constitutes in our opinion the most important subject of mathematical coding theory, the *parameter problem*.

We approach the second subject using what we have already understood about the first one. We know some upper bounds. Do there exist codes whose parameters lie on the bounds? Usually such codes have some nice property, such as, for example, perfect codes, or equidistant ones. Moreover we know rather a lot about particular classes of codes and we can put questions concerning parameters of codes belonging to a given class (such as, for example, questions about the parameters of cyclic codes, of self-dual codes, of MDS-codes, etc.). It is also interesting to know the properties of such codes, namely their weight spectra (see Section 1.1.3), their automorphism groups, their behaviour under a natural duality, and so on. This subject in general can be called the *property problem*, questions being quite various.

We would be happy if we are not only aware of the best possible parameters but also able to actually construct codes with these parameters more or less explicitly. To construct a code is just to find an algorithm producing its elements (for a linear code, for example it is enough to produce its generator matrix). Of course, there always exists a "silly" universal algorithm searching over all the subset or linear subspaces of $\mathbb{F}_q^n$; we would like to exclude "solutions" like this. This can be done using the computational complexity theory, but only when stating asymptotic problems. We just require the algorithm to be polynomial in n (precise definitions are given in Section 1.3.3). Today we know no other precise statement of the question of constructing codes explicitly, so we hold to it. This constitutes the third subject, the *constructivity problem*, we are also going to study it extensively in this book.

There is also a fourth subject, which we almost do not touch upon in this book, that of codes which are good for practical applications. Let us recall that error-correcting codes can be used to correct errors appearing in the process of data transmission over channels of certain types or data storage. To be used in practice, codes should be: usually binary (or at least 2^m-ary with m not very large), rather but sensibly long, having a fast and simple (not just polynomial) algorithm of construction, and having a simple and fast decoding algorithm. This is the *problem of practical applicability*. This subject is also connected with some interesting algebraic-geometric problems (see Chapter 3.3) and we suppose that algebraic-geometric codes should lead to significant progress also in this direction.

Last but not least is the subject which can be called the *analogue problem*. It is the subject which, in our opinion determines the place of coding theory in the whole edifice of modern mathematics. It turns out that there is a beautiful parallelism between linear codes and lattices in Euclidean spaces which in a way corresponds to one of the deepest parallelism of mathematics, that between algebraic curves and number fields (see Chapter 5.4). The hope of getting to grips with this analogy is, in fact, the core of our own interest in algebraic-geometric codes.

Advices to the reader

We do not assume any knowledge either of algebraic geometry or of error-correcting codes; therefore first two parts of our book are devoted to a brief review of these two topics. These parts are not textbooks in algebraic geometry and coding theory but we hope that the reader, ready to spend some of his time and force over them, will be able to understand the most part of modern literature on the subject with the help of these two parts. Note also that exercises constitute an essential part of our book. Their statements are of the same importance as other statements in our book. If the reader does not want to solve an exercise he must think about its statement as about proposition given without proof. Our advice is nevertheless to solve some part of exercises since it is indispensable for more profound knowledge of the subject. Mostly the exercises are not difficult and can be solved even by a non-experienced reader. The authors hope that they can solve all the exercises. Those exercises that we cannot solve are called problems.

We hope that our book will be useful for several groups of readers.

If you are a specialist in codes interested in a fast introduction to algebraic-geometric codes, we advice you to look through Chapters 1.1 and 1.2 just to get acquainted with our terminology which is not completely standard. Section 1.1.2 deserves more thorough reading. Then it is necessary to read Sections 2.1.1 - 2.1.3, 2.2.1, 2.2.2, 2.3.1 and 2.3.2. After that you can read the third part, containing the theory of AG-codes. You will be able to read Chapter 3.1 (except for the material on self-dual codes in Section 3.1.3), Chapter 3.2 (except for Section 3.2.2), and Sections 3.3.1 and 3.3.2. Before reading about self-dual codes it is necessary to read Section 2.1.4; to understand Section 3.2.2 one should read Sections 2.4.1 and 2.4.2; before reading Section 3.3.3 it is necessary to read Section 2.2.4. This course gives a basic knowledge of AG-codes. After that you can look through the rest of the book to get a general acquaintance with other topics in AG-codes.

If you are interested mostly in asymptotic problems of coding theory or in dense sphere packings you have to look through Chapters 1.1 and 1.2, and to read Chapter 1.3 more thoroughly. Then follow Sections 2.1.1.-2.1.3, 2.2.1, 2.2.2, 2.3.1 and 2.3.2. After reading Section 3.1.1. you can pass to Chapter 3.4 (note that to read Section 3.4.2 you need Sections 2.1.4 and 2.3.3). Then you can read Chapters 5.1 and 5.2. This piece of the book shows asymptotical possibilities of AG-codes and similar objects (note however that these results are in fact based on the results of the fourth part).

If you are an algebraic geometer we advice you to read thoroughly the first part and to look through Sections 2.1.2.-2.1.3, 2.2.1, 2.2.2, 2.3.1 and 2.3.2; after that you are ready to read Part 3. In any case you have to read Section 3.1.1 first and then you can follow your tastes.

If you are neither a specialist in codes nor an algebraic geometer and you wish to get quick acquaintance with the basic notions of AG-codes then you can follow this way: Sections 1.1.1, 1.1.2, 1.2.1, 2.1.1-2.1.3, 2.2.1, 2.2.2, 2.3.1, 2.3.2, 3.1.1, 3.2.1, 3.2.3, 3.2.4, 3.3.1 and 3.3.2.

Our advice is also to use tables and diagrams from Appendix.

To avoid misunderstanding we would like to remark that we use some non-standard terms and notation. Namely:

$A \subset B$ means that the set A is a *proper* subset of B, i.e. $A \neq B$. If the possibility $A = B$ is not excluded we write $A \subseteq B$. The difference of sets is denoted $A - B$;

$a \longmapsto b$ means that $a \in A$ is mapped to $b \in B$ by a map $A \longrightarrow B$;

an $[n,k,d]_q$-code can be non-linear and thus k can be a non-integral real number;

by *group codes* we mean codes which are obtained from ideals in group algebras, see Section 1.2.2;

by *modular codes* we mean codes obtained from modular curves, see Sections 4.1.2 and 4.2.2;

algebraic-geometric codes are denoted $(X,\mathcal{P},D)_L$, $(X,\mathcal{P},D)_\Omega$, etc., while in many other papers they are denoted $C(G,D)$ and $C^*(G,D)$, respectively, see Section 3.1.1;

by $\Omega(D)$ we denote the space

$$\Omega(D) = \{\omega \in \Omega(X)^* \mid (\omega) + D \geq 0\} \cup \{0\} \quad ,$$

which in some other books is denoted $\Omega(-D)$, see Section 2.1.1;

describing families of codes and code constructions we keep the general ideas but some details can be changed (see Sections 1.2.2 and 1.2.3);

$\lceil x \rceil$ is the integer part of $x \in \mathbb{R}$, i.e. $\lceil x \rceil \in \mathbb{Z}$, $\lceil x \rceil \le x < \lceil x \rceil + 1$; $\rceil x \lceil$ means an integer such that $x \le \rceil x \lceil < x + 1$;

ln means log_e ;

$exp(x) = e^x$;

the symbol ■ denotes end of proof (or its absence);

iff means "if and only if".

PART 1

CODES

This part is an introduction to the theory of error-correcting codes. Presenting principal notions and examples we always keep in mind the "algebraic-geometric philosophy", which will come out explicitly in Part 3, but here we do not touch any results using algebraic-geometric codes.

Chapter 1.1 contains principal definitions and properties of codes, and also some motivations. Chapter 1.2 is devoted to some examples, or, to be precise, to diverse interesting classes of codes and various constructions of new codes starting with those already known. In Chapter 1.3 we discuss different asymptotic problems (i.e. the rigorous way to put questions about codes of large length), the algorithmic point of view included. We end this part with some notes on history and bibliography.

CHAPTER 1.1

CODES AND THEIR PARAMETERS

In this chapter we define principal notions of the theory of error-correcting codes (coding theory).

In Section 1.1.1 we define an $[n,k,d]_q$-code (linear and non-linear) and explain why it is called error-correcting. In Section 1.1.2 we give another - more invariant - definition of a linear code, as of a system of points in a linear or a projective space; we think that this approach is natural and fruitful. In Section 1.1.3 we analyze spectra of linear codes and introduce notions of self-dual codes. We end (in Section 1.1.4) with discussing how good can the parameters of a code be, i.e. with some bounds on the parameters.

1.1.1. Definition of a code

Let A be a finite set, which we call an *alphabet*. The set $A^n = A\times\ldots\times A$ is equipped with the *Hamming metric*: the distance $d(a,b)$ is the number of coordinates in which a and b differ:

$$d((a_1,\ldots,a_n),(b_1,\ldots,b_n)) = |\{\ i\ |\ a_i \neq b_i\ \}| \quad .$$

By $q = |A|$ we denote the cardinality of the alphabet.

Any non-empty subset $C \subseteq A^n$ is called a *q-ary code of length* n . The *cardinality* of C is $M = |C| \in \mathbb{N}$, and the *log-cardinality* is $k = \log_q|C| \in \mathbb{R}$. The *minimum distance* is defined as

$$d = \min \{\ d\ (a,\ b)\ |\ a,b \in C\ ,\ a \neq b\ \}$$

A code with these parameters is called an $[n,k,d]_q$*-code*. Elements of C are called *code vectors*, and their components are called *coordinates* or *positions*. Sometimes (especially in asymptotic problems, see Section 1.3) the following relative parameters are quite useful: the *rate* $R = k/n$, and the *relative minimum distance* $\delta = d/n$. Of course $0 \le R \le 1$, $0 \le \delta \le 1$.

Remark 1.1.1. In the coding theory there is a good tradition attaching symbols C , n , k , d , q , R , δ to the defined notions. We try as far as possible to hold to this tradition.

Remark 1.1.2. Let us briefly explain why codes are called error-correcting. The notion appeared in the information theory. Start with a given message which is long enough and written in the alphabet A . When this message is transmitted over a channel it is distorted by random fluctuations (noise). Here is a way out. Take an $[n,k,d]_q$-code C . Let (for simplicity) k be an integer. Cut the message into pieces of length k each. Map every word of length k (there are q^k possibilities) to an element of C , i.e. fix an embedding $E : A^k \xrightarrow{\sim} C \subseteq A^n$, and instead of the piece $a \in A^k$ of the message let us transmit the corresponding word $E(a)$ of length n (we *encode* the message). The transmission is now R times slower, which justifies the term "rate" for R . On the other end of the channel we obtain a distorted word $E(a)' \in A^n$ and we transform it into the nearest word $E(a)'' \in C$ (i.e. we *decode* the message on the *maximum likelihood* basis); this transformation can be defined by some *decoding map* $D : A^n \longrightarrow C$. If the number of distortions is at most $\left\lceil \frac{d-1}{2} \right\rceil$, then $E(a)'' = E(a)$, i.e. the decoding is correct. The maximum likelihood decoding is almost an ideal out of reach. Usually we just give a map $D : U \longrightarrow C$, $U \subseteq A^n$ being the set of words which are at most at the distance $t \le \left\lceil \frac{d-1}{2} \right\rceil$ from C , i.e. U is the union of all balls of radius t centered in the elements of C . In this case we speak about *decoding up to* t (or correcting t errors). Usually $t = \left\lceil \frac{d-1}{2} \right\rceil$, but sometimes it is less. Let $\tau = t/n$. If the error probability per symbol transmitted is $p < \tau$ and the length n is large enough, then the probability of correct decoding is large enough (since the probability that number of errors in n transmitted symbols is at most t , is large enough).

Therefore, a "good" code is an $[n,k,d]_q$-code with large n and R and δ as large as possible.

Both for the construction of good codes and for the design of algorithms realizing coding and decoding procedures, the notion of a code over an arbitrary alphabet is too poor in structure . It is possible to enrich this structure introducing the notion of a linear code.

Now let $A = \mathbb{F}_q$ be a finite field, $q = p^m$. A *linear code* C of length n is a linear subspace $C \subseteq \mathbb{F}_q^n$; for linear codes $k = \dim C$, $d = \min \{\|a\| \mid a \in C , a \neq 0\}$, $\|a\| = |\{ i \mid a_i \neq 0\}|$ being the *weight* of a .

Any choice of basis in C yields an embedding $C : \mathbb{F}_q^k \hookrightarrow \mathbb{F}_q^n$, which we usually denote by the same symbol as the code itself. The matrix of this map, usually denoted by G , is called a *generator matrix* of the code. The map C is included into a short exact sequence

$$0 \longrightarrow \mathbb{F}_q^k \xrightarrow{C} \mathbb{F}_q^n \xrightarrow{H} \mathbb{F}_q^{n-k} \longrightarrow 0$$

(i.e. C is an embedding, H is a surjection, and kernel of H equals the image of C).

The matrix of H (which we also denote by H) is called a *parity-check matrix* of the code C (since the condition $x \in C$ is equivalent to the equality $H \cdot x = 0$).

Exercise 1.1.3. Let H be a parity-check matrix of an $[n,k,d]_q$-code C . Show that any $(d - 1)$ columns of H are linearly independent (as vectors in $\mathbb{F}_q^{n-k}$) and there exist d linearly dependent columns.

According to our definition H has n columns and $(n - k)$ linearly independent rows. Sometimes, by abuse of language, any matrix H' such that $H' \cdot x = 0$ iff $x \in C$, is also called a parity-check matrix. Any such H' has $r \geq n - k$ rows, only $(n - k)$ of which are independent.

Equivalence and automorphisms. Let $\mathcal{A}$ be defined as a subgroup in the group of linear automorphisms of $\mathbb{F}_q^n$ generated by transpositions of coordinates and by multiplications of i-th coordinate by elements of $\mathbb{F}_q^*$. The group $\mathcal{A}$ acts on subsets of $\mathbb{F}_q^n$, i.e. on codes. Two codes C and C' are called *equivalent* iff $C' = A(C)$ for some $A \in \mathcal{A}$. The subgroup $Aut\ C \subseteq \mathcal{A}$, consisting of elements preserving C, is called the *automorphism group* of the code. If is natural to consider codes up to equivalence, in many cases speaking of a code we mean rather its equivalence class. The group $\mathcal{A}$ is represented by monomial matrices (i.e. matrices such that each row and each column contains one non-zero element) and is isomorphic to a semi-direct product of $(\mathbb{F}_q^*)^n$ and S_n ; $|\mathcal{A}| = (q-1)^n \cdot n!$.

A linear $[n,k,d]_q$-code C can be defined by its generator matrix G. The choice of G corresponds to a choice of basis $\{e_1,\ldots,e_k\}$ in the k-dimensional linear space. The group $GL(k,\mathbb{F}_q)$ acts on the set of such bases, and two matrices G and G' define the same code iff $G' = B \cdot G$ for some $B \in GL(k,\mathbb{F}_q)$.

1.1.2. $[n,k,d]_q$-systems

The notion of a linear code can be reformulated in the following nice way.

Let V be a linear space over a finite field $\mathbb{F}_q$. An $[n,k,d]_q$-*system* is an ordered finite family $\mathcal{P}$ of points of V (in general, the points of $\mathcal{P}$ need not be distinct) such that $\mathcal{P}$ does not lie in a hyperplane. The parameters of a system are defined as

$$n = |\mathcal{P}| , \quad k = \dim V , \quad d = n - \max_H |\mathcal{P} \cap H| \geq 1$$

(the maximum being taken over all hyperplanes $H \subset V$).

Two $[n,k,d]_q$-systems $\mathcal{P}$ and $\mathcal{P}'$ in V and V', respectively, are called *equivalent* iff there is and isomorphism $V \simeq V'$ mapping $\mathcal{P}$ isomorphically to $\mathcal{P}'$. A natural object to consider is a class of equivalent systems.

Proposition 1.1.4. *There is a one-to-one correspondence between the set of classes of* $[n,k,d]_q$*-systems and the set of linear* $[n,k,d]_q$*-codes.*

Sketch of proof: Let V^* be the space of linear forms on V, $(V^*)^* = V$. Let $\mathcal{P} = (P_1,\dots,P_n)$. Consider a map $\varphi : V^* \longrightarrow \mathbb{F}_q^n$ defined by $\varphi(Q) = (\varphi_1(Q),\dots,\varphi_n(Q))$, $\varphi_i(Q) = Q(P_i)$. This map φ is injective, let $C = Im\,\varphi$. The other way round, for $C \subseteq \mathbb{F}_q^n$, the coordinate forms define elements of $V = C^*$. ∎

Exercise 1.1.5. Give the complete proof, checking in particular that the parameters do coincide.

Projective systems. Let is introduce a more invariant object. Let $\mathbb{P} = \mathbb{P}(V)$ be a projective space (i.e. the space of lines in a linear space V) over $\mathbb{F}_q$. A *projective* $[n,k,d]_q$*-system* is a finite unordered family $\mathcal{P}$ of points of $\mathbb{P}$ which does not lie in a (projective) hyperplane (note that $|\mathcal{P}| \geq dim\,\mathbb{P} + 1 = dim\,V$). By abuse of notation we write $\mathcal{P} \subset \mathbb{P}$ (though in general there are multiplicities). The parameters $[n,k,d]$ are defined as follows:

$$n = |\mathcal{P}|\ ,\quad k = dim\,\mathbb{P} + 1\ ,\quad d = n - \max_H |\mathcal{P} \cap H| \geq 1\ .$$

Just as for $[n,k,d]_q$-systems, $\mathcal{P} \subset \mathbb{P}$ and $\mathcal{P}' \subset \mathbb{P}'$ are called *equivalent* iff there is a (projective) isomorphism $\mathbb{P} \simeq \mathbb{P}'$ mapping $\mathcal{P}$ onto $\mathcal{P}'$.

We call a linear code $C \subset \mathbb{F}_q^n$ *degenerate* iff $C \subseteq \mathbb{F}_q^{n-1} \subset \mathbb{F}_q$, where $\mathbb{F}_q^{n-1}$ is the subspace of vectors having 0 in some fixed position.

Theorem 1.1.6. *Let $k \geq 1$ and $d \geq 1$. There is a one-to-one correspondence between the set of equivalence classes of non-degenerate linear $[n,k,d]_q$-codes and the set of equivalence classes of projective $[n,k,d]_q$-systems.* ■

Exercise 1.1.7. Prove the theorem.

Projective systems have an advantage of dispensing with a choice of some particular code in its equivalence class; they also look more natural than codes since there is no choice of basis involved. Besides the problem of possible parameters of a projective system looks quite natural, being just a question of how general a position of n points in $\mathbb{P}^{k-1}$ can be.

Remark 1.1.8. Sometimes it is indeed necessary to consider systems with multiplicities (for example when constructing codes whose parameters lie on the Griesmer bound, cf. Theorem 1.1.38).

The language of projective systems is good for the problem of spectra (cf. Section 1.1.3). It is less clear how to put in these terms the notion of duality. The following research problem looks rather important and interesting:

Problem 1.1.9. Rewrite existent books on coding theory in terms of projective systems (you can start with this chapter).

Dual systems. Let us present another approach. A *dual $[n,k,d]_q$-system* is a finite ordered family Q of points of a linear space W (multiplicities are again allowed) which does not lie in a hyperplane. The parameters are defined in the following way: $n = |Q|$, $k = n - \dim W$ (note that $k \geq 0$), d is the minimum number of linearly dependent vectors in Q (in particular, if Q includes multiplicities, then $d \leq 2$).

Exercise 1.1.10. Define a *dual projective* $[n,k,d]_q$-*system* and a proper notion of equivalence. Then prove the following theorem.

Theorem 1.1.11. *There is a one-to-one correspondence between the set of equivalence classes of dual* $[n,k,d]_q$-*systems and the set of linear* $[n,k,d]_q$-*codes, and also between the of classes of dual projective* $[n,k,d]_q$-*systems and the set of equivalence classes of non-degenerate linear* $[n,k,d]_q$-*codes.* ■

*n***-sets.** Projective systems (or dual projective systems) show immediately that $k + d \leq n + 1$. In fact, there is a hyperplane passing through any $(k - 1)$ points in $\mathbb{P}^{k-1}$, i.e. $n - d \geq k - 1$ (an argument for dual systems: no $(n - k - 1)$ points in $\mathbb{P}^{n-k-1}$ can be in general position). A projective $[n, k, n - k + 1]_q$-system is called an *n-set*. The corresponding code C is called a *maximal distance separable* code (an *MDS-code*).

In Sections 1.2.1 and 3.2.2 we give some examples of such codes, namely, Reed-Solomon codes on the projective line; their length $n \leq q + 1$.

We get a very interesting problem: For given k and q find the maximum possible length $m_q(k)$ for which their exists an $[m_q(k), k, m_q(k) + 1 - k]_q$-code.

Let us expose some known results (without proofs):

there always exist $[n,1,n]_q$, $[n,n,1]_q$, and $[n, n-1, 2]_q$-codes (see Section 1.2.1), i.e.

$$m_q(1) = \infty , \quad m_q(k) \geq k + 1 ;$$

$m_q(k) \geq q + 1$;

if q is even then

$$m_q(3) \geq q + 2 , \quad m_q(q - 1) \geq q + 2 ;$$

if $k \geq 2$ then $m_q(k) \leq q + k - 1$;

if $k \geq 3$ and q is odd then $m_q(k) \leq q + k - 2$;

if $k \geq q$ then $m_q(k) = k + 1$.

Problem 1.1.12 (the main conjecture on MDS-codes). Prove that for $2 \leq k < q$

$$m_q(k) = q + 1 ,$$

except for the case $m_q(3) = m_q(q - 1) = q + 2$ for even q.

This conjecture is proved whenever $q \leq 11$, or $k \leq 5$, or $k < (\sqrt{q} + 9)/4$). These results are obtained by the theory of finite geometries. The main conjecture is very beautiful but it seems that we are rather far from its proof.

Let $g \geq 0$. Let $m_q(g,k)$ denote the maximum length for which there exists an

$$[m_q(g,k), k, m_q(g,k) + 1 - k - g]_q\text{-code.}$$

In Chapter 3.2 we construct algebraic-geometric codes that give some lower bounds for $m_q(g,k)$ when g is small. The question about the precise value of $m_q(g,k)$ is most likely very difficult; it would be interesting to find some good bounds.

1.1.3. Spectra and duality

An important invariant of a code is its *weight enumerator* or *spectrum*. We are going to study spectra of linear codes.

Let C be a linear $[n,k,d]_q$-code, define $A_r = A_r(C)$ as the number of code vectors of weight r in C. Of course, $A_r \geq 0$, $A_r = 0$ for $0 < r < d$, and $\sum_{r=0}^{n} A_r = q^k$.

The *enumerator* is a homogeneous polynomial

$$W_C(x:y) = \sum_{r=0}^{n} A_r \cdot x^{n-r} \cdot y^r \quad ;$$

it is easy to see that

$$W_C(x:y) = \sum_{v \in C} x^{n-\|v\|} \cdot y^{\|v\|} \quad .$$

Sometimes non-homogeneous coordinates are more convenient, then we consider polynomials

$$W_C(x) = \sum_{r=0}^{n} A_r \cdot x^{n-r} \quad \text{or} \quad W'_C(y) = \sum_{r=0}^{n} A_r \cdot y^r \quad .$$

A code C has only one vector of weight 0 and has no other vectors of weight less that d. Hence, $A_0 = 1$, $A_1 = \dots = A_{d-1} = 0$, $A_d \neq 0$, i.e.

$$W_C(x:y) = x^n + y^d \cdot \sum_{i=0}^{n-d} A_{d+i} \cdot y^i \cdot x^{n-d-i} \quad .$$

Since in many cases we know not the precise value of d but only some lower estimate for it, the following form is rather convenient. Let a be some integer such that $d \geq n - a$. Then

$$W_C(x) = x^n + \sum_{i=0}^{a} A_{n-i} \cdot x^i .$$

Exercise 1.1.13. Show that if $d \geq n - a$, then

$$W_C(x) = x^n + \sum_{i=0}^{a} B_i \cdot (x - 1)^i ,$$

where

$$B_i = \sum_{j=n-a}^{n-i} \binom{n-j}{i} \cdot A_j \geq 0 , \quad A_i = \sum_{j=n-i}^{a} (-1)^{n+i+j} \cdot \binom{j}{n-i} \cdot B_j .$$

A bit later (Exercise 1.1.27) we explain conceptual meaning of B_i in terms of $[n,k,d]_q$-systems, and in Section 3.1.3 we give an interpretation of B_i for the case of algebraic-geometric codes.

For a linear code C the *dual* code $C^{\perp}$ is defined as

$$C^{\perp} = \{ x \in \mathbb{F}_q^n \mid x \cdot y = 0 \text{ for any } y \in C \}$$

(here $x \cdot y = \sum_{i=1}^{n} x_i \cdot y_i$ is the inner product).

Remark 1.1.14. Sometimes it is more convenient to define duality according to some *hermitian* product $(x,y) = \sum x_i \cdot \bar{y}_i$, where for $a \in \mathbb{F}_q$, $q = p^e$, $\bar{a}$ is defined either as $\bar{a} = a^p$ or (for even e) as $\bar{a} = a^{p^{e/2}}$. The choice of definition does not influence the results we present below. The same is also valid for the *twisted* inner product $(x,y) = \sum x_i \cdot y_i \cdot \alpha_i$, where $\alpha_i \in \mathbb{F}_q^*$.

The parameters of the dual code are $[n, n-k, d^{\perp}]_q$; a generator matrix of the code is a parity-check matrix of its dual and vice versa; the dual distance $d^{\perp}$ depends (not in an easy way) on the equivalence class of the code C (and not only on its parameters $[n,k,d]_q$). It can be calculated if we know the enumerator W_C .

Moreover, there is a beautiful relation between the spectrum of a code and that of the dual one:

Theorem 1.1.15 (the MacWilliams identity).

$$W_{C^{\perp}}(x:y) = q^{-k}\cdot W_C(x + (q-1)\cdot y : x - y) \quad .$$

We prove this theorem a bit later, right now let us introduce one notion more. Let g be a non-negative integer.

A code C is called a *code of genus at most* g iff the following relations hold:

$$k + d \geq n + 1 - g \quad ,$$

$$(n-k) + d^{\perp} \geq n + 1 - g$$

(we consider only non-negative values of g since Exercise 1.1.3 shows that for any linear code $k + d \leq n + 1$).

Of course, in this case the dual code $C^{\perp}$ is also a code of genus at most g .

Theorem 1.1.16. *Let* C *be an* $[n,k,d]_q$*-code of genus at most* g *. Let* $a = k + g - 1$ *. Then in the representation*

$$W_C(x) = x^n + \sum_{i=0}^{a} B_i\cdot(x-1)^i$$

for $a - 2g + 2 \le i \le a$ *the coefficients* B_i *satisfy inequalities*

$$\binom{n}{i}\cdot(q^{a-i+1} - 1) \ge B_i \ge \max\left\{ 0,\ \binom{n}{i}\cdot(q^{a-i-g+1} - 1) \right\},$$

and for $i \le a - 2g + 1$

$$B_i = \binom{n}{i}\cdot(q^{a-i-g+1} - 1) \quad .$$

Proof: Theorem 1.1.15 gives

$$W_{C^\perp}(x:y) = q^{-k}\cdot W_C(x + (q - 1)\cdot y : x - y) =$$

$$= q^{-k}\cdot\left((x + (q - 1)\cdot y)^n + \sum_{i=0}^{a} B_i\cdot(q\cdot y)^i\cdot(x - y)^{n-i}\right) \quad .$$

Passing to non-homogeneous coordinates and setting $a^\perp = n - k + g - 1$ and $W_{C^\perp}(x) = x^n + \sum_{i=0}^{a^\perp} B'_i\cdot(x - 1)^i$ we get

$$x^n + \sum_{i=0}^{a^\perp} B'_i\cdot(x - 1)^i =$$

$$= q^{-k}\cdot\left((x + q - 1)^n + \sum_{i=0}^{a} B_i\cdot q^i\cdot(x - 1)^{n-i}\right) \quad .$$

Expanding in powers of $z = x - 1$ we get

$$\sum_{i=0}^{a^\perp} \left(B'_i + \binom{n}{i}\right)\cdot z^i + \sum_{i=a^\perp+1}^{n} \binom{n}{i}\cdot z^i =$$

$$= \sum_{i=0}^{n-a-1} \binom{n}{i}\cdot q^{n-k-i}\cdot z^i + \sum_{i=n-a}^{n}\left(B_{n-i} + \binom{n}{i}\right)\cdot q^{n-k-i}\cdot z^i \quad .$$

Hence for $n - a \le i \le n$ we have

$$B_{n-i} = \begin{cases} \binom{n}{i} \cdot (q^{-n+k+i} - 1) & \text{for } i \ge a^{\perp} + 1 \\ \binom{n}{i} \cdot (q^{-n+k+i} - 1) + B'_i \cdot q^{-n+k+i} & \text{for } i \le a^{\perp} \end{cases}$$

i.e. (substitute $j = n - i$ using $k = a - g + 1$ and $a^{\perp} = n - k + g - 1 = n - a + 2g - 2$):

$$B_j = \begin{cases} \binom{n}{j} \cdot (q^{a-j-g+1} - 1) & \text{for } j \le a-2g+1 \\ \binom{n}{j} \cdot (q^{a-j-g+1} - 1) + B'_{n-j} \cdot q^{a-j-g+1} & \text{for } a-2g+2 \le j \le a \ . \end{cases}$$

The lower bound and the equality are proved. The upper bound will be proved below, after we interpret the values B_i in terms of $[n,k,d]_q$-systems (see Theorem 1.1.26 and Exercise 1.1.27). ■

Remark 1.1.17. It is easy to see that $B_k = B'_{n-k}$.

Exercise 1.1.18. Show that if $k + d \ge n + 1$ then $k + d = n + 1$ and $(n - k) + d^{\perp} = n + 1$, t.e. that in this case the code is of genus 0 (an MDS-code).

Example 1.1.19. Consider $[4,2,1]_2$-code C generated by the matrix

$$\begin{pmatrix} 1 & 0 & 0 & 0 \\ 0 & 1 & 0 & 0 \end{pmatrix} .$$

The dual code has the same parameters. The spectrum is

$A_0 = 1$, $A_1 = 2$, $A_2 = 1$, $A_3 = A_4 = 0$. It is a code of genus at most 2, $a = 3$; the calculation of B_i gives $B_3 = A_1 = 2$, $B_2 = 3A_1 + A_2 = 7$, $B_1 = 3A_1 + 2A_2 + A_3 = 8$, $B_0 = A_1 + A_2 + A_3 + A_4 = 3$. Let us point out that in this case $B_i > \binom{n}{i} \cdot (q^{\lceil (a-i)/2 \rceil + 1} - 1)$ for $i = 2$.

Example 1.1.20. Consider an $[n, k_0, d_0]_q$-code C_0 of genus 0. Let H_0 be its parity-check matrix. Add to H_0 one row of weight w which is not a linear combination of rows of H_0. We get a matrix H such that each $(d_0 - 1)$ columns of it are linearly independent (that is so already for H_0). The parameters of the corresponding $[n,k,d]_q$-code C are $k = k_0 - 1$ and $d \geq d_0$, i.e. $k + d \geq n$. If $w < n - d$ then $d^\perp \leq w < n - d$ (since the additional row is a code vector of $C^\perp$), therefore $d^\perp + k^\perp = d^\perp + n - k < < 2n - d - k \leq n$. If $k + d$ were more than n, then we would have got $k + d = k^\perp + d^\perp = n + 1$. Hence we have constructed a code C with $k + d = n$ which is not of genus at most 1 (since $d^\perp + k^\perp < n$).

Exercise 1.1.21. Consider a $[6,4,2]_2$-code C generated by the matrix

$$\begin{pmatrix} 1 & 1 & 0 & 0 & 0 & 0 \\ 0 & 1 & 1 & 0 & 0 & 0 \\ 0 & 0 & 0 & 1 & 1 & 0 \\ 0 & 0 & 0 & 0 & 1 & 1 \end{pmatrix} .$$

Calculate the spectra of C and $C^\perp$ in terms of B_i. (*Answer*: $B_0 = 15$, $B_1 = 42$, $B_2 = 45$, $B_3 = 24$, $B_4 = 6$, $B'_0 = 3$, $B'_1 = B'_2 = 6$, $B'_3 = 2$).

Before we pass to the proof of Theorem 1.1.15, let us recall that an additive character of $\mathbb{F}_q$ is defined as a homomorphism χ from the additive group of $\mathbb{F}_q$ to $\mathbb{C}^*$. It is easy to see that its image lies in the group μ_p of p-roots of 1.

Exercise 1.1.22. Prove that the character group of $\mathbb{F}_q$ is isomorphic to $\mathbb{F}_q$, this isomorphism mapping the trivial character $\chi_0(a) \equiv 1$ to $0 \in \mathbb{F}_q$.

Lemma 1.1.23. *For any non-trivial character* $\chi : \mathbb{F}_q \longrightarrow \mu_p$

$$\sum_{a\in\mathbb{F}_q} \chi(a\cdot b) = \begin{cases} 0 & \text{for } b \neq 0 \\ q & \text{for } b = 0 \end{cases} .$$

Proof: Let $b \neq 0$, choose a_0 such that $\chi(a_0\cdot b) \neq 1$. Then

$$\chi(a_0\cdot b)\cdot\sum_{a\in\mathbb{F}_q} \chi(a\cdot b) = \sum_{a\in\mathbb{F}_q} \chi((a_0+a)\cdot b) = \sum_{a'\in\mathbb{F}_q} \chi(a'\cdot b) ,$$

since the shift by a_0 maps $\mathbb{F}_q$ onto itself bijectively. If $b = 0$ then always $\chi(a\cdot b) = 1$. ∎

Fix a non-trivial character χ_1. For $u,v \in \mathbb{F}_q^n$ let $u\cdot v \in \mathbb{F}_q$ be their inner product. Set

$$\chi_u(v) = \chi_v(u) = \chi_1(u\cdot v) ;$$

then

$$\chi_u : \mathbb{F}_q^n \longrightarrow \mu_p$$

is an additive character of $\mathbb{F}_q^n$.

Let A be an arbitrary $\mathbb{Z}[\mu_p]$-module. For a function $f : \mathbb{F}_q^n \longrightarrow A$ define the transform

$$\hat{f}(u) = \sum_{v\in\mathbb{F}_q^n} \chi_u(v)\cdot f(v)$$

Lemma 1.1.24. *For any linear subspace* $C \subseteq \mathbb{F}_q^n$

$$\sum_{n\in C^{\perp}} f(u) = \frac{1}{|C|}\cdot \sum_{n\in C} \hat{f}(u) \quad .$$

Proof:

$$\sum_{u\in C} \hat{f}(u) = \sum_{v\in\mathbb{F}_q^n} \sum_{u\in C} \chi_u(v)\cdot f(v) = \tag{1.1.1}$$

$$= \sum_{v\in C^{\perp}} f(v) \sum_{u\in C} \chi_v(u) + \sum_{v\notin C^{\perp}} f(v) \sum_{u\in C} \chi_v(u) \quad .$$

If $v \in C^{\perp}$ then $u\cdot v = 0$ for all $u \in C$, thus

$$\sum_{u\in C} \chi_v(u) = \sum_{u\in C} \chi_1(0) = |C| \; .$$

Let v be such that there exists $u_0 \in C$, $\chi_v(u_0) \neq 1$. Then

$$\chi_v(u_0)\cdot\sum_{u\in C} \chi_v(u) = \sum_{u\in C} \chi_v(u_0 + u) = \sum_{u\in C} \chi_v(u) \quad ,$$

since $u_0 + C = C$; hence $\sum_{u\in C} \chi_v(u) = 0$.

It is left to prove that if $\chi_v(u) = 1$ for any $u \in C$ then $v \in C^{\perp}$. In fact, in this case $\chi_v(\alpha\cdot u) = \chi_1(\alpha\cdot(u\cdot v)) = 1$ for any $\alpha \in \mathbb{F}_q$, and, since there is $\beta \in \mathbb{F}_q^*$ such that $\chi_1(\beta) \neq 1$, setting $\alpha = \beta\cdot(u\cdot v)^{-1}$ we get a contradiction, i.e. $u\cdot v = 0$. The vanishing of the second term in (1.1.1) proves the lemma. ■

In the course of the proof we have also established the following fact:

Corollary 1.1.25. *For any linear subspace* $C \subseteq \mathbb{F}_q^n$

$$\sum_{u\in C} \chi_v(u) = \begin{cases} 0 & \text{for } v \notin C^{\perp} \\ |C| & \text{for } v \in C^{\perp} \end{cases}$$

■

Proof of Theorem 1.1.15: Let $A = \mathbb{C}[x,y]$,

$$f : \mathbb{F}_q^n \longrightarrow A, \quad f(u) = x^{n-\|u\|} \cdot y^{\|u\|}.$$

The left hand side of the identity of Lemma 1.1.24 equals $W_{C^\perp}(x:y)$. Let us calculate $\hat{f}(u)$. Let $u = (u_1,\ldots,u_n)$, $v = (v_1,\ldots,v_n)$; using the fact that $v_i^{q-1} = 1$ for $v_i \neq 0$ and Lemma 1.1.24 we get

$$\hat{f}(u) = \sum_{v\in\mathbb{F}_q^n} \chi_u(v) \cdot x^{n-\|v\|} \cdot y^{\|v\|} =$$

$$= \sum_{v\in\mathbb{F}_q^n} \chi_1(u_1v_1 + \ldots + u_nv_n) \cdot x^{n-\|v\|} \cdot y^{\|v\|} =$$

$$= \sum_{v\in\mathbb{F}_q^n} \prod_{i=1}^{n} \chi_1(u_iv_i) \cdot x^{1-v_i^{q-1}} \cdot y^{v_i^{q-1}} =$$

$$= \sum_{v_1\in\mathbb{F}_q} \chi_1(u_1v_1) \cdot x^{1-v_1^{q-1}} \cdot y^{v_1^{q-1}} \sum_{v_2\in\mathbb{F}_q} \ldots =$$

$$= \prod_{i=1}^{n} \sum_{v_i\in\mathbb{F}_q} \chi_1(u_iv_i) \cdot x^{1-v_i^{q-1}} \cdot y^{v_i^{q-1}} = \prod_{i=1}^{n} (x + y \cdot \sum_{v_i\in\mathbb{F}_q^*} \chi_1(u_iv_i)) =$$

$$= \prod_{i=1}^{n} (x - y + y \cdot \sum_{\alpha\in\mathbb{F}_q} \chi_1(u_i \cdot \alpha)) =$$

$$= \prod_{i=1}^{n} \left((x - y)^{u_i^{q-1}} \cdot (x + (q - 1) \cdot y)^{1-u_i^{q-1}} \right) =$$

$$= (x + (q - 1) \cdot y)^{n-\|u\|} \cdot (x - y)^{\|u\|},$$

and the theorem is proved. ■

Here is a strengthening of Theorem 1.1.16 (for an arbitrary linear code).

Theorem 1.1.26. *Let* C *be a linear* $[n,k,\geq d]_q$*-code, and let the minimum distance of the dual code be at least* $d^{\perp}$ *. Then*

$$W_C(x) = x^n + \sum_{i=0}^{n-d} B_i \cdot (x-1)^i \quad ,$$

where for $d^{\perp} - 1 \geq i \geq 0$

$$B_i = \binom{n}{i} \cdot (q^{k-i} - 1) \quad ,$$

and for $n - d \geq i \geq d^{\perp}$

$$\binom{n}{i} \cdot (q^{\min\{n-d-i+1,k-d^{\perp}+1\}} - 1) \geq B_i \geq \max \left\{0, \binom{n}{i} \cdot (q^{k-i}-1)\right\}.$$

■

Exercise 1.1.27. Check the following interpretation of B_i in terms of $[n,k,d]_q$-systems. Let $\mathcal{P} = \{P_1,\ldots,P_n\}$ be an $[n,k,d]_q$-system, $P_i \in V$. By H_i denote the hyperplane in V^* corresponding to P_i . For $\mathcal{R} \subseteq \mathcal{P}$ let $\ell(\mathcal{R}) = \dim \left(\bigcap_{P_i \in \mathcal{R}} H_i \right)$. Then

$$B_i = \sum_{\mathcal{R} \subset \mathcal{P}, |\mathcal{R}|=i} (q^{\ell(\mathcal{R})} - 1) \quad .$$

Using this interpretation prove Theorem 1.1.26.

Self-dual codes. A linear code C is called *self-dual* iff $C = C^{\perp}$. A code C is called *quasi-self-dual* iff there exists a vector $y = (y_1,\ldots,y_n) \in \mathbb{F}_q^n$, $y_i \neq 0$ for all

$i = 1,\ldots,n$, such that $y \cdot C = C^{\perp}$. Here

$$y \cdot C = \{ \, y \cdot c = (y_1 \cdot c_1, \ldots, y_n \cdot c_n) \mid c \in C \, \} \, .$$

A code C is called *formally self-dual* if $W_C = W_{C^{\perp}}$. Of course, any self-dual code is quasi-self-dual, and any quasi-self-dual code is formally self-dual. If $q = 2$ then any quasi-self-dual code is self-dual. The following fact is quite easy to prove:

Exercise 1.1.28. Let $q = 2$ or $q = 3$. Show that the weights of all code vectors of a self-dual q-ary code are divisible by q .

Note that for formally self-dual codes this statement is wrong: a $[2,1,1]_2$-code spanned by the vector $(1,0) \in \mathbb{F}_2^2$ is formally self-dual.

If an $[n,k,d]_q$-code C is (at least formally) self-dual then $n = 2k$ (since $q^k = W_C(1:1)$ is uniquely determined by the enumerator).

Everywhere below, whenever we consider some question concerning self-duality, we restrict ourselves from the very beginning to $[n,k,d]_q$-codes with $n = 2k$.

The most beautiful theorem on formally self-dual codes is proved with the help of classical invariant theory. Unfortunately we have space only for the statement:

Theorem 1.1.29. *For any formally self-dual code* C *there exists a homogeneous polynomial* $P(x:y)$ *such that*

$$W_C(x:y) = P(x^2 + (q-1) \cdot y^2 : y \cdot (x - y)) \quad . \blacksquare$$

Remark 1.1.30. Of course, not every polynomial of this form is in fact an enumerator of some linear code. For example the polynomial $y\cdot(x - y)$ cannot be an enumerator, since it lacks the term x^n corresponding to the zero code vector. Another example of the same kind is $2x^2 + 2(q - 1)\cdot y^2$ .For $q = 2$ the enumerator of any self-dual code must be an even function in x (see Exercise 1.1.28). A question which polynomials in $x^2 + (q - 1)\cdot y^2$ and $y\cdot(x - y)$ are actually enumerators of self-dual codes (or of any codes) is very subtle.

Let us present some other results (without proofs).

Theorem 1.1.31. *The enumerator of a binary self-dual code is a polynomial in* $x^2 + y^2$ *and* $x^2\cdot y\cdot(x^2 - y^2)^2$ *(the same is valid for a formally self-dual code such that all its weights are even). The enumerator of a* 3*-ary self-dual code is a polynomial in* $x^4 + 8\cdot x\cdot y^3$ *and* $y^3\cdot(x^3 - y^3)^3$ *(the same is valid for a formally self-dual code such that all its weights are divisible by* 3*). The enumerator of a* 4*-ary formally self-dual code such that all its weights are even is a polynomial in* $x^2 + 3\cdot y^2$ *and* $x^2\cdot(x^2 - y^2)^2$ *. The enumerator of a binary formally self-dual code such that all its weights are divisible by* 4 *is a polynomial in* $x^8 + 14\cdot x^4\cdot y^4 + y^8$ *and* $x^4\cdot y^4\cdot(x^4 - y^4)^4$ *.* ■

Theorem 1.1.32. *Suppose that all the weights of a formally self-dual* q*-ary code* C *are divisible by an integer* $t > 1$ *. Then either* C *is an* $[n,n/2,2]_q$*-code and* $W_C = ((q - 1)\cdot x^2 + y^2)^{n/2}$ *, or* $(q,t) = (2,2), (2,4), (3,3),$ *or* $(4,2)$*.* ■

Exercise 1.1.33. Check that if C is quasi-self-dual with respect to some $y \in (\mathbb{F}_q^*)^n$ and for any i there exists $x_i \in \mathbb{F}_q^*$ such that $y_i = x_i^2$, then there exists some $x \in (\mathbb{F}_q^*)^n$ such that $x\cdot C$ is self-dual. Check that any element of a finite field of characteristic 2 is a square, and hence if $q = 2^m$ then for any quasi-self-dual code C there exists a self-dual code equivalent to C .

1.1.4. Bounds

We have already explained that a good code should have large k and d for a given n . Let q be fixed. For which n, k, d does there exist a linear $[n,k,d]_q$-code (or just some $[n,k,d]_q$-code)? Of course, $0 \le k \le n$, $1 \le d \le n$.

We start with a rather strange but quite useful statement that having a good code we can get a lot of worse ones.

Lemma 1.1.34 (the spoiling lemma). *Suppose that there exists a non-degenerate* $[n,k,d]_q$*-code* C . *Then we can construct a non-degenerate linear code with parameters*

a) $[n + 1 , k , d]_q$;

and if $k \ge 1$ *and* $n > d \ge 2$ *then also linear codes with parameters*

b) $[n - 1 , k - 1 , d]_q$;

c) $[n - 1 , k , d - 1]_q$;

d) $[n , k - 1 , d]_q$;

e) $[n , k , d - 1]_q$.

Proof: Let $\mathcal{P} \subset \mathbb{P}^{k-1}$ and $\mathcal{Q} \subset \mathbb{P}^{n-k-1}$ be projective and dual projective systems corresponding to C (see Section 1.1.2).

a) Choose a hyperplane $H_0 \subset \mathbb{P}^{k-1}$ such that $|H_0 \cap \mathcal{P}| = \max_H |H_0 \cap \mathcal{P}|$. Add to $\mathcal{P}$ one more point from H_0 (it does not matter whether it already belongs to $\mathcal{P}$ or not).

b) Let $Q_0 \subset Q$ be a linearly dependent set of d vectors. Exclude from Q any vector which does not belong to Q_0 (it is possible since $d < n$). Then n is lessened by 1, and d and $(n - k)$ do not change.

c) Chose H_0 as in a) and exclude from $\mathcal{P}$ any point which does not belong to H_0 . Since $d \geq 2$ we get a system again (the remaining points cannot all lie in a hyperplane); k and $(n - d)$ do not change.

d) Use the operations of b) and a) .

e) Use the operations of c) and a) . ■

Exercise 1.1.35. Prove the spoiling lemma for degenerate linear codes. State and prove the spoiling lemma for non-linear codes. (*Hint*: In this case $(k - 1)$ is changed by $\log_q(\rceil q^{k-1} \lceil)$).

In many cases it is difficult to calculate precise values of parameters but it is possible to bound them. The spoiling lemma makes it possible to pass from an $[\leq n, \geq k, \geq d]_q$-code C to an $[n,k,d]_q$-code. In such a situation we say that *up to a spoiling* the code C is an $[n,k,d]_q$-code.

So we can always spoil parameters, but of course we cannot always make them better. Here are some restrictions.

Proposition 1.1.36 (the Singleton bound). *For any linear* $[n,k,d]_q$*-code*

$$n \geq k + d - 1 \quad .$$

Proof: Let us argue in terms of $[n,k,d]_q$-systems. Each set of $(k-1)$ vectors in a k-dimensional linear V lies in a hyperplane, hence $n - d = max\,|\mathcal{P} \cap H| \geq k - 1$. ■

Exercise 1.1.37. Prove that the Singleton bound is also valid for non-linear codes. (*Hint*: Different shifts of C by vectors of the form $(v_1,\ldots,v_{d-1},0,\ldots,0)$ do not intersect each other).

In the next statement linearity is essential.

Theorem 1.1.38 (the Griesmer bound). *For any linear* $[n,k,d]_q$*-code*

$$n \geq \sum_{i=0}^{k-1} \left\rceil \frac{d}{q^i} \right\lceil .$$

Proof: Consider the corresponding projective system $\mathcal{P} \subset \mathbb{P}^{k-1}$. Let

$$|\mathcal{P} \cap H_0| = \max_H |\mathcal{P} \cap H| = n - d .$$

Set $\mathbb{P}' = H_0$, $\mathcal{P}' = \mathcal{P} \cap H_0 \subset \mathbb{P}'$. This is a projective $[n',k',d']_q$-system, where $n' = n - d$, $k' = k - 1$. Let H' be a hyperplane (of dimension $k-3$) in $\mathbb{P}'$ such that $|\mathcal{P}' \cap H'| = n' - d'$. There are $(q+1)$ hyperplanes H_i of $\mathbb{P}^{k-1}$ passing through H', and $|H_i \cap \mathcal{P}| \leq n - d$. Therefore

$$(q+1)\cdot(n-d) \geq \sum_{i=1}^{q+1} |H_i \cap \mathcal{P}| = |V \cap \mathcal{P}| + q\cdot|H' \cap \mathcal{P}| =$$

$$= n + q\cdot(n - d - d') ,$$

since $\bigcup H_i = V$, $\bigcap H_i = H'$. We get $d' \geq \left]\frac{d}{q}\right[$. Iterating this operation k times we get an $[n^{(k)},0,d^{(k)}]_q$-system with $n^{(k)} = n - d - d' - \ldots \leq n - \sum_{i=0}^{k-1} \left]\frac{d}{q^i}\right[$. The condition $n^{(k)} \geq 0$ proves the theorem. ∎

The following bounds we prove for any codes.

Theorem 1.1.39 (the Plotkin bound). *For any* $[n,k,d]_q$-*code* C

$$d \leq \frac{n \cdot q^k \cdot (q - 1)}{(q^k - 1) \cdot q} .$$

Proof: The minimum distance d cannot exceed the average pairwise distance between the elements of C :

$$d \leq \frac{1}{q^k (q^k - 1)} \cdot \sum_{x,y \in C} d(x,y) .$$

Set $X_{a,i} = |\{x \in C \mid x_i = a\}|$; $\sum_{a \in \mathbb{F}_q} X_{a,i} = q^k$ for any i . Let $\delta_{a,b}$ be the Kronecker symbol.

$$\sum_{x,y \in C} d(x,y) = \sum_{i=1}^{n} \sum_{x,y \in C} (1 - \delta_{x_i,y_i}) =$$

$$= \sum_{i=1}^{n} \sum_{a,b \in \mathbb{F}_q} (1 - \delta_{a,b}) \cdot X_{a,i} \cdot X_{b,i} \leq n \cdot \max_Z Q(Z) ,$$

where Q is a quadratic from whose matrix is $(1 - \delta_{a,b})$, $Z = (Z_a)_{a \in \mathbb{F}_q}$, $Z_a \in \mathbb{R}$, and the maximum is taken over Z subject to the conditions $Z_a \geq 0$, $\sum_{a \in \mathbb{F}_q} Z_a = q^k$. Now the theorem is implied by the following fact:

Exercise 1.1.40. Prove that this maximum equals $q^{2k-1}\cdot(q-1)$ and is reached when all Z_a are equal. ■

Theorem 1.1.41 (the sphere-packing bound, or the Hamming bound). *For any* $[n,k,d]_q$*-code* C

$$n - k \geq \log_q \sum_{i=0}^{\left\lceil \frac{d-1}{2} \right\rceil} \binom{n}{i}\cdot(q-1)^i .$$

Proof: Consider spheres in $\mathbb{F}_q^n$ of radius $t = \left\lceil \frac{d-1}{2} \right\rceil$ centered at the code vectors (by definition a sphere of radius t centered at a is

$$B_i(a) = \{x \in \mathbb{F}_q^n \mid \|x - a\| \leq t\}).$$

These spheres do not intersect, hence the product of $|C| = q^k$ and the volume (the number of elements) of such a sphere is at most $|\mathbb{F}_q^n| = q^n$. Now the theorem is implied by the following statement.

Lemma 1.1.42. *The volume of a sphere of radius* t *in* $\mathbb{F}_q^n$ *equals*

$$|B_t(a)| = \sum_{i=0}^{t} \binom{n}{i}\cdot(q-1)^i . \blacksquare$$

Exercise 1.1.43. Prove the lemma.

The following theorem is obtained by combining the averaging procedure with the sphere-packing argument.

Theorem 1.1.44 (the Bassalygo-Elias bound). *For any* $[n,k,d]_q$*-code and for any integer* w *such that* $1 \le w \le n$ *and*

$$A = d - 2\cdot w + \frac{q\cdot w^2}{(q - 1)\cdot n} > 0$$

the following inequality holds:

$$n - k \ge \log_q \binom{n}{w} + w\cdot \log_q (q - 1) - \log_q d + \log_q A \quad .$$

Sketch of proof: The idea is us follows: first we establish some relation between the possible cardinality of a code in $\mathbb{F}_q^n$ and that of a spherical code on $S^n(w) = \{x \in \mathbb{F}_q^n \mid \|x\| = w\}$. Then we use the Plotkin method to estimate the cardinality of the spherical code. Let us start from the following mean inequality:

Lemma 1.1.45. *Let* $A(n,d) = q^{k_0}$, $k_0 = \max\{k\}$, *the maximum being taken over all* $[n,k,d]_q$*-codes* C *with given* n *and* d; *let* $A(n,w,d)$ *be the maximum possible number of vectors of weight* w *in* $\mathbb{F}_q^n$ *such that the distance between any two of them is at least* d. *Then for any* w

$$\frac{A(n,d,w)}{\binom{n}{w}\cdot (q - 1)^w} \ge \frac{A(n,d)}{q^n} \quad . \quad \blacksquare$$

Exercise 1.1.46. Prove the lemma. (*Hint*: use the fact that for any code C there exists a sphere of radius w such that the volume of its intersection with the code C is at least the average volume of intersection of C with such spheres).

Lemma 1.1.47. *We have*

$$A(n,d,w) \le \left\lfloor \frac{d}{d - 2\cdot w + \frac{q^2\cdot w}{(q-1)\cdot n}} \right\rfloor$$

whenever the denominator is positive. ■

Exercise 1.1.48. Prove the lemma. (*Hint*: Estimate the sum of all pairwise distances in a spherical code, as it is done in the proof of Theorem 1.1.39).

Exercise 1.1.49. Derive Theorem 1.1.44 from Lemma 1.1.45 and Lemma 1.1.47. ■

The next idea is to use spectra. Let C be a linear code. By Theorem 1.1.15 we know that

$$W_{C^\perp}(x:y) = q^{-k}\cdot W_C(x + (q-1)y : x - y)$$

or, in terms of coefficients:

$$A'_i = q^{-k}\cdot \sum_{j=0}^{n} A_j\cdot P_i(j) \quad ,$$

where $P_i(x)$ is the *Krawtchouk polynomial* defined as

$$P_i(x) = \sum_{j=0}^{i} (-1)^j\cdot (q-1)^{i-j}\cdot \binom{x}{j}\cdot\binom{n-x}{i-j} \quad .$$

Note that $P_i(0) = \binom{n}{i}\cdot(q-1)^i$. The generating function of polynomials $P_i(x)$ is

$$(1 + (q-1)\cdot z)^{n-x}\cdot(1-z)^x = \sum_{i=0}^{\infty} P_i(x)\cdot z^i \quad .$$

Since A'_i are the coefficients of $W_{C^\perp}(x:y)$ they are non-negative integers, i,e. for any i

$$\sum_{j=0}^{n} A_j \cdot P_i(j) \geq 0 \quad .$$

We want to give an upper bound for $q^k = 1 + \sum_{i=d}^{n} A_i$, i.e. to solve the following linear programming problem (note that $\sum_{j=0}^{n} A_j \cdot P_0(j) \geq 0$ for $A_j \geq 0$):

$$M = 1 + \sum_{i=d}^{n} x_i \longrightarrow max \quad .$$

$$\binom{n}{j} \cdot (q-1)^j + \sum_{i=d}^{n} P_j(j) \cdot x_i \geq 0 \quad \text{for} \quad j = 1, \ldots , n$$

$$x_i \geq 0 \quad \text{for} \quad i = d,\ d+1, \ldots , n$$

Solving the dual problem we obtain the following statement:

Theorem 1.1.50 (the linear programming bound). *For a given set of non-negative real numbers* $a_1, \ldots, a_n$ *such that for any* $j = d,\ d+1, \ldots, n$

$$1 + \sum_{i=1}^{n} a_i \cdot P_i(j) \leq 0$$

and for any $[n,k,d]_q$*-code* C *we have*

$$q^k \leq 1 + \sum_{i=1}^{n} a_i \cdot \binom{n}{i} \cdot (q-1)^i \quad .$$

■

Let us remark that this result can be strengthened if we apply the linear programming method to the constant-weight (spherical) codes and then use Lemma 1.1.45.

At last we come to an existence theorem.

Theorem 1.1.51 (the Gilbert-Varshamov bound). *Whenever*

$$q^{n-k} < \sum_{i=0}^{d-2} \binom{n-1}{i} \cdot (q-1)^i ,$$

there exists a linear $[n,k,d]_q$*-code* C .

Proof: We are going to construct a dual $[n,k,d]_q$-system $Q \subset W$, $\dim W = n-k$. For Q_1 we take any non-zero vector. Suppose that we have already constructed a system consisting of i vectors $Q_1,\dots,Q_i$ such that any $(d-1)$ of them are linearly independent. Consider the set S_i of vectors which are linear combinations of at most $(d-2)$ elements from $\{Q_1,\dots,Q_i\}$;

$$|S_i| \le M_i = \sum_{j=0}^{d-2} \binom{i}{j} \cdot (q-1)^j .$$

If $M_i < |W|$ we are able to choose $Q_{i+1} \notin S_i$. Then any $(d-1)$ vectors from the set $\{Q_1,\dots,Q_{i+1}\}$ are linearly independent. Proceeding like that we get $Q = \{Q_1,\dots,Q_n\}$ of cardinality n defined in the theorem. ∎

Remark 1.1.52. It is easy to see that there are many codes having parameters which are the best satisfying the inequality of the theorem (because of the multiple possibility of choice at each step). In a certain way (see Remark 1.3.17) we can say that the parameters of almost every code are quite near to those predicted by Theorem 1.1.51.

Exercise 1.1.53 (the Varshamov procedure). Suppose that the parameters of an $[n_0,k_0,d_0]_q$-code C_0 satisfy the inequality of Theorem 1.1.51. Prove that there exists an $[n_0,k_1,d_0]_q$-code $C_1 \supseteq C_0$ such that its parameters also satisfy the inequality and the triple $[n_0,\ k_1+1,\ d_0]$ does not satisfy it.

Remark 1.1.54 (the Gilbert bound). It is possible to prove a slightly weaker result: if

$$q^{n-k} < \sum_{i=0}^{d-1} \binom{n}{i} \cdot (q-1)^i$$

then there exists a (non-linear) $[n,k,d]_q$-code; this can be proved by choosing coding vectors in $\mathbb{F}_q^n$ one by one so that each point does not lie in the union of spheres of radius $(d-1)$ centered in the previous points.

The results we have exposed in this section (Section 1.1.4) somehow bound the possible values of parameters. Still the problem of actually finding out what is possible and what is not is very difficult.

CHAPTER 1.2

EXAMPLES AND CONSTRUCTIONS

In this chapter we present several examples of codes. Each example is in fact a method to construct some family of codes, which (in some way or other) have rather good parameters. Since in many cases these families are predecessors of algebraic-geometric codes, we try to choose constructions that are easy to generalize in that direction.

Section 1.2.1 is devoted to the example most important for us, to Reed-Solomon codes, which - as we shall see in Part 3 - are just algebraic-geometric codes of genus zero. We discuss them in detail, calculate the spectra, and give a decoding algorithm. In Section 1.2.2 we briefly discuss other interesting families of codes. Section 1.2.3 is devoted to a number of rather simple constructions which produce new codes starting with some codes we already know.

1.2.1. Codes of genus zero

Recall that a *code of genus zero* (or an *MDS-code*) is an $[n,k,d]_q$-code C such that $k + d = n + 1$; this yields $(n - k) + d^\perp = n + 1$, i.e. $C^\perp$ is also an MDS-code (see Exercise 1.1.18).

Trivial codes. For any n there are three simple q-ary codes which it is naturally to call *trivial*. These are:

$[n,n,1]_q$-code $C_1 = \mathbb{F}_q^n$,

$[n,n-1,2]_q$-code $C_2 = \{(v_1,\dots,v_n) \in \mathbb{F}_q^n \mid \sum v_i = 0\}$
(called the *parity-check code*), and

$[n,1,n]_q$-code $C_3 = \{v = (\alpha,\dots,\alpha) \in \mathbb{F}_q^n\}$, $\alpha \in \mathbb{F}_q$
(called the *repetition code*).

Now we are passing to a more conceptual construction.

Reed-Solomon codes. Let $\mathcal{P} = \{P_1,\dots,P_n\} \subseteq \mathbb{F}_q$ be a subset of cardinality n . Consider a linear space $L(a)$ of all polynomials in one variable of degree at most a with coefficients in $\mathbb{F}_q$; $\dim L(a) = a + 1$. For $n > a$ a non-zero polynomial $f(x) \in L(a)$ cannot vanish at all points of $\mathcal{P}$, moreover, it has at least $(n - a)$ non-zero values at points of $\mathcal{P}$. The "evaluation" map

$$Ev_{\mathcal{P}} : L(a) \longrightarrow \mathbb{F}_q^n$$

$$Ev_{\mathcal{P}} : f \longmapsto (f(P_1),\dots,f(P_n))$$

is injective for $n > a$ and its image C is an

$[n, a+1, n-a]_q$-code called a *Reed-Solomon code of degree* a (traditionally this name is reserved for the codes of length $n = q - 1$ with $\mathcal{P} = \mathbb{F}_q^*$, the rest are called extended or shortened Reed-Solomon codes; but for us it is natural to give one name to all of them). The parameters of such a code satisfy the condition $k + d = n + 1$ which is very good (by Proposition 1.1.36 it cannot be better), and $k = a + 1$ can be freely chosen between 1 and n. Moreover, Reed-Solomon codes form an *embedded family*, $L(a) \subset L(a+1)$. Unfortunately, their length cannot exceed q. In Section 3.1.1 we shall see that these codes can be naturally extended to codes on the projective line with $k + d = n + 1$, $n \leq q + 1$.

Choose the basis $\{1, x, x^2, \dots, x^a\}$ in $L(a)$. In this basis the generator matrix of C is (P_i^j), $j = 0, 1, \dots, a$; $i = 1, 2, \dots, n$.

The dual code. Let us start with the simplest case.

Exercise 1.2.1. Let $\mathcal{P} = \mathbb{F}_q$ or $\mathcal{P} = \mathbb{F}_q^*$. Prove that $C^\perp$ is also a Reed-Solomon code with $a^\perp = n - a - 2$ and parameters $[n, n-a-1, a+2]_q$. (*Hint*: Use the fact that $\sum_{x \in \mathbb{F}_q} x^i = 0$ for $i < q - 1$).

Then let us find the dual code for the case of an arbitrary $\mathcal{P} \subseteq \mathbb{F}_q$. Set

$$g_0(x) = \prod_{P_i \in \mathcal{P}} (x - P_i)^{-1},$$

$$g_\ell(x) = x^\ell \cdot g_0(x).$$

Consider the vector space $\Omega(a)$ spanned by the functions

$g_\ell(x)$ for $0 \le \ell \le n - a - 2$. Consider a function $F(x) = f(x) \cdot g_0(x)$, $f(x)$ being a polynomial. Recall the definition of the *residue* of F at P_i :

$$Res_{P_i} F = f(P_i) \cdot \prod_{j \neq i} (P_i - P_j)^{-1} .$$

Exercise 1.2.2 (the residue formula). Prove that for $\deg f \le n - 2$ we have $\sum_{P_i \in \mathcal{P}} Res_{P_i} F = 0$.

Proposition 1.2.3. *The dual code* $C^\perp$ *for the Reed-Solomon code* C *of degree* a *is the image of* $\Omega(a)$ *under the map*

$$Res_{\mathcal{P}} : g \longmapsto (Res_{P_1} g , \ldots , Res_{P_n} g) .$$

The code $C^\perp$ *is equivalent to some other Reed-Solomon code.*

Proof: Let $f \in L(a)$, $g \in \Omega(a)$. Then

$$\sum_{P_i \in \mathcal{P}} f(P_i) \cdot Res_{P_i} g = \sum_{P_i \in \mathcal{P}} Res_{P_i} (f \cdot g) = 0 .$$

Since $\dim L(a) + \dim \Omega(a) = n$, we get $C^\perp = Res_{\mathcal{P}}(\Omega(a))$.

Then note that if $g(x) = h(x) \cdot g_0(x)$, $h(x)$ being a polynomial, then

$$Res_{P_i} g = y_i \cdot h(P_i)$$

where

$$y_i = \prod_{j \neq i} (P_i - P_j)^{-1} ,$$

i.e. $C^\perp$ can be obtained from a Reed-Solomon code $C' = Ev_{\mathcal{P}}(L(a^\perp))$, $a^\perp = n - a - 2$ by multiplying i-th coordinate of all vectors by $y_i \in \mathbb{F}_q^*$. We write $C^\perp = y \cdot C'$, $y = (y_1, \ldots, y_n) \in (\mathbb{F}_q^*)^n$, and call such codes *generalized Reed-Solomon codes.* They are equivalent to Reed-Solomon codes in the sense of definition of Section 1.1.1. ■

Whenever $\mathcal{P} = \mathbb{F}_q$ or $\mathcal{P} = \mathbb{F}_q^*$, multiplication by y is an automorphism of C'.

Let us remark that if n is even and $a = \frac{n}{2} + 1$, then C is quasi-self-dual: $C^\perp = y \cdot C$, and if $\mathcal{P} = \mathbb{F}_q$ or $\mathcal{P} = \mathbb{F}_q^*$ it is self-dual: $C^\perp = C$.

Spectra. Applying Theorem 1.1.16 to codes of genus zero, we get a complete answer.

Proposition 1.2.4. *If C is an $[n,k,d]_q$-code of genus zero then*

$$W_C(x) = x^n + \sum_{i=0}^{k-1} \binom{n}{i} \cdot (q^{k-i} - 1) \cdot (x - 1)^i \quad .$$

■

Exercise 1.2.5. Establish this formula for Reed-Solomon codes by a direct computation. Check that for $i \neq 0$

$$A_i = \binom{n}{i} \cdot (q - 1) \cdot \sum_{j=0}^{i-d} (-1)^j \cdot \binom{i-1}{j} \cdot q^{i-d-j} \quad .$$

Decoding. Consider an $[n, n - a - 1, a + 2]_q$-code C dual to a Reed-Solomon code of degree a. Recall that decoding up to $t = \left\lceil \frac{d-1}{2} \right\rceil$ means an algorithm that makes it possible, starting with some $v \in \mathbb{F}_q^n$ which is at most at distance t from some code vector $u \in C$, to find this u. Let $v - u = e$, $\|e\| \leq t$, e is called the *error vector*.

So we are given some $v \in \mathbb{F}_q^n$. Start with a calculation of so-called *syndromes*:

$$s_j = s_j(v) = \sum_{P_i \in \mathcal{P}} v_i \cdot P_i^j , \quad 0 \le j \le 2t - 1 .$$

Note that $s_j = \sum_{i \in I} e_i \cdot P_i^j$, where $I = \{i \mid e_i \ne 0\}$ is the (unknown) set of *error locators*, since for $u \in C$ the corresponding sum equals zero by definition (the matrix (P_i^j) is a parity-check matrix of the code) .

The next step is to find the polynomial

$$g(x) = \sum_{\ell=0}^{t} y_\ell \cdot x^\ell = c \cdot \prod_{i \in I} (x - P_i) .$$

To do that we solve the system of equations

$$\sum_{\ell=0}^{t} y_\ell \cdot s_{j+\ell} = 0 , \quad 0 \le j \le t - 1 ,$$

where y_ℓ are indeterminants. The function g we are looking for is the solution of the system since $g(P_i) = 0$, hence

$$\sum_{\ell=0}^{t} y_\ell \cdot s_{j+\ell} = \sum_{\ell=0}^{t} \sum_{P_i \in \mathcal{P}} y_\ell \cdot e_i \cdot P_i^{j+\ell} =$$

$$= \sum_{i \in I} e_i \cdot P_i^j \cdot g(P_i) = 0 .$$

On the other hand if $\{y'_\ell\}$ is some other solution and

$$g'(x) = \sum_{\ell=0}^{t} y'_\ell \cdot x^\ell ,$$

then, setting

$$F_j(x) = \prod_{i\in I, i\neq j} (x - P_i) = \sum_{k=0}^{t-1} b_k \cdot x^k ,$$

for any $j \in I$ we have

$$e_j \cdot F_j(P_j) \cdot g'(P_j) = \sum_{i\in I} e_i \cdot F_j(P_i) \cdot g'(P_i) =$$

$$= \sum_{i\in I} \sum_{k=0}^{t-1} e_i \cdot b_k \cdot P_i^k \cdot g'(P_i) =$$

$$= \sum_{k=0}^{t-1} \sum_{\ell=0}^{t} b_k \cdot y'_\ell \cdot \sum_{i\in I} e_i \cdot P_i^{k+\ell} =$$

$$= \sum_{k=0}^{t-1} b_k \cdot \left(\sum_{\ell=0}^{t} y'_\ell \cdot s_{k+\ell} \right) = 0 ,$$

Therefore $g'(P_j) = 0$ for any $j \in I$, i.e. g is the only solution (up to a multiplicative constant).

Decomposing g into factors we find the set of error locators I. Let us now solve the system $\sum_{i\in I} e_i \cdot P_i^j = s_j$, $0 \le j \le a$ (in indeterminants e_i). The values of e_i we are looking for satisfy this system. If $\{e'_i\}$ is another solution, then $\sum_{i\in I} (e_i - e'_i) \cdot P_i^j = 0$, i.e. the vector $e - e' \in C$ but its weight is at most $2t \le d - 1$. The contradiction we get shows that e is found uniquely.

1.2.2. Some families of codes

Let us discuss some other interesting examples. (In our exposition, discussing some well known families of codes, we diverge in minor details from traditional constructions and definitions).

Reed-Muller codes of the first order. Consider a linear space L_m of polynomials of degree 0 and 1 in m variables; $dim\, L_m = m + 1$; let $\mathcal{P} = \{P_1, \ldots, P_n\} \subseteq \mathbb{F}_q^m$ be a subset of cardinality n such that no non-zero linear polynomial vanishes at all points $P_1, \ldots, P_n$ (it is surely so if $n > q^{m-1}$ since the number of zeroes of a linear polynomial in m variables is at most q^{m-1}). The image of the *evaluation map*

$$Ev_{\mathcal{P}} : L_m \longrightarrow \mathbb{F}_q^n$$

$$Ev_{\mathcal{P}} : f \longmapsto (f(P_1), \ldots, f(P_n))$$

is an $[n, m + 1, n - q^{m-1}]_q$-code with $n \le q^m$.

Let us generalize this construction. Consider all homogeneous linear forms in $(m + 1)$ variables. Together with zero they form a linear space L'_m of dimension $(m + 1)$ over $\mathbb{F}_q$. Then let $\mathcal{P} \subset \mathbb{F}_q^{m+1}$ be such that $P_i \ne (0, \ldots, 0)$ and $P \in \mathcal{P}$ yields $\alpha \cdot P \notin \mathcal{P}$ for any $\alpha \in \mathbb{F}_q^* - \{1\}$. Consider again the evaluation-at-$\mathcal{P}$ map. A non-zero form $f \in L'_m$ has at most q^m zeroes in $\mathbb{F}_q^{m+1}$. Recall that $0 \notin \mathcal{P}$ and if $f(P_i) = 0$ for $P_i \in \mathcal{P}$ then also $f(\alpha \cdot P_i) = 0$ for all $\alpha \ne 0, 1$. Therefore the number of zeroes of f in $\mathcal{P}$ is at most $\frac{q^m - 1}{q - 1}$. The maximum cardinality of $\mathcal{P}$ is $\frac{q^{m+1} - 1}{q - 1}$ (take for example all non-zero elements of $\mathbb{F}_q^{m+1}$ such that their first non-zero coordinate is 1). We obtain an $\left[n, m + 1, n - \frac{q^m - 1}{q - 1}\right]_q$

code for $n \le \frac{q^{m+1} - 1}{q - 1}$, in particular a code C with parameters

$$\left[\frac{q^{m+1} - 1}{q - 1}, m + 1, q^m\right]_q .$$

This is a very good code, the best possible for $d/n > \frac{q - 1}{q}$ (cf. Theorem 1.1.39, this code lies on the Plotkin bound).

Exercise 1.2.6. Show that all non-zero code vectors of C are of the same weight q^m; i.e.

$$W_C(x) = x^n - (q^{m+1} - 1)\cdot x^{q^m} .$$

Hamming codes. Consider the code $C_H = C^{\perp}$ dual to the above Reed-Muller code. Theorem 1.1.16 makes it possible to find out the spectrum.

Exercise 1.2.7. Calculate this spectrum.

The spectrum shows that $d \ge 3$. This can be seen also without knowing the spectrum. In fact, the parity-check matrix of C_H has no proportional columns (if for $P_1, P_2 \in \mathbb{F}_q^{m+1}$ all linear forms are proportional, $f(P_1) = \alpha\cdot f(P_2)$, then $P_1 = \alpha\cdot P_2$). Hence any two columns are linearly independent and (Exercise 1.1.3) $d \ge 3$.

So we have constructed codes with parameters

$$[n, n - m - 1, 3]_q , \quad n \le \frac{q^{m+1} - 1}{q - 1}$$

(and for $n = \frac{q^{m+1} - 1}{q - 1}$ we know the spectrum of such codes). These codes are good if we are interested in codes with $d = 3$. For $n = \frac{q^{m+1} - 1}{q - 1}$ they lie on the Hamming bound (cf. Theorem 1.1.41).

Reed-Muller codes of order r **.** Let $r < m\cdot(q-1)$. Consider the linear space $L_m(r)$ of all polynomials of degree at most r in m variables. Fix a subset $\mathcal{P} = \{P_1,\ldots,P_n\} \subseteq \mathbb{F}_q^m$ and consider the evaluation map:

$$Ev_{\mathcal{P}} : L_m(r) \longrightarrow \mathbb{F}_q^n$$

$$Ev_{\mathcal{P}} : f \longmapsto (f(P_1),\ldots,f(P_n)) .$$

Set $C = Ev_{\mathcal{P}}(L_m(r))$. Finding out the parameters of C is rather difficult. For simplicity let us suppose that $\mathcal{P} = \mathbb{F}_q^m$, $n = q^m$. The map $Ev_{\mathcal{P}}$ in general is not injective. In fact $Ev_{\mathcal{P}}(f) = Ev_{\mathcal{P}}(f^q)$ for any f . Let $L'_m(r)$ be the space spanned by monomials of degree at most r which are not divided by any q-power (i.e. those of the form $x_1^{\alpha_1}\cdot\ldots\cdot x_m^{\alpha_m}$, $0 \le \alpha_i \le q-1$, $\sum \alpha_i \le r$).

Exercise 1.2.8. Let

$$n = q^m , \ r = a\cdot(q-1) + b \le m\cdot(q-1) , \ 1 \le b \le q-1 .$$

Prove that then

$$C = Ev_{\mathcal{P}}(L'_m(r)),$$

$$Ev_{\mathcal{P}} : L'_m(r) \hookrightarrow \mathbb{F}_q^n ,$$

$$k = \dim L'_m(r) = \sum_{i=0}^{r} \sum_{j=0}^{\lceil i/q \rceil} (-1)^j \cdot \binom{m}{j} \cdot \binom{m-1+i-qj}{m-1} ,$$

$$d = (q-b)\cdot q^{m-a-1} .$$

In particular, for $q = 2$ we get a

$$\left[2^m , \sum_{i=0}^{r} \binom{m}{i} , 2^{m-r}\right]_2\text{-code.}$$

(*Hint*: To compute k one can calculate the number of ways to place m objects in i cells such that no cell contains more than j objects and then apply an exclusion-inclusion argument. To compute d one can just use induction over m).

Problem 1.2.9. Generalize this definition to the case of a projective space.

Cyclic codes. We can only just mention here this important topic which is mostly left out from this book.

A code $C \subseteq \mathbb{F}_q^n$ is called *cyclic* if it is invariant with respect to the cyclic shift of coordinates, i.e. $(c_1,\dots,c_n) \in C$ yields $(c_2,\dots,c_n,c_1) \in C$. Note that cyclicity is not an invariant of the equivalence class of codes.

The cyclicity condition is so strong that it is possible (in a way) to describe a cyclic code. Identify $\mathbb{F}_q^n$ with the ring $R_n = \mathbb{F}_q[x]/(x^n - 1)$. A code is cyclic iff it is invariant under the multiplication by x (in fact, if

$$c = c_0 + c_1 \cdot x + \dots + c_{n-1} \cdot x^{n-1} \in R_n$$

then

$$x \cdot c = c_{n-1} + c_0 \cdot x + \dots + c_{n-2} \cdot x^{n-1}$$

is its cyclic shift). The condition $x \cdot C \subseteq C$ is equivalent to C being an ideal in R_n. Since R_n is the principal ideal domain, $C = (g) \subseteq R_n$. Let us take for g the representative $g(x)$ of the lowest degree in $\mathbb{F}_q[x]$. It is obvious that $g(x) \mid (x^n - 1)$; let $r = deg\, g$ then the dimension of the code is $k = n - r$. An element $c(x) \in R_n$ belongs to C iff there exists $f(x) \in \mathbb{F}_q[x]$, $deg\, f \le k$ such that $c(x) = f(x) \cdot g(x)$. The polynomial $g(x)$ is called the *generator polynomial* of C, and $h(x) = (x^n - 1)/g(x)$ is called its *parity-check polynomial.*

Exercise 1.2.10. Prove that the Reed-Solomon codes with $\mathcal{P} = \mathbb{F}_q^*$ are cyclic. The same is also true for $\mathcal{P} = \mathbb{F}_p$ (for a prime p).

Let us consider the case $(n,q) = 1$, then the polynomial $(x^n - 1)$ has n distinct roots in $\overline{\mathbb{F}}_q$; let $\mathbb{F}_{q^m}$ be the splitting field of this polynomial, i.e. the smallest extension of $\mathbb{F}_q$ where $(x^n - 1)$ decomposes into a product of linear factors.

Exercise 1.2.11. Check that m is the smallest integer such that $n|(q^m - 1)$.

The group $\mathbb{F}_{q^m}^*$ is cyclic, its subgroup μ_n of n-th roots of 1 is also cyclic, fix its generator α , then

$$x^n - 1 = \prod_{i=0}^{n-1} (x - \alpha^i) .$$

Since $g(x)|(x^n - 1)$, $g(x) = \prod_{i\in I} (x - \alpha^i)$ for a subset $I \subseteq \{0,1,\ldots,n - 1\}$.

Exercise 1.2.12. Prove that the coefficients of $g(x)$ belong to $\mathbb{F}_q$ iff $q\cdot I \equiv I \pmod n$.

Now it is quite clear that $c(x) \in C$ iff $c(\alpha^i) = 0$ for every $i \in I$. Hence a parity-check matrix of C can be written as follows (here $I = \{i_1,\ldots,i_r\}$):

$$H = \begin{pmatrix} 1 & \alpha^{i_1} & \alpha^{2i_1} & \ldots & \alpha^{(n-1)i_1} \\ \cdot & \cdot & \cdot & & \cdot \\ \cdot & \cdot & \cdot & & \cdot \\ \cdot & \cdot & \cdot & & \cdot \\ 1 & \alpha^{i_r} & \alpha^{2i_r} & \ldots & \alpha^{(n-1)i_r} \end{pmatrix} .$$

Attention: H is a parity-check matrix in a Pickwick sense, its entries do not lie in $\mathbb{F}_q$. To put it precisely we should change each element $h_{ij} \in H$ by the column of its coordinates in a basis of $\mathbb{F}_{q^m}$ over $\mathbb{F}_q$, the matrix H' we obtain is in fact the parity-check matrix of dimension $m \cdot r \times n$.

We are in the situation of the field restriction, discussed below in Section 1.2.3. Suppose that we are given an $[n,k,d]_{q^m}$-code C', choose code vectors of C' all coordinates of which lie in $\mathbb{F}_q$, i.e. consider the q-ary code $C = C' \cap \mathbb{F}_q^n$; its parameters are at worst $[n,\ n - m\cdot(n - k),\ d]_q$.

It can in fact happen that some rows of H' are linearly dependent. For example if $i_1 = q \cdot i_2$ then $\mathbb{F}_q$-expansions of α^{i_1} and α^{i_2} have the same set of coordinates, hence $rk\, H' \le m \cdot s$, $s = |I_0|$, where I_0 is such that $I_0 \cup q\cdot I_0 \cup \ldots \cup q^{m-1}\cdot I_0 = I$.

Summing up we get:

Theorem 1.2.13. *Let* $(n,q) = 1$. *Any cyclic* $[n,k,d]_q$-*code* C *is the field restriction of the* $[n,k',d']_{q^m}$-*code* C' *defined by the parity-check matrix*

$$H = \begin{pmatrix} 1 & \alpha^{i_1} & \ldots & \alpha^{(n-1)i_1} \\ \cdot & \cdot & \cdot\cdot\cdot & \cdot \\ \cdot & \cdot & \cdot\cdot\cdot & \cdot \\ \cdot & \cdot & \cdot\cdot\cdot & \cdot \\ 1 & \alpha^{i_r} & \ldots & \alpha^{(n-1)i_r} \end{pmatrix}$$

where m *is the smallest integer such that* $n | (q^m - 1)$, $\langle\alpha\rangle = \mu_n \subseteq \mathbb{F}^*_{q^m}$, *the set* $I = \{i_1, \ldots, i_r\}$ *has the property*

$q\cdot I \equiv I(\mathrm{mod}\ n)$, $r = n - k'$; and $d \geq d'$, $k \geq n - m\cdot r$. Moreover, $k \geq n - m\cdot s$, where $s = \min|I_0|$ over all $I_0 \subseteq I$ such that $I_0 \cup q\cdot I_0 \cup \ldots \cup q^{m-1}\cdot I_0 = I$.

■

It would be also good to know how to estimate the minimum distance of a cyclic code. In general, this is a very subtle question. Here is an important example.

BCH-codes. Let $n|(q - 1)$, i.e. $m = 1$. Consider a cyclic code with $I = \{0,1,\ldots,r - 1\}$. This is a Reed-Solomon code, its dual (which is also a Reed-Solomon code) is built from polynomials of degree at most $(r - 1)$ being evaluated at $\mathcal{P} = \{1,\alpha,\alpha^2,\ldots,\alpha^{n-1}\}$. For such codes $k = n - r$ and we know the minimum distance: $d = n + 1 - k = r + 1$. Let us generalize this example.

A *BCH-code* C is a cyclic code such that $I \supseteq \{b , b + 1 ,\ldots, b + \Delta - 2\}$ for some $b \geq 0$ and $\Delta \geq 2$. We call Δ the *designed distance* of C. A *BCH-code in the narrow sense* is a BCH-code with $b = 1$, a *primitive BCH-code* is a BCH-code of length $n = q^m - 1$ (in this case α is a primitive element of $\mathbb{F}_{q^m}$).

Exercise 1.2.14. Prove that for a BCH-code $d \geq \Delta$. (*Hint*: This can be done at least in two ways. You can write out explicitly the condition for $(\Delta - 1)$ columns of H to be linearly dependent, and see that it is a Vandermonde determinant. You can also reduce everything to the case $b = 0$, consider a Reed-Solomon $[n,k',d']_{q^m}$-code C' , and use $d \geq d'$).

Then we are going to estimate the dimension k of C . Recall that $I \equiv q\cdot I(\mathrm{mod}\ n)$ and by Theorem 1.2.13

$n - k \le m \cdot s$, where $s = min\ |I_0|$ over $I_0 \subseteq I$ such that $I = I_0 \cup q \cdot I_0 \cup \ldots \cup q^{m-1} \cdot I_0$.

Further on we suppose that C is a primitive BCH-code in the narrow sense (i.e. $b = 1$, $n = q^m - 1$), and that $m \ge 2$. Set

$$I_1 = \{1,2,\ldots,\Delta - 1\} , \quad I_0 = I_1 - q \cdot I_1 ,$$

let $I \subseteq \mathbb{Z}/n$ be the smallest set such that $I \equiv q \cdot I\ (mod\ n)$ and $I \supseteq I_1$; $s \le |I_0| = \Delta - 1 - \left\lceil \frac{\Delta - 1}{q} \right\rceil$. We get

$$n - k \le m \cdot \left(d - 1 - \left\lceil \frac{d - 1}{q} \right\rceil \right) .$$

Summing up, we have proved

Theorem 1.2.15. *For any* $m \ge 2$ *and* $n = q^m - 1$ *there exists a primitive q-ary BCH-code in the narrow sense whose parameters are*

$$\left[n = q^m - 1,\ k \ge n - m \cdot \left(d - 1 - \left\lceil \frac{d - 1}{q} \right\rceil \right),\ \ge d \right]_q ,$$

$$d = 2,\ 3,\ \ldots,\ 1 + \frac{n \cdot q}{m \cdot (q - 1)} .$$

In particular, for $d = q \cdot \ell + 1$ *and* $n = q^m - 1$ *we get* $n - k \le m \cdot \ell \cdot (q - 1)$. ■

Of course, the most advantage when compared with the standard estimate for the field restriction we get for $q = 2$. In Section 3.4.3 we return to the idea of "throwing out q-th powers" in a more refined version.

There are many other interesting cyclic codes, for example such are *post factum* Golay codes, some Goppa codes,

Reed-Muller codes with $\mathcal{P} = \mathbb{F}_m^q - \{0\}$, etc. It is a pity we cannot answer the following important question:

Problem 1.2.16. Do there exist infinite families of cyclic codes with $n \longrightarrow \infty$ and $k \cdot d/n^2 \geq \varepsilon > 0$? (such families are called *asymptotically good*, cf. Section 1.3.1).

Let us remark that if we weaken a bit the condition of cyclicity, this problem has a positive answer: there exist asymptotically good families of *quasi-cyclic codes*, i.e. codes that are invariant under a cyclic s-shift

$$(c_1, \ldots, c_n) \longmapsto (c_{s+1}, c_{s+2}, \ldots, c_{s+n}) ,$$

$s \geq 2$ being fixed, $n \longrightarrow \infty$.

Here is another example of cyclic codes.

Quadratic-residue codes. Let $q = p$ be a prime and let ℓ be another prime such that $\ell \neq 2$ and p is a square modulo ℓ . Let I be the set of quadratic residues modulo ℓ . Recall that a quadratic residue is an element $i \in \mathbb{Z}/\ell \cdot \mathbb{Z}$ such that $i \equiv a^2 (mod\ \ell)$ for some a ; i is a quadratic residue iff $i^{(\ell-1)/2} \equiv 1 (mod\ \ell)$. Since p is a quadratic residue, $p \cdot I \equiv I (mod\ \ell)$.

Exercise 1.2.17. Prove that a cyclic code C corresponding to the set I of quadratic residues (see Theorem 1.2.13) is a $\left[\ell, \frac{\ell+1}{2}, \geq \rceil\sqrt{\ell}\lceil\right]_q$-code. Let us add to a vector $v = (v_1, \ldots, v_\ell) \in C$ one coordinate more $v_{\ell+1} = -\sum_{i=1}^{\ell} v_i$. Prove that for $\ell \equiv -1 (mod\ 4)$ the new code C' is self-dual.

Alternant codes. Let C_0 be a generalized Reed-Solomon code over $\mathbb{F}_{q^m}$ (cf. the definition given after Proposition 1.2.3). The code $C = C_0^\perp \cap \mathbb{F}_q^n$ is called an *alternant code.*

Exercise 1.2.18. Prove that the parameters of C are at worst $[n, n - m\cdot(d - 1), d]_q$.

There is a lot of alternant codes, much more than BCH-codes, that is why there are many good codes among them (e.g. there is no infinite family of BCH-codes with $k\cdot d/n^2 \geq \varepsilon > 0$, but there are such families of alternant codes). Here is an important subclass of these codes:

Goppa codes. Let $G(x)$ be a polynomial over $\mathbb{F}_{q^m}$, and let $\mathcal{P} = \{P_1, \ldots, P_n\} \subseteq \mathbb{F}_{q^m}$ be such that $G(P_i) \neq 0$. The alternant code constructed from the generalized Reed-Solomon code corresponding to $\mathcal{P}$ and $y = (G(P_1)^{-1}, \ldots, G(P_n)^{-1})$ is called a *Goppa code.*

Consider another definition. For any

$$v = (v_1, \ldots, v_n) \in \mathbb{F}_q^n$$

let

$$R_v(z) = \sum_{i=1}^{n} \frac{v_i}{z - P_i} .$$

The code

$$C = \{v \in \mathbb{F}_q^n \mid R_v(z) \equiv 0 \ (mod\ G(z))\}$$

is called a *Goppa code.*

Exercise 1.2.19. Check that these definition are equivalent and prove that the parameters are at worst $[n, n - m\cdot r, r + 1]_q$, where $r = deg\ G(z)$.

Exercise 1.2.20. Let $q = 2$. Show that if $G(z)$ has no multiple roots then $d \ge 2r + 1$. (*Hint*: In this situation $R_V(z) = f'_V(z)/f_V(z)$ for some polynomial $f_V(z)$, and $f'_V(z)$ is a square for any polynomial $f_V(z)$).

Justesen codes. Consider a q^m-ary Reed-Solomon $[n, k, n + 1 - k]_{q^m}$-code C_0 , and let

$$y = (1,\alpha,\alpha^2,\ldots,\alpha^{n-1}) ,$$

α being a primitive element of $\mathbb{F}_{q^m}$. Let

$$C = C_0 \oplus y\cdot C_0 \subseteq \mathbb{F}_{q^m}^{2n} \simeq \mathbb{F}_q^{2mn} ,$$

i.e.

$$C = \{(c_1,\ldots,c_n;c_1,\alpha\cdot c_2,\ldots,\alpha^{n-1}\cdot c_n) \mid (c_1,\ldots,c_n) \in C_0\})$$

and consider C as a q-ary code.

Exercise 1.2.21. Prove that C is an

$$\left[2m\cdot n,\ m\cdot k,\ \sum_{i=1}^{\ell} i\cdot\binom{2m}{i}\cdot(q - 1)^i\right]_q\text{-code,}$$

ℓ being the largest integer such that

$$\sum_{i=1}^{\ell}\binom{2m}{i}\cdot(q - 1)^i \le n - k + 1 .$$

(*Hint*: The line $(c_1,c_1;c_2,c_2\cdot\alpha_2;\ldots;c_n,\alpha^{n-1}\cdot c_n)$ includes at least $n + 1 - k$ different q-ary vectors $(c_i,\alpha^{i-1}\cdot c_i)$ of length $2m$; estimate the total weight).

Golay codes. Speaking of interesting examples it is impossible not to mention such a phenomenon as Golay codes. These codes can be described in several different ways.

Consider the field $\mathbb{F}_{2^{11}}$; $2^{11} - 1 = 2047 = 23 \cdot 89$, hence the roots of 1 of degree 23 lie in this field. Let $\alpha \in \mathbb{F}_{2^{11}}$ be such a primitive root. Let

$$C_{23} = \{x \in \mathbb{F}_2^{23} \mid \sum_{i=1}^{23} x_i \cdot \alpha^i = 0\} \quad .$$

The code C_{23} is in fact a $[23,12,7]_2$-code. It is easy to check that this code is given by Theorem 1.2.13 for $q = 2$, $m = 11$, $n = 23$, $I_0 = \{1\}$; in this case

$$I = \{1, 2, 3, 4, 6, 8, 9, 12, 13, 16, 18\} \ .$$

Its minimum distance is much larger than it is natural to expect. This code is, of course, cyclic. Its generator polynomial is either

$$g_1(x) = 1 + x + x^5 + x^6 + x^7 + x^9 + x^{11} \ ,$$

or

$$g_2(x) = 1 + x + x^4 + x + x^6 + x^{10} + x^{11}$$

(it depends on the choice of α ; in fact

$$x^{23} - 1 = (x - 1) \cdot g_1(x) \cdot g_2(x) \quad).$$

The automorphism group M_{23} of the Golay $[23,12,7]_2$-code C_{23} is of order $23 \cdot 22 \cdot 21 \cdot 20 \cdot 48 = 10200960$. Its spectrum is

$$W_{23}(x) = x^{23} + 253\ x^{16} + 506\ x^{15} + 1288\ x^{12} +$$
$$+ 1288\ x^{11} + 506\ x^8 + 253\ x^7 + 1 = x^{23} \cdot W_{23}(1/x) \quad .$$

This code is a quadratic-residue code: the roots of its generator polynomial are α^i , i being a quadratic residue modulo 23 (the set of quadratic residues is I). This code has also something to do with quadratic residues modulo 11. Consider the vector

$$v = (1,0,1,0,0,0,1,1,1,0,1) ,$$

$v_i = 1$ iff i is a quadratic residue modulo 11. Let A be the matrix whose rows are formed by all cyclic shifts of this vector. Adding to A the row of ones we get a matrix B . Then the code C_{23} is given by the generator matrix

$$G = \boxed{E \mid B} ,$$

$E = E_{12}$ being the unity matrix.

The code C_{24} obtained from C_{23} by adding the parity-check (i.e. the column $(1,1,\ldots,1,0)$) is even more wonderful. Its parameters are $[24,12,8]_2$, its automorphism group M_{24} is 24 times larger:

$$|M_{24}| = 24\cdot 23\cdot 22\cdot 21\cdot 20\cdot 48 = 244823040 .$$

This group M_{24} is 5-transitive. The spectrum has only weights 24, 16, 12, 8, and 0:

$$W_{24}(x) = x^{24} + 759\, x^{16} + 2576\, x^{12} + 759\, x^{8} + 1 =$$

$$= x^{24}\cdot W_{24}(1/x) .$$

A similar situation takes place over $\mathbb{F}_3$. Consider $\mathbb{F}_{3^5}$; $3^5 - 1 = 242 = 11\cdot 22$. Let $\alpha \in \mathbb{F}_{3^5}$ be a primitive root of degree 11. Let

$$C_{11} = \{x \in \mathbb{F}_3^{11} \mid \sum_{i=1}^{11} x_i\cdot\alpha^i = 0\} .$$

This is an $[11,6,5]_3$-code. For this code $I_0 = \{1\}$, $I = \{1,3,4,5,9\}$ is the set of quadratic residues modulo 11.

It is cyclic and quadratic-residue:

$$x^{11} - 1 =$$

$$= (x - 1)\cdot(x^5 + x^4 - x^3 + x^2 - 1)\cdot(x^5 - x^3 + x^2 - 1) .$$

The generator matrix of C_{11} can be written out by a similar method, starting with $(0,1,-1,-1,1)$, i.e. using quadratic residues modulo 5.

Adding a parity-check (which in this case is the column $(-1,-1,-1,-1,-1,0)$) we get a $[12,6,6]_2$-code C_{12} . The order of its automorphism group M_{12} is $12\cdot 11\cdot 10\cdot 9\cdot 8 = 95040$. The spectrum of C_{12} is

$$W_{12}(x) = x^{12} + 264\, x^6 + 440\, x^3 + 24 .$$

Exercise 1.2.22. Prove all the above statements. (*Hint*: This is rather difficult, use some book on codes).

Perfect codes. A code C is called *perfect* if $d = 2t + 1$ and $\mathbb{F}_q^n$ is the union of spheres of radius t centered at code vectors. It is clear that this property depends only on the parameters $[n,k,d]_q$.

Theorem 1.2.23. *Let an* $[n,k,d]_q$*-code* C *(linear or non-linear) be perfect. Then either* $k = 0$ *, or* $k = n$ *, or* $q = 2$ *,* $k = 1$ *and* $n = d$ *is odd, or the parameters of* C *are exactly those of Hamming or Golay codes, i.e. either*

$$\left[n = \frac{q^m - 1}{q - 1} , n - m , 3\right]_q ,$$

or $[23,12,7]_2$ *, or* $[11,6,5]_3$. ■

Group codes. The notion of a cyclic code can be generalized to the case of any group G.

Recall that a *group algebra* of G is an algebra of functions

$$\mathbb{F}_q[G] = \{f\colon G \longrightarrow \mathbb{F}_q\} ,$$

where the multiplication is defined as convolution

$$(f_1 * f_2)(g) = \sum_{h\in G} f_1(h)\cdot f_2(h^{-1}g) .$$

The group G acts on $\mathbb{F}_q[G]$ from the right: $(fg)(h) = f(hg)$. For a subgroup $H \leq G$ the invariant space is defined as

$$\mathbb{F}_q[G/H] = \{f \in \mathbb{F}_q[G] \mid f(gh) = f(g) \text{ for any } g \in G,\ h \in H\} .$$

Any module $\mathbb{F}_q[G/H]$ has a natural basis $f_1,\dots,f_\ell$, where $\ell = [G:H]$, and the functions f_i have the property

$$f_i(g_j\cdot H) = \begin{cases} 1 & \text{if } i = j \\ 0 & \text{if } i \neq j \end{cases} ,$$

$G = g_1\cdot H \cup \dots \cup g_\ell\cdot H$ being the decomposition of G into disjoint right cosets over H.

Thus any subspace $C \leq \mathbb{F}_q[G/H]$ can be naturally viewed as a linear code, since the choice of basis $\{f_1,\dots,f_\ell\}$ gives an identification $\mathbb{F}_q[G/H] \simeq \mathbb{F}_q^\ell$. The same is valid for any permutational G-module $M = \mathbb{F}_q[G/H]\oplus\dots\oplus\mathbb{F}_q[G/H_m]$, where $H_1,\dots,H_m$ are arbitrary subgroups in G, so that any subspace $C \leq M$ is a linear code over $\mathbb{F}_q$.

Then let $G \leq (S_n \cap \mathit{Aut}\ C)$ be a subgroup of the automorphism group of $C \leq \mathbb{F}_q$ acting by permutations of coordinates. Let $B = \{e_1,\dots,e_n\}$ be the basis of $\mathbb{F}_q^n$. The group G acts on B, hence B is a disjoint union of G-orbits $B = O_1 \cup \ldots \cup O_m$, where $O_i = Gb_i$ is an orbit of some $b_i \in B$. Let H_i be the stabilizer of b_i, so that there is an isomorphism of G-sets $O_i \simeq G/H_i$. Then $\mathbb{F}_q^n$ is identified with a G-module $\mathbb{F}_q[G/H_1] \oplus \ldots \oplus \mathbb{F}_q[G/H_m]$. In particular, if G acts transitively, then $\mathbb{F}_q^n$ is identified with $\mathbb{F}_q[G/H]$, and C is a G-submodule in $\mathbb{F}_q[G/H]$. Since $\mathbb{F}_q[G/H]$ is a right ideal in the group algebra $\mathbb{F}_q[G]$, the code C is in this case also a right ideal in this algebra.

For example, if $Aut\ C$ contains a cyclic subgroup of order n permuting $e_1,\ldots,e_n$ then C is embedded into the group algebra $R_n = \mathbb{F}_q[x]/(x^n - 1)$ as an ideal, i.e. C is a cyclic code.

1.2.3. Some constructions

There are many ways to construct new codes starting with some known ones. The simplest were used above to prove the spoiling lemma (Lemma 1.1.34).

Exercise 1.2.24. Given an $[n,k,d]_q$-code C, construct $[n+\ell,k,d]_q$-, $[n-\ell,k-\ell,d]_q$- and $[n-\ell,k,d-\ell]_q$-codes. (*Attention*: There is an abuse of notation here, if $k \notin \mathbb{Z}$ one should write $log_q(\rceil q^{k-\ell}\lceil)$ instead of $(k-\ell)$). If C is linear new codes should be also linear.

Here we give some constructions more, starting with those that do not change q.

Direct sum. Let us start with two given codes

$$C_1 \subseteq \mathbb{F}_q^{n_1} \quad \text{and} \quad C_2 \subseteq \mathbb{F}_q^{n_2} .$$

Their *direct sum*

$$C = C_1 \oplus C_2 \subseteq \mathbb{F}_q^{n_1+n_2}$$

is the set of vectors $v = (v_1, v_2)$ where $v_1 \in C_1$, $v_2 \in C_2$. It is easy to see that C is a linear $[n_1 + n_2, k_1 + k_2, d]_q$-code with $d = min\,\{d_1, d_2\}$. We can also consider direct sums of any finite number of codes. If all these codes are equal, we get the *power* C^ℓ of the original code C; if C was an $[n,k,d]_q$-code we get an $[\ell\cdot n, \ell\cdot k, d]_q$-codes for all $\ell = 1, 2, \ldots$.

Tensor product. Consider the *tensor* (or *Kroneckerian*) product of codes

$$C_1 \subseteq \mathbb{F}_q^{n_1} \quad \text{and} \quad C_2 \subseteq \mathbb{F}_q^{n_2} \quad ,$$

i.e. the subspace

$$C = C_1 \otimes C_2 \subseteq \mathbb{F}_q^{n_1 n_2} \, ,$$

consisting of matrices such that all their rows are elements of C_1 and all their columns of C_2 ($\mathbb{F}_q^{n_1 n_2}$ is identified with the space of $n_1 \times n_2$ matrices).

Exercise 1.2.25. Prove that the obtained code C is an

$$[n_1 \cdot n_2, k_1 \cdot k_2, d_1 \cdot d_2]_q\text{-code.}$$

We can, of course, consider the tensor product of any finite number of codes, in particular the *tensor power* $C^{\otimes \ell}$ of a code. Its parameters are $[n^\ell, k^\ell, d^\ell]_q$.

Pasting. Start with an $[n_1, k, d_1]_q$-code

$$C_1 : \mathbb{F}_q^k \longrightarrow \mathbb{F}_q^{n_1}$$

and an $[n_2,k,d_2]_q$-code

$$C_2 : \mathbb{F}_q^k \longrightarrow \mathbb{F}_q^{n_2} .$$

It is natural to consider the diagonal map

$$\mathbb{F}_q^k \xrightarrow{(C_1,C_2)} \mathbb{F}_q^{n_1} \oplus \mathbb{F}_q^{n_2} = \mathbb{F}^{n_1+n_2} .$$

Its image $C = (C_1|C_2)$ is called the *pasting* of C_1 and C_2. It is easy to see that C is an

$$[n_1 + n_2,\ k,\ d_1 + d_2]_q\text{-code.}$$

Applying this construction several times to one and the same code we get ℓ-time *repetition* of the code. Its parameters are $[\ell\cdot n,k,\ell\cdot d]_q$.

A code from an embedded pair. Let codes $C_1 \supset C_2$ have parameters, respectively, $[n,k,d]_q$ and $[n,\ k-1,\ d+1]_q$. Chose $v \in C_1$, $v \notin C_2$. Any vector of C_1 is of the form $w + x\cdot v$, $w \in C_2$, $x \in \mathbb{F}_q$. Consider the extension of C_1 adding to $w + x\cdot v$ the $(n + 1)$-coordinate equal to x . We obtain an $[n + 1,\ k,\ d + 1]_q$-code.

$(u|u+v)$-construction. Let $C_1 \subseteq \mathbb{F}_q^n$ and $C_2 \subseteq \mathbb{F}_q^n$ lie in the same space $\mathbb{F}_q^n$. Consider the subspace $C = (\ C_1 \mid C_1 + C_2)$ of vectors of the form $(u,\ u + v)$, where $u \in C_1$, $v \in C_2$. It is easy to see that C is a $[2n,\ k_1 + k_2,\ d]_q$-code with $d = min\{2d_1,d_2\}$.

Shortening by the distance. In the proof of Theorem 1.1.38 it is shown that starting with a linear $[n,k,d]_q$-code C we can get an

$$\left[n - d,\ k - 1,\ \geq \left]\frac{d}{q}\right[\ \right]_q\text{-code.}$$

Shortening by the dual distance. Let C be a linear $[n,k,d]_q$-code such that its dual code has the minimum distance $d^{\perp} \le k$. Choose a parity-check matrix H such that one of its rows v is of weight $d^{\perp}$. Cross out this row v and $d^{\perp}$ columns where v has non-zero coordinates. We get an $[n - d^{\perp}, k - d^{\perp} + 1, d]_q$-code.

Parity-check. Let $q = 2$. Then we can lengthen $[n,k,d]_2$-code with odd d putting the sum of n coordinates for the additional one. Since now every code vector has an even weight, we get an

$$[n + 1, k, d + 1]_2\text{-code}\ .$$

Exercise 1.2.26. Which of these constructions can be applied to non-linear codes?

Now we pass to the other type of constructions, where we change q. We start with linear ones. Let C be a linear $[n,k,d]_q$-code and $q = r^m$.

Subfield restriction. Let $C' = C \cap \mathbb{F}_r^n \subseteq \mathbb{F}_q^n$ (recall that $\mathbb{F}_r$ is naturally embedded into $\mathbb{F}_q$); $C' \subseteq \mathbb{F}_r^n$ is a linear $\mathbb{F}_r$-subspace. This is an $[n,k',d']_r$-code and clearly

$$d' \ge d\ ,\ k' \ge n - m\cdot(n - k)\ .$$

This construction makes sense when applied to high-rate codes (i.e. when k/n is not too far from 1).

Concatenation. Let C_0 be an $[N,m,D]_r$-code (we call it the *inner code*, and C the *outer code*). We define a new code C' as the composition of maps

$$\mathbb{F}_r^{mk} \xrightarrow{\sim} \mathbb{F}_q^k \xhookrightarrow{C} \mathbb{F}_q^n \xrightarrow{\sim} (\mathbb{F}_r^m)^n \xhookrightarrow{(C_0,\ldots,C_0)} \mathbb{F}_r^{Nn}\ ,$$

and call it the *concatenation* of C and C_0 .

Exercise 1.2.27. Prove that the minimum distance of C' is at least $d \cdot D$.

So we have constructed (up to a spoiling) an $[N \cdot n, k \cdot m, D \cdot d]_r$-code.

When for the inner code we take the $[m,m,1]_r$-code, this construction is called the *field descent*, and the parameters are $[n \cdot m, k \cdot m, d]_r$.

Generalized concatenation. We are now starting with a family of $[n, k_i, d_i]_{q_i}$-codes C_i , $i = 1, \dots, \ell$, and an embedded family of $[N, K_i, D_i]_r$-codes

$$\{0\} = C'_0 \subset C'_1 \subset \dots \subset C'_\ell$$

such that $K_i = \sum_{j=1}^{i} m_j$, $q_i = r^{m_i}$. Consider C'_ℓ as a direct sum $C'_\ell = \bigoplus_{j=1}^{\ell} V_j$ such that $C'_i = \bigoplus_{j=1}^{i} V_j$. Let W_i be the concatenation of V_i and C_i (note that $\dim V_i = m_i$). Set $C = \bigoplus_{i=1}^{\ell} W_j$.

Exercise 1.2.28. Show that the code C is an

$$\left[N \cdot n, \sum_{i=1}^{\ell} m_i \cdot k_i, \geq \min_{1 \leq i \leq \ell} \{d_i \cdot D_i\}\right]_r\text{-code.}$$

Remark 1.2.29. It is possible to generalize both concatenation and generalized concatenation to the case of non-linear codes. The same is true about subfield restriction (subalphabet restriction), but instead of C' we should take "the best" of its shifts.

Now we are going to present two essentially non-linear constructions. Let C be a given arbitrary $[n,k,d]_q$-code, $k \in \mathbb{R}$, $q \in \mathbb{Z}$, $q \geq 2$.

Alphabet extension. Let $r \geq q$. Let us embed an alphabet A of cardinality q into an alphabet B of cardinality r. Considering an embedding $C \hookrightarrow A^n \hookrightarrow B^n$; we obtain an $[n, k\cdot log_r q, d]_r$-code.

Alphabet restriction. Let, vice versa, $r \leq q$. Embed $B \subseteq A$ and make A an abelian group (setting, for example, $A \simeq \mathbb{Z}/q$). Consider all q^n shifts C_v of C by vectors $v \in A^n$, $C_v = \{v + c \mid c \in C\}$. In the totality of sets C_v each word of A^n appears M times. Consider all intersections $B^n \cap C_v$, there are q^n of them and their total cardinality is $M\cdot r^n$, hence there exist C_v such that $|B^n \cap C_v| \geq M\cdot \left(\frac{r}{q}\right)^n$.

Since the shift does not change the minimum distance, we have obtained an $[n, \geq n - (n - k)\cdot log_r q, \geq d]_r$-code.

Decoding. When we construct a code using one or several others, there is a question how to decode the code we get, supposing that we know decoding algorithms for the codes we start with.

For the most part of the above constructions there is no problem to find a decoding algorithm, the interesting question is how to construct a "good" algorithm (from the point of view of the complexity theory).

For example, if there is a decoding algorithm of a q-ary code $C \subseteq \mathbb{F}_q$, the same algorithm decodes the field restriction $C' = C \cap \mathbb{F}_r^n \subseteq \mathbb{F}_r^n$.

It is more difficult with the concatenation. We need the following notion: *erasure* is an error whose position we know. Suppose that we have transmitted $u \in C$ and received $v \in \mathbb{F}_q^n$, $v = u + e + \varepsilon$, $I = \{i \mid e_i \neq 0\}$, $J = \{j \mid \varepsilon_j \neq 0\}$, I being unknown to us (it is called the *set of error locators*) and J being known (the *set of erasure locators*). The vector e is called the *error vector*, and ε the *erasure vector*. If there is an algorithm which finds out the nearest code vector u (starting with v) for any e and ε such that $|I| \le t$ and $|J| \le s$, we say that the code *corrects* t *errors and* s *erasures*.

Exercise 1.2.30. Suppose that there is given an algorithm decoding any shortening $C' \subseteq \mathbb{F}_q^{n'}$ of the code C up to $t' \le \left\lceil \frac{d'-1}{2} \right\rceil$, where C is an $[n,k,d]_q$-code and $d' = d - (n - n')$. Construct a decoding algorithm for C correcting any t errors and s erasures for $2t + s \le d - 1$.

Then we turn to the decoding of concatenated codes.

Proposition 1.2.31. *Let C be a linear $[n,k,d]_q$-code, $q = r^m$, which is given together with a decoding algorithm correcting t errors and s erasures for any t and s such that $2t + s \le d' - 1$, $d' \le d$. Let C_0 be an $[N,m,D]_r$-code given with a decoding algorithm correcting any $T \le \left\lceil \frac{D'-1}{2} \right\rceil$ errors, $D' \le D$. Then the concatenated $[n{\cdot}N, k{\cdot}m, d{\cdot}D]_r$-code C' has a decoding algorithm correcting any $u \le \left\lceil \frac{d'{\cdot}D' - 1}{2} \right\rceil$ errors.* ■

Exercise 1.2.32. Prove the proposition. (*Hint*: Decode the given vector first with the decoding algorithm for C_0, then decode the obtained vector and all vectors got from it by erasing one symbol using the algorithm for C; among the vectors you get there is one you want).

CHAPTER 1.3

ASYMPTOTIC PROBLEMS

The parameters of codes of large length n are of particular interest. As it happens quite often, if a problem has no easy answer for each particular n - just as our problem of finding out the best possible parameters of a code - passing to the limit for $n \longrightarrow \infty$ helps us to avoid "deviations" and to understand the behaviour of parameters better. In this chapter we state the problems rigorously and discuss those results that do not use algebraic-geometric codes. We shall return to asymptotic problems in Chapter 3.4, since asymptotic results are the best to demonstrate the power of algebraic-geometric methods.

In Section 1.3.1 we put the main asymptotic problem ($n, k, d \longrightarrow \infty$; $k/n, d/n \longrightarrow const$). In Section 1.3.2 we discuss known results in the direction of its solution (except those applying algebraic geometry), i.e. upper and lower asymptotic bounds. Section 1.3.3 is devoted to the problem of effective construction of asymptotically good codes, and Section 1.3.4 to the results in this direction (i.e. to the polynomial bounds). In Section 1.3.5 we discuss other asymptotic problems (e.g. $n, k \longrightarrow \infty$, $d = const$) and present some results.

1.3.1. The principal asymptotic problem

Fix the alphabet cardinality q. Each q-ary code C defines a pair of its relative parameters $(\delta(C), R(C))$, i.e. a point in the unit square $[0,1]^2$ on the plane with coordinates (δ, R). All such points form the family of code points V_q. Let U_q denote the set of limit points of this family. We call V_q and U_q *code domains*. In other terms, $(\delta, R) \in U_q$ iff there exists an infinite sequence of different codes C_i such that

$$\lim_{i \longrightarrow \infty} (\delta(C_i), R(C_i)) = (\delta, R) ;$$

since there is only a finite number of codes of each given length, for each sequence $n(C_i) \longrightarrow \infty$. If $\delta > 0$ and $R > 0$ such a sequence (a family) of codes C_i is called *asymptotically good*. Sometimes, by abuse of language, we say just *good codes*.

Theorem 1.3.1. *There exists a continuous function* $\alpha_q(\delta)$, $\delta \in [0,1]$, *such that*

$$U_q = \{(\delta, R) \mid 0 \le R \le \alpha_q(\delta)\} ;$$

moreover $\alpha_q(0) = 1$, $\alpha_q(\delta) = 0$ *for* $(q-1)/q \le \delta \le 1$, *and* $\alpha_q(\delta)$ *decreases on the segment* $[0, (q-1)/q]$.

Proof: Let $(\delta_0, R_0) \in U_q$. Consider the corresponding family of codes C and apply the spoiling lemma to each of them. For a given $[n,k,d]_q$-code Lemma 1.1.34.b yields an $[n-\ell, k-\ell, d]_q$-code for any $\ell \le k$. The corresponding code points lie on the segment of the line $R = 1 - \delta\cdot\frac{n-k}{d}$ passing through the points $\left(\frac{d}{n},\frac{k}{n}\right)$ and $(0,1)$ lower and to the left of the point $(d/n,k/n)$. If $[n_i,k_i,d_i]_q$ are the parameters of C_i, $n_i \longrightarrow \infty$, $k_i/n_i \longrightarrow R_0$, $d_i/n_i \longrightarrow \delta_0$, then the points we obtain fill this segment in more and more dense manner (at each $1/n_i$), and at the limit we get the statement that each point of the segment of the line passing through (δ_0,R_0) and $(0,1)$ for $\delta \ge \delta_0$ belongs to U_q (see Fig. 1.1).

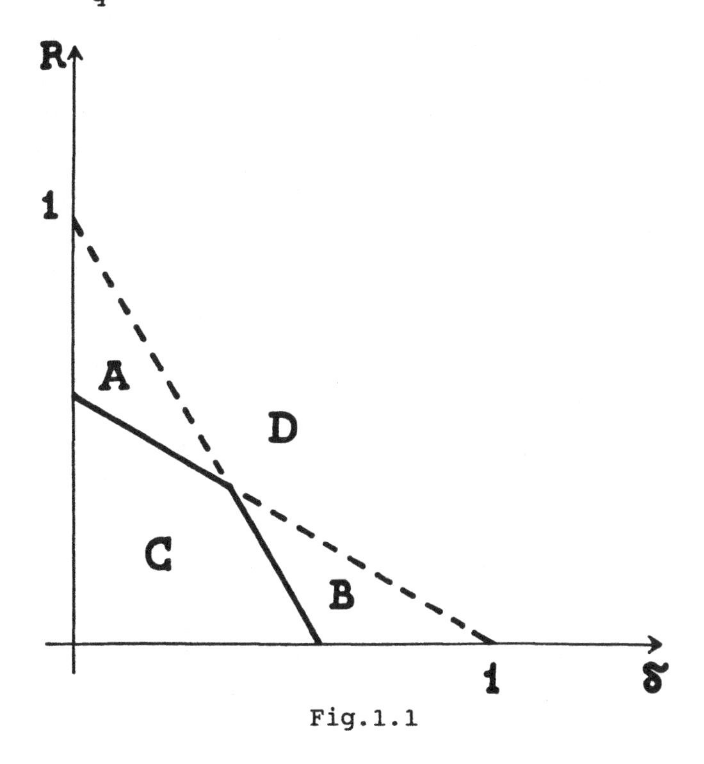

Fig.1.1

Similarly, by Lemma 1.1.34.c we get a segment of the line through (δ_0,R_0) and $(1,0)$ for $\delta \le \delta_0$. This line is given by $R = R_0\cdot(1-\delta)/(1-\delta_0)$.

Let $\alpha_q(\delta) = sup \{R \mid (\delta, R) \in U_q\}$, this function satisfies the first condition of the theorem. Let us check that it is continuous.

By $A = A(\delta_0, R_0)$ we denote the upper and by $B = B(\delta_0, R_0)$ the lower triangle between the lines $L_1 = L_1(\delta_0, R_0)$ given by $R = 1 - (1 - R_0)\cdot\delta/\delta_0$ and $L_2 = L_2(\delta_0, R_0)$ given by $R = R_0\cdot(1 - \delta)/(1 - \delta_0)$ (see Fig. 1.1). Let $R_0 = \alpha_q(\delta_0)$, i.e. (δ_0, R_0) is a point of the boundary of the code domain, and let $\delta_0 \cdot R_0 \neq 0$. We are going to show that all the rest points of the boundary belong to the triangles A and B (this immediately yields the continuity of the boundary at the point (δ_0, R_0)). In fact, if a point (δ_1, R_1) of the boundary lies in $C(\delta_0, R_0)$ (which is defined as the domain below both lines), then, since the point with the same δ_1 and larger R, lying on the segments of $L_1(\delta_0, R)_0$ and $L_2(\delta_0, R_0)$, lies in the code domain, we come to a contradiction: (δ_1, R_1) cannot lie on the boundary. If the boundary point (δ_1, R_1) lies in $D(\delta_0, R_0)$ (i.e. above both lines), then (δ_0, R_0) lies in $C(\delta_1, R_1)$ and cannot be a boundary point. The fact that the boundary lies in sectors A and B also yields its decreasing (up to the point where $\alpha_q(\delta) = 0$).

Let us now show that $\alpha_q(\delta) = 0$ for $\delta \geq (q - 1)/q$. Use the Plotkin bound (Theorem 1.1.39): for any code

$$\delta \leq \frac{q^k \cdot (q - 1)}{(q^k - 1)\cdot q} .$$

If (δ, R) is a code point with $R > 0$ which is the limit one for the family of $[n_i, k_i, d_i]_q$-codes then $k_i \longrightarrow \infty$, hence $\delta \leq (q - 1)/q$. Thus for $\delta > (q - 1)/q$ we have $\alpha_q(\delta) = 0$, by continuity $\alpha_q((q - 1)/q) = 0$ as well.

The only thing we have not proved yet is that $\alpha_q(\delta) > 0$ for $0 \leq \delta < (q - 1)/q$. This follows from Theorem 1.3.15 which we shall prove a bit later. ■

Now let q be a power of a prime.

Define the code domains V_q^{lin} and U_q^{lin} for linear codes similarly to the definition of V_q and U_q (considering only points (δ,R) corresponding to linear q-ary codes).

Exercise 1.3.2. Check that Theorem 1.3.1 is also valid for U_q^{lin} and α_q^{lin} (instead of U_q and α_q).

Of course, $\alpha_q^{lin}(\delta) \leq \alpha_q(\delta)$.

To find out the functions α_q and α_q^{lin} is doubtlessly one of the central problems of the coding theory. Nowadays there are no ideas how to answer these questions. Moreover, we are even unable to solve the following problems:

Problem 1.3.3. Are these functions differentiable in the interval $(0,(q-1)/q)$?

It is not difficult to prove differentiability at 0 and at $(q-1)/q$ (an *Exercise*!).

Problem 1.3.4. Are these functions convex?

Problem 1.3.5. Is it true that $\alpha_q(\delta) = \alpha_q^{lin}(\delta)$, or not?

1.3.2. Asymptotic bounds

Upper bounds. Having no knowledge of exact values of α_q and α_q^{lin} we have nothing to do but to satisfy ourselves with upper and lower bounds for these functions.

Let us first consider the form the results of Section 1.1.4 take when $n \longrightarrow \infty$.

The Singleton bound yields

$$\alpha_q(\delta) \le 1 - \delta \quad .$$

The Griesmer bound gives

$$\alpha_q^{lin}(\delta) \le 1 - \frac{q}{q - 1}\cdot\delta \quad .$$

The same inequality is also true for non-linear codes:

Theorem 1.3.6 **(the asymptotic Plotkin bound).**

$$\alpha_q(\delta) \le R_P(\delta) = 1 - \frac{q}{q - 1}\cdot\delta \quad .$$

■

Exercise 1.3.7. Prove the theorem. (*Hint*: Cf. the proof of Theorem 1.3.1; the Plotkin bound yields $\alpha_q((q - 1)/q) = 0$, the rest follows from the spoiling lemma).

Let us introduce the q-ary entropy function

$$H_q(x) = x\cdot log_q(q - 1) - x\cdot log_q x - (1 - x)\cdot log_q(1 - x) \quad .$$

Theorem 1.3.8 **(the Hamming bound).**

$$\alpha_q(\delta) \le R_H(\delta) = 1 - H_q(\delta/2) \quad .$$

■

Exercise 1.3.9. Prove that

$$\frac{1}{n}\cdot log_q\left(\sum_{i=0}^{t}\binom{n}{i}\cdot(q - 1)^i\right) = H_q\left(\frac{t}{n}\right) + o(1)$$

for $n \longrightarrow \infty$, $t/n \longrightarrow const$. Using this fact derive

Theorem 1.3.8 from Theorem 1.1.41. Prove that

$$\frac{1}{n}\cdot log_q\left(\binom{n}{t}\cdot(q-1)^t\right) = H_q\left(\frac{t}{n}\right) + o(1) .$$

(*Hint*: Use the Stirling formula).

Theorem 1.3.10 (the Bassalygo-Elias bound).

$$\alpha_q(\delta) \le R_{BE}(\delta) = 1 - H_q\left(\frac{q-1}{q} - \frac{q-1}{q}\cdot\sqrt{1 - \frac{q\cdot\delta}{q-1}}\right) .$$

Proof: For $d/n \longrightarrow \delta$, $\frac{1}{n}\cdot log_q A(n,d) = k/n \longrightarrow R$, and $w/n \longrightarrow \omega$ Theorem 1.1.44 yields

$$R \le 1 - H_q(\omega) + \frac{1}{n}\cdot log_q\left[\frac{\delta}{\delta - 2\omega + \frac{q}{q-1}\cdot\omega^2}\right] \sim 1 - H_q(\omega) ,$$

subject to the condition that

$$\omega^2 - 2\cdot\frac{q-1}{q}\cdot\omega + \frac{q-1}{q}\cdot\delta \ge \varepsilon > 0 ;$$

tending $\varepsilon \longrightarrow 0$ and choosing the largest $\omega \le 1$ with this property, i.e.

$$\omega = \frac{q-1}{q} - \sqrt{\left(\frac{q-1}{q}\right)^2 - \frac{q-1}{q}\cdot\delta} ,$$

we get the statement of the theorem. ■

The linear programming bound (Theorem 1.1.50) can be also used to get upper bounds, but one has to apply rather subtle technic which does not fit into the frames of this book. Here is the result:

Theorem 1.3.11 (the McEliece-Rodemich-Ramsey-Welch bound).

$$\alpha_q(\delta) \le R_4(\delta) =$$

$$= H_q\left(\frac{(q-1) - \delta\cdot(q-2) - 2\sqrt{(q-1)\cdot\delta\cdot(1-\delta)}}{q}\right) . \blacksquare$$

Linear programming applied to the constant-weight codes (cf. Remark 1.1.51) for $q = 2$ leads to the following result:

Theorem 1.3.12 (the second McEliece-Rodemich-Ramsey-Welch bound).

$$\alpha_2(\delta) \le R_{4(2)}(\delta) = \min_{0<u\le 1-2\delta} (1 + h(u^2) - h(u^2 + 2\delta\cdot u + 2\delta)) ,$$

where

$$h(x) = H_2\left(\frac{1-\sqrt{1-x}}{2}\right) . \blacksquare$$

Functions R_4 and $R_{4(2)}$ are sometimes called *the bounds of four*. Let us point out that for $q = 2$ and for $\delta \ge 0.273$ the bound $R_{4(2)}$ equals R_4 (since the minimum is achieved for $u = 1 - 2\delta$).

Remark 1.3.13. For q large enough the bound R_4 is not convex. Moreover, it can be ameliorated by the help of the spoiling lemma (cf. Exercise 1.3.7). More precise argument leads to the bound

$$\alpha_q(\delta) \le R_{ALZ}(\delta) =$$

$$= \min \{ \tau + (1-\tau)\cdot(1 - H_q(w) + f_q(\xi,\eta)) - \tau\cdot H_q(\gamma/\tau)\} ,$$

where the function $f_q(\xi,\eta)$ is defined as follows:

$$f_q(\xi,\eta) = H_q(\eta) + (1-\eta)\cdot H_q\left(\frac{\xi-\eta}{1-\eta}\right) -$$

$$- \xi\cdot log_q(q-1) + \eta\cdot log_q(q-2) \quad ,$$

the minimum being taken over τ, w, ξ, η, and γ subject to the conditions

$$0 \le \tau \le 1 \; , \; 0 \le \gamma/\tau \le 1/2 \; , \; 0 \le w \le 1,$$

$$0 \le \eta \le (q-2)\cdot w/(q-1) \quad ,$$

$$0 \le \xi - \eta \le min\ \{w-\eta,\ 1-w\} \quad ,$$

$$\beta = (1-\eta)\cdot h\left(\frac{w-\eta}{1-\eta}\ ,\ \frac{\xi-\eta}{1-\eta}\right) \le w - \frac{q-1}{q-2}\cdot\eta \quad ,$$

$$\delta \ge 2\gamma + \left(2\beta + (w-\beta)\cdot K_{q-1}\left(\frac{\eta}{w-\beta}\right)\right)\cdot(1-\tau) \quad ,$$

where

$$h(x,y) = \frac{x\cdot(1-x) - y\cdot(1-y)}{1 + 2\sqrt{y\cdot(1-y)}} \quad , \; 0 \le x,\ y \le 1 \quad ,$$

$$K_{q-1}(x) = \frac{q-2}{q-1} - \frac{q-3}{q-1}\cdot x - \frac{2}{q-1}\cdot\sqrt{(q-2)\cdot x\cdot(1-x)} \quad ,$$

$$0 \le x \le 1 \quad .$$

Here we stop to discuss upper (i.e. possibility) bounds, and pass to existence bounds.

Lower bounds. Let

$$R_{GV}(\delta) = 1 - H_q(\delta) \quad .$$

This curve is called the *Gilbert-Varshamov curve.*

Exercise 1.3.14. Check the following facts. On the segment $[0,(q-1)/q]$ the curve $R_{GV}(\delta)$ is differentiable (of class C^{∞}) and convex; $R_{GV}(0)=1$, $R_{GV}((q-1)/q)=0$. For $\delta \longrightarrow 0$ there is the asymptotical equality

$$R_{GV}(\delta) = 1 + \delta\cdot \log_q\delta + o(\delta\cdot\log_q\delta) .$$

In particular, the tangent at $\delta = 0$ is vertical. For $\delta \longrightarrow (q-1)/q$ there is the asymptotical equality

$$R_{GV}\left(\frac{q-1}{q} - x\right) = \frac{q^2}{2(q-1)\cdot \ln q}\cdot x^2 + o(x^2) .$$

The tangent at $(q-1)/q$ is horizontal and the tangent order is two. Tangents to $R_{GV}(\delta)$ are of the form:

$$R_t(\delta) = 1 - (\log_q(q^t + q - 1) - t) - t\cdot\delta , \quad 0 \le t \le \infty .$$

Each $R_t(\delta)$ is tangent to $R_{GV}(\delta)$ at the point

$$\delta_0 = \frac{q-1}{q^t + q - 1}$$

and

$$R(\delta_0) = 1 + \frac{t\cdot q^t}{q^t + q - 1} - \log_q(q^t + q - 1) .$$

Now it is time to explain the role of $R_{GV}(\delta)$ in the coding theory.

Theorem 1.3.15 (the Gilbert-Varshamov bound).

$$\alpha_q^{lin}(\delta) \ge R_{GV}(\delta) = 1 - H_q(\delta) .\blacksquare$$

Exercise 1.3.16. Prove the theorem using Exercise 1.3.9 and Theorem 1.1.51.

Remark 1.3.17. The Gilbert-Varshamov bound has remarkable statistical properties. The following facts, which can be easily stated rigorously, are valid:

a) The parameters of almost all linear codes lie on the curve R_{GV}.

b) Let us allow "to correct" each non-linear $[n,k,d]_q$-code, crossing out at most $\frac{n-1}{n}\cdot q^k$ code vectors. Then almost every code can be corrected so that the parameters of the correction lie on the curve R_{GV}. Note that $log_q(q^k/n) = k - log_q n \sim k$, i.e. correction does not change asymptotic parameters.

Exercise 1.3.18. State rigorously and prove the above facts.

Thus codes whose parameters lie higher than the Gilbert-Varshamov bound are quite rare. "Bad" codes are also quite rare. Nevertheless it is more than easy to construct a bad code, and very difficult to construct a good one. It would be nice to understand the reason. Here is an *ad homini* argument: if it were vice versa, the mathematical coding theory would not exist.

Connections between different q**'s.** Let us derive some corollaries from the constructions of Section 1.2.3.

Theorem 1.3.19.

$$\alpha_q(\delta) \geq max\ \{ \max_{q' \geq q} \{1 - (1 - \alpha_{q'}(\delta))\cdot log_q q'\}\ ,$$

$$\max_{q' \leq q} \{\alpha_{q'}(\delta)\cdot log_q q'\}\}\ . \blacksquare$$

Exercise 1.3.20. Prove the theorem. (*Hint*: Apply alphabet restriction and alphabet extension, cf. Section 1.2.3).

Concatenation helps us to prove the following important fact.

Theorem 1.3.21. *If there exists an* $[n,k,d]_q$*-code* C_0 *,* q *being a prime power and* k *being an integer, then*

$$\alpha_q(\delta) \geq \frac{k}{n}\cdot \alpha_{q^k}\left(\frac{n}{d}\cdot\delta\right) .$$

If C_0 *is linear, the same is valid also for* α_q^{lin} . ■

Corollary 1.3.22.

$$\alpha_q(\delta) \geq \max_C \left(\frac{k}{n}\cdot \alpha_{q^k}\left(\frac{n}{d}\cdot\delta\right)\right) .$$

where max *is taken over all* $[n,k,d]_q$*-codes such that* k *is an integer. The same is also valid for linear codes.* ■

Exercise 1.3.23. Prove the theorem and the corollary.

Other problems. Besides the main functions α_q and α_q^{lin} it is reasonable to consider asymptotic problems with some algorithmic constraints (as we do later, cf. Sections 1.3.3 and 1.3.4).

Self-dual problem. We can also put an asymptotic problem for self-dual codes. For such codes $R = 1/2$. Set

$$\delta_q^{sd} = \limsup \frac{d}{n} ,$$

the limit being taken over all self-dual codes (and,

similarly, δ_q^{qsd} for quasi-self-dual ones). There is a "Gilbert-Varshamov bound" in this case:

Theorem 1.3.24. *$\delta_q^{sd} \geq H_q^{-1}\left(\frac{1}{2}\right)$, where $H_q^{-1}(x)$ is the inverse function to the q-ary entropy.* ■

1.3.3. Polynomial families of codes

Effectiveness problems play an important role in the coding theory. It would be nice to be able to construct good codes explicitly (effectively). The only rigorous notion of effectiveness we know is given by the computational complexity theory. The results of this theory are mostly asymptotic themselves, i.e. we should consider sequences of codes of growing length, and not separate code with given parameters. There are three main questions about a code (or a class of codes): construction, encoding, and decoding. Therefore we arrive at three corresponding questions on the *complexity of construction*, the *complexity of encoding*, and the *complexity of decoding* of some class of codes. We first consider the question of the complexity of construction, and then briefly discuss two others. Let us remark that for all codes considered in this book there is no real problem in encoding, since the most part of them is linear and (knowing the generator matrix) their encoding procedure is trivial, and all the considered non-linear codes differ from linear ones "in a finite number of times", and they are also easy from the point of encoding algorithms.

Let $\{C_i\}$ be a family of q-ary codes of growing length. This means that for any $i = 1,2,\ldots$ there is an $[n_i,k_i,d_i]_q$-code C_i (the alphabet cardinality $q = |A|$ is always fixed). We also suppose that the family is ordered in such a way that $n_{i+1} \geq n_i$ for any i, and that each two codes in the family differ (as subsets in A^n). The last condition yields $n_i \longrightarrow \infty$ for $i \longrightarrow \infty$. We also suppose for simplicity that $q = p^a$, p being a prime, $A = \mathbb{F}_q$, and that all codes C_i are linear (later on we are going to comment on the non-linear case as well). Let $G_i \in Mat(n_i \times k_i, \mathbb{F}_q)$ be a generator matrix of C_i. The family C_i is called *polynomial* (or having a *polynomial construction complexity*), iff there exists an algorithm $\mathfrak{a}$, constructing matrices G_i, whose complexity (the number of elementary operations in $\mathbb{F}_q$ needed) is bound by a polynomial in the length of the code. "Algorithm" here means any reasonable definition of this notion you like, since the particular computational model does not influence the polynomial class.

In general, a construction algorithm for a family of codes $\{C_i\}$ is an algorithm $\mathfrak{a}$ which for each i produces an algorithm A_i of encoding for C_i (to be precise a text $\bar{A}_i$ of this algorithm written in some language). The produced algorithm A_i applied to a vector $x \in \mathbb{F}_q^k$ (for simplicity we consider the case of $k \in \mathbb{Z}$, otherwise instead of $x \in \mathbb{F}_q^k$ one should consider $x \in \{1,2,\ldots,q^k\}$) gives an element $A_i(x) \in C_i \subseteq \mathbb{F}_q^n$ which is different for different values of x. It is clear that a generator matrix G_i of a linear code (together with the rule of multiplying a vector by a matrix) is a particular case of such A_i.

Saying that a family of codes $\{C_i\}$ is polynomial we mean that not only the construction algorithm $\mathfrak{a}$ is polynomial in n_i, but also that the encoding algorithms A_i are polynomial, in the sense that for any x the calculation of $A_i(x)$ needs a polynomial in n_i number of operations in $\mathbb{F}_q$.

We define families of codes having a polynomial decoding procedure in a similar way. Decoding of a code C

up to $t \le \left\lceil \frac{d-1}{2} \right\rceil$ is given by a map $D : U \longrightarrow C$, U being the set of vectors in $\mathbb{F}_q^n$ whose distance to the nearest point of C is at most t, such that $D(x) = x$ for $x \in C$, and $D(x)$ is the nearest to x vector of C in the general case. Let there be given a family $\{C_i\}$ of codes equipped by decoding algorithms $\{B_i\}$ defining maps $D_i : U_i \longrightarrow C_i$ such that $D_i(x)$ is the point of C_i nearest to $x \in U_i$. Then we call $(\{C_i\},\{B_i\})$ a *polynomially decodable* family (or a family with a *polynomial decoding algorithm*) iff there exists a universal algorithm $\mathfrak{b}$ generating all B_i which is polynomial in n_i and the number of operations needed to apply each B_i to a vector is also polynomial in n_i. If $t_i/n_i \longrightarrow \tau$ we say that the family is polynomially decodable up to τ.

1.3.4. Polynomial bounds

Now we are going to expose some definitions and results which are the polynomial analogues of those of Section 1.3.1. They are all proved by thorough study of corresponding proofs of Sections 1.3.1 and 1.1.4, therefore we mostly abstain from giving any proofs here. Of course, not every result can be transferred to the polynomial situation; we shall point this out each time we come over such a case.

Define $U_q^{pol} \subset [0,1]^2$ as a set of limit points of relative parameters for polynomial families of codes, i.e. $(\delta,R) \in U_q^{pol}$ iff there exists a polynomial (in $n_i \longrightarrow \infty$) family of $[n_i,k_i,d_i]_q$-codes with $d_i/n_i \longrightarrow \delta$, $k_i/n_i \longrightarrow R$. The definition of $U_q^{pol,lin}$ (for linear codes) is similar.

Theorem 1.3.25. *There exist continuous functions $\alpha_q^{pol}(\delta)$ and $\alpha_q^{pol,lin}(\delta)$, $\delta \in [0,1]$ such that*

$$U_q^* = \{(\delta,R) \mid 0 \le R \le \alpha_q^*(\delta)\} \quad .$$

Moreover, $\alpha_q^*(0) = 1$, $\alpha_q^*(\delta) = 0$ for $(q-1)/q \le \delta \le 1$, and on the segment $[0,(q-1)/q]$ the functions $\alpha_q^*(\delta)$ are decreasing (here the upper $*$ means either pol , or pol,lin). ■

Exercise 1.3.26. Prove the theorem, supposing that we already know that $\alpha_q^{pol,lin}(\delta) > 0$ for $0 \le \delta < (q-1)/q$ (this is really so, see Theorem 1.3.31 and Exercise 1.3.32 below).

Of course,

$$\alpha_q^{pol,lin}(\delta) \le \alpha_q^{pol}(\delta) \le \alpha_q(\delta) \ , \quad \alpha_q^{pol,lin}(\delta) \le \alpha_q^{lin}(\delta) \ .$$

The main problems about the polynomially constructable codes are the following.

Problem 1.3.27. Find out $\alpha_q^{pol}(\delta)$ and $\alpha_q^{pol,lin}(\delta)$.

Problem 1.3.28. Is it true that $\alpha_q^{pol}(\delta) = \alpha_q(\delta)$, $\alpha_q^{pol,lin}(\delta) = \alpha_q^{lin}(\delta)$, or not?

Problem 1.3.29. Is this at least true for the asymptotic behaviour of these bounds for $\delta \longrightarrow 0$ and $\delta \longrightarrow (q-1)/q$?

We do not know a single specifically polynomial upper bound. All the known upper bounds are those for $\alpha_q(\delta)$.

Codes on the Gilbert-Varshamov bound are constructed by an essentially non-polynomial method, on the way one uses an exponential (in n) search out.

Concatenation is polynomial in the following sense:

Exercise 1.3.30. Start with two polynomial families of codes with parameters $[N_i, K_i, D_i]_{Q_i}$ and $[n_i, k_i, d_i]_q$, $Q_i = q^{k_i}$. Prove that their pairwise concatenations form a polynomial family of $[N_i \cdot n_i, K_i \cdot k_i, D_i \cdot d_i]_q$-codes, and that the same is even true if the family of $[n_i, k_i, d_i]_q$-codes is polynomial only in N_i (it can be exponential in n_i).

The last point is used to prove the following bound:

Theorem 1.3.31 (the Zyablov bound).

$$\alpha_q^{pol,lin}(\delta) \geq R_Z(\delta) =$$

$$= \max_{\delta \leq \delta_0 \leq (q-1)/q} \left\{ \left(1 - \frac{\delta}{\delta_0}\right) \cdot (1 - H_q(\delta_0)) \right\} .$$

Proof: Consider a family of Reed-Solomon $[N_i, K_i, D_i]_{q_i}$-codes, $q_i = q^{k_i}$, such that $k_i \longrightarrow \infty$, $D_i/N_i \longrightarrow \delta_1$, $K_i/N_i \longrightarrow R_1$, $\delta_1 + R_1 = 1$, $N_i = q_i$, and a family of Gilbert-Varshamov $[n_i, k_i, d_i]_q$-codes with $n_i \longrightarrow \infty$, $d_i/n_i \longrightarrow \delta_0$, $k_i/n_i \longrightarrow R_0$, $R_0 = 1 - H_q(\delta_0)$. Concatenation gives us $[n_i \cdot N_i, k_i \cdot K_i, d_i \cdot D_i]_q$-codes with $R = R_0 \cdot R_1 = (1 - H_q(\delta_0)) \cdot (1 - \delta_1)$, $\delta = \delta_0 \cdot \delta_1$. We get the line $(1 - \delta/\delta_0) \cdot (1 - H_q(\delta_0))$, where the parameters of concatenated codes (asymptotically) lie. Since we can choose any δ_0 such that $\delta \leq \delta_0 \leq \frac{q-1}{q}$ (then $R_0 \geq 0$) and $0 \leq \delta_1 = \delta/\delta_0 \leq 1$ (then $R_1 \geq 0$), we get the right-hand side of the inequality. We could have chosen the codes to be linear, so the only point left is to prove that the construction is polynomial in the length $n_i \cdot N_i$ of the codes obtained. Indeed, the Gilbert-Varshamov codes are exponential in n_i and $n_i \sim log(n_i \cdot N_i)$, all the rest being polynomial. ■

Remark 1.3.32. Justesen codes considered in Section 1.2.2 yield a bit worse bound (for $q = 2$)

$$\alpha_2^{pol,lin}(\delta) \geq \frac{1}{2} \cdot \left(1 - \frac{\delta}{H_2^{-1}\left(\frac{1}{2}\right)}\right) ,$$

where H_2^{-1} is the inverse function to $H_2(\delta)$; $H_2^{-1}(1/2) \approx 0.110$. By some elaboration of this construction (using so called punctured Justesen codes) this inequality can be made a bit better.

Remark 1.3.33 (the Blokh-Zyablov bound). Using generalized concatenation one can do better than in Theorem 1.3.31. The bound obtained is

$$\alpha_q^{pol,lin}(\delta) \geq R_{BZ}(\delta) = R_{GV}(\delta) - \delta \cdot \int_0^{R_{GV}(\delta)} \frac{dx}{R_{GV}^{-1}(x)}$$

where

$$R_{GV}(\delta) = 1 - H_q(\delta) =$$

$$= 1 - \delta \cdot log_q(q - 1) + \delta \cdot log_q \delta + (1 - \delta) \cdot log_q(1 - \delta),$$

and R_{GV}^{-1} is the inverse function to R_{GV}). This bound is the best lower polynomial bound known which does not use algebraic-geometric codes.

Exercise 1.3.34. Prove that both Zyablov and Bloch-Zyablov bounds are smooth curves passing through (0,1) and

$((q-1)/q, 0)$. Find out their asymptotics at the ends. (*Answer*:

$$1 - R_{BZ}(\delta) \sim \frac{\ln q}{2} \cdot \delta \cdot (\log_q \delta)^2 \quad \text{for} \quad \delta \longrightarrow 0 ,$$

$$R_{BZ}(x) \sim \frac{q^3}{6(q-1)^2 \cdot \ln q} \cdot x^3 \quad \text{for} \quad x = \frac{q-1}{q} - \delta \longrightarrow 0 \text{).}$$

Theorem 1.3.31 uses concatenation with outer codes over growing alphabets. Let us see, what can be done over a fixed field $\mathbb{F}_{q^k}$.

Theorem 1.3.35. *If there exists an* $[n,k,d]_q$*-code* C_0 *, then*

$$\alpha_q^{pol}(\delta) \geq \frac{k}{n} \cdot \alpha_{q^k}^{pol}\left(\frac{n}{d} \cdot \delta\right) \quad ;$$

if the code C_0 *is linear then*

$$\alpha_q^{pol,lin}(\delta) \geq \frac{k}{n} \cdot \alpha_{q^k}^{pol,lin}\left(\frac{n}{d} \cdot \delta\right) \quad .$$

■

Corollary 1.3.36. $\alpha_q^{pol}(\delta) \geq \max\limits_{C} \left\{ \frac{k}{n} \cdot \alpha_{q^k}^{pol}\left(\frac{n}{d} \cdot \delta\right) \right\}$,

where max *is taken over all* q*-ary codes* C ($[n,k,d]_q$ *being the parameters of* C). *The same is valid for* $\alpha_q^{pol,lin}$ (*max being taken over all linear codes*). ■

Exercise 1.3.37. Prove the theorem and the corollary.

For non-linear codes (and arbitrary alphabets) we have the analogue of the second half of Theorem 1.3.19:

Exercise 1.3.38. Show that

$$\alpha_q^{pol}(\delta) \geq \max_{q' \leq q} \{ \alpha_{q'}^{pol}(\delta) \cdot \log_q q' \} .$$

Remark 1.3.39. On the contrary, the alphabet restriction uses an exponential search out of different shifts of the code and is not suitable to bound α_q^{pol} .

We shall return to bounds for α_q^{pol} in Chapter 3.4, after we learn the main properties of algebraic-geometric codes.

1.3.5. Some other asymptotics

The problem we have studied up to this moment can be called the main asymptotic problem of the coding theory: $n, k, d \longrightarrow \infty$, $k/n \longrightarrow R$, $d/n \longrightarrow \delta$, find the relations between δ and R . There are several other natural asymptotic questions. In any asymptotic problem $n \longrightarrow \infty$, but for k and d it is not necessarily so (in all the problems we suppose that q is fixed).

Fix the minimum distance d , and let $n \longrightarrow \infty$, how can one express the relation between n and the largest possible k ? Set

$$\kappa_q(d) = \lim\inf \left(\frac{n - k}{\log_q n} \right) ,$$

the limit being taken over all $[n,k,d]_q$-codes with fixed q and d . The definition of $\kappa_q^{lin}(d)$ (for linear codes) is similar.

Theorem 1.3.40.

$$\left\lceil \frac{d-1}{2} \right\rceil \le \kappa_q(d) \le \kappa_q^{lin}(d) \le \left] \frac{(d-1)\cdot(q-1)}{q} \right[\quad .$$

Proof: Spheres of radius $t = \left\lceil \frac{d-1}{2} \right\rceil$ centered in vectors of C do not intersect each other. The volume of such a sphere is $V_t^n = \sum_{i=0}^{t} \binom{n}{i}\cdot(q-1)^i \ge \binom{n}{t}\cdot(q-1)^t$. The total volume $V \ge q^k\cdot\binom{n}{t}\cdot(q-1)^t$ is at most the volume of the whole space $\mathbb{F}_q^n$, i.e. $q^k\cdot\binom{n}{t}\cdot(q-1)^t \le q^n$. Hence

$$n - k \ge t\cdot log_q(q-1) + log_q\binom{n}{t} \quad .$$

The first term of the sum disappears when we divide by $log_q n$, and for $n \longrightarrow \infty$ (using the asymptotical equality $\binom{n}{t} \sim n^t$ for a fixed t) we get:

$$\kappa_q(d) \ge \left\lceil \frac{d-1}{2} \right\rceil \quad .$$

Essentially, this is the same Hamming bound as in Theorem 1.1.41.

To prove the existence bound (which in this case is the upper bound) consider BCH-codes with parameters

$$\left[n = q^m - 1 \ , \ k = n - m\cdot\left(d - 1 - \left\lceil \frac{d-1}{q} \right\rceil\right) \ , \ d \right]_q \quad ,$$

Using $m = log_q(n+1) \sim log_q n$ we get

$$\kappa_q^{lin}(d) \le \left] \frac{(d-1)\cdot(q-1)}{q} \right[\quad . \qquad \blacksquare$$

Remark 1.3.41. For $q = 2$ and an odd d the inequalities of Theorem 1.3.40 give the answer:

$$\kappa_q(d) = \kappa_q^{lin}(d) = \frac{d - 1}{2} \ .$$

For $q = 2$ and an even d we can use the

$$[n = 2^m \ , \ k = n - m \cdot \frac{d - 2}{2} \ , \ d]_2\text{-codes}$$

obtained by adding the parity-check position to the BCH

$$[n = 2^m - 1 \ , \ k = n - m \cdot \frac{d - 2}{2} \ , \ d - 1]_2\text{-codes.}$$

These codes give us $\kappa_q(d) = \frac{d - 2}{2} = \left\lceil \frac{d - 1}{2} \right\rceil$.

Let us also remark that for any q and $d = 3$, using Hamming $\left[n = \frac{q^m - 1}{q - 1} \ , \ n - m \ , \ 3 \right]_q$-codes, we get $\kappa_q(3) = 1$.

There are some nice results also for $d = 4, 5, 6$.

Problem 1.3.42. Find out the precise value of $\kappa_q(d)$ for $q > 2$.

Consider another natural asymptotical question. Set

$$\delta_q(k) = \lim \sup \frac{d}{n} \ ,$$

the limit being taken over all $[n,k,d]_q$-codes with fixed q and k ; $\delta_q^{lin}(k)$ is defined similarly.

Theorem 1.3.43.

$$\delta_q(k) = \delta_q^{lin}(k) = \frac{q^k}{q^k - 1} \cdot \frac{q - 1}{q} \ .$$

Proof: The inequality $\delta_q(k) \le \frac{q^k}{q^k - 1} \cdot \frac{q - 1}{q}$ is given by the Plotkin bound (Theorem 1.1.39). The opposite estimate (the existence bound) is given by the Reed-Muller codes of order 1, their parameters being

$$\left[\frac{q^m - 1}{q - 1}, m, q^{m-1}\right]_q . \blacksquare$$

There are other possible asymptotic questions. Since $n \longrightarrow \infty$, and we have already considered the situations when either k or d is constant, the only cases left are those when $n, k, d \longrightarrow \infty$ and either $k/n \longrightarrow 0$, or $d/n \longrightarrow 0$ (if neither, we come to the main asymptotic problem). Here are some examples.

Exercise 1.3.44. Let $\delta_0 > \frac{q - 1}{q}$. Prove that in the class of codes with $\frac{d}{n} \longrightarrow \delta_0$ the dimension is bounded:

$$\mathit{lim\ sup}\ (k) \le \left\lceil \log_q \left(\frac{q \cdot \delta_0}{q \cdot \delta_0 - q + 1}\right)\right\rceil .$$

Thus there are two classes of asymptotic problems left: either $\delta \longrightarrow 0$ and $R \longrightarrow 1$, or $\delta \longrightarrow (q - 1)/q$ and $R \longrightarrow 0$. Here is a way to put problems. Let $\varphi(n)$ be some increasing function with $\varphi(n)/n \longrightarrow 0$.

Set

$$\rho_\varphi(n) = \mathit{inf}(n - k) ,$$

inf being taken over all $[n,k,d]_q$-codes with the given n and $d \ge \varphi(n)$. What is the asymptotic behaviour of $\rho_\varphi(n)$ for $n \longrightarrow \infty$?

The other problem is to find out the asymptotic

behaviour of the best possible distance

$$d_\varphi(n) = sup(d) \quad ,$$

sup being taken over $[n,k,d]_q$-codes with $k \geq \varphi(n)$.

Sometimes we can answer these two questions using results on the asymptotic problem and on those about $\kappa_q(d)$ and $\delta_q(k)$.

Exercise 1.3.45. Prove that for $\varphi = n^\alpha$, $0 < \alpha < 1$,

$$\frac{1-\alpha}{2} \lesssim \frac{\rho_\varphi(n)}{n^\alpha \cdot log_q n} \lesssim min \left\{1 - \alpha \; , \; \frac{q-1}{q}\right\}$$

(*Hint*: The lower estimate is given by the Hamming bound, the upper by the Gilbert-Varshamov bound and by BCH-codes, cf. Theorems 1.1.41, 1.1.51, and 1.2.15).

Exercise 1.3.46. Prove that for $\varphi = n^\alpha$, $0 < \alpha < 1$,

$$d_\varphi(n) \sim \frac{q-1}{q} \cdot n \quad .$$

(*Hint*: Use the bounds for the main asymptotic problem). What is it possible to say about the second asymptotic term?

Remark 1.3.47. Up to now, studying existence bounds in all asymptotics we considered limits over subsequences $\{n_i\} \subseteq \mathbb{N}$ for which the parameters are as good as possible. The other way round also gives rather interesting problems: what is it possible to say about *any* infinite subsequence $\{n_i\} \subseteq \mathbb{N}$? For example, set

$$\kappa'_q(d) = \lim_n sup \left(inf \frac{n-k}{log_q n}\right) \quad ,$$

inf being taken over all codes of length n and minimum distance d (recall that before we have considered $\kappa_q(d) = \liminf_n \left(\inf \frac{n-k}{\log_q n} \right)$). What can one say about $\kappa'_q(d)$? The same type of questions is also interesting for the main asymptotic problem, and for other asymptotics.

To finish this section, here is a question more, which is of particular interest because of algebraic-geometric codes.

Let us call a family of $[n,k_i,d_i]_q$-codes, $i = 0,1,\ldots,n+1-g$, a *g-family* for some $g \in \mathbb{N}$ iff

$$k_i \geq i \quad ,$$

$$k_i + d_i \geq n + 1 - g \quad ,$$

$$(n - k_i) + d_i^{\perp} \geq n + 1 - g \quad .$$

Problem 1.3.48. Set

$$\nu_q = \limsup_{g \longrightarrow \infty} \frac{n_g}{g} \quad ,$$

n_g being the maximum possible length of a q-ary g-family. Find out or bound the value ν_q .

Historical and bibliographic notes to Part 1

The theory of error-correcting codes was born about forty years ago. Though the theory is rather young it is possible to say that the most part of this chapter is quite classical. We are not planning to write a history of the coding theory here, an interested reader should turn to the books [MW/Sl], [Pe/We], [Ber 2], [Lin] and to the collection of papers [Ber 1] . Our exposition of classical results mostly follows an excellent book by F.J.MacWilliams and N.J.A.Sloane [MW/Sl] with the difference that we always try to work over an arbitrary finite field (not only over $\mathbb{F}_2$).

Let us briefly list the authors of the principal achievements. The notion of an error-correcting code was discovered by R.V.Hamming, and of a linear code by D.Slepian. Cyclic codes are due to E.Prange, I.S.Reed, and G.Solomon; BCH codes to A.Hocquenghem, R.C.Bose, and D.K.Ray-Chaudhuri. The spectrum of a dual code was calculated by F.J.MacWilliams. The most part of classes of codes, and of asymptotic bounds are named after those who discovered them. Let us remark that the Gilbert-Varshamov bound was first established by E.N.Gilbert for non-linear codes, and then by R.R.Varshamov for linear ones. The Bassalygo-Elias bound was independently discovered by

P.Elias and L.A.Bassalygo. The statement of the first bound of four (due itself to R.J.McEliece, E.R.Rodemich, H.C.Rumsey Jr., L.R.Welch) for an arbitrary q is due to V.I.Levenshtein. Concatenation is due to G.D.Forney, generalized concatenation to E.L.Blokh, V.V.Zyablov, and V.A.Zinoviev. First asymptotically good codes of polynomial construction were discovered by V.V.Zyablov, almost immediately the construction by J.Justesen followed. The fact that the list of perfect codes is complete was proved by A.Tietäväinen (basing on results of S.P.Lloyd, J.H.van Lint, V.A.Zinoviev and V.K.Leontiev). Applications of the invariant theory to self-dual codes is due to A.M.Gleason. Computational complexity in coding theory was studied by L.A.Bassalygo, V.V.Zyablov, M.S.Pinsker, and others.

We must apologize that in this brief list we are not able to review many excellent advantages (including those we have mentioned in this chapter), and proceed to a bit more detailed review of those results of this chapter which appeared directly or indirectly because of algebraic-geometric codes.

The importance of asymptotic questions was understood from the very beginning of the coding theory. The rigorous approach to these problems (Theorem 1.3.1, Exercise 1.3.2) is due to Yu.I.Manin (cf.[Ma], [Ma/Vl]), and independently to M.J.Aaltonen [Aa 1].

Results on spectra and duality (Exercise 1.1.13, Theorems 1.1.16 and 1.1.26) are taken from a paper by G.L.Katsman and M.A.Tsfasman [Ka/Ts 1] (statements and proofs are slightly changed).

Remark 1.3.13 is due to M.J.Aaltonen [Aa 2], S.N.Litsyn and V.A.Zinoviev. Theorem 1.3.21 and Corollary 1.3.22, as well as Theorem 1.3.35 and Corollary 1.3.36 are taken from papers by G.L.Katsman and the authors [Ka/Ts/Vl 1], [Ka/Ts/Vl 2]; the results of Theorem 1.3.19 and Exercise 1.3.38 come from a note by S.N.Litsyn and M.A.Tsfasman [Li/Ts 1].

Our main goal, when writing this part, was to understand the coding theory results from the point of view of mathematics as a whole. To do that we introduce a simple and natural notion of $[n,k,d]_q$-systems (cf. Section 1.1.2, especially Proposition 1.1.4, Theorems 1.1.6 and 1.1.11) and try to use it in proofs. Let us remark that earlier mostly dual systems were used. The same attempt of understanding concerns our point of view on asymptotic problems (Chapter 1.3, cf. also Introduction and Section A.2 of the Appendix).

PART 2

CURVES

This part contains the algebraic geometry we need. Almost all constructions of our book use only algebraic curves, that is the reason why in this part we concentrate mostly on the theory of algebraic curves. Multi-dimensional algebraic-geometric objects appear as instruments for the study of curves. We try to use the possible minimum of technical devices, and to work mostly with objects lucid for one's geometric intuition. For example schemes appear rather late when we are just absolutely unable to avoid them. Notwithstanding our strife to restrict ourselves to elementary means, the scope of information we really need for algebraic-geometric codes is such that this part may happen to be rather difficult for a non-algebraic-geometric reader. However we hope that (using may be some extra books and papers quoted in our bibliographic notes) the reader with any background will be able to read this part, or at least sections he needs most.

Chapter 2.1 presents principle definitions and tools concerning curves (Jacobians included); for curves over the field $\mathbb{C}$ of compex numbers we also discuss the connections with Riemann surfaces. Chapter 2.2 is consentrated around differential forms and the Riemann-Roch theorem; we also discuss the Hurwitz formula and some properties of the Cartier operator. In Chapter 2.3 the curves are mostly over a finite field, we are interested in their points and divisors defined over the ground field. Chapter 2.4 is devoted to a particular non-trivial case - to elliptic curves - and contains many important and beautiful results specific for such curves. Up to this point all the curves are smooth. Chapter 2.5 deals with singular curves. Chapter 2.6 studies reductions of curves; to study reductions one needs some scheme theory which we introduce (up to representable functors we need to study modular curves in Part 4). We end the part with some historical and bibliographic notes.

CHAPTER 2.1

ALGEBRAIC CURVES

In our book we are mainly concerned with algebraic curves. This chapter contains a description of their basic properties. We use the geometric language but the algebraic approach is also considered; over $\mathbb{C}$ we use also analysis and topology. We do not consider arithmetical questions in this chapter; the ground field k is assumed here to be algebraically closed.

Section 2.1.1 contains elementary properties of quasi-projective varieties; the next Section 2.1.2 is devoted to quasi-projective curves. Section 2.1.3 contains definitions and properties of divisors, linear systems, and of line bundles on curves. Section 2.1.4 is devoted to Jacobians. In Section 2.1.5 we describe how complex algebraic curves are connected with Riemann surfaces.

2.1.1. Quasi-projective varieties

By $\mathbb{A}^n$ we denote here and below the n-dimensional *affine space* over k ; its points $P \in \mathbb{A}^n$ are n-tuples $P = (x_1,\dots,x_n)$, $x_i \in k$. The *projective space* of dimension n is denoted by $\mathbb{P}^n$; its points $Q \in \mathbb{P}^n$ are equivalence classes of n-tuples $Q = (y_0:y_1:\dots:y_n)$, where not all of y_i vanish, and n-tuple $(y_1:\dots:y_n)$ is equivalent to $(\lambda\cdot y_0:\lambda\cdot y_1:\dots:\lambda\cdot y_n)$ for $\lambda \in k$, $\lambda \neq 0$. Note that there is a natural embedding $\mathbb{A}^n \hookrightarrow \mathbb{P}^n$, given by $(x_1,\dots,x_n) \longmapsto (1:x_1:\dots:x_n)$. Hear and below we use this inclusion of $\mathbb{A}^n$ into $\mathbb{P}^n$ unless the contrary is mentioned explicitly. When it is necessary to stress the dependence of $\mathbb{A}^n$ and $\mathbb{P}^n$ on k , we write $\mathbb{A}^n(k)$ and $\mathbb{P}^n(k)$.

Sometimes it is convenient to use a coordinateless definition of $\mathbb{P}^n$ which follows. Let V be a vector space over k of dimension $(n + 1)$. By $\mathbb{P}(V)$ we denote the set of lines in V , i.e. the set of equivalence classes $\{v\}$, $v \in V - \{0\}$, v being equivalent to $\lambda\cdot v$ for $\lambda \in k - \{0\}$. Fixing a basis in V one obtains identifications $V \simeq \mathbb{A}^{n+1}$, $\mathbb{P}(V) \simeq \mathbb{P}^n$.

Closed sets. By a *closed set* X in $\mathbb{A}^n$ (an *affine closed set*) we mean the set of mutual zeroes of a finite number of polynomials $F_1,\dots,F_s \in k\,[T_1,\dots,T_n]$:

$$X = \{P = (x_1,\dots,x_n) \in \mathbb{A}^n \mid$$

$$\mid F_1(x_1,\dots,x_n) = \dots = F_s(x_1,\dots,x_n) = 0\} .$$

(Note that if $s = 0$ then $X = \mathbb{A}^n$). A subset $U \in \mathbb{A}^n$ is *open* in $\mathbb{A}^n$ iff its complement $\mathbb{A}^n - U$ is closed in $\mathbb{A}^n$. A *closed* subset $Y \subseteq \mathbb{P}^n$ in $\mathbb{P}^n$ (a *projective closed set*) is the set of mutual zeroes of a finite number of homogeneous polynomials (*forms*) $G_1, \ldots, G_r \in k[T_0, \ldots, T_n]$:

$$Y = \{Q = (y_0 : \ldots : y_n) \in \mathbb{P}^n \mid$$

$$\mid G_1(y_0, \ldots, y_n) = \ldots = G_r(y_0, \ldots, y_n) = 0\}$$

(here we need forms since the vanishing of G_i should depend only on the equivalence class of Q). A complement $\mathbb{P}^n - Y$ of a closed set is called *open* in $\mathbb{P}^n$. From any closed affine set $X \subseteq \mathbb{A}^n$ one can obtain a projective closed set $\tilde{X} \subseteq \mathbb{A}^n$ which is called the *projective closure* of X : let $F_1 = \ldots = F_s = 0$ be the equations defining X where $F_1, \ldots, F_s \in k[T_1, \ldots, T_n]$, and let the forms $\tilde{F}_i \in k\ [T_0, \ldots, T_n]$ be such that

$$\tilde{F}_i(T_0, \ldots, T_n) = \sum_{j=0}^{deg\ F_i} T_0^{deg\ F_i - j} \cdot F_{ij}(T_1, \ldots, T_s)$$

for $i = 1, \ldots, s$, $F_{ij}(T_0, \ldots, T_n)$ being forms of degree j with

$$\sum_{j=0}^{deg\ F_i} F_{ij}(T_1, \ldots, T_n) = F_i(T_1, \ldots, T_n)\ .$$

Then $\tilde{X} \subseteq \mathbb{P}^n$ is defined by

$$\tilde{F}_1(T_0 : \ldots : T_n) = \ldots = \tilde{F}_s(T_0 : \ldots : T_n)\ .$$

For an arbitrary subset S of $\mathbb{P}^n$ we call $T \subseteq S$ *closed in* S , iff $T = S \cap X$ for a closed $X \subseteq \mathbb{P}^n$. A subset $U \subseteq S$ is called *open in* S iff $S - U$ is closed in S .

Exercise 2.1.1. Check that the above definitions make S a topological space. This topology on S is call the *Zariski topology*.

Exercise 2.1.2. Show that $\mathbb{P}^n$ and $\mathbb{A}^n$ are irreducible in the Zariski topology (recall that a topological space is called *irreducible* iff any two its non-empty open subsets U and V have a non-empty intersection).

Thus $\mathbb{A}^n$ and $\mathbb{P}^n$ are non-separable in the Zariski topology.

Quasi-projective sets. An open subset U of a closed projective set X is called a *quasi-projective set*. In particular, any affine closed set is quasi-projective. A quasi-projective set $X \subseteq \mathbb{P}^n$ is called *reducible* if there exist a pair of non-empty proper closed subsets $X_1 , X_2 \subset X$ such that $X = X_1 \cup X_2$. If it is not the case the set X is called *irreducible*. An irreducible quasi-projective set is called a *quasi-projective variety*. We often omit "quasi-projective" and just speak about varieties. If a variety is closed in $\mathbb{P}^n$ then it is called *projective*.

Theorem 2.1.3. *Any quasi-projective set* X *can be uniquely represented as a union of a finite number of non-empty varieties* X_i *such that* X_i *does not contain* X_j *for* $i \neq j$. ∎

The varieties X_i are called *irreducible components* of X.

Exercise 2.1.4. Check that $\mathbb{A}^1 - \{0\}$ is a quasi-projective variety which is not projective.

Exercise 2.1.5. Check that the set $X = \{T_1 \cdot T_2 = 0\}$, $(T_0:T_1:T_2)$ being the homogeneous coordinates on $\mathbb{P}^2$, is a reducible closed projective set. Find its irreducible components.

Dimension. Let X be a variety. The *dimension* $n = \dim X$ is the largest integer n such that there exists a strictly descending chain of varieties

$$X = X_0 \supset X_1 \supset \ldots \supset X_n \neq \emptyset ,$$

X_i being closed in X_{i-1} for $i = 1,\ldots,n$. If Y is closed in X, then we define its *codimension* in X as $\operatorname{codim}_X Y = \dim X - \dim Y$.

Exercise 2.1.6. Show that $\dim \mathbb{A}^n = \dim \mathbb{P}^n = n$.

Exercise 2.1.7. Let $X \subset Y$ be projective varieties. Show that $\dim X < \dim Y$. In particular, if X is closed in $\mathbb{P}^n$ then $\dim X < n$.

The dimension of a quasi-projective set is, by definition, the maximum of the dimensions of its irreducible components.

Rational functions. Let $X \subseteq \mathbb{P}^n$ be a quasi-projective variety, let $F, G \in k[T_0,\ldots,T_n]$ be forms of the same degree in homogeneous coodinates on $\mathbb{P}^n$, and let $G(P) \neq 0$ for some $P \in X$. We say that the ratio $F/G \in k(T_0,\ldots,T_n)$

defines a *rational function*. Another ratio F'/G' defines the same function as F/G iff $(F'\cdot G - F\cdot G')$ vanishes on X. The set of rational functions on X is denoted by $k(X)$.

Exercise 2.1.8. Check that $k(X)$ is a field (rational functions are added and multiplied in the usual way:

$$F/G + F'/G' = (F\cdot G' + F'\cdot G)/(F\cdot G) \ ,$$

$$(F/G)\cdot(F'/G') = = (F\cdot F')/(G\cdot G') \ ).$$

This field is called the *field of rational functions* on X.

A function $f \in k(X)$ is called *regular* at $P \in X$ iff there exists a representation $f = F/G$ with $G(P) \neq 0$. If this is the case, $f(P) = F(P)/G(P)$ is called the *value* of f at P.

Exercise 2.1.9. Show that for any $f \in k(X)$ the set of $P \in X$ such that f is regular at P is open and non-empty.

If $f \in k(X)$ is regular at any $P \in X$ then it is called *regular* on X. The set $k[X]$ of regular functions on X is a k-algebra.

Exercise 2.1.10. Show that $k[X]$ is an integral ring (i.e. that it has no non-trivial divisors of zero).

Exercise 2.1.11. Show that $k[\mathbb{P}^n] = k$ and that $k[\mathbb{A}^n] = k[T_1,\ldots,T_n]$ is a polynomial ring.

Exercise 2.1.12. Let X be an affine variety. Then the fraction field of $k[X]$ is $k(X)$. (*Hint*: Consider the restrictions of coordinate functions to X).

Exercises 2.1.11 and 2.1.12 show that the properties of regular functions on affine varieties and on $\mathbb{P}^n$ are quite different (cf. also Corollary 2.1.23 below).

Rational maps. A *rational map* f from a variety $X \subseteq \mathbb{P}^m$ to $\mathbb{A}^n$ is given by an n-tuple $f_1,\dots,f_n$, $f_i \in k(X)$; we write $f = (f_1,\dots,f_n)$. If all f_i are regular at $P \in X$ then f is called *regular at* P . If f is regular at any $P \in X$ then it is called a *regular map* from X to $\mathbb{A}^n$.

A *rational map* f from $X \subseteq \mathbb{P}^n$ to $\mathbb{P}^n$ is given by an $(n+1)$-tuple $(F_0:F_1:\dots:F_n)$, F_i being forms of the same degree on $\mathbb{P}^m$ such that there exist $i \in \{1,\dots,n\}$ and $P \in X$ with $F_i(P) \neq 0$. We say that $(n+1)$-tuples $(F_0:F_1:\dots:F_n)$ and $(F'_0:F'_1:\dots:F'_n)$ define the same rational map iff all the forms $(F_i\cdot F'_j - F'_i\cdot F_j)$ vanish on X . If there exists a representation $(F_0:F_1:\dots:F_n)$ of f such that $F_i(P) \neq 0$ for some $i \in \{1,\dots,n\}$ then f is called *regular* at $P \in X$.

Exercise 2.1.13. Show that for any rational map $f : X \longrightarrow \mathbb{P}^n$ the set U_f , formed by $P \in X$ such that f is regular at P , is open in X . If f is regular at any $P \in X$ then f is called a *regular map* (or a *morphism*) of varieties. Thus a rational map of X is a regular map of its open subset. If there exists a variety $Y \subseteq \mathbb{P}^n$ such that $F(P) \in Y$ for any $P \in U_f$ then we write $f : X \longrightarrow Y$ and call f a *rational map from* X *to* Y .

Exercise 2.1.14. Show that a regular map $f : X \longrightarrow Y$ is continuous in the Zariski topology (i.e. that $f^{-1}(U)$ is open in X for any open $U \subseteq Y$).

Exercise 2.1.15. Show that a rational map $f : X \longrightarrow \mathbb{A}^n$ defines a rational map $\overline{f} : X \longrightarrow \mathbb{P}^n$. If f is regular then $\overline{f}$ is also regular. The converse of the last statement is false (give a counter-example).

Let $f : X \longrightarrow Y$ be a rational map. The set $f(U_f) = Im\, f$ is called the *image* of f. If it is dense in Y (i.e. its closure equals Y) then f is called *dominant*. Clearly, any rational map $f : X \longrightarrow \mathbb{P}^n$ can be considered as a dominant map $f : X \longrightarrow Y$, Y being the projective closure of $Im\, f$.

Let $f : X \longrightarrow Y$ be a dominant rational map, and let $g \in k(Y)$. Define $f^*(g)$ by

$$f^*(g)(P) = g(f(P)) \quad , \qquad (2.1.1)$$

P being a point such that $P \in U_f$, and $f(P) \in U_g$.

Exercise 2.1.16. Check that $f^*(g) \in k(X)$.

Thus a rational dominant map $f : X \longrightarrow Y$ defines an embedding $f^* : k(Y) \hookrightarrow k(X)$ which is identical on k. (Check that f^* is an embedding!).

Exercise 2.1.17. Let $\varphi : k(Y) \hookrightarrow k(X)$ be an arbitrary embedding over k. Show that there exists a rational dominant map $f : X \longrightarrow Y$ such that $\varphi = f^*$.

If $f : X \longrightarrow Y$ and $f' : Y \longrightarrow Z$ are rational maps and f is dominant, then one can define the composition $f' \circ f$ in the usual way, namely $(f' \circ f)(P) = f'(f(P))$ for $P \in X$. Since f is dominant, $U_{f'} \cap Im\, f \neq 0$ and $f' \circ f$ is a rational map.

Exercise 2.1.18. Let f and f' be dominant. Show that $(f' \circ f)^* = f^* \circ f'^*$.

Let $f : X \longrightarrow Y$ be a regular map (we do not assume it to be dominant). Then (2.1.1) defines a homomorphism of k-algebras $f^* : k[X] \longrightarrow k[Y]$.

Exercise 2.1.19. Let $f : X \longrightarrow Y$ and $f' : Y \longrightarrow Z$ be regular maps. Show that $(f' \circ f)^* = f^* \circ f'^*$.

If $k(X)$ is isomorphic to $k(Y)$ as a k-algebra then X and Y are called *birationally isomorphic*. According to Exercises 2.1.17 and 2.1.18 this is equivalent to the existence of dominant rational maps $f : X \longrightarrow Y$ and $f' : Y \longrightarrow X$ such that $f' \circ f$ and $f \circ f'$ are (as rational maps) identical automorphisms of X and Y , respectively. If f and f' are regular then X and Y are called *isomorphic*. It is possible to consider isomorphic varieties as just one variety with different projective embeddings.

Exercise 2.1.20. Show that a variety is birationally isomorphic to any its open non-empty subset.

Exercise 2.1.21. Let X be a hyperbola in $\mathbb{A}^2$ given by $x_1 \cdot x_2 = 1$ and let $Y = \mathbb{A}^1 - \{0\}$. Show that the projection $f : X \longrightarrow Y$ defines an isomorphism from X to Y .

If a variety is birationally isomorphic to $\mathbb{A}^n$ (or, which is the same, to $\mathbb{P}^n$), then it is called a *rational variety;* for example Exercise 2.1.21 shows that hyperbola is a rational curve. If a variety is isomorphic to an affine closed set then it is called an *affine variety*. Note that Exercise 2.1.21 shows that an affine variety can be non-closed while embedded into an affine space. For projective varieties the situation is completely different.

Theorem 2.1.22. *Let* $f : X \longrightarrow \mathbb{P}^n$ *be a regular map,* X *being a projective variety. Then* $f(X)$ *is closed in* $\mathbb{P}^n$. ■

Corollary 2.1.23. $k[X] = k$ *for a projective* X. ■

Corollary 2.1.24. *An affine variety of positive dimension is never isomorphic to a projective variety.* ■

In contrast with Corollary 2.1.23 an affine variety is completely determined by its algebra of regular functions.

Exercise 2.1.25. Let X and Y be affine varieties. Show that for any k-linear morphism $\varphi : k[X] \longrightarrow k[Y]$ there exists a unique regular map $f : Y \longrightarrow X$ such that $\varphi = f^*$. (*Hint:* $k[X] = k[\mathbb{A}^n]/I_X$, I_X being an ideal formed by polynomials vanishing on X).

Exercise 2.1.26. Deduce from the last exercise that affine varieties X and Y are isomorphic iff $k[X]$ and $k[Y]$ are k-isomorphic.

Local ring of a point. Let P be a point of X. The set of rational functions which are regular at P is denoted by O_P; it is clear that O_P is a ring. When we want to stress its dependence on X we denote it by $O_{X,P}$.

Exercise 2.1.27. Check that O_P has a unique maximal ideal m_P formed by $f \in O_P$ with $f(P) = 0$ (recall that a commutative ring with unique maximal ideal is called *local*).

The ring O_P is called the *local ring* of P.

Exercise 2.1.28. Show that $\mathcal{O}_P/m_P = k$. (*Hint:* $\mathcal{O}_P/m_P$ is an algebraic extension of k).

The factor-group m_P/m_P^2 is an $\mathcal{O}_P/m_P$-module, i.e. it is a k-vector space. The space $\theta_P = (m_P/m_P^2)^*$ dual to m_P/m_P^2 is called the *tangent space* to X at P (and m_P/m_P^2 itself is called the *co-tangent space* to X at P).

Exercise 2.1.29. Let $X \subseteq \mathbb{P}^n$. Show that for any $P \in X$

$$dim\ X \le dim_k\ \theta_P \le n\ .$$

Smooth and singular points. A point P is called *smooth* (synonims: *non-singular*, *regular*, *simple*) iff $dim_k\ \theta_P = dim\ X$. If it is not the case P is called *singular*.

Exercise 2.1.30. Show that $P = (0,0,0)$ is singular on the surface $X \subset \mathbb{A}^3$ defined by $x_1 \cdot x_2 = x_3^2$. All the other points $Q \in X$ are smooth.

If each point of X is smooth, X is called a *smooth* (or a *non-singular*) *variety*.

Exercise 2.1.31. Show that $\mathbb{P}^n$ and $\mathbb{A}^n$ are smooth varieties.

Proposition 2.1.32. *The set* X_{smooth} *formed by non-singular points of* X *is open and non-empty.* ■

Product of varieties. Let $X \subseteq \mathbb{A}^r$ and $Y \subseteq \mathbb{A}^s$. The set of pairs $(P,Q) \in \mathbb{A}^{r+s}$ for $P \in X$ and $Q \in Y$ is called the *product* of X and Y .

Exercise 2.1.33. Check that if X and Y are closed sets then $X \times Y$ is also closed in $\mathbb{A}^{r+s}$.

Exercise 2.1.34. Show that if X and Y are varieties then $X \times Y$ is also a variety.

Now let $X \subseteq \mathbb{P}^n$ and $Y \subseteq \mathbb{P}^m$ be varieties, and let $X \times Y$ be the set of pairs (P,Q) for $P \in X$ and $Q \in Y$. We want to find an embedding $\varphi : X \times Y \hookrightarrow \mathbb{P}^N$ (for some N) such that $\varphi(X \times Y)$ is a quasi-projective subvariety of $\mathbb{P}^N$. It is sufficient to consider the case $X = \mathbb{P}^n$, $Y = \mathbb{P}^m$. Indeed, if $\psi : \mathbb{P}^n \times \mathbb{P}^m \hookrightarrow \mathbb{P}^N$ is already constructed then $\varphi : X \times Y \hookrightarrow \mathbb{P}^N$ is obtained by restriction of ψ to $X \times Y \subseteq \mathbb{P}^n \times \mathbb{P}^m$.

To construct $\psi : \mathbb{P}^n \times \mathbb{P}^m \hookrightarrow \mathbb{P}^N$ we set $N = (m+1)\cdot(n+1) - 1 = m\cdot n + m + n$. Let T_{ij} be homogeneous coordinates on $\mathbb{P}^N$ for $i = 0,1,\dots,n$ and $j = 0,1,\dots,m$. For $P = (x_0: \dots :x_m) \in X$ and $Q = (y_0: \dots :y_n) \in Y$ we set $\varphi(P, Q) = (x_i \cdot y_j)$.

Exercise 2.1.35. Show that $\varphi(\mathbb{P}^n \times \mathbb{P}^m)$ is closed in $\mathbb{P}^N$. (*Hint:* $\varphi(\mathbb{P}^n \times \mathbb{P}^m)$ is defined by the equations $T_{ij}\cdot T_{k\ell} = T_{i\ell}\cdot T_{kj}$).

Proposition 2.1.36. *For any varieties* X *and* Y

$$\dim X \times Y = \dim X + \dim Y .$$

■

Line bundles. Let X be a variety. A *family of vector spaces* on X is a regular map $p : \mathcal{E} \longrightarrow X$ such that for any $P \in X$ its *fibre* $\overline{\mathcal{E}}_P = p^{-1}(P)$ is a vector space over k , and two structures of a variety on $\overline{\mathcal{E}}_P$ coincide (the first structure being that of a vector space over k and the second being that of a closed subset in $\mathcal{E}$; note that $\overline{\mathcal{E}}_P$ is closed by Exercise 2.1.14).

Morphism from a family $p : \mathcal{E} \longrightarrow X$ to a family $p' : \mathcal{E}' \longrightarrow X$ is a regular map $f : \mathcal{E} \longrightarrow \mathcal{E}'$ such that $p = p' \circ f$ and for any $P \in X$ the induced map of fibres $f_P : \overline{\mathcal{E}}_P \longrightarrow \overline{\mathcal{E}}'_P$ is k-linear. Isomorphism of families is defined in a natural way.

The simplest example of family of vector spaces is the product $X \times \mathbb{A}^m$ with its natural projection $p_X : X \times \mathbb{A}^m \longrightarrow X$. This family is called *trivial*. If U is open in X and $p : \mathcal{E} \longrightarrow X$ is a family then the restriction $p|_{p^{-1}(U)} : p^{-1}(U) \longrightarrow U$ is also a family of vector spaces over U, which is called the *restriction of* $\mathcal{E}$ *to* U, and is denoted $\mathcal{E}|_U$.

Further on we consider only line families (i.e. with $dim\, \overline{\mathcal{E}}_P = 1$ for any $P \in X$). A line family $p : \mathcal{L} \longrightarrow X$ is called a *line bundle* on X iff for any $P \in X$ there exists an open set U, $P \in U \subset X$, such that $\mathcal{E}|_U$ is isomorphic to the trivial family. We denote line bundles by $\mathcal{L}, \mathcal{M}, \mathcal{N}$ etc., the trivial line bundle is denoted by $\mathcal{O}$. A regular map $s : X \longrightarrow \mathcal{L}$ such that $p \circ s = id_X$ is called a *section* of the bundle $p : \mathcal{L} \longrightarrow X$. In particular any line bundle $\mathcal{L}$ has the zero-section s_0 such that $s_0(P) = 0 \in \overline{\mathcal{L}}_P$. The set of sections $s : X \longrightarrow \mathcal{L}$ is a vector space over k with operations

$$(s + s')(P) = s(P) + s'(P) \;, \quad (\alpha \cdot s)(P) = \alpha \cdot s(P)$$

for $P \in X$ and $\alpha \in k$. This vector space is denoted by $H^0(X,\mathcal{L})$. In particular, one easily checks that $H^0(X,\mathcal{O}) = k[X]$. Any $s \in H^0(X,\mathcal{L})$ defines a map $\tilde{s} : \mathcal{O} \longrightarrow \mathcal{L}$ which is uniquely determined by the condition $\tilde{s}(1) = s(P)$, 1 being a fixed constant section of $\mathcal{O}$. The set $T_s = \{P \in X \mid s(P) = 0\}$ is called the *set of zeroes* of s. From the definition of a line bundle and Exercise 2.1.14 one can deduce that T_s is a closed subset of X.

Let $p : \mathcal{L} \longrightarrow X$ be a line bundle, and let $X = \bigcup_{\alpha} U_\alpha$ be an open covering of X such that $\mathcal{L}|U_\alpha$ is trivial for any α. Let $\varphi_\alpha : \mathcal{L}|_{U_\alpha} \xrightarrow{\sim} U_\alpha \times \mathbb{A}^1$ be corresponding isomorphisms. Let us consider the map

$$\varphi_\alpha \circ \varphi_\beta^{-1} : (U_\alpha \cap U_\beta) \times \mathbb{A}^1 \longrightarrow (U_\alpha \cap U_\beta) \times \mathbb{A}^1 .$$

It is an isomorphism of trivial families over $U_\alpha \cap U_\beta$. Let $P \in U_\alpha \cap U_\beta$ then $(\varphi_\alpha \circ \varphi_\beta^{-1})_P$ is a k-automorphism of $\overline{\mathcal{L}}_P$ and since $dim\, \overline{\mathcal{L}}_P = 1$ it is given by an element $\lambda_P \in k^*$. Therefore $(\varphi_\alpha \circ \varphi_\beta^{-1})$ defines a function $f_{\alpha\beta} \in k[U_\alpha \cap U_\beta]$ which vanished nowhere on $U_\alpha \cap U_\beta$. For any α one can check that $f_{\alpha,\alpha} = id$, and for any triple (α,β,γ) one can check that $f_{\alpha\gamma} = f_{\alpha\beta} \cdot f_{\beta\gamma}$ on $U_\alpha \cap U_\beta \cap U_\gamma$. Conversely any family of functions $f_{\alpha\beta} \in k[U_\alpha \cap U_\beta]$ satisfying these requirements determines a line bundle.

Let $\mathcal{L}$ and $\mathcal{M}$ be line bundles given by families $\{f_{\alpha\beta}\}$ and $\{g_{\alpha\beta}\}$, respectively, with $f_{\alpha\beta}$ and $g_{\alpha\beta} \in k[U_\alpha \cap U_\beta]$ (one can easily show that there exists an open covering $\{U_\alpha\}$ of X trivializing both $\mathcal{L}$ and $\mathcal{M}$). The family $\{f_{\alpha\beta} \cdot g_{\alpha\beta}\}$ defines a line bundle $\mathcal{L} \otimes \mathcal{M}$ which is called the *(tensor) product* of $\mathcal{L}$ and $\mathcal{M}$. The family $f_{\alpha\beta}^{-1}$ defines a line bundle $\mathcal{L}^{-1} = \mathcal{L}^{\otimes -1}$ such that $\mathcal{L}^{-1} \otimes \mathcal{L} = \mathcal{O}$. The tensor product of $m > 0$ copies of $\mathcal{L}$ is denoted $\mathcal{L}^{\otimes m}$ or $\mathcal{L}^m$; for $m < 0$ we set $\mathcal{L}^m = \mathcal{L}^{\otimes m} = (\mathcal{L}^{-1})^{\otimes(-m)}$. By definition, $\mathcal{L}^0 = \mathcal{L}^{\otimes 0} = \mathcal{O}$. One can check that the set $Pic(X)$ of line bundles on X is a group which is called the *Picard group* of X.

Here is the most important line bundle on $\mathbb{P}^n$. Let V be a vector space of dimension $n + 1$ such that $\mathbb{P}^n = \mathbb{P}(V)$. Let

$$\mathcal{E} = \{(P,v) \in \mathbb{P}^n \times V \mid v \in \ell_P\}$$

ℓ_P being the line in V corresponding to P, and let $p : \mathcal{E} \longrightarrow \mathbb{P}^n$ be the restriction of the natural projection $\mathbb{P}^n \times V \longrightarrow \mathbb{P}^n$. One can show that $\mathcal{E}$ is a line bundle on $\mathbb{P}^n$ which is denoted by $\mathcal{O}(-1)$. Let $\mathcal{O}(1) = \mathcal{O}(-1)^{-1}$ and let $\mathcal{O}(m) = \mathcal{O}(1)^{\otimes m}$. For any $s \in H^0(\mathbb{P}^n, \mathcal{O}(1))$ its zero-set T_s is a hyperplane in $\mathbb{P}^n$. Conversely, for any hyperplane $H \subset \mathbb{P}^n$ there exists $s \in H^0(\mathbb{P}^n, \mathcal{O}(1))$ with $T_s = H$. Moreover, there is a natural isomorphism $H^0(\mathbb{P}^n, \mathcal{O}(1)) = V^*$, V^* being the space dual to V.

Exercise 2.1.37. Prove all the above statements about line bundles.

2.1.2. Quasi-projective curves

Quasi-projective variety of dimension 1 is called a *quasi-projective curve* (*algebraic curve* or just a *curve*). Since for any non-empty variety X any $P \in X$ is closed in X, one can define a curve to be a variety such that any its closed subvariety is a point. If a curve X is a closed subset in $\mathbb{P}^n$ it is called *projective* (or *complete*); note that by Theorem 2.1.22 this property depends only on the isomorphism class of X.

Example 2.1.38. The zero set in $\mathbb{P}^2$ of an irreducible form F is a complete curve; $\mathbb{A}^1$ is not complete.

Sometimes we call the union $X = \bigcup_{i=1}^{n} X_i$ of several curves a (*reducible*) *curve*; in this case curves X_i are called *irreducible components*.

Non-singular points on curves. From Proposition 2.1.32 it follows that a curve X has only a finite number of singular points. Let $P \in X$ be an arbitrary point.

Exercise 2.1.39. Show that $P \in X$ is non-singular iff the ideal m_P is principal (i.e. $m_P = t_P \cdot O_P$ with $t_P \in O_P$).

If P is non-singular then any $t_P \in O_P$ with $m_P = t_P \cdot O_P$ is called a *local parameter* at P. To check non-singularity of a point $P \in X$ it is convenient to use the *differential criterion of non-singularity.* Let $P \in X \subseteq \mathbb{P}^n$. Without loss of generality we can assume that $P \in X \cap \mathbb{A}^n$ (since $\mathbb{P}^n$ is a union of $(n+1)$ copies of $\mathbb{A}^n$). Let $(x_1, \dots, x_n)$ be coordinates on $\mathbb{A}^n$, and let $F_1, \dots, F_N \in k[x_1, \dots, x_n]$ be generators of the ideal $I_{X \cap \mathbb{A}^n}$ (recall that $I_{X \cap \mathbb{A}^n}$ is the ideal of polynomials vanishing on $X \cap \mathbb{A}^n$). Thus $F_1(x) = \dots = F_N(x) = 0$ is a complete system of equations defining $X \cap \mathbb{A}^n$. Consider the matrix

$$\frac{\partial F}{\partial x}(P) = \left(\frac{\partial F_i}{\partial x_j}(P) \right)_{\substack{i=1,\dots,N \\ j=1,\dots,n.}}$$

Proposition 2.1.40. *P is non-singular iff the rank of $\frac{\partial F}{\partial x}(P)$ equals $n-1$.* ■

In particular, if $N = 1$, $n = 2$ (the case of a plane curve with affine equation $F(x_1, x_2) = 0$) a point P is non-singular on X iff either $\frac{\partial F}{\partial x_1}(P) \neq 0$ or $\frac{\partial F}{\partial x_2}(P) \neq 0$.

Exercise 2.1.41. Which of the points listed below are non-singular?

a) $P = (0,0)$ on X given by $x_1^2 = x_2^3$;

b) $P = (0,1)$ on $X : \{x_1^2 + 1 = x_2^3\}$;

c) $P = (0,0,0)$ on $X : \{x_1^2 = x_2^2 + x_3 \ , \ x_1^2 = x_3^2 + x_2\}$.

Power series expansions. Non-singular points have the following important property: a function which is regular at a non-singular point P has a unique power series expansion provided that a local parameter t_P is fixed.

Let P be a non-singular point of X. Fix a local parameter t_P at P. For any power series $F(t) = \sum_{i=0}^{\infty} a_i \cdot t^i \in k[[t]]$ and for a positive integer m by $F(t)_m$ we denote the polynomial $\sum_{i=0}^{m} a_i \cdot t^i$ of degree m .

Proposition 2.1.42. *There exists a unique embedding of k-algebras* $\tau_P : \mathcal{O}_P \longrightarrow k[[t_P]]$ *such that* $f - \tau_P(f)_n \in m_P^{n+1}$ *for any* $n > 0$. ∎

Exercise 2.1.43. Let $X = \mathbb{A}^1$, $P = \{0\}$, $t_P = t$, $f = (1 - t)^{-1}$. Find $\tau_P(f)$. (*Hint*: This is a geometric progression).

Note that any rational function f can be uniquely expanded into a Laurent series at any non-singular $P \in X$ since there exists an integer m such that $t_P^m \cdot f \in \mathcal{O}_P$, and we can set $\tau_P(f) = t_P^{-m} \cdot \tau_P(t_P^m \cdot f)$. Thus we obtain a field embedding $\tau_P : k(X) \hookrightarrow k((t_P))$ for any nonsingular $P \in X$.

Smooth complete curves. Smooth complete curves are the most important. They have the following useful property.

Theorem 2.1.44. *Smooth complete curves are isomorphic iff they are birationally isomorphic.* ∎

Exercise 2.1.45. Let $X \subset \mathbb{P}^2$ be the curve defined by $x_0 \cdot x_1^2 = x_2^3$. Show that X is birationally isomorphic to $\mathbb{P}^1$ (i.e. X is a rational curve). Is X isomorphic to $\mathbb{P}^1$?

Since it is clear that $\mathbb{A}^1$ and $\mathbb{P}^1$ are birationally isomorphic, but are not isomorphic, we see that for non-complete or singular curves Theorem 2.1.44 is not valid.

Remark 2.1.46. For higher-dimensional varieties Theorem 2.1.44 is also wrong. For example smooth projective surfaces $\mathbb{P}^2$ and $\mathbb{P}^1 \times \mathbb{P}^1$ are not isomorphic, being birationally isomorphic.

One can describe fields of the form $k(X)$ for a complete non-singular curve X.

Proposition 2.1.47. *A field K containing k is isomorphic to $k(X)$ for a complete non-singular curve X iff K is of transcendence degree one and is finitely generated over k.* ∎

These results make it possible to use both equivalent languages - algebraic and geometric - when studying smooth complete curves.

Let as explain how one can algebraically express the notion of a point on a curve.

Let V be an arbitrary proper subring in $k(X)$, and let $V \supset k$. The ring V is called a *valuation ring* iff the condition $f \notin V$ implies $f^{-1} \in V$. Let as show that the local ring $\mathcal{O}_P$, P being a non-singular point of X, is a valuation ring. To prove this we introduce the following important notion. For $f \in \mathcal{O}_P$, $f \neq 0$ we set

$$ord_P(f) = max \{ k \mid f \in m_P^k , f \notin m_P^{k+1} \} .$$

For any $f = g/h$, $g,h \in \mathcal{O}_P$ we set

$$ord_P(f) = ord_P(g) - ord_P(h) .$$

Since $\mathcal{O}_P \supset k[U]$ for any open neighbourhood U of P , the fraction field of $\mathcal{O}_P$ is $k(X)$. Thus $ord_P(f)$ is defined for any $f \in k(X) - \{0\}$. It is called the *order (of zero)* of f at P . If $ord_P(f) < 0$ then $|ord_P(f)|$ is called the *order of pole* of f at P . Hence, ord_P is a surjective group homomorphism $ord_P : k(X)^* \longrightarrow \mathbb{Z}$ such that for any $f,g \in k(X)^*$

$$ord_P(f\cdot g) = ord_P(f) + ord_P(g) ;$$

$$ord_P(f + g) \geq min (ord_P(f) , ord_P(g)) .$$

Such a homomorphism is called a *discrete valuation* on $k(X)$. Note that

$$\mathcal{O}_P = \{f \in k(X)^* \mid ord_P(f) \geq 0\} \cup \{0\} .$$

From this description it is obvious that $f \in k(X)$, $f \notin \mathcal{O}_P$ implies $f^{-1} \in \mathcal{O}_P$. Hence $\mathcal{O}_P$ is a valuation ring. Moreover, one can prove

Theorem 2.1.48. *Let X be smooth projective curve. Then $P \longmapsto \mathcal{O}_P$ is a bijection of the set of points $P \in X$ onto the set of valuation subrings in $k(X)$.* ∎

Degree of a map. Note that if $f : X \longrightarrow Y$ is a non-constant rational map then f is dominant. Hence

$f^* : k(Y) \hookrightarrow k(X)$ is an embedding. Since $k(X)$ and $k(Y)$ are of transcendence degree 1 over k and are finitely generated, the degree $[k(X) : f^*(k(Y))]$ is finite; it is called the *degree* of f and is denoted by $\deg f$. In the next section we give another interpretation of $\deg f$.

2.1.3. Divisors

Let X be a smooth projective curve over k (recall that we assume k to be algebraically closed and X to be irreducible). A *divisor* D on X is a finite formal sum $D = \sum a_P \cdot P$, $P \in X$, a_P being integers. The set $\{P \in X \mid a_P \neq 0\}$ is called the *support* of D and is denoted by $Supp\, X$. Sometimes we write $D = \sum a_i \cdot P_i$ instead of $D = \sum a_{P_i} \cdot P_i$, where $a_i = a_{P_i}$. The set of divisors on X is denoted by $Div(X)$; it is an abelian group since we can add and subtract divisors: if $D = \sum_P a_P \cdot P$ and $E = \sum_P b_P \cdot P$, then $D \pm E = \sum_P (a_P \pm b_P) \cdot P$. The *degree* $\deg D$ of D is $\sum a_P$, the degree map $Div(X) \longrightarrow \mathbb{Z}$ is surjective, its kernel is denoted by $Div^0(X)$. If $a_P \geq 0$ for any P then we call $D = \sum a_P \cdot P$ an *effective* divisor and write $D \geq 0$. If moreover $D \neq 0$ we call it *positive*. The set of effective divisors is denoted by $Div^+(X)$. This definition induces a partial order on $Div(X)$: $D \geq F$ iff $D - F \in Div^+(X)$. Note that any divisor is a difference of two effective ones.

For a function $f \in k(X)^*$ let

$$(f) = \sum ord_P(f) \cdot P \quad ;$$

(f) is called the *divisor of* f. Note that (f) is indeed a divisor since any $f \in k(X)^*$ has only a finite number of zeroes and poles and thus $ord_P(f) \neq 0$ only for a finite number of points $P \in X$. Note also that

$$(f) = (f)_0 - (f)_\infty ,$$

where

$$(f)_0 = \sum_{ord_P(f)>0} ord_P(f) \cdot P$$

and

$$(f)_\infty = \sum_{ord_P(f)<0} (-ord_P(f)) \cdot P$$

are effective; $(f)_0$ is called the *divisor of zeroes*, and $(f)_\infty$ is called the *divisor of poles* of f.

Exercise 2.1.49. Find (f) on $\mathbb{P}^1$ for a polynomial $f = F(x) \in k[x]$, $x = T_1/T_0$ being a coordinate on $\mathbb{A}^1 \subset \mathbb{P}^1$.

Divisors of the form (f) are called *principal*. Since it is clear that $(f \cdot g^{\pm 1}) = (f) \pm (g)$, principal divisors form a subgroup $P(X)$ in $Div(X)$. If $D_1 - D_2 \in P(X)$ then D_1 and D_2 are called *linearly equivalent* (or just *equivalent*); in this case we write $D_1 \sim D_2$.

Theorem 2.1.50. *The degree of a principal divisor equals zero.* ■

Thus, $P(X) \subseteq Div^0(X)$. The theorem implies that we can speak about the degree of an equivalence class of divisors. Factor-groups $Cl(X) = Div(X)/P(X)$ and $Cl^0(X) = Div^0(X)/P(X)$ play a crucial role in the study of curves. One can check that the group $Cl(X)$ is canonically isomorphic to $Pic(X)$ (we shall discuss it later). Further on we use the notation $Pic(X)$ for $Cl(X)$ and $Pic^0(X)$ for $Cl^0(X)$.

Let D be a divisor on X. Let

$$L(D) = \{f \in k(X)^* \mid (f) + D \geq 0\} \cup \{0\} \quad ;$$

$L(D)$ is a vector space over k called the *space associated to* D. Its dimension is denoted $\ell(D)$.

Theorem 2.1.51. *$L(D)$ is finite-dimensional for any $D \in Div(X)$.*

Proof: Let $D = D_1 - D_2$, where $D_1 \geq 0$, $D_2 \geq 0$. Since $L(D) \subseteq L(D_1)$ we can assume D to be effective so that $D = \sum a_P \cdot P$ and $a_P \geq 0$. Choose a local parameter t_P at some $P \in Supp\, D$. Since $t_P^{a_P} \cdot f$ is regular at P for any $f \in L(D)$, we can define a linear functional $\varphi_P : L(D) \longrightarrow k$ setting $\varphi_P(f) = (t_P^{a_P} \cdot f)(P)$. It is clear that the kernel of φ_P lies in $L(D')$ where $D' = D - P$ is an effective divisor with $deg(D') < deg\, D$. Since $\ell(D) \leq \ell(D') + 1$ we can apply the induction on $deg\, D$. ∎

Corollary 2.1.52. *If $D \geq 0$ then $\ell(D) \leq deg\, D + 1$.* ∎

Exercise 2.1.53. Show that for any $D \in Div(X)$

$$\ell(D) \leq max\, \{0\ ,\ deg\, D + 1\} \quad .$$

Exercise 2.1.54. Let $X = \mathbb{P}^1$, $D = n \cdot \infty$. Show that $L(D)$ is the space of polynomials of degree at most n. Thus $\ell(D) = n + 1$.

Proposition 2.1.55. *$\ell(D)$ depends only on the linear equivalence class of D.*

Proof: If $D_1 - D_2 = (f)$ with $f \in k(X)^*$ then

multiplication by f defines an isomorphism of $L(D_1)$ onto $L(D_2)$. ■

Linear systems. Let $M \neq \{0\}$ be a subspace in $L(D)$. The set of effective divisors of the form $(f) + D$, $f \in M - \{0\}$, is called a *linear system* and is denoted by $|M|$. If $M = L(D)$, $|M|$ is called a *complete linear system* and is denoted by $|D|$.

Exercise 2.1.56. Show that $|M| = \mathbb{P}(M)$.

The dimension of $\mathbb{P}(M)$ is denoted by $dim\,|M|$. Thus $dim\,|M| = dim\,M - 1$; in particular, $dim\,|D| = \ell(D) - 1$.

There exists a close connection between linear systems on a curve X and rational maps from X to projective spaces. Indeed, let $\varphi : X \longrightarrow \mathbb{P}^n$ be a rational map ,

$$\varphi : P \longmapsto (f_0(P):\ldots:f_n(P)) \quad , \qquad (2.1.2)$$

and suppose that $Im(\varphi)$ is not contained in a hyperplane $H \subset \mathbb{P}^n$ (if the last condition does not hold φ can be considered as a map from X to $\mathbb{P}^m$ for $m < n$). Let

$$(f_i) = \sum a_{i,P} \cdot P \ , \quad i = 0,1,\ldots,n \ ,$$

and let

$$D = inf \ \{(f_0),\ldots,(f_n)\} \ ,$$

i.e. $D = \sum a_P \cdot P$ for $a_P = \min\limits_{0 \le i \le n} a_{i,P}$. It is clear that $(f_i) - D \ge 0$, thus $f_i \in L(-D)$. Let M_φ be a linear subspace in $L(D)$ generated by f_i . So we have associated with φ a linear system $|M_\varphi|$. Conversely, let $|M| \subseteq |D|$ and $|M| \neq \varnothing$. Let $\{f_0,\ldots,f_n\}$ be a basis in M , the formula (2.1.2) defines a rational map $\varphi : X \longrightarrow \mathbb{P}^n$ for

$n = dim\,|M|$. In this situation we have $(f_i) - D = \varphi^*(H_i)$, where H_i is a hyperplane defined by $x_i = 0$. Moreover if $\lambda = (\lambda_0:\ldots:\lambda_n)$ and H_λ is the hyperplane with the equation $\sum \lambda_i \cdot x_i = 0$ then $f^*(H_\lambda) = (\sum \lambda_i \cdot f_i) - D$.

One can also describe φ in invariant terms. Indeed for any $P \in X - Supp\, D$ there is a functional φ_P on M given by $\varphi_P(f) = f(P)$. Thus we can define

$$\varphi = \varphi_M : X - Supp\, D \longrightarrow \mathbb{P}(M^*)$$

by $\varphi_M(P) = \varphi_P$ for $P \in X - Supp\, D$, i.e. a rational map from X to $\mathbb{P}(M^*)$. Further on we use the notation φ_M for the map defined by a linear system $|M|$; in the case $|M| = |D|$ we write φ_D for φ_M . Using the connection between linear systems and rational maps one can prove the following important fact:

Theorem 2.1.57. *Any rational map of a smooth complete curve* X *to a projective space is regular.*

Proof: Let $\varphi : X \longrightarrow \mathbb{P}^n$ be given by (2.1.2). It is clear that φ is regular at any $P \notin Supp\, D$. Let $Q \in Supp\, D$ and let t_Q be a local parameter at Q . Let $\varphi' : X \longrightarrow \mathbb{P}^n$ be defined by

$$\varphi' : P \longmapsto (f'_0(P) : \ \ldots \ : f'_n(P))$$

where $f'_i = t_Q^{-a_Q} \cdot f_i$. By the definition of D any f'_i is regular at Q and there exists i_0 with $f'_{i_0}(Q) \neq 0$. Hence φ' is regular at Q . It is sufficient to note that φ and φ' coincide as rational maps. ■

Corollary 2.1.58. *Let* $\varphi : X \longrightarrow Y$ *be a rational map ,* X *and* Y *being smooth complete curves. Then* φ *is regular.* ■

One can also obtain Theorem 2.1.44 as a corollary of Theorem 2.1.57.

Lattices. Now we shall describe how one can speak about divisors on X in terms of the function field $k(X)$.

For any $P \in X$ let there be given a *lattice*, i.e. a free $\mathcal{O}_P$-submodule $\mathcal{L}_P$ in $k(X)$ or rank 1 ; $\mathcal{L}_P = \ell_P \cdot \mathcal{O}_P$ for $\ell_P \in k(X)$. In this situation we say that there is given a *family of lattices* $(\mathcal{L}_P)_{P \in X}$. If for all but a finite number of point we have $\mathcal{L}_P = \mathcal{O}_P$ then the family $(\mathcal{L}_P)_{P \in X}$ is called an *asymptotically standard family of lattices*.

To any $D \in Div(X)$ we can associate an asymptotically standard family of lattices. Let $D = \sum a_P \cdot D$, then

$$\mathcal{L}_P = \begin{cases} \mathcal{O}_P & \text{for } P \notin Supp\ D \\ t_P^{-a_P} \cdot \mathcal{O}_P & \text{for } P \in Supp\ D \end{cases} \tag{2.1.3}$$

t_P being a local parameter at P.

Proposition 2.1.59. *The formula* (2.1.3) *defines a bijection of* $Div(X)$ *onto the set of asymptotically standard lattice families on* X.

Proof: Let us construct an inverse map $(\mathcal{L}_P) \longmapsto D$. For any $P \in X$ we have $\mathcal{L}_P = t_P^{a_P} \cdot \mathcal{O}_P$, a_P being an integer; since for all but a finite number of P, $\mathcal{L}_P = \mathcal{O}_P$, we can define a divisor $D = \sum(-a_P) \cdot P$. It is clear that maps $D \longmapsto (\mathcal{L}_P)_{P \in X}$ and $(\mathcal{L}_P) \longmapsto D$ are inverse to each other. ■

Let $\mathcal{L} = (\mathcal{L}_P)_{P \in X}$ be an asymptotically standard lattice family. We call the intersection $\bigcap_{P\in X} \mathcal{L}_P \subset k(X)$ the *space of sections of* $\mathcal{L}$ and we denote it by $H^0(\mathcal{L})$; its elements are called *sections of* $\mathcal{L}$. This definition immediately implies

Proposition 2.1.60. *Let* $\mathcal{L} = (\mathcal{L}_P)_{P\in X}$ *and* $D \in Div\ X$ *correspond to each other. Then* $H^0(\mathcal{L}) = L(D)$. ■

By the *fibre* of a family $(\mathcal{L}_P)_{P\in X}$ at $P \in X$ we mean the one-dimensional space $\overline{\mathcal{L}}_P = \mathcal{L}_P/t_P\cdot\mathcal{L}_P$ over $k = \mathcal{O}_P/t_P\cdot\mathcal{O}_P$. The image of $s \in H^0(\mathcal{L})$ in $\overline{\mathcal{L}}_P$ is called the *value* of s at P .

If $s \in H^0(\mathcal{L}) - \{0\}$, $\mathcal{L}$ being an asymptotically standard lattice family, then we call

$$D_s = \sum_P (ord(s) - a_P)\cdot P$$

the *divisor of zeroes of* s , where $\mathcal{L}_P = t_P^{a_P}\cdot\mathcal{O}_P$ for any P.

Exercise 2.1.61. Show that D_s is well-defined and effective.

We call asymptotically standard lattice families $\mathcal{L} = (\mathcal{L}_P)$ and $\mathcal{M} = (\mathcal{M}_P)$ *linearly equivalent* iff there exists $f \in k(X)^*$ such that $\mathcal{L}_P = f\cdot\mathcal{M}_P$ for any $P \in X$.

Exercise 2.1.62. Show that asymptotically standard lattice families are linearly equivalent iff corresponding divisors are.

Cartier divisors. One can give an alternative definition for divisors (*Cartier divisors*), which is very useful. Let $D \in Div(X)$, $D = \sum a_P\cdot P$, and let U be an open subset in X . We denote by $D|_U$ the following divisor on U :

$$D|_U = \sum_{P\in U} a_P \cdot P \quad .$$

Exercise 2.1.63. Show that for any $D \in Div(X)$ and any $P \in X$ there exists an open $U = U(P,D)$ such that $P \in U$ and $D|_U$ is a principal divisor on U , i.e. there exists $f \in k(X)$ with $D|_U = (f)|_U$.

Exercise 2.1.64. Deduce from the previous exercise that for any $D \in Div(X)$ there exists a finite open covering $\{U_i\}$ of X such that $D|_{U_i} = (f_i)|_{U_i}$. Show that $(f_i \cdot f_j^{-1}) \in k[U_i \cap U_j]^*$. Conversely for any family $(\{U_i\},\{f_i\})$, where $\{U_i\}$ is a finite open covering of X , $\{f_i\}$ are rational functions, and $(f_i \cdot f_j^{-1}) \in k[U_i \cap U_j]^*$, there exists a unique divisor $D \in Div(X)$ with $D|_{U_i} = (f_i)|_{U_i}$.

We call $(\{U_i\},\{f_i\})$ a *Cartier divisor.*

Behaviour under maps. Using Cartier divisors we can define the *inverse image* $\varphi^*(D) \in Div(X)$ of a divisor $D \in Div(Y)$ under a regular map $\varphi : X \longrightarrow Y$ of smooth irreducible complete curves. Let $(\{U_i\},\{f_i\})$ be a Cartier divisor on Y corresponding to $D \in Div(Y)$. Then $\varphi^*(D)$ is defined on the covering $\{\varphi^{-1}(U_i)\}$ of X by the family $\{\varphi^*(f_i)\}$. One can easily show that φ^* is a group homomorphism $\varphi^* : Div(Y) \longrightarrow Div(X)$, note also that $\varphi^*(P(Y)) \subseteq P(X)$ since $\varphi^*((f)) = (\varphi^*(f))$ for any $f \in k(Y)$. Thus we get a homomorphism $\varphi^* : Pic(Y) \longrightarrow Pic(X)$. Moreover, $\varphi^*(Div^0(Y)) \subseteq Div^0(X)$ (see Corollary 2.1.65 below), and thus $\varphi^* : Pic^0(Y) \longrightarrow Pic^0(X)$.

Using Cartier divisors we can also define the *divisor* $(F) \in Div(X)$ *of a form* F of degree s on $\mathbb{P}^n$. Indeed let $U_i = \{T_i \neq 0\} \cap X$ and let $f_i = F/T_i^s$.

Exercise 2.1.65. Check that $(\{U_i\},\{f_i\})$ is a Cartier divisor.

In particular if $F = L$ is a linear form then (L) is called the *divisor of a hyperplane section*; all such divisors are linearly equivalent. The degree $deg(L)$ is called the *degree* of X, $deg(X) = deg(L)$. If a map $f : X \longrightarrow \mathbb{P}^m$ is given by $M \subseteq L(-D)$ then effective divisors $D' \in |M|$ are precisely inverse images of hyperplane section divisors $(L) \in Div(f(X))$.

Connection with line bundles. Let us show that divisors and line bundles are closely connected. If $D = (\{U_\alpha\},\{f_\alpha\})$ is a Cartier divisor we can set $f_{\alpha\beta} = f_\alpha \cdot f_\beta^{-1}|_{U_\alpha \cap U_\beta}$; since $f_{\alpha\beta} \in k[U_\alpha \cap U_\beta]^*$, functions $f_{\alpha\beta}$ define a line bundle on X which is denoted by $O(D)$.

Exercise 2.1.66. Show that the correspondence $D \longmapsto O(D)$ is a bijection of the set of linear equivalence classes of divisors on X onto the set of isomorphism classes of line bundles on X.

Thus we have a canonical isomorphism $Cl(X) \simeq Pic(X)$. Further on we use only the notation $Pic(X)$ and $Pic^0(X)$. By Exercise 2.1.66 we can speak about the degree $deg(\mathcal{L})$ of a line bundle $\mathcal{L}$ on X; if $\mathcal{L} \simeq O(D)$ we denote the map φ_D by $\varphi_{\mathcal{L}}$.

Exercise 2.1.67. Let $\mathcal{L} = O(D)$. Show that $H^0(X,\mathcal{L})$ is isomorphic to $L(D)$.

Let $s \in H^0(X,\mathcal{L})$ and let $X = \bigcup_\alpha U_\alpha$ be an open covering such that $\mathcal{L}|_{U_\alpha}$ is trivial for any α. Since $\mathcal{L}|_{U_\alpha} \simeq \mathcal{O}|_{U_\alpha}$ the restriction $s|_{U_\alpha}$ of s can be regarded as a regular function f_α on U_α. Let $D_{s,\alpha} = (f_\alpha)_0|_{U_\alpha}$.

Exercise 2.1.68. Show that there exists a unique divisor D_s such that $D_s|_{U_\alpha} = D_{s,\alpha}$ for any α.

Divisor $D_s \geq 0$ is called the *divisor of zeroes* of s ; $D_s \sim D$ for any $D \in Div(X)$ with $\mathcal{L} \simeq \mathcal{O}(D)$.

Inverse images of points. To study non-constant maps $f : X \longrightarrow Y$ it is important to consider divisors of the form $f^*(P)$, P being a point of Y.

Proposition 2.1.69. *For any* $P \in Y$

$$deg\ f^*(P) = deg\ f ,$$

and for any $D \in Div\ Y$ *we have*

$$deg\ f^*(D) = (deg\ D)\cdot(deg\ f) . \blacksquare$$

Exercise 2.1.70. Deduce Theorem 2.1.50 from Proposition 2.1.69. (*Hint*: Apply this proposition to the map $f : X \longrightarrow \mathbb{P}^1$ and to $D = (0) - (\infty)$).

Ramification points. Let $f : X \longrightarrow Y$ be a regular map, X and Y being smooth complete curves. Let $P \in X$ and $f(P) = Q$. It is clear that $P \in Supp(f^*(Q))$. If $f^*(Q) = P + D'$ with $P \notin Supp\ D'$, we call f *unramified* at P. If it is not the case, and thus $f^*(Q) = e_P \cdot P + D'$ with $e_P \geq 2$ and $P \notin Supp\ D'$, then f is called *ramified* at P, and P is called a *ramification point* of f ; e_P

is called the *ramification index* of f at P. From the definition of $f^*(Q)$ it follows that $f^*(t_Q) = t_P^{e_P} \cdot u$ with $u \in \mathcal{O}_P^*$, t_P and t_Q being local parameters at P and Q, respectively. If for $Q \in Y$ the map f is unramified at any $P \in f^{-1}(Q)$, we say that f is *unramified* at Q. If it is not the case Q is called a *ramification point* of f.

Remark 2.1.71. One can define divisors and related objects for any variety X (which we assume to be smooth and projective). Simple divisor F on X is a subvariety in X of codimension one (F can be singular). Divisor D on X is a finite sum

$$D = \sum n_i \cdot F_i \quad ,$$

n_i being integers, and F_i being simple divisors on X. The set $Div(X)$ of divisors on X is an abelian group; the set $Div^+(X)$ of effective divisors $D \geq 0$ (i.e. with all $n_i \geq 0$) is a subsemigroup in $Div(X)$. If $f \in k(X)^*$ and F is a simple divisor on X one can define an integer $ord_F(f)$ as follows. It can be shown that there exists an open affine set $U \subset X$ such that $F \cap U \neq \emptyset$ and $F \cap U$ is defined by $g = 0$ with $g \in k[U]$. Set

$$ord_F(f) = \{min\ \ell \mid f = g^{\ell} \cdot h \text{ with } h \in k[U]\} .$$

One can check that $ord_F(f)$ is well defined (i.e. it does not depend on U) and that

$$ord_F(f_1 \cdot f_2) = ord_F(f_1) + ord_F(f_2)$$

for any $f_1, f_2 \in k(X)^*$, and

$$ord_F(f_1 + f_2) \geq min\ \{ord_F(f_1)\ ,\ ord_F(f_2)\}$$

for $f_1 + f_2 \neq 0$. For $ord_F(f) \geq 0$ we call $ord_F(f)$ the *order of zero* of f at F. If $ord_F(f) < 0$ then $|ord_F(f)|$ is called the *order of pole* of f at F. We can define the divisor (f) of f by

$$(f) = \sum ord_F(f) \cdot F \quad .$$

For any $D \in Div(X)$ let

$$L(D) = \{f \in k(X)^* | \ (f) + D \geq 0\} \cup \{0\} \quad .$$

Using properties of $ord_F(f)$ one can easily show that $L(D)$ is a vector subspace in $k(X)$; one can also prove that it is finite-dimensional for any D ; its dimension is denoted by $\ell(D)$. The connection between divisors and rational maps takes place for arbitrary smooth projective varieties. The theory of Cartier divisors and the connection between divisor classes and linear bundles is also quite general.

2.1.4. Jacobians

In the study of algebraic curves an important role is played by their Jacobians. The Jacobian of a curve is an algebraic group, i.e. an algebraic variety which is a group such that the group structure is compatible with the structure of the algebraic variety.

Algebraic groups. Let G be quasi-projective variety which is a group. Then G is called an *algebraic group* iff the maps

$$\psi : G \times G \longrightarrow G \ , \quad \psi(g_1, g_2) = g_1 \cdot g_2$$

and

$$\varphi : G \longrightarrow G \ , \quad \varphi(g) = g^{-1}$$

are regular.

Exercise 2.1.72. Check that the following varieties are algebraic groups:

a) $\mathbb{A}^1$ with standard addition. This algebraic group is called additive and is denote by $\mathbb{G}_a$.

b) $\mathbb{A}^1 - \{0\}$ with multiplication of coordinates. This algebraic group is called multiplicative and is denoted by $\mathbb{G}_m$.

c) The set

$$GL(n) = GL(n,k) =$$

$$= \{A \text{ is } n \times n\text{-matrix over } k \mid det\ A \neq 0\}$$

with standard matrix multiplication. This algebraic group is called the general linear group (over k). Note that $GL(1) = \mathbb{G}_m$.

Note that for a fixed $h \in G$ the map $L_h : G \longrightarrow G$, $L_h(g) = h \cdot g$ is an isomorphism (of varieties but not of groups), since $L_h^{-1} = L_{h^{-1}}$.

Proposition 2.1.73. *If G is an algebraic group then G is a smooth variety.*

Proof: On G there exists a non-singular point. Applying L_h , where h runs over all elements of G , we obtain the proposition. ■

Abelian varieties. If G is an algebraic group and G is projective then G is called an *abelian variety*.

Theorem 2.1.74. *An abelian variety is a commutative group.* ■

Proposition 2.1.75. *Let* $\psi : G \longrightarrow H$ *be a regular map from an abelian variety to an algebraic group. Then there exists a morphism of algebraic groups* $\varphi : G \longrightarrow H$ *(i.e. a regular map which is a group homomorphism) such that* $\psi = L_h \circ \varphi$ *, where* $h = \psi(e) \in H$ *.* ■

Corollary 2.1.76. *If* A *and* B *are abelian varieties which are isomorphic as varieties then they are also isomorphic as algebraic groups.* ■

Thus on abelian varieties "algebra is determined by geometry".

Jacobian of a curve. Let X be a smooth projective curve. Recall that $Pic^0(X) = Div(X)/P(X)$ is the group of equivalence classes of divisors of degree zero.

Theorem 2.1.77. *For any smooth projective curve* X *there exists a unique abelian variety* J_X *such that*

a) J_X *is isomorphic to* $Pic^0(X)$ *as a group;*

b) *the map*

$$i_{P_0} : X \longrightarrow J_X , \quad i_{P_0} : P \longmapsto P - P_0$$

is regular for any $P_0 \in X$ *;*

c) *for any regular map* $\varphi : X \longrightarrow A$ *from* X *to an abelian variety such that* $\varphi(P_0)$ *is the neutral element of* A *there exists a morphism of abelian varieties* $\lambda : J_X \longrightarrow A$ *, with* $\varphi = \lambda \circ i_{P_0}$ *.* ■

The abelian variety J_X is called the *Jacobian* of X .

The dimension of the Jacobian is called the *genus* of X and is denoted $g(X)$. One can show that this definition coincides with the definition of genus in terms of differentials given in Section 2.2.1 below; for $k = \mathbb{C}$ it also coincides with topological definition of genus given in Section 2.1.5.

Example 2.1.78. a) If $X = \mathbb{P}^1$ then J_X is trivial since $Pic^0(X) = 0$. Therefore $g(\mathbb{P}^1) = 0$.

b) If X a smooth plane curve of degree 3 then $J_X \simeq X$ (see Section 2.4.1 below). Hence $g(X) = 1$.

Functoriality. For a regular map $f : X \longrightarrow Y$ of smooth projective curves one can define two morphisms of Jacobians $f_* : J_X \longrightarrow J_Y$ and $f^* : J_Y \longrightarrow J_X$; for $f : X \longrightarrow Y$ and $g : Y \longrightarrow Z$ one has $(g \circ f)_* = g_* \circ f_*$ and $(g \circ f)^* = f^* \circ g^*$.

Exercise 2.1.79. Deduce the existence of the map $f_* : J_X \longrightarrow J_Y$ from the last statement of Theorem 2.1.77.

Theorem 2.1.80. *Let* $f : X \longrightarrow Y$ *be a regular map. Then the map* $f^* : J_Y \longrightarrow J_X$, *defined by the inverse image of a divisor, is a morphism of abelian varieties.* ∎

Embedding into Jacobian. It should be remarked that any curve which is not isomorphic to $\mathbb{P}^1$ can be embedded into its Jacobian:

Proposition 2.1.81. *If* $i_{P_0} : X \longrightarrow J_X$ *is not injective then* $X \simeq \mathbb{P}^1$.

Proof: If i_{P_0} is not injective then there exist P and $Q \in X$, $P \neq Q$ such that $P = Q + (f)$ for some $f \in k(X) - k$. Consider f as a regular map $f : X \longrightarrow \mathbb{P}^1$. Since $(f) = P - Q$ we have $(f)_0 = P$ and by Proposition 2.1.69 $deg\, f = 1$; hence f is an isomorphism. ■

Note that J_X can be also identified with the set of isomorphism classes of line bundles of degree zero. If $\mathcal{L}$ is a line bundle of degree a , the map $\mathcal{M} \longmapsto \mathcal{M} \otimes \mathcal{L}$ is a bijection of J_X onto the set of isomorphism classes of line bundles of degree a .

Proposition 2.1.82. *The set of isomorphism classes of line bundle of a fixed degree is in a natural bijection with* J_X *.* ■

2.1.5. Riemann surfaces

If the ground field is $\mathbb{C}$, any smooth algebraic curve can be considered as a Riemann surface, which is compact in the case of a complete curve.

Recall that a *Riemann surface* is a one-dimensional complex-analytic manifold. To be more precise, a Riemann surface is a connected Hausdorff topological space T with an atlas S . By an *atlas* S we mean a family $S = \{(U_\alpha, p_\alpha) \mid \alpha \in A\}$, A being an arbitrary index set, $\{U_\alpha\}_{\alpha\in A}$ being an open covering of T , and $p_\alpha : U_\alpha \longrightarrow V_\alpha$ for $\alpha \in A$ being a homeomorphism of U_α onto an open set $V_\alpha \subseteq \mathbb{C}$ such that the map

$$p_\beta \circ p_\alpha^{-1} : p_\alpha(U_\alpha \cap U_\beta) \longrightarrow p_\beta(U_\alpha \cap U_\beta)$$

is complex analytic for $U_\alpha \cap U_\beta \neq \emptyset$. Usually one requires S to be a maximal set satisfying these conditions. In fact this requirement is not essential since for any S there exists a maximal set $S' \supseteq S$. Further on we do not require S to be maximal.

Let X be a smooth quasi-projective curve over $\mathbb{C}$. Let us define a complex topology on X . We begin with some elementary remarks on the complex topology of $\mathbb{A}^N$ and $\mathbb{P}^N$. It is clear that $\mathbb{A}^N(\mathbb{C}) \simeq \mathbb{C}^N$ is homeomorphic to $\mathbb{R}^{2N}$. We define the complex topology on $\mathbb{P}^N(\mathbb{C})$ as the factor-topology under the canonical projection

$$\mathbb{A}^{N+1} - \{0\} \longrightarrow \mathbb{P}^N(\mathbb{C}) .$$

Since the restriction of this map to the $(2N+1)$-sphere

$$S^{2N+1} = \{z = (z_0,\ldots,z_N) \in A^{N+1}(\mathbb{C}) - \{0\} \mid \sum_{i=0}^{N} |z_i|^2 = 1\}$$

gives a surjection $S^{2N+1} \longrightarrow \mathbb{P}^N(\mathbb{C})$, we see that $\mathbb{P}^N(\mathbb{C})$ is compact in the complex topology. Now we define the complex topology on $X \subset \mathbb{P}^N$ as the restriction of the complex topology of $\mathbb{P}^N(\mathbb{C})$. Note that if X is projective then X is a closed set in $\mathbb{P}^N(\mathbb{C})$, X being the intersection of zero sets of continuous functions on $\mathbb{P}^N(\mathbb{C})$. Therefore a projective curve is compact in the complex topology.

Now let us check that the complex topology on X is well-defined, i.e. that it does not depend on the embedding of X into a projective space. It is sufficient to define a fundamental basis of neighbourhoods for any $P \in X$. Let $P \in X$ be an arbitrary point (by our assumptions it is non-singular). Let U be an affine open neighbourhood of P so that $U \subset \mathbb{A}^N(\mathbb{C})$ and let t_P be a local parameter at P . Since t_P is a rational function which is regular at

P , there exists $\varepsilon > 0$ such that $U_\varepsilon = \{Q \in X \mid t_P(Q) < \varepsilon\}$ is continuously mapped to $\mathbb{C}$ by t_P . Consider the restrictions $t_1,\ldots,t_N$ to X of coordinate functions $T_1,\ldots,T_N$ on $\mathbb{A}^N$, and let $\Phi_i(t) = \tau_P(t_i)$ be the power series expansions of t_i at P . Using the implicit function theorem one can prove

Lemma 2.1.83. *There exists* $\varepsilon > 0$ *such that power series* $\Phi_i(t)$ *converge in a disc of radius* ε . ∎

Let us define a map $u : D_\varepsilon \longrightarrow \mathbb{A}^N(\mathbb{C})$ by $u(t) = (\Phi_1(t),\ldots,\Phi_N(t))$, D_ε being the disc of radius ε in $\mathbb{C}$. One proves that the maps $t_P \circ u$ and $u \circ t_P$ are identical on some neighbourhoods of 0 and P in $\mathbb{C}$ and U_ε , respectively. Therefore t_P is a homeomorphism of a neighbourhood U_P of P in X onto a neighbourhood V_P of 0 in $\mathbb{C}$. Since this argument is valid for any choice of t_P the complex topology on X depends neither on the choice of t_P nor on the choice of an embedding $X \subset \mathbb{P}^N$. Moreover Lemma 2.1.83 implies that the family (U_P, t_P) is an atlas on X . To prove that X is a Riemann surface one should prove

Theorem 2.1.84. *Any algebraic curve is connected in the complex topology.*

Proof: For simplicity we assume X to be complete. We use here the immediate corollary of the Riemann-Roch theorem which follows (note that the proof of the Riemann-Roch theorem in Chapter 2.2 below does not depend on this section). Let $P_0 \in X$, then there exists a non-constant rational function $f \in k(X)$ which is regular at all points of X except P_0. Suppose that $X = X_1 \cup X_2$, X_1 and X_2 being non-empty non-intersecting closed subsets of X. Let $P_0 \in X_1$ and let f be a non-constant rational function which is regular everywhere except P_0. Since f is regular on X_2 and X_2 is compact, $f|_{X_2}$ is constant; hence f itself is constant, a contradiction. ■

Since Riemann surfaces can be considered as two-dimensional orientable differentiable manifolds, i.e. as orientable surfaces, any smooth complex algebraic curve X is an orientable surface. If X is complete then this surface is compact. Recall the following fact.

Theorem 2.1.85. *Let S be a compact orientable connected surface. Then S is diffeomorphic to the sphere with g handles, g being a non-negative integer.* ■

This integer g is called the *genus* of S (and if S corresponds to an algebraic curve X it is also called the *genus* of X).

In particular, for $g = 0$ the surface S is diffeomorphic to the sphere S^2, for $g = 1$ it is diffeomorphic to the torus $T^2 = S^1 \times S^1$.

One can show that this topological definition of the genus coincides with the above denition in terms of the Jacobian and with the definition in terms of differentials given below.

In fact in the compact case the construction of Riemann surfaces from algebraic curves can be inverted. This deep statement is called the Riemann existence theorem:

Theorem 2.1.86. *Let* Y *be a compact Riemann surface. There exists a unique (up to an isomorphism) smooth projective curve* X *with* $Y = X_{an}$ *, where* X_{an} *is the Riemann surface associated to* X *.* ■

Moreover one can show that for any curves X and X' the set of complex analytic maps from X_{an} to X'_{an} coincides with the set of regular maps from X to X' .

Therefore the notions of a smooth complete complex curve and of a compact Riemann surface are essentially equivalent. This makes it appropriate to use powerful technique of complex analysis to the study of complex algebraic curves. We shall use the connection between curves and Riemann surfaces to construct some classes of algebraic curves which arise naturally as Riemann surfaces (see Part 4 below).

CHAPTER 2.2

RIEMANN-ROCH THEOREM

Since many important questions of the geometry of curves are reduced to the calculation of $\ell(D)$ for various divisors D, an explicit expression of $\ell(D)$ plays an essential role in the theory of curves. Such an expression is given by the Riemann-Roch theorem which is the crucial result of the theory. To state it one should study differential forms on curves which are also useful in many other questions.

In this chapter we give a brief exposition of this subject. In Section 2.2.1 the definition and basic properties of differential forms are given; Section 2.2.2 contains the statement and the scheme of proof of the Riemann-Roch theorem and of some of its corollaries. Next section is devoted to the Hurwitz formula which describes the behaviour of genus under regular maps; this result is rather useful to compute the genus of many curves. In Section 2.2.4 we describe basic properties of special divisors and Weierstrass points on curves. Section 2.2.5 is devoted to the Cartier operator, useful to study curves over fields of positive characteristic.

In this chapter by X we mean a smooth irreducible curve over an algebraically closed field k.

2.2.1. Differential forms

Let $P \in X$ and let $f \in O_P$. Let $d_P f$ be the image of $f - f(P) \in m_P$ in m_P/m_P^2; note that m_P/m_P^2 is a one-dimensional vector space over k. We call $d_P f$ the *differential* of f at P.

Exercise 2.2.1. Check that the map $d_P : O_P \longrightarrow m_P/m_P^2$ is k-linear and for any $f, g \in O_P$

$$d_P(f \cdot g) = f(P) \cdot d_P g + g(P) \cdot d_P f \quad .$$

Let U be an open subset of X and let $f \in k[U]$. Let $\Phi[U]$ be the set of maps φ which send each $P \in U$ to some $\varphi(P) \in m_P/m_P^2$; it is clear that $\Phi[U]$ is a $k[U]$-module. Any $f \in k[U]$ defines $df \in \Phi[U]$ by $(df)(P) = d_P f$. We call $\varphi \in \Phi[U]$ a *differential form regular on* U iff for any $P \in U$ there exist an open neighbourhood V of P in U such that $\varphi|_V$ lies in the $k[V]$-submodule $\sum_{f \in k[V]} k[V] \cdot df$ of $\Phi[V]$. Differential forms which are regular on U form a $k[U]$-module $\Omega[U]$; in particular for $U = X$ we obtain a k-vector space $\Omega[X]$. Sometimes we write Ω for $\Omega[X]$.

Theorem 2.2.2. *$\dim_k \Omega[X]$ is finite.*

A proof of the theorem will be given later (after the proof of Corollary 2.2.8); $g(X) = dim_k\Omega[X]$ is called the *genus* of X. One can show that this definition coincides with definitions given above in Section 2.1.4 and 2.1.5.

Proposition 2.2.3. *Let* $P \in X$ *and let* $t = t_P$ *be a local parameter at* P. *Then there exists an open neighbourhood* U *of* P *such that* $\Omega[U] = k[U]\cdot dt$.

Sketch of proof: Note that for any $F \in k[T_1,\ldots,T_N]$ and any $f_1,\ldots,f_N \in k[U]$ one has

$$d(F(f_1,\ldots,f_N)) = \sum_{i=1}^{N} \frac{\partial F}{\partial T_i}(f_1,\ldots,f_N)\cdot df_i \quad ,$$

which easily follows from $d(f\cdot g) = f\cdot dg + g\cdot df$. Let now V be an open neighbourhood of P which is an affine curve, let $V \subset \mathbb{A}^N$, and let $F_1,\ldots,F_m \in k[T_1,\ldots,T_N]$ be a basis of the ideal I_V, formed by $F \in k[T_1,\ldots,T_N]$ vanishing on V. Since $F_i|_V = 0$ for $i = 1,\ldots,m$ we obtain

$$\sum_{j=1}^{N} \frac{\partial F_i}{\partial T_j}\cdot dt_j = 0 \quad \text{for} \quad i = 1,\ldots,m \qquad (2.2.1)$$

where $t_j = T_j|_V \in k[V]$. Since P is non-singular, the rank of $(\partial F_i/\partial T_j(P))$ equals $N - 1$. Without loss of generality we can assume that $t = t_1$. One can express dt_j for $j = 2,\ldots,N$ from (2.2.1):

$$dt_j = f_j\cdot dt \quad ,$$

f_j being rational functions regular at P. Let U be an open subset in V such that $f_j \in k[U]$. Then $\Omega[U] = k[U]\cdot dt$ since we can express any $\omega \in \Omega[U]$ in $dt_1,\ldots,dt_N$ and hence in dt. ∎

Corollary 2.2.4. *Let* $\omega \in \Omega[U]$ *. Then the set* F_ω *of zeroes of* ω *is closed in* U *.* ∎

Let U be an open subset in X, and let $\omega \in \Omega[U]$. We say that ω defines a *rational differential form* on X; if $\omega \in k[U]$ and $\omega' \in k[U']$ satisfy $\omega|_{U\cap U'} = \omega'|_{U\cap U'}$ then we say that ω and ω' define the same rational differential form on X. The set of rational differential forms on X is denoted by $\Omega(X)$; clearly $\Omega(X)$ is a $k(X)$-vector space. Moreover Proposition 2.2.3 implies

Proposition 2.2.5. $dim_{k(X)}\Omega(X) = 1$. ∎

Canonical class. From the definition of a differential form ω it follows that for any $P \in X$ there exists an open neighbourhood U of P such that $\omega = f\cdot dt$, for $f \in k(X)$, where $t - t(P)$ is a local parameter at any $P \in U$. It follows that for any $\omega \neq 0$ there exists an open covering $\{U_i\}$ of X such that $\omega|_{U_i} = f_i\cdot dt_i$ for any i. Since $f_i\cdot dt_i = f_j\cdot dt_j$ on $U_i \cap U_j$, $t_i - t_i(Q)$ and $t_j - t_j(Q)$ being local parameters for any $Q \in U_i \cap U_j$, we see that f_i/f_j and $f_j/f_i \in (k[U_i \cap U_j])^*$. Therefore $\{U_i, f_i\}$ is a Cartier divisor which is denoted by (ω); it is called the divisor associated to ω. Since any $\omega' \in \Omega(X)$ can be written as $\omega' = f\cdot\omega$ with $f \in k(X)$ (Proposition 2.2.5) we see that $(\omega') = (\omega) + (f)$. Hence the class $K = K_X$ of linear equivalence of (ω) does not depend on the choice of ω; it is called the *canonical class* of X. Sometimes we denote by K_X a divisor from the canonical class of X.

Example 2.2.6. a) Let $X = \mathbb{P}^1$, $\omega = dt$, t being a coordinate on $\mathbb{P}^1$, let $u = t^{-1}$, and let

$$U_0 = \{x \in \mathbb{P}^1 \mid u \neq 0\}, \qquad U_1 = \{x \in \mathbb{P}^1 \mid t \neq 0\},$$

then $\mathbb{P}^1 = U_0 \cup U_1$; $t - t(P)$ is a local parameter at any $P \in U_0$, $u - u(Q)$ is a local parameter at any $Q \in U_1$. In U_0 we have $\omega = d(u^{-1}) = -u^{-2} \cdot du$. Thus $(\omega) = -2 \cdot \infty$, ∞ being a point of $\mathbb{P}^1$ with $u(\infty) = 0$. Therefore $K_{\mathbb{P}^1}$ consists of divisors of degree -2.

b) Let X be the curve in $\mathbb{P}^2$ defined by $x_0^3 + x_1^3 + x_2^3 = 0$, and let $U_{ij} = \{P \in X \mid x_i(P) \cdot x_j(P) \neq 0\}$ for $i = 0,1,2$; $X = U_{01} \cup U_{02} \cup U_{12}$. Let $x = x_1/x_0$, $y = x_2/x_0$, and let $\omega = dy/x^2 = dx/y^2$ in U_{01}. Then $\omega = -dv/u^2$ in U_{12}, where $u = x_2/x_1$, $v = x_0/x_1$, and $\omega = -ds/t^2$ in U_{02}, where $s = x_0/x_2$, $t = x_1/x_2$. Therefore $\omega \in \Omega[X]$ and one easily checks that $(\omega) = 0$. Hence $K_X = 0$.

Smooth plane curves. Let $F(x_0 : x_1 : x_2) = 0$ be a homogeneous equation of degree m, which defines a smooth curve $X \subset \mathbb{P}^2$. Let $x = x_1/x_0$, $y = x_2/x_0$ be coordinates on $\mathbb{A}^2 \subset \mathbb{P}^2$, and let $G(x,y) = F(1:x:y) = 0$ be the equation of the affine curve $X' = X \cap \mathbb{A}^2$. Let us consider rational differential forms on X given by

$$\omega = \frac{P \cdot dy}{G'_X}. \tag{2.2.2}$$

where $G'_X = \partial G/\partial x$, $P \in k[x,y]$.

Proposition 2.2.7. *A form given by (2.2.2) is regular iff* $\deg P \leq m - 3$. *Conversely any* $\omega \in \Omega[X]$ *can be* *written in the form (2.2.2) with* $P \in k[x,y]$ *,* $\deg P \le m-3$ *. Hence*

$$\Omega[X] = \left\{ \frac{P\cdot dy}{G'_x} \;\middle|\; P \in k[x,y],\ \deg P \le m-3 \right\} .$$

Moreover $K_X = (m-3)\cdot L$ *,* L *being the class of a hyperplane section.*

Proof: Let $\omega_0 = dy/G'_x$; clearly it is regular at any $P = (x,y) \in X' = X \cap \mathbb{A}^2$ with $G'_x(x,y) \neq 0$. Moreover since $G(x,y) = 0$ for $P = (x,y) \in X$, $\omega_0 = -\omega'_0$, where $\omega'_0 = dx/G'_y$. Therefore, ω_0 is regular at any $P = (x,y)$ with $G'_y(x,y) \neq 0$. Since X is smooth for any point, either G'_x or G'_y does not vanish and ω_0 is regular on X' . It is clear that $(\omega)|_{X'} = 0$. Let us consider the behaviour of ω_0 at a point of $X \cap (\mathbb{P}^2 - \mathbb{A}^2)$. Let $X'_1 = X \cap \mathbb{A}^2_1$, where $\mathbb{A}^2_1 = \{(x_0:x_1:x_2) \in \mathbb{P}^2 \mid x_1 \neq 0\}$. Note that $u = 1/x$ and $v = y/x$ are coordinates on $\mathbb{A}^2_1$ and one has $x = 1/u$, $y = v/u$, $dx = -du/u^2$, $dy = (u\cdot dv - v\cdot du)/u^2$. The equation of $X'_1 = X \cap \mathbb{A}^2_1$ is $H(u,v) = u^m\cdot G(1/u,v/u)$, and $H'_v(u,v) = u^{m-1}\cdot G'_y(x,y)$. Thus $\omega_0 = u^{m-3}\cdot du/H'_v$ and we get $(\omega_0)|_{X'_1} = (m-3)\cdot(u) = (m-3)\cdot(x_0)$. Therefore $K_X = (m-3)\cdot L$, L being the class of a hyperplane section.

From the above argument it is clear that $\Omega[X'] = \{P\cdot\omega_0 \mid P \in k[x,y]\}$. Indeed, the condition $(f\cdot\omega_0) \ge 0$ is equivalent to $(f)|_{X'} \ge 0$, i.e. $(f) = D - m\cdot(x_0)$ with $D \ge 0$, $m \ge 0$. Let now $\omega = P\cdot\omega_0$ with $P \in k[x,y]$, $\deg P = s$. Then on X'_1 one has $\omega = -\tilde{P}(u,v)\cdot u^{m-3-s}\cdot du$, where $\tilde{P}(u,v) = u^s\cdot P(1/u,v/u)$. Thus ω is regular in X'_1 iff $s \le m-3$. Applying the same argument to the other neighbourhood $X'_2 = X \cap \mathbb{A}^2_2$, where $\mathbb{A}^2_2 = \{(x_0:x_1:x_2) \in \mathbb{P}^2 \mid x_2 \neq 0\}$ one obtains the proposition. ■

Therefore $D \in |K_X|$ iff it is of the form (F), F being a form of degree $(m - 3)$. It is possible to say that these divisors are intersections of X with curves $F = 0$, for $deg\, F = m - 3$. Such curves are called *adjoint* for X; for $m \leq 3$ there are no adjoint curves.

Corollary 2.2.8. *For a smooth plane curve* X

$$g(X) = \frac{(m - 1)\cdot(m - 2)}{2} \quad .$$

Proof: Since the space of polynomials in 2 variables of degree at most $m - 3$ has the dimension $(m - 1)\cdot(m - 2)/2$, it is sufficient to note that if P and Q are polynomials of degree at most $m - 3$ and $P\cdot\omega_0 = Q\cdot\omega_0$ then $P - Q$ is divisible by G, thus $P = Q$. ■

Note that the condition $(\omega) \geq 0$ is equivalent to the regularity of ω. Hence $\Omega[X] \simeq L(K_X)$ for $K_X = (\omega)$, $\omega \neq 0$ being an arbitrary rational differential form on X. This remark and Theorem 2.1.51 imply Theorem 2.2.2. We also need the following space $\Omega(D)$ for $D \in Div(X)$:

$$\Omega(D) = \{\omega \in \Omega(X) - \{0\} \mid (\omega) + D \geq 0\} \cup \{0\} .$$

From this definition it is clear that $\Omega(D) \simeq L(K + D)$; in particular, $\Omega(D)$ is finite-dimensional for any $D \in Div(X)$. The line bundle $\mathcal{O}(K + D)$ is sometimes denoted by $\tilde{\Omega}(D)$, thus $\Omega(D) \simeq H^0(\tilde{\Omega}(D))$.

Functoriality. Let $\varphi : X \longrightarrow Y$ be a regular map of smooth projective curves, and let $\omega = \sum_{i=1}^{m} f_i \cdot dg_i \in \Omega(Y)$. We set

$$\varphi^*(\omega) = \sum_{i=1}^{m} \varphi^*(f_i)\cdot d\varphi^*(g_i) \in \Omega(X) \quad .$$

Exercise 2.2.9. Check that $\varphi^*(\omega)$ is well defined, i.e. it does not depend on the choice of representation $\omega = \sum f_i \cdot dg_i$. Show that if ω is regular at $Q \in Y$ then the inverse image $\varphi^*(\omega)$ is also regular at any $P \in \varphi^*(Q)$.

Therefore, there exist k-linear maps

$$\varphi^* : \Omega(Y) \longrightarrow \Omega(X) \quad ,$$

$$\varphi^* : \Omega[Y] \longrightarrow \Omega[X] \quad .$$

Exercise 2.2.10. Show that if φ is a *separable map* (i.e. the extension $k(X)/\varphi^* k(Y)$ is separable) then φ^* is an inclusion. If it is not the case φ^* is trivial. Show that if $\varphi : X \longrightarrow Y$, $\psi : Y \longrightarrow Z$ are regular maps then $(\psi \circ \varphi)^* = \varphi^* \circ \psi^*$.

Remark 2.2.11. Note that in general $\varphi^*((\omega)) \neq (\varphi^*(\omega))$ (consider any map $\varphi : X \longrightarrow \mathbb{P}^1$ of degree more than 1). Hence, the maps φ^* for divisors and differential forms do not commute with each other. The precise commutation rule is given below in Section 2.2.4 (the Hurwitz formula).

Automorphisms. An isomorphism $g : X \longrightarrow X$ of a curve X onto itself is called its *automorphism*. The set of automorphisms of X is denoted $Aut(X)$, or $Aut_k(X)$ if it is necessary to indicate its dependence on k ; it is clear that $Aut(X)$ is a group.

Exercise 2.2.12. Show that $Aut(\mathbb{P}^1) = PGL(2,k)$, $PGL(2,k)$ being the factor-group of $GL(2,k)$ over its center (recall that the center of $GL(2,k)$ consist of matrices of the form $\begin{pmatrix} a & 0 \\ 0 & a \end{pmatrix}$ for $a \in k^*$).

Thus $Aut(\mathbb{P}^1)$ is an infinite group. In Section 2.4.1 we shall see that if $g(X) = 1$ then X is an abelian variety (of dimension one) and hence $Aut(X)$ contains X as a subgroup; this subgroup is normal in $Aut(X)$ and $Aut(X)/X$ is a finite group of order 2, 4, or 6 (for $char\ k \neq 2,3$; for $p = 3$ the order of G is a divisor of 12 and for $p = 2$ it is a divisor of 24). Therefore for $g(X) = 1$ the group $Aut(X)$ is also infinite. On the other hand we have

Theorem 2.2.13. *If* $g(x) \geq 2$ *then* $Aut(X)$ *is finite.* ∎

Remark 2.2.14. If $p = char\ k = 0$ then

$$|Aut(X)| \leq 84 \cdot (g - 1) .$$

The same is true for $p \geq g + 2$ with the unique exception of the curve $y^2 = x^p - x$ (for $p \neq 2$) with $g = (p - 1)/2$ and $|Aut(X)| = 2p \cdot (p^2 - 1)$. For $p \leq g + 1$ we shall see (in Chapter 4.2) that the group $Aut(X)$ can be very large also for some other curves.

Let H be a finite subgroup of $Aut(X)$; it is clear that H also acts on $k(X)$. Consider the field $K = k(X)^H$ of invariants of this action; since K is finitely generated and has the transedence degree 1 over k, by Proposition 2.1.47 and Theorem 2.1.44 there exists a unique smooth projective curve Y such that $k(Y) = K$. The embedding $k(Y) \hookrightarrow k(X)$ defines a regular map $X \longrightarrow Y$ of degree $h = |H|$. In this situation Y is denoted X/H. In particular, if $H = \langle g \rangle$ is a cyclic subgroup in $Aut(X)$ generated by $g \in Aut(X)$ then X/H is denoted $X/\langle g \rangle$ or X^g.

Exercise 2.2.15. Show that $Y = X/H$ as a set.

Exercise 2.2.16. Let $Y = X/H$, let $f : X \longrightarrow Y$ be the natural map, and let $D \in Div(Y)$. Show that H acts on $L(f^*D)$ and $L(f^*D)^H$ is naturally isomophic to $L(D)$.

2.2.2. Riemann-Roch theorem

Theorem 2.2.17 (the Riemann-Roch theorem). *Let* X *be a smooth projective curve,* $K = K_X$ *being its canonical class. Then for any* $D \in Div(X)$

$$\ell(D) - \ell(K - D) = deg\, D - g + 1 \quad . \blacksquare \tag{2.2.3}$$

We do not prove this profound result; we just describe the main points of its proof. By the way we introduce some important notions of geometry of algebraic curves. Note that (2.2.3) is valid for $D = 0$ since $\ell(0) = 1$, $deg(0) = 0$, and $\ell(K) = g$ by the definition of g.

Let $R = R(X)$ be the *algebra of distributions* on X, i.e. of families $\{r_P\}_{P \in X}$, $r_P \in k(X)$ such that $r_P \in \mathcal{O}_P$ for almost all $P \in X$ (i.e. for all but a finite number). For $D = \sum a_P \cdot P \in Div(X)$ by $R(D)$ we denote the set of distributions $r = \{r_P\}_{P \in X}$ such that $ord_P(r_P) \geq -a_P$ for any $P \in X$. Since for any $f \in k(X)$ we have $f \in \mathcal{O}_P$ for almost all $P \in X$, $k(X)$ is a subalgebra of $R(X)$. Let us consider a k-vector space

$$I(D) = R/(R(D) + k(X)) \ ;$$

we shall see below that $I(D)$ is finite-dimensional. Let

$$i(D) = dim_k I(D) \ .$$

Here is a preliminary version of the Riemann-Roch theorem.

Theorem 2.2.18. *For any* $D \in Div\ X$

$$\ell(D) - i(D) = deg\ D - g + 1 \quad . \tag{2.2.4}$$

Proof: To prove (2.2.4) for $D = 0$ it is sufficient to check that $i(0) = g$; this follows from Theorem 2.2.22 below. Now it is sufficient to prove that (2.2.4) is valid for D iff it is valid for $D' = D + P$, P being an arbitrary point of X . Indeed, any divisor can be obtained from the trivial one by successive addition and subtraction of points. We have

$$deg(D') - g + 1 = (deg(D) - g + 1) + 1 \ .$$

Therefore one has to show that

$$\ell(D') - i(D') = (\ell(D) - i(D)) + 1 \ .$$

In fact we shall prove that either $\ell(D') = \ell(D)$ and $i(D') = i(D) - 1$, or $\ell(D') = \ell(D) + 1$ and $i(D') = i(D)$.

To begin with, we note that the argument in the proof of Theorem 2.1.51 shows that $\ell(D')$ equals either $\ell(D)$ or $\ell(D) + 1$.

Since $dim\ R(D')/R(D) = 1$ it is clear that $i(D')$ equals $i(D)$ or $i(D) - 1$. From the definition of $I(D)$ and $I(D')$ it follows that the sequences

$$0 \longrightarrow R(D) + k(X) \longrightarrow R \longrightarrow I(D) \longrightarrow 0$$

$$0 \longrightarrow R(D') + k(X) \longrightarrow R \longrightarrow I(D') \longrightarrow 0$$

are exact. (Recall that a sequence of abelian groups $0 \longrightarrow A \xrightarrow{f} B \xrightarrow{g} C \longrightarrow 0$ is called *exact* iff f is injective, g is surjective and $Im(f) = Ker(g)$, i.e. g

induces an isomorphism $B/f(A) \xrightarrow{\sim} C$). It follows that the sequences

$$0 \longrightarrow (R(D) \ + k(X))/k(X) \longrightarrow R/k(X) \longrightarrow I(D) \longrightarrow 0$$

$$0 \longrightarrow (R(D') + k(X))/k(X) \longrightarrow R/k(X) \longrightarrow I(D') \longrightarrow 0$$

are also exact. Using the "modular" isomorphism $(A + B)/B \simeq A/(A \cap B)$ which is valid for any subgroups A and B of an abelian group C, we obtain the exact sequences

$$0 \longrightarrow (R(D)/L(D)) \longrightarrow R/k(X) \longrightarrow I(D) \longrightarrow 0$$

$$0 \longrightarrow (R(D')/L(D)) \longrightarrow R/k(X) \longrightarrow I(D') \longrightarrow 0$$

since $L(D) = k(X) \cap R(D)$ for any $D \in Div(X)$. Since $dim\, R(D)/R(D') = 1$, $\ell(D) = \ell(D')$ implies $i(D') = i(D) - 1$ and $\ell(D) = \ell(D') + 1$ implies $i(D) = i(D')$. ■

Residues. To deduce Theorem 2.2.17 from Theorem 2.2.18 it is necessary to consider the following important construction. Let $\omega \in \Omega(X)$, let t be a local parameter at $P \in X$, and let $\omega = f \cdot dt$ for $f \in k(X)$. Expanding f into a Laurent power series in t we get

$$f = \sum_{f=-M}^{\infty} a_i \cdot t^i$$

We call the coefficient a_{-1} the *residue* of ω at P and denote it by $Res_P(\omega)$. The basic properties of residues can be summed up as follows

Proposition 2.2.19. a) $Res_P(\omega)$ *does not depend on the choice of local parameter* $t = t_P$;

b) Res_P *is a k-linear functional on* $\Omega(X)$;

c) *If* ω *is regular at* P *then* $Res_P(\omega) = 0$;

d) $Res_P(df) = 0$ *for any* $f \in k(X)^*$;

e) $Res_P(df/f) = ord_P(f)$ *for any* $f \in k(X)^*$.

Proof: Properties b), c), and d) are obvious. Let us prove e): if $f = t^n \cdot u$ with $ord_P(u) = 0$ then $df/f = n \cdot dt/t + du/u$ hence $Res_P(\omega) = n$ since du/u is regular at P . We shall prove the property a) only in the case $char\ k = 0$; the general case can be deduced from this case. Let us write

$$\omega = \sum a_n \cdot dt/t^n + \omega_0 ,$$

ω_0 being regular at P . Let us choose another local parameter u at P and let $Res'_P(\omega)$ be the residue of ω at P defined by this choice. By the properties b) and c) we have

$$Res'_P(\omega) = \sum_{n \geq 1} a_n \cdot Res'_P(dt/t^n) .$$

Since $char\ k = 0$ the rational function $g_n = t^{n-1}/(n-1)$ is well defined for $n \geq 2$ and we obtain $Res'_P(dt/t^n) = Res'_P(dg_n) = 0$ by the property d). Hence $Res'_P(\omega) = Res_P(\omega)$. ∎

Here is the most important result on residues.

Proposition 2.2.20 (the residue formula). *For any* $\omega \in \Omega(X)$

$$\sum_{P \in X} Res_P(\omega) = 0 \qquad (2.2.5)$$

∎

Note that the formula (2.2.5) makes sense since $Res_P(\omega) = 0$ for all but a finite number of points $P \in X$. We do not prove this proposition. Note that if $k = \mathbb{C}$ then X has a natural structure of a Riemann surface, and one can prove that $Res_P(\omega) = \oint_P \omega/2\pi i$. Hence (2.2.5) for $k = \mathbb{C}$ follows from the Stokes theorem.

Let us now define a pairing

$$< \, , \, > \, : \Omega(X) \times R \longrightarrow k \quad ,$$

$$<\omega, r> = \sum_{P \in X} Res_P(r_P \cdot \omega) \quad .$$

Exercise 2.2.21. Show that

a) if $r \in k(X) \subset R$ then $<\omega, r> = 0$;

b) $<\omega, r> = 0$ for any $\omega \in \Omega(-D)$ and any $r \in R(D)$;

c) $<f \cdot \omega, r> = <\omega, f \cdot r>$ for any $f \in k(X)$.

Let $\omega \in \Omega(-D)$. Then $r \longmapsto <\omega, r>$ defines (by Exercise 2.2.21.b) a linear functional $\theta(\omega)$ on the space $I(D) = R/(k(X) + R(D))$.

Theorem 2.2.22 (the duality theorem). *The map* $\omega \longmapsto \theta(\omega)$ *defines an isomorphism of* $\Omega(-D)$ *onto the space dual to* $I(D)$. ∎

From Theorem 2.2.22 and Theorem 2.2.18 the Riemann-Roch theorem follows since $\Omega(-D) \simeq L(K - D)$. Now let us give some of its corollaries.

Exercise 2.2.23. Let X be a smooth projective curve, let K_X be its canonical class, and let $g = g(X)$ be its genus. Show that:

a) $\deg K_X = 2g - 2$;

b) $\ell(K - D) = g - 1 - \deg D$ for $\deg D < 0$;

c) $\ell(D) = \deg D - g + 1$ for $\deg D > 2g - 2$.

(*Hint*: $\ell(D) = 0$ for $\deg D < 0$).

Sometimes it is convenient to write the Riemann-Roch theorem using the language of line bundles. Let $\mathcal{L}$ be a line bundle on X . Let $\mathcal{L} = \mathcal{O}(D)$ and thus

$$h^0(\mathcal{L}) = \dim_k H^0(\mathcal{L}) = \ell(D) .$$

Theorem 2.2.24. *Let $\mathcal{L} = \mathcal{O}(D)$ be a line bundle on X , and let $\mathcal{K}_X = \mathcal{O}(K_X)$ be the canonical line bundle on X . Then*

$$h^0(\mathcal{L}) - h^0(\mathcal{K}_X \otimes \mathcal{L}^{-1}) = \deg \mathcal{L} - g + 1 .$$

■

Corollary 2.2.25. a) *$\deg \mathcal{K}_X = 2g - 2$;*

b) *$h^0(\mathcal{L}) = \deg(\mathcal{L}) - g + 1$ for $\deg(\mathcal{L}) > 2g - 2$.* ■

Note that the uniqueness of the curve of genus zero also easily follows from the Riemann-Roch theorem.

Exercise 2.2.26. Show that if $g(X) = 0$ then $X \simeq \mathbb{P}^1$. (*Hint*: Apply the Riemann-Roch theorem to $D = P$, P being a point of X).

Embeddings into projective spaces. The Riemann-Roch theorem makes it possible in many cases to prove that the map $\varphi : X \longrightarrow \mathbb{P}^n$ defined by a complete linear system $|D|$ is an embedding, i.e. defines an isomorphism of X onto its image in $\mathbb{P}^n$. This can be done using the following result:

Proposition 2.2.27. *Let D be a divisor on X such that for any $P, Q \in X$*

$$\ell(D - P - Q) = \ell(D) - 2 \quad . \tag{2.2.6}$$

Then $\varphi_D : X \longrightarrow \mathbb{P}^n$ is an embedding ($n = \dim |D|$).

Sketch of proof: For $P \neq Q$ the equality (2.2.6) implies that $\varphi_D(P) \neq \varphi_D(Q)$, i.e. φ_D is injective. For $P = Q$ (2.2.6) implies that φ_D^* defines an isomorphism of $m_{P'}/m_{P'}^2$ onto m_P/m_P^2 , where $P' = \varphi_D(P)$ and $m_{P'}$ is the maximal ideal of the local ring of P' on the image of X .. Thus φ_D is an injection which induces isomorphisms of all tangent spaces. Any such injection is an imbedding. Indeed, the injectivity of φ_D (together with $\varphi_D^* \neq 0$) implies that $\deg \varphi_D = 1$. Hence there exists an inverse rational map $\varphi_D^{-1} : Y = \varphi_D(X) \longrightarrow X$. Since the map $\varphi_D^* : m_{P'}/m_{P'}^2 \longrightarrow m_P/m_P^2$ is an isomorphism for any P , the curve Y is smooth and hence φ_D is an isomorphism. ∎

Exercise 2.2.28. Give a complete proof of the proposition.

Applying Exercise 2.2.23.c we obtain

Corollary 2.2.29. *If $\deg D \geq 2g + 1$ then φ_D is an embedding of X into $\mathbb{P}^n$.* ∎

The maps φ_{mK} defined by multiplicities of the canonical class $K = K_X$ are of particular interest. For these maps we get

Corollary 2.2.30. *If X is a curve of genus at least 2 then φ_{3K} is an embedding. If the genus of X is at least 3 then φ_{2K} is an embedding.* ∎

Remark 2.2.31. There exist curves of arbitrary genera such that φ_K is not an embedding. These curves are called *hyperelliptic*. Let X be a smooth projective curve of genus $g(X) > 0$ with the canonical class K. If φ_K is not an embedding then there exist points P and Q such that (2.2.6) is not valid, hence

$$\ell(K - P - Q) \geq \ell(K) - 1 = g - 1 .$$

Since $\ell(K - P) \geq \ell(K - P - Q)$, the equality $\ell(K - P) = g$ implies that

$$\ell(P) = 1 + 1 - g + \ell(K - P) = 2$$

and hence there exists a non-constant map $f : X \longrightarrow \mathbb{P}^1$ such that $deg\, f = deg\, P = 1$, i.e. $X = \mathbb{P}^1$ and $g(X) = 0$. Therefore

$$\ell(K - P - Q) = \ell(K - P) = g - 1$$

for some $P, Q \in X$. By Riemann-Roch

$$\ell(P + Q) = 2 - g + 1 + \ell(K - P - Q) = 2 ,$$

hence there is a map $f : X \longrightarrow \mathbb{P}^1$ of degree 2. Conversely, if $f : X \longrightarrow \mathbb{P}^1$ is a regular map, $deg\, f = 2$, and $g(X) > 0$ then $f^*(Q) = P + P'$, $Q \in \mathbb{P}^1$, $P, P' \in X$ and by Riemann-Roch

$$\ell(K - P - P') = 2g - 2 - 2 + 1 - g + \ell(P + P') = g - 1 ,$$

$$\ell(K - P) = 2g - 2 - 1 + 1 - g + \ell(P) = g - 1$$

($\ell(P) = 1$ since X is not rational and $\ell(P + P') = 2$

since $P + P' = f^*(Q)$); thus φ_K is not an embedding. It can be proved that hyperlliptic curves are in some (precise) sense rare exceptions, but we do not explain this here.

Noether theorem. The canonical class of a non-hyperelliptic curve has an important property which can be proved using the canonical embedding of X into $\mathbb{P}^{g-1}$. For any $D, D' \in Div(X)$ there exists a natural map

$$L(D) \otimes L(D') \longrightarrow L(D + D') \quad ,$$

$$f \otimes f' \longmapsto f \cdot f' \quad .$$

In particular, for any $m \geq 1$ we have the map

$$L(D)^{\otimes m} \longrightarrow L(m \cdot D) \quad .$$

Moreover, since for any permutation π of the set $\{1,\ldots,m\}$ the images of $f_1 \otimes \ldots \otimes f_m$ and $f_{\pi(1)} \otimes \ldots \otimes f_{\pi(m)}$ in $L(m \cdot D)$ coincide, there exists a natural map of m-th symmetrical power $S^m L(D)$ to $L(m \cdot D)$.

Theorem 2.2.32. *For a non-hyperellitic* X *the map* $S^m L(K) \longrightarrow L(m \cdot K)$ *is surjective.* ∎

A curve Y which is the image $\varphi_K(X)$ in $\mathbb{P}^{g-1}$ of some X under the canonical embedding is called a *canonical curve*.

Exercise 2.2.33. Show that Theorem 2.2.32 can be restated as follows: a canonical curve in $\mathbb{P}^{g-1}$ is *projectively normal*, i.e. for any $D \in |m(L)|$, (L) being a hyperplane section divisor, there exists a form F of degree m such that $(F) = D$.

2.2.3. Hurwitz formula

Let $f : X \longrightarrow Y$ be a non-constant regular map of smooth complete curves. A formula due to Hurwitz gives an expression of $g(X)$ in terms of $g(Y)$ and some invariants of f. Let us consider the field extension $k(X)/f^*(k(Y))$, and let K' be the maximal separable over $f^*(k(Y))$ subfield of $k(X)$; thus, $f^*(k(Y)) \subseteq K' \subseteq k(X)$. By Proposition 2.1.47 there exists a smooth projective curve X' such that $k(X') = K'$. Therefore the map f can be factored,

$$f : X \xrightarrow{f'} X' \xrightarrow{f''} Y ,$$

f' being purely inseparable, and f'' being separable. Let us show that $g(X) = g(X')$. To do this indroduce the following definition. Let X be a smooth projective curve over the field k of positive characteristic $p > 0$, let $K = k(X)$ and let $K_p = K^{1/p}$ be the field

$$K_p = \{a \in \overline{K} \mid a^p \in K\}$$

Again by Proposition 2.1.47 there exists a unique smooth projective curve X_p with $K_p = k(X_p)$ and the inclusion $K \subset K_p$ defines the map $\varphi_p : X_p \longrightarrow X$ which is called the *p-linear Frobenius map* (or the *Frobenius map over* $\mathbb{F}_p$). Since the trancendence degree of K equals 1, $deg\ \varphi_p = p$.

Proposition 2.2.34. *Let $f : X \longrightarrow Y$ be a non-constant purely inseparable map. Then $g(X) = g(Y)$.*

Sketch of proof: Let $deg\ f = p^r$. By induction one can assume that $r = 1$. In this case $k(X) \subseteq f^*(k(Y)^{1/p})$

and since $[k(X):f^*(k(Y))] = p$ the map f coincides with φ_p and $X = Y_p$. Let us show that the dimension of $\Omega[Y]$ equals that of $\Omega[Y_p]$. Indeed, for $\omega = f\cdot dg \in \Omega[Y]$ the form $\omega' = f'\cdot dg'$ belongs to $\Omega[Y_p]$ where f' and g' are obtained from f and g by raising all their coefficient to the p-th power; conversely, for $\omega' = f'\cdot dg' \in \Omega[Y_p]$ the form $\omega = f\cdot dg$ lies in $\Omega[Y]$, here f and g are obtained from f' and g' by exracting the p-th power root from all their coefficients. ■

By Proposition 2.2.34 computing $g(X)$ we can assume that f is separable.

Proposition 2.2.35. *The set of ramification points of a separable map* $f : X \longrightarrow Y$ *is finite.*

Proof: Deleting finite sets of points from X and Y we can assume X and Y to be affine curves. Let $k[X] = A$ and $k[Y] = A'$, let $a \in A$ be a primitive element of the extension $k(X)/f^*(k(Y))$ and let $F(T)$ be the minimal polynomial of a , $D(F) \in A'$ being its discriminant. Since f is separable, $D(F) \neq 0$ and hence $D(F)$ vanishes only on a finite set $S \subset Y$; at any $P \notin S$ the map f is unramified. ■

Let $f : X \longrightarrow Y$ be a non-constant separable map of smooth projective curves, let $P \in X$, let $f(P) = Q$, and let t_P and t_Q be local parameters at P and Q , respectively. Then $f^*(dt_Q) = g\cdot dt_P$ for some $g \in \mathcal{O}_P$. Let $ord_P(g) = b_P$; it is clear that $b_P \neq 0$ only for ramification points P of f . Let

$$B = B_f = \sum b_P\cdot P \in Div(X) \ ;$$

B_f is called the *ramification divisor* of f .

Theorem 2.2.36. *Let $f : X \longrightarrow Y$ be a separable non-constant map of degree n. Then*

$$2g(X) - 2 = n\cdot(2g(Y) - 2) + deg\ B_f \quad .$$

Proof: By Exercise 2.2.23.a and Proposition 2.1.69 it is sufficient to show that $K_X = f^*K_Y + B_f$. Let $\omega \in \Omega(Y) - \{0\}$ be such that $Supp((\omega))$ is disjoint from the finite set where f is ramified. Let us compute the divisor $K_X = (f^*(\omega))$. If f is unramified at $Q \in Y$ and $\omega = g\cdot dt_Q$, t_Q being a local parameter at Q then $f^*(t_Q)$ is a local parameter at any $P \in X$ with $f(P) = Q$ and hence $K_X|_U = f^*K_Y|_U$ where $U = X - Supp\ B_f$. If f is ramified at $Q \in Y$ then $\omega = h\cdot dt_Q$ with $ord_Q h = 0$ and hence we see that $(f^*(\omega)) = (g)_P$ where $f(P) = Q$, and $f^*(dt_Q) = g\cdot dt_Q$, since $ord_P(f^*(h)) = 0$. ■

We call $f : X \longrightarrow Y$ *tamely ramified* iff $p \nmid e_P$ for all $P \in X$ (here $p = char\ k$).

Corollary 2.2.37 (the Hurwitz formula). *Let a map $f : X \longrightarrow Y$ be seperable, non-constant, and tamely ramified. Then*

$$2g(X) - 2 = n\cdot(2g(Y) - 2) + \sum_{P\in X}(e_P - 1) \quad . \qquad (2.2.8)$$

Proof: It is sufficient to show that

$$deg\ B_f = \sum_{P\in X}(e_P - 1)$$

provided f is tamely ramified. Indeed, if $f^*(t_Q) = h\cdot t_P^{e_P}$ with $ord_P h = 0$ then

$$f^*(dt_Q) = h\cdot e_P\cdot t_Q^{e_P-1}\cdot dt_P + t_P^{e_P}\cdot dh$$

and since $e_P \neq 0$ in k we obtain the formula. ■

Note that for $p = char\ k = 0$ any f is separable and tamely ramified. A form of the Hurwitz formula which follows is of particular interest.

Corollary 2.2.38. *Let $f : X \longrightarrow \mathbb{P}^1$ be a non-constant map of degree n which is tamely ramified. Then*

$$g(X) = 1 - n + \frac{1}{2} \cdot \sum_{P \in X} (e_P - 1) \quad . \tag{2.2.9}$$

■

2.2.4. Special divisors

We call a divisor D *special* iff $\ell(K - D) > 0$.

Exercise 2.2.39. Show that D is special iff $D \sim K - D'$ for an effective D' .

Exercise 2.2.40. Show that any divisor of degree at least $2g - 1$ is non-special and any divisor of degree at most $g - 2$ is special.

Exercise 2.2.41. For any divisors D and D' show that

$$\ell(D + D') \geq \ell(D) + \ell(D') - 1 ,$$

i.e.

$$dim\ |D| + dim\ |D'| \leq dim\ |D + D''| .$$

(*Hint*: Consider the natural map $|D| \times |D'| \longrightarrow |D + D'|$).

Using the result of the last exercise one can prove

Theorem 2.2.42 (the Clifford theorem). *For any special effective divisor* D

$$\dim |D| \le \frac{1}{2}\cdot \deg D \quad .$$

Proof: By Exercise 2.2.39 $D' = K - D$ is equivalent to an effective divisor. Applying the result of Exercise 2.2.41 to D and D' we obtain

$$\dim |D| + \dim |K - D| \le \dim |K| = g - 1 \quad .$$

By Riemann-Roch

$$\dim |D| - \dim |K - D| = \deg D + 1 - g \quad .$$

Adding these formulae we obtain the theorem. ■

Remark 2.2.43. Usually one includes into the statement of the Clifford theorem the following complement. If $\dim |D| = \frac{1}{2}\cdot \deg D$ then either $D = 0$, or $D = K_X$, or X is hyperelliptic. We do not need this fact in the book.

Weierstrass points. The simplest example of an effective divisor is $D = a\cdot P$, $P \in X$. Considering divisors of such a form it is possible to define Weierstrass points which play an essential role in the study of geometry of curves.

Let $P \in X$, let $g = g(X) \ge 2$ and let $a \ge 1$ be an integer. We call a a *gap* at P if $\ell(a\cdot P) = \ell((a - 1)\cdot P)$ and a *non-gap* if it is not the case. In other words a is a non-gap at P iff there exists $f \in k(X)$ with $ord_P(f) = a$. If such f does not exist then a is a gap at P .

Exercise 2.2.44. Show that for any $P \in X$:

a) If a and b are non-gaps at P then $a + b$ is also a non-gap;

b) 1 is a gap at P ;

c) The number of gaps at P equals g ;

d) If $a \geq 2g$ then a is a non-gap at P .

We call $P \in X$ a *Weierstrass point* iff the gap sequence $\{a_1,\dots,a_g\}$ at P with $a_1 < \dots < a_g$ does not coincide with $\{1,\dots,g\}$.

Exercise 2.2.45. Let $P \in X$. Show that the following conditions are equivalent

a) P is a Weierstrass point;

b) Divisor $g \cdot P$ is special;

c) Divisor $a \cdot P$ is special for some $a \geq g$.

Let $\{a_1,\dots,a_g\}$, $1 \leq a_1 < a < \dots < a_g \leq 2g - 1$, be the gap-sequence at $P \in X$. We call

$$w(P) = \sum_{i=1}^{g} (a_i - i)$$

the (*Weierstrass*) *weight* of P ; $w(P) > 0$ iff P is a Weierstrass point.

We have a *mass-formula* which follows:

$$\sum_{P \in X} w(P) = g \cdot (g - 1) \cdot (g + 1) + \alpha(X) \quad ,$$

where $\alpha(X) \geq 0$; $\alpha(X) = 0$ if $p = char\ k = 0$, and for $p > 0$ the value $\alpha(X)$ vanishes for all but a finite number of "exceptional" curves.

Therefore the number of Weierstrass points on X if finite and does not exceed $g\cdot(g-1)\cdot(g+1)+\alpha(X)$.

Exercise 2.2.46. Show that $w(P) \le g\cdot(g-1)/2$ and $w(P) = g\cdot(g-1)/2$ iff 2 is a non-gap at P .

A point P with $w(P) = g\cdot(g-1)/2$ is called a *hyperelliptic point*.

Exercise 2.2.47. Show that X has hyperelliptic points iff X is a hyperelliptic curve.

For $p = char\ k \ne 2$ any hyperelliptic curve has $2\cdot(g+1)$ Weierstrass points all of them being hyperelliptic.

To construct Weierstrass points one can use the following result.

Proposition 2.2.48. *Let* $\varphi : X \longrightarrow Y$ *be a regular map of smooth projective curves of degree* n *, let* $Q \in Y$ *be such that* $\varphi^{-1}(Q)$ *is a single point* P *, and let* $g(Y) \le \lceil g(X)/n \rceil - 1$ *. Then* P *is a Weierstrass point.*

Proof: By Riemann-Roch we have $\ell((g(Y)+1)\cdot Q) \ge 2$ hence there exists a non-constant $f \in k(Y)$ having the pole of order at most $g(Y)+1$ at Q and regular outside Q . Let us consider $\varphi^{*}(f)$. Its unique pole is P and

$$|ord_P(\varphi^{*}(f))| = n\cdot|ord_Q(f)| \le n\cdot(g(Y)+1) ,$$

since the ramification index of P equals n. Since $n\cdot(g(Y)+1) \le g(X)$ we see that $\ell(g(X)\cdot P) \ge 2$, thus P is a Weierstrass point. ∎

We shall use this result in the following situation:

Corollary 2.2.49. *Let* $h \in Aut(X)$ *be of order* n , $h(P) = P$, *and let* $g(X^h) \le \lceil g(X)/n \rceil - 1$. *Then* P *is a Weierstrass point.* ■

2.2.5. Cartier operator

If the characteristic p of the ground field k is positive, one can define an operator on $\Omega(X)$ which plays an important role in the study of curves in a positive characteristic. Let $K = k(X)$, X being a smooth projective curve over k , let $t \in K$ be such that $t \notin K^p$ and thus $K = K^p(t)$. We can write any $\omega \in \Omega(X)$ in the form

$$\omega = (y_0^p + y_1^p \cdot t + \ldots + y_{p-1}^p \cdot t^{p-1}) \cdot dt \ , \ y_i \in K \ . \qquad (2.2.10)$$

Set

$$C(\omega) = y_{p-1} \cdot dt \ .$$

One can show that this definition does not depend on the choice of t . The formula (2.2.10) implies that any $\omega \in \Omega(X)$ can be uniquely written in the form $\omega = df + g^p \cdot dt/t$; then it is easy to cheek that $C(\omega) = g \cdot dt/t$.

Exercise 2.2.50. Let $z \in K$. Show that:

a) $C(z^p \cdot \omega) = z \cdot C(\omega)$;

b) $C(dz) = 0$;

c) $C(dz/z) = dz/z$;

d) if ω is regular at $P \in X$ then $C(\omega)$ is also regular at P ;

e) for any $P \in X$

$$(Res_P(C(\omega)))^p = Res_P \omega \ .$$

We can now give a characterization of *exact* (i.e. of the form $\omega = df$) and of *logarithmic* (i.e. of the form $\omega = df/f$) differential forms.

Proposition 2.2.51. *Let* $\omega \in \Omega(X)$. *Then*

a) $C(\omega) = 0$ *iff* $\omega = df$ *for some* $f \in K$;

b) $C(\omega) = \omega$ *iff* $\omega = df/f$ *for some* $f \in K$. ■

Note that the property a) is obvious.

The operator $C = C_p$ is called the *Cartier operator over* $\mathbb{F}_p$; if $q = p^m$ then we call its m-th power $C_q = C_p^m$ the *Cartier operator over* $\mathbb{F}_q$.

In Part 3 we shall need the following fact.

Exercise 2.2.52. Let $D = P_1 + \ldots + P_n \in Div(X)$, where $P_1,\ldots,P_n$ are distinct points of X , let $P \in X$, $P \neq P_i$, $i = 1,\ldots,n$, and let $\omega \in \Omega(D - a\cdot P)$. Show that $C_q(\omega) \in \Omega(D - \lceil\frac{a}{q}\rceil\cdot P)$.

CHAPTER 2.3

Rational points

In Chapters 2.1 and 2.2 we have assumed the ground field k to be algebraically closed. In fact there are many cases in which the consideration of non-closed field such as $\mathbb{Q}$ or $\mathbb{F}_q$ is indispensable. For example, applying algebraic geometry to coding theory one should study curves over $\mathbb{F}_q$ and their points with coordinates in $\mathbb{F}_q$ (such points are called $\mathbb{F}_q$-*rational*). This section is devoted to the basic properties of curves over a non-closed (mainly finite) field and of their points. In Section 2.3.1 we give the definition and basic properties of rational points and divisors on a curve over a non-closed field; Section 2.3.2 contains basic properties of $\mathbb{F}_q$-rational points on a curve over $\mathbb{F}_q$, including the question about the number of points. In Section 2.3.3 we study the asymptotic behaviour of the number of $\mathbb{F}_q$-rational points on a curve of high genus.

2.3.1. Rational points and divisors

Recall that a field k is called *perfect* iff its algebraic closure $\bar{k}$ is separable over k.

Rational points. Let k be a perfect field (one can assume $k = \mathbb{F}_q$; we need only this case), let $\bar{k}$ be its algebraic closure and let $G = Gal(\bar{k}/k)$ be its Galois group. It is not convenient to define a curve over k as a set of points. For instance some non-trivial systems of equations have no solutions in k. We shall define a smooth projective curve over k in terms of its field of rational functions.

Let K be a finitely generated field of transcendence degree one over k which does not contain any non-trivial algebraic extension of k. If k is algebraically closed then by Proposition 2.1.47 and Theorem 2.1.44 there exists a unique smooth projective curve X with $K = k(X)$. Let now k be a non-closed field. We say that in this case $K = k(X)$ is also *the field of rational functions on a smooth projective curve* X over k. It should be pointed out that we do not give any other definition of X except this one. Let $K' = \bar{k}\cdot k(X)$ where $\bar{k}$ and $k(X)$ are considered as subfields of an algebraic closure $\bar{K}$ of $K = k(X)$. Let $\bar{X}$ be a smooth projective curve over $\bar{k}$ such that $K' = \bar{k}(\bar{X})$. Note that $\bar{X}$ is indeed a curve, i.e. it is irreducible; to emphasize this fact we say that X is *absolutely irreducible;* sometimes we denote $\bar{X}$ by $X\otimes\bar{k}$. One can show (in a way similar to the proof of Proposition 2.3.4 below) that the curve $\bar{X} \subset \mathbb{P}^N$ can be defined by a system of equations $F_1 = \ldots = F_m = 0$, F_i being forms with coefficients in k. For any algebraic extension k' of k

the subset of $\overline{X}$ consisting of points with coordinates in k' is denoted by $X(k')$ and is called the *set of k'-rational points of* X. In particular, $X(\overline{k}) = \overline{X}$ and $X(k)$ is the set of *k-rational points* of X. The points of X are described in terms of K as follows. Let V be a proper valuation ring in K, containing k, $V \neq k$, and let m be its maximal ideal.

Exercise 2.3.1. Show that the extension $k(V)/k$ is of finite degree (here $k(V) = V/m$).

Let $s = [k(V):k]$; we say that V determines a *point* P *of* X *of degree* s and write $V = \mathcal{O}_P$, $m = m_P$, $s = deg\, P$.

Let $P \in \overline{X}$ and let $k(P)$ be the normal extension of k generated by the coordinates of P (we assume an embedding $X \hookrightarrow \mathbb{P}^n$ to be fixed); it is easy to check that $k(P)$ is the intersection of all normal extensions k' of k with $P \in X(k')$. Let $s = deg\, P$; one can show that $s = [k(P):k]$. For any $\sigma \in Gal(\overline{k}/k)$ the conjugate point $\sigma(P) \in X(k(P))$ since $\overline{X}$ can be defined by equations with coefficients in $k \subseteq k(P)$. Since $s = [k(P):k]$ the set $\{\sigma(P) \mid \sigma \in Gal(\overline{k}/k)\}$ has s elements $P_1,\dots,P_s$. Let $\overline{V}_i = \mathcal{O}_{P_i}$ be the local ring of P_i in $K' = \overline{k}(\overline{X})$.

Proposition 2.3.2. *Let* $V = K \cap \overline{V}_1$. *Then* $K \cap \overline{V}_i = V$ *for any* $i = 1,\dots,s$; V *is a valuation subring of* $K = k(X)$ *which defines a point of* X *of degree* s,

$$V = \bigcap_{i=1}^{s} \overline{V}_i , \quad \overline{k}\cdot V = \bigcup_{i=1}^{s} \overline{V}_i .$$

Vice versa for any valuation ring V *with* $k \subset V \subset K$ *which defines a point of* X *of degree* s *there exists* $P \in X(k')$, k' *being an extension of* k *of degree* s, *such that* $V = K \cap \mathcal{O}_P$.

Proof: Note that if $\sigma(P_1) = P_r$, $\sigma \in Gal(\bar{k}/k)$, then $\sigma(\bar{V}_1) = \bar{V}_r$, $r = 1,\ldots,s$. Hence $K \cap \bar{V}_1 = K \cap \bar{V}_r$ for $r = 1,\ldots,s$ since $K = (K')^G$; $\bar{V}_1 \cap K$ is a valuation ring since $f \in K$, $f \notin V = \bar{V}_1 \cap K$ implies $f^{-1} \in V_1$ and hence $f^{-1} \in V$. Clearly $\bigcap_{i=1}^{s} \bar{V}_i \supseteq V$; since $(\bigcap_{i=1}^{s} \bar{V}_i) \subset (K')^G = K$ we have $\bigcap_{i=1}^{s} \bar{V}_i = (\bigcap_{i=1}^{s} V_i) \cap K = V$. The proof of the second part of the proposition is similar. ■

Exercise 2.3.3. Prove the second part of Proposition 2.3.2.

Therefore any point of X of degree s defines a unique s-tuple of points of $\bar{X}$ which are conjugate to each other under the action of the Galois group and vice versa, any such s-tuple defines a unique point of X of degree s.

Rational divisors. We call a *divisor* on X (a *k-rational divisor* or a *k-divisor*) a finite formal sum $D = \sum a_P \cdot P$, a_P being integers and P being points of X (in general, of different degrees). We call

$$deg\ D = \sum a_P \cdot deg\ P$$

the *degree* of D. Since any P of degree s defines an s-tuple $\{P_1,\ldots,P_s\}$, $P_i \in \bar{X}$, D determines the divisor

$$\bar{D} = \sum a_P \cdot (\sum_{i=1}^{s} P_i)$$

on $\bar{X}$ with

$$deg\ \bar{D} = \sum a_P \cdot deg\ P = deg\ D\ .$$

Conversely, let

$$\bar{D} = \sum a_P \cdot P \in Div(\bar{X})$$

and let $a_P = a_{\sigma P}$ for any $\sigma = Gal(\bar{k}/k)$. Then there exists a unique k-rational divisor D on X such that $\bar{D}$ is obtained from D by the above construction.

Hence we can look at a k-rational divisor on X in two ways: as at a divisor D on X or as at a divisor $\bar{D}$ on $\bar{X}$ which is invariant under the Galois action. Further on we identify the divisors D and $\bar{D}$ and use these two points of view simultaneously.

Proposition 2.3.4. *Let $\bar{D}$ be a Galois invariant divisor on $\bar{X}$ (i.e. corresponding to a k-rational divisor D on X). Then $L(\bar{D})$ has a basis whose elements belong to $L(\bar{D}) \cap k(X)$.*

Proof: Let $f \in L(\bar{D})$. It is clear that $f \in k'(X)$ for a finite Galois extension k' of k . Let $Gal(k'/k) = \{\sigma_1,\ldots,\sigma_n\}$, let $\{\omega_1,\ldots,\omega_n\}$ be a basis of the extension k'/k and let

$$g_j = \sum_{i=1}^{s} \sigma_i(\omega_j \cdot f) = \sum_{i=1}^{s} \sigma_i(\omega_j)\cdot\sigma_i(f)$$

for $j = 1,\ldots,n$. By the Galois theory the determinant of the matrix $\|\sigma_i(\omega_j)\|$ does not vanish. Therefore, if

$$L(\bar{D}) = \bigoplus_{i=1}^{m} f_i \cdot \bar{k}$$

then the corresponding functions $g_{11},\ldots,g_{1n},\ldots,g_{mn}$ generate $L(\bar{D})$ over $\bar{k}$, on the other hand $g_{ij} \in k(X)$. ■

Define $L(D)$ for a k-divisor D as $L(\bar{D}) \cap k(X)$. From Proposition 2.3.4 it follows that

$$L(D) = \{f \in k(X) \mid (f) + \bar{D} \geq 0\}$$

We set $\ell(D) = dim_k L(D)$.

Corollary 2.3.5. *Let D be a k-rational divisor on X. Then $L(\overline{D}) = L(D)\otimes\overline{k}$; in particular, $\ell(D) = \ell(\overline{D})$.* ■

Exercise 2.3.6. Let D and D' be divisors on X and let $\overline{D} \sim \overline{D}'$. Show that there exists $f \in k(X)$ such that $\overline{D} = \overline{D}' + (f)$.

Line bundles. By an *asymptotically standard family of lattices* $\mathcal{L} = (\mathcal{L}_P)_{P\in X}$ on X we mean a family $(\mathcal{L}_P)_{P\in X}$ of free $\mathcal{O}_P$-submodules $\mathcal{L}_P \subset k(X)$ of rank one such that $\mathcal{L}_P = \mathcal{O}_P$ for all but a finite number of P . The fibre $\overline{\mathcal{L}}_P$ of the family $\mathcal{L} = (\mathcal{L}_P)_{P\in X}$ at $P \in X$ is the one-dimensional space $\overline{\mathcal{L}}_P = \mathcal{L}_P/m_P\cdot\mathcal{L}_P$ over $k(P) = \mathcal{O}_P/m_P$. As above we denote the intersection $\bigcap_{P\in X}\mathcal{L}_P \subset k(X)$ by $H^0(\mathcal{L})$. The family $\mathcal{L} = (\mathcal{L}_P)_{P\in X}$ on X defines a family $\tilde{\mathcal{L}} = (\tilde{\mathcal{L}}_Q)_{Q\in\overline{X}}$ on $\overline{X}$, where $\tilde{\mathcal{L}}_Q = \overline{k}\cdot\mathcal{L}_P \subset \overline{k}(\overline{X})$ for any point $Q \in \overline{X}$ corresponding to $P \in X$.

Exercise 2.3.7. Show that the correspondence between asymptotically standard families of lattices and divisors takes place on X and is also a bijection. Define line bundles on X and show that line bundles on X can be identified with line bundles on $\overline{X}$ which are invariant under the Galois action.

Exercise 2.3.8. Deduce from Corollary 2.3.5 that $h^0(\mathcal{L}) = h^0(\overline{\mathcal{L}})$, $\mathcal{L}$ being a line bundle on X and $\overline{\mathcal{L}}$ being the corresponding bundle on $\overline{X}$.

Therefore the Riemann-Roch theorem and its corollaries are valid for divisors and line bundles on X .

Remark 2.3.9. In the case of a multi-dimensional smooth projective variety X we can also consider the notion of rationality. In particular, X is called *defined over* k iff it can be given in $\mathbb{P}^N$ by a system $F_1 = \ldots = F_m = 0$, where all the coefficients of F_i lie in k; for any algebraic extension k'/k the set $X(k')$ is defined similarly to the case of curves. A *k-rational divisor* $D = \sum a_F \cdot F \in Div(X)$ can be defined as a Galois invariant divisor on $\overline{X}$ (where $\overline{X}$ is the variety over $\overline{k}$ defined by $F_1 = \ldots = F_m = 0$). The spaces $L(D)$ for $D \in Div(X)$ and line bundles on X are defined similarly to the case of curves. All the above results on k-rational divisors, linear systems, and line bundles are valid for any (smooth projective) variety X defined over k.

k-rational maps. Let X and Y be smooth projective curves over k. We say that a field embedding $\varphi : k(Y) \hookrightarrow k(X)$ which is identical on k defines a *(regular) map* $f : X \longrightarrow Y$ *over* k (a *k-morphism*) and write $\varphi = f^*$ (note that since X and Y are smooth and projective every rational map is regular). The set of k-maps from X to Y is denoted by $Hom_k(X,Y)$.

Exercise 2.3.10. Show that

$$Hom_k(X,Y) = (Hom_{\overline{k}}(\overline{X},\overline{Y}))^{Gal(\overline{k}/k)} .$$

Exercise 2.3.11. Let $f : X \longrightarrow Y$ be a k-map, let $P \in X$, let $V = \mathcal{O}_P$, and let

$$V' = (f^*)^{-1}(f^*(k(Y)) \cap V) .$$

Show that there exists a unique point $Q = f(P) \in Y$ such that $V' = \mathcal{O}_Q$. Show that for any algebraic extension k' of k if $P \in X(k')$ then $f(P) \in Y(k')$. In particular, $f(X(k)) \subseteq Y(k)$.

2.3.2. Curves over a finite field

Weil theorems. Let now $k = \mathbb{F}_q$ be a finite field of characteristic p with $q = p^a$ elements. For any integer $s \geq 1$ the set $X(\mathbb{F}_{q^s})$ is finite; let N_s be its cardinality. In particular, $N = N_1$ is the number of $\mathbb{F}_q$-points of X. To study N_s it is useful to consider the following formal power series with rational coefficients

$$Z(t) = Z(X;t) = exp(\sum_{r=1}^{\infty} N_r \cdot t^r / r)$$

which is called the *zeta-function* of X over $\mathbb{F}_q$. Here is the main result concerning $Z(t)$.

Theorem 2.3.12 (the Weil theorem). *$Z(t)$ is (a power series expansion of) a rational function of the form*

$$Z(t) = \frac{P_X(t)}{(1 - t)(1 - qt)} ,$$

where $P_X(t) = q^g \cdot t^{2g} + p_1 \cdot t^{2g-1} + \ldots + p_{2g-1} \cdot t + 1 \in \mathbb{Z}[t]$ *is a polynomial with integral coefficients. Moreover*

$$P_X(t) = \prod_{i=1}^{2g} (1 - \omega_i \cdot t) ,$$

where $\omega_{g+i} = \bar{\omega}_i$, *and* $|\omega_i| = \sqrt{q}$ *for any* $i = 1, \ldots, 2g$. ■

The very last statement of the theorem is called the *Riemann hypothesis* (since it is equivalent to the statement that $Z(q^{-s}) = 0$ only for $Re(s) = \frac{1}{2}$).

From this theorem we can easily obtain formulae for N_s which follow.

Exercise 2.3.13. Show that

$$N_s = \frac{1}{(s-1)!} \cdot \frac{d^s}{dt^s}(\log Z(t))|_{t=0} \quad .$$

Exercise 2.3.14. Deduce from the last exercise that

$$N_s = q^s + 1 - \sum_{i=1}^{s} \omega_i^s \tag{2.3.1}$$

In particular,

$$|N_s - (q \;\; + 1)| \le 2 \cdot q^{s/2} \cdot g \tag{2.3.2}$$

since $|\omega_i| \le \sqrt{q}$. In terms of N_s we can give also an expression for the number of line bundles on X of degree zero (and hence of line bundles of any fixed degree).

Theorem 2.3.15. *Let $M = M(X)$ be the number of line bundles of degree zero on X (i.e. the cardinality of the set of $\mathbb{F}_q$-rational points on the Jacobian J_X of X). Then*

$$M = P_X(1) = \prod_{j=1}^{2g}(\omega_j - 1) \quad .$$

In particular,

$$(\sqrt{q} - 1)^{2g} \le M(X) \le (\sqrt{q} + 1)^{2g} \quad .$$

■

The number $N_q(g)$. For any X over $\mathbb{F}_q$ the inequality (2.3.2) implies

$$|N - q - 1| \le \lceil 2g \cdot \sqrt{q} \rceil \tag{2.3.3}$$

It is interesting to know whether this inequality can be ameliorated.

Let us consider the number $N_q(g) = \max_X |X(\mathbb{F}_q)|$ where X runs over all curves over $\mathbb{F}_q$ of genus g.

Theorem 2.3.16.

$$N_q(g) \le q + 1 + g\cdot\lceil 2\sqrt{q}\rceil \quad . \tag{2.3.4}$$

■

Small genera. For $g = 1$ and 2 we know the value of $N_q(g)$.

Theorem 2.3.17. *Let $q = p^a$, p being a prime. Then*

$$N_q(1) = q + 1 + \lceil 2\sqrt{q}\rceil$$

with the exception of the case when $p \mid \lceil 2\sqrt{q}\rceil$, $a \ge 3$, and a is odd. In this case $N_q(1) = q + \lceil 2\sqrt{q}\rceil$. ■

Theorem 2.3.18. *Let $q = p^a$. Then*

a) *if a is even and $q \ne 4$ or 9 then*

$$N_q(2) = q + 1 + 4\cdot\sqrt{q} \quad .$$

b) $N_4(2) = 10$, $N_9(2) = 20$;

c) *let a be odd; we call q* special *iff either $p \mid \lceil 2\sqrt{q}\rceil$ or $q = m^2 + 1$, $q = m^2 + m + 1$, or $q = m^2 + m + 2$ for some integer m. If q is not special then*

$$N_q(2) = q + 1 + 2\lceil 2\sqrt{q}\rceil \quad ;$$

if q *is special then*

$$N_q(2) = \begin{cases} q + 2\lceil\sqrt{q}\rceil & \text{for } 2\sqrt{q} - \lceil 2\sqrt{q}\rceil > \frac{\sqrt{5} - 1}{2} \\ q + 2\lceil\sqrt{q}\rceil - 1 & \text{for } 2\sqrt{q} - \lceil 2\sqrt{q}\rceil < \frac{\sqrt{5} - 1}{2} \end{cases} .$$

■

For $g = 3$ we know $N_q(g)$ only for some values of q .

Theorem 2.3.19. *The value of* $N_q(3)$ *for* $q \le 25$ *is given by*

$q =$	2	3	4	5	7	8	9	11	13	16	17	19	25
$N_q(3) =$	7	10	14	16	20	24	28	28	32	38	40	44	56

■

Proofs of these results are rather delicate; they have two parts: the first gives estimates for $N_q(g)$ which are based on so called "explicit formulae" for the number of rational points of a curve over a finite field (cf. Remark 2.3.22) and uses some technique from algebraic number theory; the second part is a construction of curves with required number of points and uses various methods of algebraic geometry.

Another form of $Z(t)$ **.** In many questions it is important to study the numbers B_s of points on X of degree s . Since any point of degree s on X gives exactly s points of $X(\mathbb{F}_{q^s})$, we see that

$$N_r = \sum_{d|r} d \cdot B_d .$$

One can express $Z(t)$ in terms of B_s :

Exercise 2.3.20. Show that

$$Z(t) = \prod_{d=1}^{\infty}(1 - t^d)^{-B_d} \quad . \tag{2.3.5}$$

(*Hint*: Consider $\ln Z(t)$). Show that

$$Z(t) = 1 + \sum_{i=1}^{\infty} R_i \cdot t^i \quad ,$$

R_i being the number of effective divisors on X of degree i .

Canonical class. Curves over finite fields have a special rationality property which concerns their canonical class.

Proposition 2.3.21. *Let* X *be a complete projective curve over* $\mathbb{F}_q$ *and let* K_X *be its canonical class. Then* $K_X \sim 2 \cdot D$ *for some* $\mathbb{F}_q$*-divisor* D *on* X . ∎

In the general case the proof of this proposition is based on the class field theory. Nevertheless for $p = char\, \mathbb{F}_q = 2$ it can be proved in a quite elementary way. Indeed let $f \in \mathbb{F}_q(X) - \mathbb{F}_q$, let $\omega = df$, let $P \in Supp(\omega)$ and let $t = t_P$ be a local parameter at P . Then $df = \sum_{i=i_0}^{\infty} a_{2i+1} \cdot t^{2i} \cdot dt$ where $\tau_P(f) = \sum_{i=m}^{\infty} a_i \cdot t^i$ is the power series expansion at P , and i_0 is the least integer i with $a_{2i+1} \neq 0$. Hence $(\omega) = (df) = 2i_0 \cdot P + (\omega')$ with $P \notin Supp(\omega')$ and the induction gives $(\omega) = 2D$ for some $\mathbb{F}_q$-divisor D .

2.3.3. Asymptotics

The number of points on a curve. Let us fix a finite field $\mathbb{F}_q$. Theorem 2.3.16 gives an upper bound for $N_q(g)$. As we have seen above in many cases this bound is exact for curves of low genera ($g = 1$, 2, and 3). Let us now consider the behaviour of $N_q(g)$ for $g \longrightarrow \infty$. To state the corresponding results it is necessary to introduce the following definition

$$A(q) = \limsup_{g \longrightarrow \infty} \frac{N_q(g)}{g} .$$

Theorem 2.3.16 implies the inequality

$$A(q) \le \lceil 2\sqrt{q} \rceil .$$

In fact it can be ameliorated:

Theorem 2.3.22 (the Drinfeld-Vladuţ bound).

$$A(q) \le \sqrt{q} - 1 .$$

Proof: Let X be a curve of genus g and let $\alpha_i = \omega_i/\sqrt{q}$ for $i = 1,\dots,2g$ in the notation of Theorem 2.3.12; hence $|\alpha_i| = 1$. Since $N_1 \le N_r$ for any $r \ge 1$ we obtain

$$N_1 \cdot q^{-r/2} \le N_r \cdot q^{-r/2} = q^{r/2} + q^{-r/2} - \sum_{i=1}^{2g} \alpha_i^r ,$$

hence

$$\sum_{i=1}^{2g} \alpha_i^r \le q^{r/2} + q^{-r/2} - N_1 \cdot q^{-r/2} . \qquad (2.3.6)$$

On the other hand for any $\alpha_i \in \mathbb{C}$ and any positive integer m we have

$$0 \le \left| \sum_{r=0}^{m} \alpha_i^r \right|^2 = \sum_{r,s=0}^{m} \alpha_i^{r-s} =$$
$$= m + 1 + \sum_{r=1}^{m} (m + 1 - r)\cdot(\alpha_i^r + \alpha_i^{-r}) \ . \tag{2.3.7}$$

Summing (2.3.7) over $i = 1,\ldots,2g$ and applying (2.3.6) we obtain

$$0 \le 2g\cdot(m + 1) + 2\cdot \sum_{r=1}^{m} (m + 1 - r)\cdot \sum_{i=1}^{2g} \alpha_i^r \le$$
$$\le 2g\cdot(m + 1) + 2\cdot \sum_{r=1}^{m} (m + 1 - r)\cdot(q^{r/2} + q^{-r/2} - N_1\cdot q^{-r/2}) \ . \tag{2.3.8}$$

Hence

$$\frac{N_1}{g}\cdot \sum_{r=1}^{m} \frac{m + 1 - r}{m + 1}\cdot q^{-r/2} \le 1 + \frac{1}{g}\cdot \sum_{r=1}^{m} \frac{m + 1 - r}{m + 1}\cdot(q^{r/2} + q^{-r/2}) \ .$$

If g and m tend to ∞ in such a way that $m/\log_q g \longrightarrow 0$ then it follows that for any $\varepsilon > 0$

$$\frac{N_1}{g}\cdot \frac{1}{\sqrt{q} - 1} \le 1 + \varepsilon$$

and we are done. ■

Remark 2.3.23. Developing the method of the proof of Theorem 2.3.22 we come to an "explicit formula" for the number of rational points of a curve over $\mathbb{F}_q$.

Let $f(\varphi) = 1 + 2\cdot \sum_{m=1}^{N} c_m\cdot \cos(m\cdot\varphi)$ for $c_m \in \mathbb{R}$ be an even periodic function in φ, and let $\psi_1(t) = \sum_{m=1}^{N} c_m\cdot t^m$,

$\psi_d(t) = \sum_{m=1}^{\lceil N/d \rceil} c_{md}\cdot t^m$. Then we have an "explicit formula" which follows

$$\sum_{k=1}^{g} f(\theta_k) + \sum_{d\ge 1} d\cdot B_d\cdot \psi_d(q^{-1/2}) = \psi_1(q^{1/2}) + \psi_1(q^{-1/2}) + g \;,$$

where $\alpha_k = e^{\sqrt{-1}\cdot\theta_k}$ and $\alpha_1,\ldots,\alpha_{2g}$ are ordered in such a way that $\alpha_k\cdot\alpha_{k+g} = 1$. If $c_m \ge 0$ for any $m = 1,\ldots,N$ and $f(\varphi) \ge 0$ for any φ then the "explicit formula" implies that

$$g \ge \sum_{d=1}^{N} d\cdot B_d\cdot\psi_d(q^{-1/2}) - \psi_1(q^{1/2}) - \psi_1(q^{-1/2}) \quad .$$

For $f(\varphi) = 1 + \cos\varphi$ we obtain the Weil bound $N \le 1 + q + 2g\cdot\sqrt{q}$. The bounds which are used to prove Theorems 2.3.18 and 2.3.19 are obtained for

$$f(\varphi) = (a_0 + a_1\cdot\cos\varphi + \ldots + a_n\cdot\cos(n\cdot\varphi))^2$$

or for

$$f(\varphi) = (a_0 + a_1\cdot\cos\varphi + \ldots + a_n\cdot\cos(n\cdot\varphi))^2\cdot(1 + \cos\varphi) \quad .$$

The bound of Theorem 2.3.22 is tight provided that q is a square, i.e. $q = p^{2m}$ for a prime p . This fact follows from an explicit construction of the corresponding family of curves which is described below in Part 4 (Theorems 4.1.52 and 4.2.38). Hence we have

Theorem 2.3.24. *For* $q = p^{2m}$

$$A(q) = \sqrt{q} - 1 \quad . \blacksquare$$

In the general case (for odd powers of a prime) only the following facts are known:

Theorem 2.3.25. a) *There exists an absolute constant* $c \geq 0$ *such that*

$$A(q) \geq c \cdot \log_2 q \quad ,$$

b)
$$A(q^3) \geq 2(q^2 - 1)/(q + 2) \quad ,$$

c) *let* $r > 4\ell + 1$, ℓ *being an odd prime, and either* $q = r^\ell$ *and* $r \equiv 1 \pmod{\ell}$, *or* $r \not\equiv 1 \pmod{\ell}$ *and* $q = r^k \equiv 1 \pmod{\ell}$ *then*

$$A(q) \geq \frac{\sqrt{\ell \cdot (r - 1)} - 2\ell}{\ell - 1} \quad .$$

The proof of a) consists in an explicit construction of a family of curves such that $N_1/g \geq c \cdot \log_2 q$ for $g \longrightarrow \infty$. It is done by the class field theory which provides a family $\{X_n\}$ of curves of genus at least 2, $n = 0,1,\ldots$ with maps $\pi_n\colon X_n \longrightarrow X_0$ which are unramified and such that any $\mathbb{F}_q$-point P of X_0 is completely split by any π_n (i.e. $\pi_n^{-1}(P) \subseteq X_n(\mathbb{F}_q)$) . Hence by the Hurwitz formula

$$\lim_{n \longrightarrow \infty} \frac{|X_n(\mathbb{F}_q)|}{g(X_n) - 1} \geq \frac{|X_0(\mathbb{F}_q)|}{g(X_0) - 1} \quad .$$

The proof of c) is based on similar considerations.

The proof of b) is based on the study of some special curves on Shimura varieties which asymptotically have the required ratio N_1/g .

Asymptotic for Jacobians. We are also interested in the asymptotical behaviour of $M = M(X) = |J_X(\mathbb{F}_q)|$ when $g(X) \longrightarrow \infty$. Theorem 2.3.15 implies that

$$2\cdot log_q(\sqrt{q} - 1) \le \frac{log_q M}{g} \le 2\cdot log_q(\sqrt{q} + 1) \quad .$$

One can ameliorate it in the following situation. Let X be a curve from a family of curves of growing genus such that

$$\lim_{g\longrightarrow\infty} \frac{N_1(X)}{g(X)} = A > 0$$

Then we have

Proposition 2.3.26.

$$\liminf_{g(X)\longrightarrow\infty} \frac{log_q M(X)}{g(X)} \ge 1 + A\cdot log_q\left(\frac{q}{q-1}\right) \quad .$$

Proof: By Theorem 2.3.15, since $\prod_{i=1}^{2g} \omega_i = |\omega_1|^{2g} = q^g$, we have

$$M = \prod_{i=1}^{2g}(\omega_i - 1) = q^g\cdot\prod_{i=1}^{2g} (1 - \alpha_i\cdot q^{-1/2}) \quad .$$

Hence

$$ln\ M = g\cdot ln\ q + \sum_{i=1}^{2g} ln(1 - \alpha_i\cdot q^{-1/2}) =$$

$$= g\cdot ln\ q - \sum_{i=1}^{2g} \sum_{m=1}^{\infty} \alpha_i^m\cdot q^{-m/2}/m =$$

$$= g\cdot ln\ q - \sum_{m=1}^{\infty} \frac{q^{-m/2}}{m}\cdot \sum_{i=1}^{2g} \alpha_i^m \quad .$$

Note that (2.3.1) can be rewritten as

$$-\sum_{i=1}^{2g} \alpha_i^m = N_m \cdot q^{-m/2} - (q^{m/2} + q^{-m/2}) \quad . \tag{2.3.9}$$

Replacing the left-hand side of (2.3.9) by its right hand side in all terms of the sum with $m \le b = \lceil log_q g \rceil$ we see that

$$ln\ M = g \cdot ln\ q + \sum_{m=1}^{b} \frac{q^{-m/2}}{m} \cdot (N_m \cdot q^{-m/2} - q^{m/2} - q^{-m/2}) -$$

$$- \sum_{m=b+1}^{\infty} \frac{q^{-m/2}}{m} \cdot \sum_{i=1}^{2g} \alpha_i^m \quad .$$

Note that

$$\sum_{m=1}^{b} \frac{q^{-m/2}}{m} \cdot (N_m \cdot q^{-m/2} - q^{m/2} - q^{-m/2}) =$$

$$= \sum_{m=1}^{b} \frac{N_m \cdot q^{-m}}{m} - \sum_{m=1}^{b} \frac{1}{m} - \sum_{m=1}^{b} \frac{q^{m}}{m}$$

and

$$\sum_{m=1}^{\infty} q^{-m}/m = ln \frac{q}{q-1} \quad .$$

Since $b + 1 > log_q g$, we have

$$\sum_{m=b+1}^{\infty} \frac{q^{-m}}{m} < q^{-(b+1)} \sum_{j=0}^{\infty} q^{-j} = \frac{q}{q-1} \cdot q^{-(b+1)} < \frac{1}{g} \cdot \frac{q}{q-1} \quad .$$

Note also that

$$\sum_{m=1}^{b} \frac{N_m}{m} \cdot q^{-m} \ge N_1 \cdot \sum_{m=1}^{b} \frac{q^{-m}}{m} = N_1 \cdot \left(\sum_{m=1}^{\infty} \frac{q^{-m}}{m} - \sum_{m=b+1}^{\infty} \frac{q^{-m}}{m} \right) \quad .$$

Therefore we see that

$$\sum_{m=1}^{b} q^{-m/2}\cdot(N_m\cdot q^{-m/2} - q^{m/2} - q^{-m/2})/m >$$

$$> N_1\cdot ln\,\frac{q}{q-1} - \frac{N_1\cdot q}{g(q-1)} - b - ln\,\frac{q}{q-1} =$$

$$= N_1\cdot ln\,\frac{q}{q-1} - o(g) \ .$$

On the other hand,

$$\left|\sum_{m=b+1}^{\infty}\frac{q^{-m}}{m}\cdot\sum_{i=1}^{2g}\alpha_i^m\right| \le 2g\cdot\sum_{m=b+1}^{\infty}\frac{q^{-m/2}}{m} <$$

$$< 2g\cdot q^{-(b+1)/2}\cdot\sum_{j=0}^{\infty} q^{-j/2} < 2\cdot\sqrt{g}\cdot\frac{\sqrt{q}}{\sqrt{q}-1} \ .$$

Hence

$$ln\,M \ge g\cdot ln\,q + N_1\cdot ln\,\frac{q}{q-1} - o(g) \ .$$

Dividing by $g\cdot ln\,q$, we obtain the proposition. ■

Exercise 2.3.27. Show that for $A = \sqrt{q} - 1$ we have

$$\lim_{g\longrightarrow\infty}\frac{log_q M}{g} = 1 + A\cdot log_q\left(\frac{q}{q-1}\right) \ .$$

Show that if for a family $\{X\}$ of curves of growing genus for $m = 1, \ldots, r$ there exist limits

$$\lim_{g\longrightarrow\infty}\frac{B_m(X)}{g(X)} = A_m \ge 0 \ ,$$

(recall that $B_m = B_m(X)$ is the number of points of degree m on X) then

$$\liminf_{g\longrightarrow\infty} \frac{\log_q M}{g} \geq 1 + \sum_{m=1}^{r} A_m \cdot \log_q\left(\frac{q^m}{q^m - 1}\right) .$$

Moreover, if $\sum_{m=1}^{r} m\cdot A_m/(q^{m/2} - 1) = 1$ then

$$\lim_{g\longrightarrow\infty} \frac{\log_q M}{g} = 1 + \sum_{m=1}^{r} A_m \cdot \log_q\left(\frac{q^m}{q^m - 1}\right) .$$

Points of degree ≥ 2 . The proof of Theorem 2.3.22 implies that for any fixed r and any family $\{X\}$ of curves of growing genus such that

$$\lim_{g\longrightarrow\infty} \frac{N_1(X)}{g(X)} = \sqrt{q} - 1 ,$$

we have

$$\lim_{g\longrightarrow\infty} \frac{N_r(X) - N_1(X)}{g(X)} = 0 .$$

Since $r\cdot B_r \leq N_r - N_1$, we obtain

$$\lim_{g\longrightarrow\infty} \frac{B_r(X)}{g(X)} = 0 .$$

In fact there exists a more precise form of this relation:

Proposition 2.3.28. *Let* $q \geq 4$ *; let* X *be a curve over* $\mathbb{F}_q$ *with* $N_1(X) \geq (\sqrt{q} - 1)\cdot g(X)$ *and let* $c = \lceil 2\cdot \log_q g\rceil$ *. Then for any* s *such that* $2 \leq s \leq c$ *we have*

$$B_s \leq \frac{9g\cdot q^{s/2}}{s\cdot(c + 1 - s)} .$$

Proof: Note that for any $\alpha \in \mathbb{C}^* - \{1\}$ we have

$$\sum_{j=0}^{n-1} (j+1)\cdot\alpha^j = \frac{n\cdot\alpha^{n+1} - (n+1)\cdot\alpha^n + 1}{(\alpha - 1)^2} \tag{2.3.10}$$

$$\sum_{j=0}^{n-1} (n-j)\cdot\alpha^{-j} = \frac{n\cdot\alpha^{n+1} - (n+1)\cdot\alpha^n + 1}{\alpha^{n-1}\cdot(\alpha - 1)^2} \tag{2.3.11}$$

$$\sum_{j=0}^{n-1} (n-j)\cdot\alpha^j = \frac{\alpha^{n+1} - (n+1)\cdot\alpha + n}{(\alpha - 1)^2} \tag{2.3.12}$$

The formulae (2.3.11) and (2.3.12) easily follow from (2.3.10), and (2.3.10) can be obtained from the formula

$$\sum_{j=0}^{n} \alpha^j = \frac{1 - \alpha^{n+1}}{1 - \alpha}$$

by taking its derivative.

We use the notation of the proof of Theorem 2.3.22. We can rewrite the first inequality of (2.3.8) in the form

$$\sum_{r=1}^{m} (m+1-r)\cdot(N_r\cdot q^{-r/2} - q^{r/2} - q^{-r/2}) \le g\cdot(m+1) \quad .$$

Since $B_1 = N_1$ and $r\cdot B_r \le N_r - N_1$ for $r \ge 2$ it follows that

$$\sum_{r=1}^{m} (m+1-r)\cdot q^{-r/2}\cdot r\cdot B_r - \sum_{r=1}^{m} (m+1-r)\cdot(q^{r/2} + q^{-r/2}) + \tag{2.3.13}$$

$$+ N_1\cdot\sum_{r=1}^{m} (m+1-r)\cdot q^{-r/2} \le g\cdot(m+1) \quad .$$

Using (2.3.11) and (2.3.12) for $\alpha = \sqrt{q}$, the inequality $N_1 \ge (\sqrt{q} - 1)\cdot g$ and the non-negativity of B_r for any

r , for any s with $2 \le s \le m$ we obtain

$$B_s \le \frac{q^{s/2}}{s\cdot(m+1-s)}\cdot\left(\frac{q^{(m+2)/2}}{(\sqrt{q}-1)^2} + \frac{g}{\sqrt{q}-1} + \frac{m\cdot(\sqrt{q}+1)}{(\sqrt{q}-1)^2}\right). \tag{2.3.14}$$

Setting $m = c \ge s$ we end the proof. ∎

Remark 2.3.29. Note that in Proposition 2.3.28 we have assumed that $N_1 \ge (\sqrt{q}-1)\cdot g$ rather than $\limsup_{g\to\infty} \frac{N_1}{g} = \sqrt{q}-1$. In fact this inequality holds for any family of curves with $N_1/g \longrightarrow \sqrt{q}-1$ known to us. Hence Proposition 2.3.28 will be sufficient for our purposes. Note also that the proposition is valid for $q = 2$ and 3 with the right-hand side multiplied by a suitable constant but we do not need this fact.

Exercise 2.3.30. Check that (2.3.14) implies the proposition and that (2.3.13) implies (2.3.14).

CHAPTER 2.4

ELLIPTIC CURVES

Theory of curves of genus one which are called *elliptic* is of special importance. Elliptic cuves have many remarkable properties; the most important is that any elliptic curve is an abelian variety. In this chapter we present some basic notions and results concerning elliptic curves.

In Section 2.4.1 we give a description of the group law on an elliptic curve; Section 2.4.2 is devoted to isomorphism classes of elliptic curves and to the j-invariant. In Section 2.4.3 we consider morphisms of elliptic curves. Section 2.4.4 contains the theory of elliptic curves over a finite field; Section 2.4.5 is devoted to complex elliptic curves and elliptic functions.

In this chapter all the curves are smooth and projective defined over an algebraically closed field k (with the only exception of Section 2.4.4 where the ground field k is finite; in Section 2.4.5 $k = \mathbb{C}$).

2.4.1. Group law

The group structure on an elliptic curve is essentially unique; nevertheless it can be defined in different ways. We begin with the most general (and most abstract) way and then translate it into more down-to-earth terms.

Let E be an elliptic curve, i.e a curve of genus one. Let us fix a point $P_0 \in E$ and let us consider the map $\lambda_{P_0} : E \longrightarrow Pic^0(E)$, $\lambda_{P_0}(P) = \{P - P_0\}$, $\{P - P_0\}$ being the class of $P - P_0$ in $Pic^0(E)$. In Section 2.1.4 we stated (without proof) that $Pic^0(E)$ is an abelian variety and that λ_{P_0} is an isomorphism. We do not use these facts here; on the contrary, we obtain them as corollaries of our argument.

Proposition 2.4.1. *The map* $\lambda_{P_0} : E \longrightarrow Pic^0(E)$ *is bijective.*

Proof: Let D be a divisor of degree 0 on E . Appling the Riemann-Roch theorem to the divisor $D + P_0$ we see that $\ell(D + P_0) - \ell(K - D - P_0) = 1$; note that $deg\, K = 2g - 2 = 0$, hence $\ell(K - D - P_0) = 0$ and thus $\ell(D + P_0) = 1$. It follows that there exists a unique point $P \in E$ which is equivalent to $D - P_0$. Therefore for any $D \in Div^0(E)$ there exists a unique $P \in E$ such that $D \sim P - P_0$ and we are done. ■

Since $Pic^0(E)$ is a group, E is also a group whose zero is P_0 . Let as describe the group law by geometric means. To do this we consider the map $\varphi_{3P_0} : E \longrightarrow \mathbb{P}^2$

which is defined by the divisor $3\cdot P_0$ (note that $\ell(3\cdot P_0) = 3$ by Riemann-Roch). Applying Corollary 2.2.29 we get

Proposition 2.4.2. *The map* φ_{3P_0} *is an embedding of* E *into* $\mathbb{P}^2$. ■

Therefore any elliptic curve is isomorphic to a plane cubic. Conversely, Corollary 2.2.8 implies that any smooth irreducible plane cubic is an elliptic curve. Further on we assume E to be a plane cubic. Let P, Q, and R be any points of E. The condition $P + Q + R = 0$ (where $+$ is the group law on E) can be written as $P + Q + R \sim 3P_0$. Since $3P_0$ is a line section divisor (i.e. it is of the form (L), L being a linear form) it means that P, Q, and R belong to a line ℓ (with the equation $L = 0$). If $P = Q$ it means that ℓ is tangent to E at P and if $P = Q = R$ it means that P is a flex point of E (if the ground field is $\mathbb{R}$, the notions of a tangent line and a flex point are well-known; in general one can consider the above statements as definitions).

Coordinate expression. To express the group law on an elliptic curve in coordinates it is useful to write its equation in a special form. Note that by Riemann-Roch $\ell(n\cdot P_0) = n$ for $n \geq 1$. Let

$$x \in L(2\cdot P_0) - k\ , \quad y \in L(3\cdot P_0) - L(2\cdot P_0)\ .$$

Then the functions $1, x, y, x\cdot y, x^2, y^2, x^3$ lie in $L(6\cdot P_0)$ and thus are linearly dependent. Since only y^2 and x^3 have a pole of order 6 at P_0 their coefficients in this linear equation do not vanish (if they did vanish then either the functions $1, x, y, x\cdot y, x^2, y^2$ or the functions $1, x, y, x\cdot y, x^2, x^3$ would be linearly dependent which is not the case since E is not rational). Multiplying x and y by appropriate elements of k we can assume that the linear equation is of the form

$$y^2 + a_1 \cdot x \cdot y + a_3 \cdot y = x^3 + a_2 \cdot x^2 + a_4 \cdot x + a_6 \qquad (2.4.1)$$

where $a_i \in k$. Assume for simplicity that $char\, k \neq 2$ (note that similar results are also valid for $char\, k = 2$). Making the substitution $y \longmapsto y + (a_1 \cdot x + a_2)/2$ we obtain $y^2 = (x - a) \cdot (x - b) \cdot (x - c)$ for some a, b, and c from k. Now we substitute $x \longmapsto (x - a)/(b - a)$ and obtain the equation

$$y^2 = x \cdot (x - 1) \cdot (x - \lambda) \qquad (2.4.2)$$

with $\lambda \in k$; λ is called the *Legendre modulus* of E.

In homogeneous coordinates (2.4.2) can be rewritten as

$$x_0 \cdot x_2^2 = x_1 \cdot (x_1 - x_0) \cdot (x_1 - \lambda \cdot x_0) ,$$

where $x = x_1/x_0$, $y = x_2/x_0$. Note that $P_0 = (0:0:1)$. Let $y_0^2 = a \cdot (a - 1) \cdot (a - \lambda)$. The line $x_1 = a \cdot x_0$ intersects E at points P_0, $Q = (1:a:y_0)$, and $Q' = (1:a:-y_0)$. The points Q and Q' are opposite to each other (i.e. $Q = -Q'$). Now it is easy to describe the addition law in geometric terms and to write it out. If $P_1 = (1:x_1:y_1)$ and $P_2 = (1:x_2:y_2)$ then $Q = P_1 + P_2$ can be obtained as follows: let $Q' = (1:x_3:y_3)$ be the third point of intersection of the line joining P_1 and P_2 with E, then $P_1 + P_2 = Q = (1:x_3:y_3)$ is the reflection of Q' (see Fig. 2.1).

Here is a picture, illustrating the addition law on an elliptic curve.

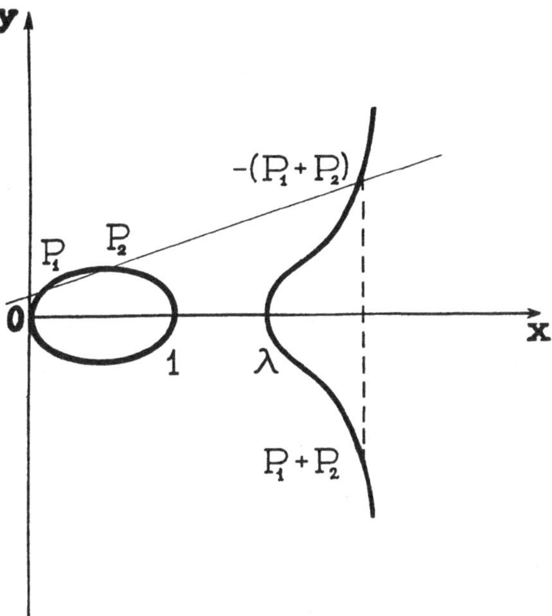

Fig. 2.1

Exercise 2.4.3. Express x_3 and y_3 in x_1, x_2, y_1, y_2, where $P_i = (1:x_i:y_i)$ for $i = 1, 2, 3$, and $P_3 = P_1 + P_2$. (*Answer*:

$$x_3 = -x_1 - x_2 - \left(\frac{y_1 - y_2}{x_1 - x_2}\right)^2 + \lambda + 1$$

$$(2.4.3)$$

$$y_3 = \frac{x_2 \cdot y_1 - x_1 \cdot y_2}{x_1 - x_2} - x_3 \cdot \left(\frac{y_1 - y_2}{x_1 - x_2}\right) \quad).$$

The formulae (2.4.3) immediately imply that the addition map $\mu : E \times E \longrightarrow E$, $\mu(P_1,P_2) = P_1 + P_2$ is regular (note that (2.4.3) is valid only for $x_1 \neq x_2$ but it is easy to get a similar expression for $P_1 = P_2$, and in the case $P_1 \neq P_2$ and $x_1 = x_2$ we have $y_1 + y_2 = 0$,

$P_1 = -P_2$). It is obvious that the map $P \longmapsto -P$ is also regular. Hence E is an abelian variety which can be identified with its Jacobian. Usually this group is denoted by $E(k)$ to stress its dependence on k.

Divisibility of $E(k)$. Here is an important property of $E(k)$.

Proposition 2.4.4. *The group* $E(k)$ *is* divisible, *i.e. for any positive integer* n *and any* $P \in E(k)$ *there exists* $Q \in E(k)$ *such that* $n \cdot Q = P$ *in this group.*

Sketch of proof: We can assume that $P = (1:x:y)$ since if it not the case then $P = P_0$ and $Q = P_0$. Let $Q = (1:z:w)$, (z,w) being unknown coordinates of Q. The group law makes it possible to express x and y in z and w. We obtain two equations $F_n(z,w) = 0$ and $G_n(z,w) = 0$ where F_n and G_n are polynomials whose coefficients depend on x, y, and λ. It can be shown that since k is algebraically closed this system has a solution. ■

For $k = \mathbb{C}$ the group $E(k)$ has a more precise description which is given below in Section 2.4.5.

Automorphisms. Since E is an abelian variety we have

Proposition 2.4.5. *For any fixed* $P \in E$ *the map* $L_P : Q \longmapsto Q + P$ *is an automorphism of* E *(as of an algebraic variety, not as of a group).* ■

Corollary 2.4.6. *The group* $Aut(E)$ *acts on* E *transitively.* ■

2.4.2. Isomorphisms and j-invariant

Let us assume that $char\ k \neq 2$. Note that the value λ from (2.4.2) can be different for isomorphic elliptic curves. In particular, the equations of the form (2.4.2) with λ and $\lambda' = 1/\lambda$ define isomophic curves. Let us set

$$j = j(E) = 2^8 \cdot \frac{(\lambda^2 - \lambda + 1)^3}{\lambda^2 \cdot (1 - \lambda)^2} . \qquad (2.4.4)$$

This value is called the *j-invariant* (or the *absolute invariant*) of E . Note that the coefficient 2^8 is introduced to make sure that j has integer coefficients being expanded into a power series in some natural variable q (cf. Section 4.1.1 below).

The importance of j-invariant in the theory of elliptic curves is shown by

Theorem 2.4.7. a) *The value* $j = j(\lambda) = j(E) \in k$ *is well-defined, i.e. it depends only on the isomorphism class of* E .

b) *Elliptic curves* E *and* E' *are isomorphic iff* $j(E) = j(E')$.

c) *For any* $j \in k$ *there exists an elliptic curve* E *over* k *with* $j(E) = j$.

Therefore the rule $E \longmapsto j(E)$ *defines a bijection of the set of isomorphism classes of elliptic curves over* k *onto* $\mathbb{A}^1(k)$.

Sketch of proof: a) Note that $j(\lambda) = j(\lambda')$ for any

$\lambda' \in \Lambda = \{\lambda, 1/\lambda, 1 - \lambda, 1/(1 - \lambda), \lambda/(\lambda - 1), (\lambda - 1)/\lambda\}$
which can be verified by brute force. If we write an equation of E in the form (2.4.2) then the projection $f(x,y) = x$ defines a map $f : E \longrightarrow \mathbb{P}^1$ of degree 2 with 4 ramification points $0, 1, \lambda$, and ∞. Let

$$y^2 = x\cdot(x - 1)\cdot(x - \lambda')$$

be another equation of E; we obtain another map $f' : E \longrightarrow \mathbb{P}^1$ of degree 2 with ramification points $0, 1, \lambda'$, and ∞. Let P and $P' \in E$ be such that $f(P) = \lambda$ and $f(P') = \lambda'$. By Corollary 2.4.6 there exists $\sigma \in Aut(E)$ with $\sigma(P) = P'$. Since f is defined by $|2\cdot P|$ and f' by $|2\cdot P'|$ the maps f and $f' \circ \sigma$ are defined by the same linear system and thus differ by an automorphism of $\mathbb{P}^1$. Such an automorphism (being an element of $PGL(2,k)$) sends the 4-tuple $(0,1,\lambda,\infty)$ to the 4-tuple $(0,1,\lambda',\infty)$ iff $\lambda' \in \Lambda$ and a) is proved.

b) Let E and E' be elliptic curves, let λ and λ' be their Legendre moduli and let $j(\lambda) = j(\lambda')$. Considering λ' as a variable and λ as a parameter, we obtain an equation of degree 6 in λ', vanishing on Λ. Therefore it has no other roots and hence E and E' are isomorphic.

c) Let $j \in k$ and let λ be a root of the equation

$$2^8\cdot(\lambda^2 - \lambda + 1)^3 = j\cdot(1 - \lambda)^2\cdot\lambda^2 \quad .$$

Then (2.4.2) defines an elliptic curve E with $j(E) = j$. ■

Remark 2.4.8. For *char* $k \neq 2, 3$ another form of the equation of an elliptic curve E is quite useful. Making the substitution $x \longmapsto x - \dfrac{\lambda + 1}{3}$ in (2.4.2) we obtain an equation of E of the form

$$y^2 = x^3 + a\cdot x + b \quad . \qquad (2.4.5)$$

Usually one makes the substitution $y \longmapsto 4y$, $x \longmapsto 4x$ in (2.4.5) and writes the equation in the form

$$y^2 = 4x^3 - g_2 \cdot x - g_3 \qquad (2.4.6)$$

which is called the *Weierstrass normal form* of E . In Section 2.4.5 we shall explain why this form is useful (this is connected with the theory of elliptic functions). It is easy to verity that

$$j(E) = 1728 \cdot g_2^3/(g_2^3 - 27g_3^2) \quad .$$

Remark 2.4.9. Theorem 2.4.7 also remains valid for *char* $k = 2$. We do not give its proof for this case. Let us only define $j(E)$ for a curve E over a field of *char* $k = 2$. Making the substitution $x \longmapsto x + a_2$ we get

$$y^2 + a_1 \cdot x \cdot y + a_3 \cdot y = x^3 + a_4 \cdot x + a_6 \qquad (2.4.7)$$

Then $j(E) = a_1^{12}/\Delta$, where

$$\Delta = a_1^6 \cdot a_6 + a_1^5 \cdot a_3 \cdot a_4 + a_1^4 \cdot a_4^2 + a_3^4 + a_1^3 \cdot a_3^3 \quad .$$

Exercise 2.4.10. Let $F : E \longrightarrow E^{(p)}$ be the Frobenius morphism of E . Show that $j(E^{(p)}) = j(E)^p$.

2.4.3. Isogenies

Let $f : E \longrightarrow E'$ be a non-constant map of elliptic curves, let P_0 be the zero of the group law on E and let $P_0' = f(P_0)$ be the zero of the group law on E' . Then by Proposition 2.1.75 f is a morphism of abelian varities. We call such f an *isogeny*; since E and E' are curves we can speak about its *degree*. Since $E \simeq J_E$ any isogeny $f : E \longrightarrow E'$ gives rise to the *dual isogeny* (or the *transpose* of f) $f^t = f^* : E' = J(E') \longrightarrow J(E) = E$.

Exercise 2.4.11. Show that $f^{tt} = f$ for any f .

Proposition 2.4.12. *Let* $f : E \longrightarrow E'$ *be an isogeny of degree* n *. Then* $\deg f^t = n$, $f \circ f^t = n_{E'}$ *and* $f^t \circ f = n_E$, *where* n_E *and* $n_{E'}$ *are morphisms of multiplication by* n *on* E *and* E' , *respectively* . ■

In the case of complex elliptic curves this will be proved in the next section.

Corollary 2.4.13. *The degree of* n_E *equals* n^2 . ■

We denote $Ker(n_E) \subset E$ by E_n or $E[n]$.

Exercise 2.4.14. Let $p \nmid n$. Show that

$$E[n] \simeq (\mathbb{Z}/n)^2$$

(*Hint*: Study the case of a prime n and then use the induction on the number of divisors of n).

Exercise 2.1.15. Let f be an isogeny of degree $n = n' \cdot n''$, where n' and n'' are coprime. Show that there exists isogenies f' and f'' such that $f = f' \circ f''$, $\deg f' = n'$ and $\deg f'' = n''$.

Now let us describe $E[p^a]$ for $p = char\ k$, $a \geq 1$. To being with, we assume that $a = 1$. Let $F : E \longrightarrow E^{(p)}$ be the Frobenius morphism and $V = F^t$ its transpose. Then by Proposition 2.4.12 we have $F \circ V = V \circ F = p_E$ and since F

is purely inseparable, p_E is not separable. Let us consider two cases: a) V is a separable morphism, b) V is not separable. In the first case we call E an *ordinary* elliptic curve, in the second case we call E a *supersingular* elliptic curve (in this case p_E is purely inseparable). Note that for supersingular E we have $V = F$, since F is the only purely inseparable map of degree p.

Proposition 2.4.16. *The kernel of the multiplication by* p^a *on* $E(k)$ *is trivial for a supersigular* E *and is isomorphic to* $\mathbb{Z}/p^a$ *for an ordinary* E.

Sketch of proof: From the above argument it follows that the proposition holds for $a = 1$. For any $a \geq 1$ it can be easily deduced by induction over a. ∎

Supersingular curves. For a fixed p there exists only a finite number of non-isomorphic supersingular curves in characteristic p. Their moduli can be found out:

Theorem 2.4.17. *Let* $p = char\ k > 2$ *and let* E *be the curve defined by* $y^2 = x\cdot(x-1)\cdot(x-\lambda)$. *Then* E *is supersingular iff*

$$\sum_{i=0}^{(p-1)/2} \binom{(p-1)/2}{i}^2 \cdot \lambda^i = 0 . \qquad (2.4.8)$$

∎

Therefore all supersingular values of the modulus λ and of the j-invariant lie in a finite field. Moreover, one has

Proposition 2.4.18. *Let* $p = char\ k > 2$, *and let* j *be a supersingular value of* j-*invariant (i.e.* $j = j(E)$

for a supersingular curve E *). Then* $j \in \mathbb{F}_{p^2}$.

Proof: Theorem 2.4.17 implies $j \in \overline{\mathbb{F}}_p$. Hence it is sufficient to show that $j^{p^2} = j$. Indeed, for a supersingular curve $V = F$ and since $j(E^{(p)}) = j^p$ we see that $(j^p)^p = j$. ■

Example 2.4.19. For $p = 2$ there exists only one supersingular curve E which can be given by $y^2 + y = x^3$. Note that $j(E) = 0$.

The group $Hom(E,E')$. Let now E and E' be elliptic curves. The set of *algebraic group morphisms* $f : E \longrightarrow E'$ (i.e. of regular maps which are group homomorphisms) is denoted by $Hom(E,E')$; if $E = E'$ it is denoted by $End(E)$. Note that $Hom(E,E')$ is an abelian group since we can add its elements: $(f + g)(x) = f(x) + g(x)$; moreover $End(E)$ is a ring: multiplication is the composition of maps. It is clear that $Hom(E,E')$ has no torsion since the condition $n \cdot f = 0$ implies that $f(E)$ is contained in the finite set $E[n]$ and hence is trivial. Studying the behaviour of morphisms at torsion points of E one can prove the following

Proposition 2.4.20. *The rank of* $Hom(E,E')$ *equals* 0, 1, 2, *or* 4 . *If it is equal to* 4 *then* E *and* E' *are supersingular (and hence the characteristic of* k *is positive).* ■

If $Hom(E,E') \neq 0$ then we call the curves E and E' *isogenous*. Note that if E and E' are isogenous then $Hom(E,E') \otimes \mathbb{Q} = End(E) \otimes \mathbb{Q}$, which follows from Proposition 2.4.12. Note that $End(E)$ is embedded into $End^0(E) = End(E) \otimes \mathbb{Q}$ since they have no torsion. Moreover from Proposition 2.4.12 it follows that $End^0(E)$ is a division algebra.

Theorem 2.4.21. *There are the following possibilities for the division algebra* $End^0(E)$:

a) $End^0(E) = \mathbb{Q}$;

b) $End^0(E)$ *is a complex quadratic field;*

c) $End^0(E)$ *is a quaternion algebra over* $\mathbb{Q}$ *which is ramified at* p *and at* ∞ *; this is the case if* $p = char\, k > 0$ *and* E *is a supersingular curve over* k . ∎

Recall that a *quaternion algebra over* $\mathbb{Q}$ *ramified at* p *and at* ∞ is a division algebra $H_{p,\infty}$ of degree 4 over $\mathbb{Q}$ such that for any prime $\ell \neq p$ the algebra $H_{p,\infty} \otimes \mathbb{Q}_\ell \simeq M_2(\mathbb{Q}_\ell)$, $M_2(\mathbb{Q}_\ell)$ being the matrix algebra of degree 2 over ℓ-adic field $\mathbb{Q}_\ell$, whereas $H_{p,\infty} \otimes \mathbb{Q}_p$ and $H_{p,\infty} \otimes \mathbb{R}$ are division algebras over $\mathbb{Q}_p$ and $\mathbb{R}$, respectively. It is well-known that there is a unique division algebra over $\mathbb{Q}$ with these properties.

Therefore $End(E)$ is an *order* in the division algebra $End(E)$ of degree 1, 2, or 4 over $\mathbb{Q}$ (this means that $End(E)$ is a free $\mathbb{Z}$-module generating $End^0(E)$ over $\mathbb{Q}$).

Theorem 2.4.22. *There are the following possibilities for the order* $End(E)$:

a) $End(E) = \mathbb{Z}$;

b) $End(E) = \mathbb{Z} + c \cdot \mathcal{O}_k$ *where* $c \in \mathbb{Z}$, $p \nmid c$, $\mathcal{O}_k$ *being the maximal order in the complex quadratic field* $k = End^0(E)$ *(in this case* c *is called the* conductor *of* $End(E)$) ;

c) $End(E)$ *is a maximal order in the quaternion algebra* $End^0(E)$. ∎

Proofs of Proposition 2.4.10 and of Theorems 2.4.21 and 2.4.22 are rather cumbersome. For $k = \mathbb{C}$ there are more simple proofs which are given in Section 2.4.5 below (note howeover that the most complicate case c) of Theorems 2.4.21 and 2.4.22 does not occure for the complex ground field).

Automorphisms. Theorem 2.4.22 makes it possible to determine the group $Aut_0(E)$ of automorphisms of E as of an algebraic group (i.e. of those preserving the initial point P_0) which is isomorphic to the group $End(E)^*$ of units of $End(E)$. Here is the answer:

Theorem 2.4.23. a) *If* $j(E) \neq 0$ *or* 1728 *then*

$$Aut_0(E) = \{\pm 1\} .$$

b) *For* $p \neq 2, 3$:

if $j(E) = 0$ *then* $Aut_0(E) = \mu_6$,

if $j(E) = 1728$ *then* $Aut_0(E) = \mu_4$,

where $\mu_n = \{\zeta \in \mathbb{C}^* \mid \zeta^n = 1\}$ *is the cyclotomic group of order* n .

c) *If* $p = 2$ *and* $j(E) = 0 = 1728$ *then* $Aut_0(E) = SL(2,\mathbb{F}_3)$ *is of order* 24 .

d) *If* $p = 3$ *and* $j(E) = 0 = 1728$ *then* $Aut_0(E)$ *is the semi-direct product of* $\mathbb{Z}/3$ *by* $\mathbb{Z}/4$ *of order* 12 . ■

Remark 2.4.24. a) Since the group $Aut(E)$ of automorphisms of the curve E is a semidirect product of $E(k)$ by $Aut_0(E)$, Theorem 2.4.23 gives also a description of $Aut(E)$.

b) One has a noteworthy *mass-formula*:

$$\sum |Aut(E)|^{-1} = (p - 1)/24 \qquad (2.4.9)$$

where the sum is taken over the set of isomorphism classes of supersingular curves in characteristic $p > 0$.

Exercise 2.4.25. Find the number of isomorphism classes of supersingular curves in characteristic p . Find supersingular values of j-invariant for $p \le 11$. (*Hint*: Use (2.4.9) which implies that this number is of the form $\lceil p/12 \rceil + \delta_p$, where $\delta_p \in \{0,1,2\}$ depends on the residue of p modulo 12).

2.4.4. Elliptic curves over finite fields

The theory of elliptic curves exposed above concerns the case of an algebraically closed ground field, while we are mainly interested in elliptic curves over a finite field $k = \mathbb{F}_q$ with $q = p^a$ elements. To study this case one should make some changes in the theory. The definition of an elliptic curve over a finite field is the same as in the case of an algebraically closed field. For a finite ground field it is necessary to check that an elliptic curve has an $\mathbb{F}_q$-point. This follows from (2.3.2), since

$$|E(\mathbb{F}_q)| \ge q + 1 - 2\cdot\sqrt{q} = (\sqrt{q} - 1)^2 > 0$$

and hence $|E(\mathbb{F}_q)| \ge 1$. Let $P_0 \in E(\mathbb{F}_q)$. If we use P_0 as the zero we obtain a group structure on the finite set $E(\mathbb{F}_q)$. (Note that, say, for an algebraic number field k there exist curves of genus one with no k-point). We can write down an equation of E in the form

$$y^2 + a_1\cdot x\cdot y + a_3\cdot y = x^3 + a_2\cdot x^2 + a_4\cdot x + a_6$$

since it can be done using the Riemann-Roch theorem which is valid over an arbitrary ground field. We have seen above that for an algebraically closed ground field j-invariant classifies isomophism classes of elliptic curves. This is not the case over a finite field (nor for the most part of non-closed fields). From Theorem 2.4.23 one can deduce

Proposition 2.4.26. *Let $j(E) = j(E')$. Then E and E' are isomorphic over a finite extension K of the ground field k such that $[K:k]$ divides 24.*

If $p = \mathrm{char}\, k \neq 2, 3$ then $[K:k]$ divides 4 or 6. If $j(E) \neq 0$ or 1728 then $[K:k] = 1$ or 2. ■

Note also that there exist elliptic curves E and E' over $\mathbb{F}_q$ such that $j(E) = j(E') \in \mathbb{F}_q$ and $Hom_{\mathbb{F}_q}(E,E') = 0$, i.e. E and E' are not isogenous over $\mathbb{F}_q$.

Theorem 2.4.27. *Let E and E' be elliptic curves over a finite field $\mathbb{F}_q$. Then E is isogenous to E' iff $|E(\mathbb{F}_q)| = |E'(\mathbb{F}_q)|$.* ■

Exercise 2.4.28. Let

$$E : y^2 + y = x^3 + x$$

and

$$E' : y^2 + y = x^3 + x + 1$$

be elliptic curves over $\mathbb{F}_2$. Show that $j(E) = j(E')$. Compute $|E(\mathbb{F}_{2^r})|$ and $|E'(\mathbb{F}_{2^r})|$ for $r = 1,2,\ldots$. Find the least field over which E and E' are isomorphic.

Endomorphisms. Elliptic curves over finite fields have sufficiently many endomorphisms. To be more precise, let $End_{\mathbb{F}_q}(E)$ be the subring of $End(E)$ which is formed by maps defined over $\mathbb{F}_q$.

Proposition 2.4.29. $End_{\mathbb{F}_q} E \neq \mathbb{Z}$.

Sketch of proof: We prove this statement for ordinary curves and supersingular curves E with $j(E) \in \mathbb{F}_p$. Indeed, let $q = p^a$ and let $F^a : E \longrightarrow E^{(p^a)}$ be the power of the Frobenius morphisms. Since E is defined over $\mathbb{F}_q$, $E^{(p^a)} = E$ and hence $F^a \in End_{\mathbb{F}_q}(E)$. If E is an ordinary curve then $F^a \notin \mathbb{Z}$, since no $f \in \mathbb{Z}$ is purely inseperable. If E is defined over $\mathbb{F}_p$ then $a = 1$ and $F \notin \mathbb{Z}$ since its degree equals p (the degree of $n \in \mathbb{Z}$ equals n^2). For supersingular curves E with $j(E) \notin \mathbb{F}_p$ one needs slightly more elaborate argument. ■

Therefore $End_{\mathbb{F}_q}(E)$ contains an order in a complex quadratic field.

Zeta-function. By Theorem 2.3.12 the zeta-function $Z_E(t)$ of an elliptic curve E over $\mathbb{F}_q$ is of the form

$$Z_E(t) = \frac{1 - m\cdot t + q\cdot t^2}{(1 - t)\cdot(1 - qt)} ,$$

where $m \in \mathbb{Z}$ and $|E(\mathbb{F}_q)| = q + 1 - m$; the integer m is called *the trace of Frobenius*. This name is due on the following fact: for any prime $\ell \neq p$ there exist a two-dimensional vector space $V_\ell(E)$ over ℓ-adic field $\mathbb{Q}_\ell$ and an embedding $\varphi_\ell : End_{\mathbb{F}_q}(E) \longrightarrow End_{\mathbb{Q}_\ell}(V_\ell(E))$ such that the trace of $\varphi_\ell(F^a)$ equals m . The endomorphism $Frob = F^a \in End_{\mathbb{F}_q}(E)$ is called the *Frobenius endomorphism over* $\mathbb{F}_q$.

If E is an ordinary curve then there exists a two-dimensional vector space $V(E)$ over $\mathbb{Q}$ and an embedding $\varphi : End_{\mathbb{F}_q}(E) \longrightarrow End_{\mathbb{Q}}(V(E))$ such that $Tr(\varphi(Frob)) = m$ (and hence one can set $V_\ell(E) = V(E)\otimes\mathbb{Q}_\ell$, $\varphi_\ell = \varphi\otimes\mathbb{Q}_\ell$). Indeed we can set $V(E) = End^0_{\mathbb{F}_q}(E)$ which is a complex quadratic field.

Structure of $E(\mathbb{F}_q)$. One can describe all the possible types of the group $E(\mathbb{F}_q)$. We begin with a description of their order (note than $E(\mathbb{F}_q) \subseteq E_N \simeq (\mathbb{Z}/N)^2$, where $N = |E(\mathbb{F}_q)|$).

Theorem 2.4.30. *The set of isogeny classes of elliptic curves over* $\mathbb{F}_q$ *is an a natural bijection with the set of integers* m *such that* $|m| \le 2\cdot\sqrt{q}$ *and one of the following condition holds*

(1) $(q,m) = 1$;

(2) q *is a square and* $m = \pm 2\cdot\sqrt{q}$;

(3) q *is a square,* $p \not\equiv 1 (mod\ 3)$, *and* $m = \pm\sqrt{q}$;

(4) q *is not a square ,* $p = 2$ *or* 3 , *and* $m = \pm\sqrt{p\cdot q}$;

(5) q *is not a square and* $m = 0$;
or q *is a square,* $p \not\equiv 1 (mod\ 4)$, *and* $m = 0$.

Moreover $E(\mathbb{F}_q) = q + 1 - m$ *for any curve* E *from the isogeny class which corresponds to* m . ∎

Now we can give a description of all possible types of groups $E(\mathbb{F}_q)$.

Theorem 2.4.31. *A group* G *of order* $N = q + 1 - m$ *is isomorphic to* $E(\mathbb{F}_q)$ *for some elliptic* E *over* $\mathbb{F}_q$ *iff one of the following conditions holds*

(1) $(q,m) = 1$, $|m| \le 2\cdot\sqrt{q}$ *and* $G \simeq \mathbb{Z}/A \times \mathbb{Z}/B$, *where* $B|A$, $B|(m - 2)$.

(2) *q is a square,* $m = \pm 2\cdot\sqrt{q}$ *, and* $G \simeq (\mathbb{Z}/A)^2$ *, where* $A = \sqrt{q} \mp 1$ *;*

(3) *q is a square,* $p \not\equiv 1 (mod\ 3)$ *,* $m = \pm\sqrt{q}$ *, and* G *is cyclic;*

(4) *q is not a square,* $p = 2$ *or* 3 *,* $m = \pm\sqrt{p\cdot q}$ *, and* G *is cyclic;*

(5a) *q is not a square and* $p \not\equiv 3 (mod\ 4)$ *, or* q *is a square and* $p \not\equiv 1 (mod\ 4)$*,* $m = 0$ *, and* G *is cyclic;*

(5b) *q is not a square,* $p \equiv 3 (mod\ 4)$ *,* $m = 0$ *, and* G *is either cyclic or* $G \simeq \mathbb{Z}/M \times \mathbb{Z}/2$ *where* $M = (q + 1)/2$ *.*

Sketch of proof: Since $E(\mathbb{F}_q) \subseteq (\mathbb{Z}/N)^2$ we see that $E(\mathbb{F}_q) \simeq \mathbb{Z}/A \times \mathbb{Z}/B$, where $B|A$, $A\cdot B = N$. Let us choose a basis in E_N such that $E(\mathbb{F}_q)$ is generated by $\begin{pmatrix} A \\ 0 \end{pmatrix}$ and $\begin{pmatrix} 0 \\ B \end{pmatrix}$. Let M_F be the matrix of the Frobenius endomorphism $Frob = F^a$ in this basis. Then $M_F = \begin{pmatrix} a & b \\ c & d \end{pmatrix}$, where $a,b,c,d \in \mathbb{Z}/N$. By the above remark on the zeta-function of E we have $Tr\ M_F \equiv m\ (mod\ N)$, $det\ M_F = q\ (mod\ N)$. Since $E(\mathbb{F}_q)$ is the subgroup of E_N formed by elements invariant under M_F , we see that $a\cdot A \equiv A\ (mod\ N)$, $d\cdot B \equiv B\ (mod\ N)$. Hence $(m - 2) = (a - 1) + (d - 1)$ is divisible by B , since $B|A$. Note also that if $B|(m - 2)$ then for $M'_F = \begin{pmatrix} m - 1 & -A \\ B & 1 \end{pmatrix}$ we have $Tr\ M'_F \equiv m$, $det\ M'_F \equiv q$, $a\cdot A \equiv A$, $d\cdot B \equiv B\ (mod\ N)$.

Let us now consider the cases of Theorem 2.4.30 one by one.

In case (1) it can be shown that for any matrix M'_F such that $Tr\ M'_F \equiv m$, $det\ M'_F \equiv q\ (mod\ N)$ there exists an elliptic curve E over $\mathbb{F}_q$ such that M'_F is the matrix of the Frobenius endomorphism (it follows from the fact that any endomorphism of E can be lifted to characteristic zero, which can be proved along the same lines as Theorem 2.4.22 and Proposition 2.4.29). Applying this result to $M'_F = \begin{pmatrix} m-1 & -A \\ B & 1 \end{pmatrix}$ we obtain (1) of Theorem 2.4.31.

In case (2) of Theorem 2.4.30 it is not hard to show that M_F is a scalar matrix (one can just use the fact that any ℓ-adic lifting of M_f is semi-simple) and (2) of Theorem 2.4.31 follows.

In case (3) we have $q \equiv N - 1 + m \equiv 1 (mod\ B)$ since $B | (m-2)$ and $B | N$. Further on, $q = m^2 \equiv 4 (mod\ B)$. Note that $B \neq 3$, since $B = 3$ would imply $m \equiv 2, 5$ or $8 (mod\ 9)$ and $q \equiv 4, 7$ or $1 (mod\ 9)$, respectively; thus $N \equiv 3 (mod\ 9)$ which contradicts the condition $B^2 | N$. Nence $B = 1$ and thus $E(\mathbb{F}_q)$ is cyclic.

In case (4) we have either $p = 2$, $m \equiv 2 (mod\ B)$, N and B being odd, $2q = m^2 \equiv 4$, $q \equiv 2$, $N \equiv 1 (mod\ B)$ and hence $B = 1$, or $p = 3$, $(B,3) = 1$, $m \equiv 2$, $3q = m^2 \equiv 4$, $3N \equiv 1 (mod\ B)$, and hence $B = 1$. Therefore $E(\mathbb{F}_q)$ is cyclic.

In case (5) $m = 0$ and thus $B = 1$ or 2. If $B = 2$ then $4 | N$ and hence $q \equiv 3 (mod\ 4)$.

Since in cases (2), (3), (4), and (5a) the group $E(\mathbb{F}_q)$ is uniquely determined by its order and Theorem 2.4.30 implies the existence of a corresponding curve in these cases we are done. Let us consider the case (5b) and show that both possibilities $B = 1$ and $B = 2$ can be realized.

Let us first show that in this case for a curve E with a cyclic $E(\mathbb{F}_q)$ there exists an isogeny $E \longrightarrow E'$ of degree 2 such that $E'(\mathbb{F}_q)$ is not cyclic. Indeed, let $E(\mathbb{F}_q)$ be cyclic, then $E_4 \cap E(\mathbb{F}_q)$ is also cyclic, let v be its generator. Let $E_2 = \{0, 2v, e, f = e + 2v\}$. Since E_2 does not lie in $E(\mathbb{F}_q)$, the Frobenius endomorphism permutes

e and f . Let $\varphi : E \longrightarrow E'$ be the isogeny whose kernel is generated by $2v$. It is clear that the Frobenius endomorphism of E' preserves $\varphi(v)$ and $\varphi(e) = \varphi(f)$ and thus $E'_2(\mathbb{F}_q) = (\mathbb{Z}/2)^2$.

Conversely, let $E(\mathbb{F}_q) \supseteq (\mathbb{Z}/2)^2$. Then $E_4(\mathbb{F}_q)$ equals $(\mathbb{Z}/2)^2$ or $(\mathbb{Z}/2)\times(\mathbb{Z}/4)$ (since $E_4(\mathbb{F}_q) \subseteq E(\mathbb{F}_q) \subseteq \mathbb{Z}/2\times\mathbb{Z}/N$). If $E_4(\mathbb{F}_q) = (\mathbb{Z}/2)^2$ then $E'(\mathbb{F}_q)$ is cyclic for any isogeny $E \longrightarrow E'$ of degree 2 . If $E_4(\mathbb{F}_q) = \mathbb{Z}/2\oplus\mathbb{Z}/4 = \langle u,v\rangle$ and $\varphi : E \longrightarrow E'$ is the isogeny whose kernel is generated by $u + 2v$, then $E'(\mathbb{F}_q)$ is cyclic. ∎

2.4.5. Complex elliptic curves

The theory of complex elliptic curves is closely connected with the theory of elliptic functions which are very useful in the study of such curves.

Elliptic functions. Let Λ be a *lattice* in $\mathbb{C}$, i.e. a free subgroup in $\mathbb{C}$ of rank 2 which generates $\mathbb{C}$ over $\mathbb{R}$. Therefore if $\Lambda = \mathbb{Z}\cdot\omega_1\oplus\mathbb{Z}\cdot\omega_2$ then $\tau = \omega_2/\omega_1 \notin \mathbb{R}$. Without loss of generality we can assume that $Im(\tau) > 0$.

By an *elliptic function* f with the *period lattice* Λ we mean a meromorphic on $\mathbb{C}$ function f such that $f(z + \omega) = f(z)$ for any $z \in \mathbb{C}$ with $f(z) \neq \infty$ and for any $\omega \in \Lambda$. Note that any integral (i.e. analytic on the whole plane $\mathbb{C}$) elliptic function is constant by the Liouville theorem.

Elliptic functions really do exist. Let

$$\mathcal{P}(z) = \frac{1}{z^2} + \sum_{\omega\in\Lambda-\{0\}} \left(\frac{1}{(z-\omega)^2} - \frac{1}{\omega^2}\right) . \qquad (2.4.10)$$

Exercise 2.4.32. Show that the series (2.4.10) converges absolutely and uniformly on any compact K such that $K \cap \Lambda = \emptyset$.

Therefore (2.4.10) defines a meromorphic function which is called the *Weierstrass* *$\mathcal{P}$-function* (or just a *$\mathcal{P}$-function*).

Exercise 2.4.33. Show that $\mathcal{P}(z)$ is an even elliptic function with the period latiice Λ which has a double pole at any $\omega \in \Lambda$ and no other poles. Show also that

$$\mathcal{P}'(z) = -2 \cdot \sum_{\omega \in \Lambda} \frac{1}{(z-\omega)^3}$$

is an odd elliptic function with the period lattice Λ which has a pole of order 3 at any $\omega \in \Lambda$ and no other poles.

Theorem 2.4.34.

$$\mathcal{P}'(z)^2 = 4 \cdot \mathcal{P}(z)^3 - g_2 \cdot \mathcal{P}(z) - g_3$$

where $g_2 = 60 \cdot \sum_{\omega \in \Lambda - \{0\}} \omega^{-4}$ and $g_3 = 140 \cdot \sum_{\omega \in \Lambda - \{0\}} \omega^{-6}$. ■

Exercise 2.4.35. Prove Theorem 2.4.34. (*Hint*: Consider expansions of $\mathcal{P}(z)$ and $\mathcal{P}'(z)$ into Laurent series at the origin).

Uniformization. Theorem 2.4.34 shows that points of the form $(\mathcal{P}(z), \mathcal{P}'(z))$ lie on an elliptic curve with the equation

$$y^2 = 4 \cdot x^3 - g_2 \cdot x - g_3 \ .$$

Moreover one has

Theorem 2.4.36. *The map*

$$\varphi : \mathbb{C} \longrightarrow \mathbb{P}^2(\mathbb{C})$$

$$\varphi : z \longmapsto \begin{cases} (1:\mathcal{P}(z):\mathcal{P}'(z)) & , \text{ if } z \notin \Lambda \\ (0:0:1) & , \text{ if } z \in \Lambda \end{cases}$$

defines an isomorphism of Riemann surfaces $\mathbb{C}/\Lambda$ *and* $E(\mathbb{C})$, *which is also a group homomorphism, where* E *is the elliptic curve with the homogeneous equation*

$$x_0 \cdot x_2^2 = 4x_1^3 - g_2 \cdot x_0^2 \cdot x_1 - g_3 \cdot x_0^3 \quad .$$

■

It is said that elliptic functions *uniformize* elliptic curves.

Corollary 2.4.37. $E(\mathbb{C})$ *is isomorphic to the torus* $(\mathbb{R}/\mathbb{Z})^2$ *as a real Lie group.* ■

Corollary 2.4.37 implies Propositions 2.4.4 and 2.4.12 for complex elliptic curves. Theorems 2.4.21 and 2.4.22 can be also deduced from Theorem 2.4.36 using the following result.

Exercise 2.4.38. Show that for any $f \in End(E)$ there exists $\alpha_f \in \mathbb{C}$ such that $\alpha_f \cdot \Lambda \subseteq \Lambda$ and the endomorphis $\varphi^{-1} \circ f \circ \varphi$ of $\mathbb{C}/\Lambda$ coincides with the multiplication by α_f . (*Hint*: Consider the Tailor expansion of $\varphi^{-1} \circ f \circ \varphi$ at the origin).

Exercise 2.4.39. Deduce Theorems 2.4.21 and 2.4.22 for $k = \mathbb{C}$ from the previous exercise and Theorem 2.4.36.

CHAPTER 2.5

SINGULAR CURVES

The study of non-singular curves is most important for the theory of curves. Nevertheless the study of singular curves is in many cases indispensable, since many smooth curves have singular models which are useful to prove some of their properties. For example, any curve has a plane (singular) model. In this chapter we discuss some properties of singular curves and describe their connections with smooth curves.

Section 2.5.1 is devoted to the normalization of a curve which is the simplest way to obtain a smooth curve from a singular one. In Section 2.5.2 we define the divisor of double points of a curve which is essential for the theory of singular curves. In Section 2.5.3 we give some properties of plane singular curves, in particular a description of differential forms on these curves; Section 2.5.4 is devoted to a desingularization technique for plane singular curves.

In this chapter we assume the ground field k to be algebraically closed.

2.5.1. Normalization

Let A be a *domain* (i.e. a commutative ring without zero-divisors) let K be its fraction field, and let $x \in L$, where L is an extension of K. We call x *integral* over A iff there exist elements $a_0, a_1, \ldots, a_{n-1} \in A$ such that $x^n + a_{n-1} \cdot x^{n-1} + \ldots + a_0 = 0$. The set $\bar{A}_L$ formed by $x \in L$ which are integral over A is called the *integral closure* of A in L. The set $\bar{A}_K$ is denoted by $\bar{A}$. If $\bar{A} = A$ we call A *integrally closed.*

Exercise 2.5.1. Show that for any A and any L the set $\bar{A}_L$ is an integrally closed domain.

We call a curve X *normal* iff O_P is integrally closed for any $P \in X$.

Exercise 2.5.2. Show that X is normal iff $k[U]$ is integrally closed for any open $U \subset X$.

Exercise 2.5.3. Show that a smooth curve is normal.

Moreover, one has:

Proposition 2.5.4. *Any normal curve is smooth.* ∎

Hence the normality and non-singularity conditions for curves are equivalent.

Remark 2.5.5. These conditions are equivalent over an arbitrary perfect field. If the ground field is not perfect there exist normal singular curves.

Exercise 2.5.6. Let X be a affine curve, and let $A = k[X]$. Show that there exists a unique affine curve X^{ν} such that $k[X^{\nu}] = \bar{A}$.

The embedding $k[X] \subseteq k[X^{\nu}]$ defines a regular surjective map $X^{\nu} \longrightarrow X$ which is called the *normalization* map.

Theorem 2.5.7. *Let X be a quasi-projective curve. There exist a unique smooth curve X^{ν} , such that there is a regular surjective map $\nu : X^{\nu} \longrightarrow X$ which is a birational isomorphism. If X is a projective curve then X^{ν} is also projective.* ■

We call ν the *normalization map* and X^{ν} the *normalization* of X .

Exercise 2.5.8. Let U be an affine open subset in X , let $\nu : X^{\nu} \longrightarrow X$ be the normalization map, and let $V = \nu^{-1}(U)$. Show that $V = U^{\nu}$.

2.5.2. Divisor of double points

Local invariants. To study singular curves it is important to consider some characteristics of a point P on a curve X which depend only on the local ring $\mathcal{O}_P$. Such characteristics are called *local invariants* of P . Let $\nu : X^{\nu} \longrightarrow X$ be the normalization map, and let $\mathcal{O}'_P$ be the integral closure of $\mathcal{O}_P$.

Exercise 2.5.9. Show that the set $S(P) = \nu^{-1}(P)$ is finite for any $P \in X$ and that $\mathcal{O}'_P = \mathcal{O}_{S(P)}$, where $\mathcal{O}_{S(P)}$ is a ring of rational functions on X^ν which are regular at any $Q \in S(P)$. Show that $\mathcal{O}_{S(P)} = \bigcap_{Q \in S(P)} \mathcal{O}_P$ and that $\mathcal{O}_{S(P)}$ is a *semi-local* ring, i.e. it has a finite number of maximal ideals (this number equals the cardinality of $S(P)$).

Let c_P be the *annihilator* of the $\mathcal{O}_P$-module $\mathcal{O}'_P/\mathcal{O}_P$, i.e. the set $c_P = \{a \in \mathcal{O}_P \mid a \cdot \mathcal{O}'_P \subseteq \mathcal{O}_P\}$. It is clear that c_P is an ideal both in $\mathcal{O}_P$ and in $\mathcal{O}'_P$. It can be shown that c_P is the largest ideal in $\mathcal{O}_P$ which is also an ideal in $\mathcal{O}'_P$. We call c_P the *conductor* of the extension $\mathcal{O}'_P/\mathcal{O}_P$.

Exercise 2.5.10. Show that c_P is nontrivial. (*Hint*: $\dim_k \mathcal{O}'_P/\mathcal{O}_P < \infty$).

Set

$$n_P = \dim_k \mathcal{O}'_P/c_P \quad , \quad \delta_P = \dim_k \mathcal{O}'_P/\mathcal{O}_P \quad ;$$

we call n_P and δ_P *fundamental local invariants* of P.

Exercise 2.5.11. Show that the following conditions are equivalent:

a) P is a non-singular point;

b) $\delta_P = 0$;

c) $n_P = 0$.

Hence δ_P and n_P do not vanish only for singular points.

Exercise 2.5.12. Show that $\delta_P \neq 0$ implies

$$\delta_P \leq n_P - 1 \ .$$

(*Hint*: $\mathcal{O}_P \supseteq k + c_P$).

One can also prove

Proposition 2.5.13. *For any* $P \in X$

$$n_P \le 2 \cdot \delta_P \quad .$$

■

Exercise 2.5.14. a) Let $X \subset \mathbb{A}^2$ be defined by $y^2 = x^2 + x^3$ for $char(k) \neq 2$ and let $P = (0,0)$. Show that $X^\nu = \mathbb{A}^1$ and that ν can be given by $\nu(t) = (t^2 - 1, t\cdot(t^2 - 1))$, t being a coordinate on $\mathbb{A}^1$. Show that $S(P) = \{Q,Q'\}$ where $Q = \{t = 1\}$, $Q' = \{t' = -1\} \in \mathbb{A}^1$. Show that $\mathcal{O}'_P$ consists of quotients $P(t)/Q(t)$ where $Q(t)$ is divisible neither by $(t + 1)$ nor by $(t - 1)$ and that $c_P = (t^2 - 1)\cdot\mathcal{O}'_P$, $\mathcal{O}_P = k + c_P$, $n_P = 2$, $\delta_P = 1$.

b) Let $X \subset \mathbb{A}^2$ be defined by $y^2 = x^3$ and let $P = (0,0)$. Show that then $X^\nu = \mathbb{A}^1$, $\nu(t) = (t^2,t^3)$, $S(P) = Q = \{t = 0\}$,

$$\mathcal{O}'_Q = \{P(t)/Q(t) \mid P,Q \in k(t) \, , \, Q(0) \neq 0\} \quad ,$$

and $c_P = t^2\cdot\mathcal{O}'_P$, $\mathcal{O}_P = k + c_P$, $n_P = 2$, $\delta_P = 1$.

A point $P \in X$ is called a *simplest singularity* iff $\delta_P = 1$ (and hence $n_P = 2$). For such a point $S(P)$ has either one or two elements. If $S(P)$ has two elements then P called a *simplest double point*, if $S(P)$ consist of a single point then P is called a *simplest cuspidal point*. Thus P is a simplest double point in Exercise 2.5.14.a, and P is a simplest cuspidal point in Exercise 2.5.14.b.

Divisor of double points. Let X be a projective curve, let $P \in X$ and let $Q \in S(P)$. Since Q is a non-singular point we have $c_P \cdot \mathcal{O}_P = m_Q^{n_Q}$ for a non-negative integer n_Q . The divisor

$$D = \sum_P \sum_{Q \in S(P)} n_Q \cdot Q$$

is called the *divisor of double points* (or the *conductor*) of X . Note that D is well-defined since for a non-singular P we have $|S(P)| = 1$ and $n_Q = 0$ for $Q \in S(P)$. Thus D is a divisor on X^ν . It plays a crucial role in the study of the connection between X and X^ν .

Exercise 2.5.15. Show that $n_P = \sum_{Q \in S(P)} n_Q$.

Exercise 2.5.16. Find out divisors of double points for projective closures of curves from Exercise 2.5.14.

Exercise 2.5.17. Let $X \subset \mathbb{P}^2$ be defined by

$$x_0 \cdot x_2^4 = x_0 \cdot x_1^4 + x_1^5 ,$$

and let $P = (1:0:0)$. Compute n_P , δ_P , and D . Is P a simplest singularity?

2.5.3. Plane curves

Note that for any smooth projective curve X there exists a projective plane curve X' which is birationally isomorphic to X . We call such X' a plane model of X . Indeed, let $f \in k(X) - k$ be such that $k(X)/k(f)$ is a separable extension (it is not hard to show that f does exist). By the primitive element theorem there exists $g \in k(X)$ such that $k(X) = k(f,g)$. It is clear that

$F(f,g) = 0$ for some $F \in k[T_1,T_2]$. Let G be a polynomial such that $G(f,g) = 0$ and which has the least possible degree. Then G is irreducible and we can define X' as the projective closure of the plane affine curve defined by $G(x,y) = 0$.

As a rule, it is impossible to choose a nonsingular X' . Indeed, the genus of a smooth projective plane curve equals $(m - 1)\cdot(m - 2)/2$ where m is its degree. Since the genus of a curve can be arbitrary non-negative integer (for example, for $char\ k \neq 2$ the genus of the curve $y^2 = f(x)$, $f \in k[x]$ equals $\lceil (deg\ f - 1)/2 \rceil$) many smooth projective curve are not isomorphic to smooth projective plane ones. Therefore the study of singular plane curves is indispensable. Let X be a smooth projective curve, X' its plane model, and let f and $g \in k(X)$ be as above. Then the rational map $\varphi : X \longrightarrow X'$ defined by $x \longmapsto (f(x),g(x))$ is regular by Theorem 2.1.57. Since the normalization of the curve is unique we see that $X = (X')^{\nu}$ and the map φ is the normalization map for X' . The main question is to express main invariants of X in terms of invariants of X' or rather of its equation in $\mathbb{P}^2$.

Genus. By definition, the genus of a singular curve is the genus of its normalization.

Theorem 2.5.18. *Let X' be a projective plane (singular) curve of degree m , let X be its normalization and let $g = g(X') = g(X)$ be its genus. Then*

$$g = \frac{(m - 1)(m - 2)}{2} - \sum_P \delta_P \quad , \tag{2.5.1}$$

where the sum is taken over all singular points of X' . ∎

The number $p_a(X') = (m-1)\cdot(m-2)/2$ is sometimes called the *arithmetic genus* of X'. Thus for smooth plane curves (and only for them)

$$p_a(X') = g(X') \quad .$$

Example 2.5.19. The curve $x_0\cdot x_2^2 = x_1^3 - x_0\cdot x_1^2$ (or $y^2 = x^3 - x^2$ in affine coordinates) is of arithmetic genus 1 (its genus being equal to 0). The same is true for the curve $x_0\cdot x_2^2 = x_1^3$.

Differentials. In fact one can describe $\Omega[X]$ in a way similar to that of Proposition 2.2.7 for smooth plane curves. Let $F(x_0:x_1:x_2) = 0$ be a homogeneous equation of X' and let $G(x,y) = F(1:x:y)$ be its affine equation. For any P and $Q \in k[x,y]$ the expression $P(x,y)\cdot dy/Q(x,y)$ defines a rational differential form on X .

Proposition 2.5.20.

$$\Omega[X] \subseteq \left\{ \frac{P(x,y)\cdot dy}{G'_x(x,y)} \;\middle|\; deg\, P(x,\, y) \le m-3 \right\} \quad . \quad \blacksquare$$

A polynomial $P(x,y)$ is called *adjoint* (for X') iff $P(x,y)\cdot dy/G'_x(x,y) \in \Omega[X]$; if $X = X'$ then any P with $deg\, P \le m-3$ is adjoint by Proposition 2.2.7. Let us describe adjoint polynomials in the general case. Note that to any $P(x,y)$ with $deg\, P = s$ one can associate a form F_P of degree s , namely

$$F_P(x_0:x_1:x_2) = x_0^s\cdot P(x_1/x_0, x_2/x_0) \quad .$$

In Section 2.1.2 we have seen that any form F defines a Cartier divisor $(F) = (\{U_i, f_i\})$ on a smooth plane curve X . In fact one can notice that this definition does not use non-singularity of X and thus is valid for any plane curve X. The Cartier divisor $(\{\nu^{-1}(U_i), \nu^*(f_i)\})$ on X, where $\nu : X \longrightarrow X'$ is the normalization map, is called the *divisor of* F on X and is denoted also by (F). Recall that by D we denote the divisor of double points of X'. In the above notation we have

Proposition 2.5.21. *A polynomial* $P \in k[x,y]$ *of degree* $s \leq m - 3$ *is adjoint for* X' *iff* $(F_P) \geq D$ *in* $Div(X)$. ■

Thus for an adjoint P the curve $F_P(x_0:x_1:x_2) = 0$ contains all the singular points of X'. If X' has only simplest singularities then this condition is sufficient for P to be adjoint for X'; in this case Theorem 2.5.18 follows from Propositions 2.5.20 and 2.5.21.

Noether theorem. In the study of plane curves an essential role is played by the following result which is called the *Noether* $(A\cdot\varphi+B\cdot\psi)$*-theorem.*

Theorem 2.5.22. *Let* F *be a homogeneous equation of* X' *and let* G *and* H *be forms in* (x_0,x_1,x_2). *Then the condition* $(G) \geq (H) + D$ *implies that there exist forms* A *and* B *such that*

$$G = A\cdot H + B\cdot F \quad .$$

in the ring $k[x_0,x_1,x_2]$. ■

In other words, the condition $(G) \geq (H) + D$ is sufficient for the divisibility of $\overline{G}$ by $\overline{H}$ in the homogeneous coordinate ring $R_X = k[x_0,x_1,x_2]/(F)$ of X', where $\overline{G}$ and $\overline{H}$ are the images of G and H in R_X.

Corollary 2.5.23. *Let X' be a smooth complete plane curve. Then $(G) \geq (H)$ implies that $\overline{G}$ is divisible by $\overline{H}$ in R_X.* ■

Auxiliary formulae. Let in the above notation the degree of G as a polynomial in x be equal to m and as a polynomial in y be equal to n, let its total degree equal r and let g be the genus of X'.

Proposition 2.5.24.

$$deg\ D = (r - 1)\cdot(r - 2) - 2g \quad .$$

■

We also need expressions for the divisors (dx) and (G'_X) of the differential form dx and of the function G'_X, respectively. Let B_X (respectively B_Y) be the ramification divisor of the map $X \longrightarrow \mathbb{P}^1$ defined on X' by $(x,y) \longmapsto x$ (respectively by $(x,y) \longmapsto y$).

Proposition 2.5.25.

$$(dx) = B_X - 2\cdot(x)_\infty \quad ,$$

$$(G'_X) = B_Y - (m - 2)\cdot(x)_\infty - n\cdot(y)_\infty - D \quad .$$

where, as usual, $(f)_\infty$ is the divisor of poles of $f \in k(X)$. ■

Exercise 2.5.26. Prove Proposition 2.5.25 and deduce Proposition 2.5.24 from it.

2.5.4. Desingularization

Theorem 2.5.7 states the existence of the normalization of a curve but it gives no idea of how to construct the normalization. For a plane curve X there exists a techinique to construct X^{ν} explicitly. To describe it we have to start with the following construction from the theory of surfaces.

Monoidal transformation. Recall that an *algebraic surface* is a two-dimensional algebraic variety. Let us begin with an example. Let $\mathbb{P}^2$ be the projective plane with homogeneous coordinates $(x_0:x_1:x_2)$ and let $\mathbb{P}^1$ be the projective line with coordinates $(y_1:y_2)$. Consider the closed subset W in $\mathbb{P}^2 \times \mathbb{P}^1$ defined by

$$x_1 \cdot y_2 = x_2 \cdot y_1 \quad . \tag{2.5.2}$$

It is clear that W is a surface, i.e. that W is two-dimensional and irreducible. The map $\sigma : W \longrightarrow \mathbb{P}^2$ which is the restriction of the natural projection $\mathbb{P}^2 \times \mathbb{P}^1 \longrightarrow \mathbb{P}^2$ is called the *monoidal transformation* centered at the point $P_0 = (1:0:0)$. Note that for $P = (x_0:x_1:x_2) \neq P_0$ from (2.5.2) it follows that $(y_1:y_2) = (x_1:x_2)$ and hence the map

$$(x_0:x_1:x_2) \longmapsto (x_0:x_1:x_2 \ ; \ x_1:x_2)$$

$$\mathbb{P}^2 - \{P_0\} \longrightarrow W - \{\sigma^{-1}(P_0)\}$$

is inverse to σ on $\mathbb{P}^2 - \{P_0\}$. Thus the restriction of σ defines an isomorphism of $W - \{\sigma^{-1}(P_0)\}$ with $\mathbb{P}^2 - \{P_0\}$. For $(x_0:x_1:x_2) = P_0$ from (2.5.2) it follows that $\sigma^{-1}(P_0) = \{P_0\} \times \mathbb{P}^1$. One says that the monoidal transformation blows up P_0 into the projective line $\{P_0\} \times \mathbb{P}^1$. Let us now consider an open subset U_i in W defined by $x_0 \neq 0$ and $y_i \neq 0$ for $i = 1$ or 2 . We can set $x_0 = 1$ and $y_i = 1$ and thus the equation (2.5.2) takes the form $x_1 = x_2 \cdot y_1$ on U_2 , and $x_2 = x_1 \cdot y_2$ on

U_1. Thus U_1 is isomorphic to $\mathbb{A}^2$ with coordinates (x_1, y_1) and U_2 is isomorphic to $\mathbb{A}^2$ with coordinates (y_1, x_2). To clarify the picture of the monoidal transformation it is useful to consider the restriction of the rational map $\sigma^{-1} : \mathbb{P}^2 \longrightarrow W$ to the line L with the equation $\alpha \cdot x_1 = \beta \cdot x_2$ containing P_0. Since L is a smooth curve and W is projective, this restriction is a regular map. In fact one easily shows that

$$\sigma^{-1}|_L : (x_0:x_1:x_2) \longmapsto (x_0:x_1:x_2;\beta:\alpha)$$

Therefore the intersection point of the curves $\sigma^{-1}(L)$ and $\sigma^{-1}(P_0)$ runs over $\sigma^{-1}(P_0)$ when L runs over all lines containing P_0. Let X be a plane curve with the equation $F(x_0:x_1:x_2) = 0$, and let $P_0 \in X$. Let X_1 be the closure of $\sigma^{-1}(X - \{P_0\})$ in W; it is called the *proper inverse image* of X under σ. Note that if P_0 is a non-singular point of X then the restriction of σ^{-1} to X is regular and hence X_1 is isomorphic to X. We shall see below that if P_0 is a singular point of X then X_1 is never isomorphic to X.

Example 2.5.27. Let $X : \{x_0 \cdot x_2^2 = x_1^3\}$. Then the point $P_0 = (1:0:0)$ is singular on X. The curve $\sigma^{-1}(X - \{P_0\})$ is defined by $x_0 \cdot x_2^2 = x_1^3$, $x_1 \cdot y_2 = x_2 \cdot y_1$ and $(x_1, x_2) \neq (0,0)$. Let $U_1 = \{x_0 \neq 0, y_1 \neq 0\}$ be an open subset of W; in U_1 the curve $\sigma^{-1}(X - \{P_0\})$ is defined by $x_2^2 = x_1^3$, $x_2 = y_2 \cdot x_1$ and $(x_1, x_2) \neq (0,0)$ which is equivalent to $y_2^2 = x_1$, $x_2 = y_2^2 \cdot x_1$ and $(x_1, x_2) \neq (0,0)$. The closure $\overline{\sigma^{-1}(X - \{P_0\})}$ is defined in U_1 by $y_2^2 = x_1$, $x_2 = y_2 \cdot x_1$. Hence $\overline{\sigma^{-1}(X - \{P_0\})} \cap U_1$ is a nonsingular curve. Similarly, for $U_2 = \{x_0 \neq 0, y_2 \neq 0\}$ the curve

$\overline{\sigma^{-1}(X - \{P_0\})} \cap U_2$ is also non-singular. Therefore the map $\sigma|_{X_1} : X_1 \longrightarrow X$ is the normalization map.

Local monoidal transformation. Let now $\mathbb{A}^2$ be the affine plane with coordinates (x_1,x_2), and let $\mathbb{P}^1$ be the projective line with coordinates $(y_1:y_2)$. Then the equation (2.5.2) defines a close subset W' of $\mathbb{A}^2 \times \mathbb{P}^1$, which is an open subset of W. The map $\sigma|_{W'} : W' \longrightarrow \mathbb{A}^2$ is called the *local monoidal transformation centered at* P_0.

Desingularization. Let V be a projective surface such that any its point has an open neighbourhood isomorphic to $\mathbb{A}^2$ (for example, the surfaces $\mathbb{P}^2$ and W have this property). One can show that for any $P_0 \in V$ there exists a unique projective smooth surface V_1 and a morphism $\sigma : V_1 \longrightarrow V$ such that σ gives an isomorphism of $V_1 - \sigma^{-1}(P_0)$ with $V - \{P_0\}$, $\sigma^{-1}(P_0)$ is isomorphic to $\mathbb{P}^1$ and the restriction of σ to $\sigma^{-1}(U)$ coincides with the local monoidal transformation centered at P_0, where U is a neighbourhood of P_0 isomorphic to $\mathbb{A}^2$. The map σ is called the *monoidal transformation centered at* P_0 or the *blow-up* of P_0. Note that V_1 has the above property and hence any its point can be also blown up.

If X is a curve on V which contains P_0 then its *proper inverse image* is defined as the curve

$$X_1 = \overline{\sigma^{-1}(X - \{P_0\})} .$$

Here is the basic fact on the desingularization of plane curves:

Theorem 2.5.28. *Let $X \subset \mathbb{P}^2$ be a plane curve. Then there exists a sequence of monoidal transformations*

$$V_n \longrightarrow V_{n-1} \longrightarrow \dots \longrightarrow V_1 \longrightarrow V_0 \longrightarrow \mathbb{P}^2$$

such that the proper inverse image X_n of X on V_n is a smooth curve. ∎

Note that a similar result is valid for curves on any surface but we do not need it here.

Multiplicity of a point on a curve. Let P be a point of $\mathbb{A}^2$ and let X be an affine plane curve. Let $F_X(x,y) = 0$ be an equation of X. Let m_P be the maximal ideal of the local ring $\mathcal{O} = \mathcal{O}_{\mathbb{A}^2,P}$ of P on $\mathbb{A}^2$. The *multiplicity* $\mu_P(X)$ of P on X is defined as the largest integer r such that $F_X \in m_P^r$. Thus $\mu_P(X) = 0$ for $P \notin X$ and $\mu_P(X) > 0$ for $P \in X$. Note that $\mu_P(X) = 1$ iff P is a non-singular point of X. Indeed, we have $\mathcal{O}_{X,P} = \mathcal{O}/F_X \cdot \mathcal{O}$ and the condition $F_X \in m_P - m_P^2$ is equivalent to $m_{X,P}$ being principal. Therefore $\mu_P(X) \geq 2$ iff X is singular at P. Note that since the definition of $\mu_P(X)$ depends only on the local ring $\mathcal{O}$ of P, $\mu_P(X)$ can be defined for any curve X on a surface V satisfying the above condition. The proof of Theorem 2.5.28 uses the fact that for the monoidal transformation centered at P and for any $Q \in V_1$ with $\sigma(Q) = P$ we have $\mu_Q(X_1) < \mu_P(X)$; this makes it possible to use the induction on $\mu_P(X)$.

The genus formula. Let

$$V_n \longrightarrow V_{n-1} \longrightarrow \dots \longrightarrow V_1 \longrightarrow V_0 = \mathbb{P}^2$$

be the sequence of monoidal transformations which desingularizes a plane curve X, and let $P \in X$. The set of *infinitely close points* corresponding to P consists, by definition, of points Q on some V_i lying over P (including P itself). This set can be represented as a tree whose root corresponds to P, offsprings of the first order correspond to points obtained by blow-up of P, ..., its offsprings of the r-th order for $r \geq 2$ correspond to points which are obtained by blow-up of points corresponding to offsprings of the $(r - 1)$-th order.

Proposition 2.5.29. *Let X be a plane curve, and let $P \in X$. Then*

$$\delta_P = \sum_{Q \longmapsto P} r_Q \cdot (r_Q - 1)/2 \tag{2.5.3}$$

where $r_Q = \mu_Q(X_i)$ for $Q \in X_i$, and the sum is taken over the set of infinitely close points corresponding to P (including P itself). ∎

Note that points Q with $r_Q = 1$, i.e. smooth points make no contribution into δ_P.

Corollary 2.5.30. *Let X be a plane curve of degree m. Then*

$$g = g(X^\nu) = \frac{(m - 1) \cdot (m - 2)}{2} - \sum_Q \frac{r_Q \cdot (r_Q - 1)}{2} \tag{2.5.4}$$

where the sum is taken over the set of all infinitely close points of X. ∎

In fact the techique of monoidal transformations makes it possible to desingularize a plane curve efficiently. A desingularization algorithm will be given below in Part 4 (for a class of curves we need). Here we give some auxiliary constructions.

Desingularization tree. Let X be a plane curve with a homogeneous equation $F(x_0:x_1:x_2) = 0$, and let $P = (a_0:a_1:a_2)$ be a singular point of X. Without loss of generality we can assume that $a_0 \neq 0$. Let $G(x,y) = F(1:x:y) = 0$ be an affine equation of X, and let $\alpha = a_1/a_0$, and $\beta = a_2/a_0$ be affine coordinates of P. We associate with P the *desingularization tree* Γ_P, which is the labeled tree of infinitely close points of P, whose labels make it possible to construct it inductively beginning from the root. Any its node s is labeled by the triple $(P(x_s,y_s), (X_s,Y_s), \Lambda)$ where x_s and y_s are local coordinates on V_i in a neighbourhood of the point Q_s, corresponding to s, $P(x_s,y_s) = 0$ is a local equation of the proper inverse image X_i of X, i.e. a polynomial which generates the ideal of functions vanishing on X_i in the local ring $\mathcal{O}_{V_i,Q_s}$, X_s and Y_s are polynomials in x_s and y_s which give expressions of local coordinates on $\mathbb{P}^2$ in a neighbourhood of P, Λ_s is the smoothness mark of Q_s:

$$\Lambda_s = \begin{cases} 1 & \text{if } X_i \text{ is smooth at } Q_s \\ 0 & \text{otherwise} \end{cases} .$$

Example 2.5.31. Let us construct the desingularization tree for $P = (0,0)$ on the curve $X : y^2 = x^3$. Its root is labeled by $(y^2 - x^3, (x,y), 0)$. After blowing-up of P which is given by $y = t\cdot x$, (t,x) being coordinates on V_1 in a neighbourhood of Q lying over P, we get an equation of $X_1 : x = t^2$ and hence Q is non-singular. Thus Γ_P looks like this:

	free node	$(t^2 - x\ ,\ (x, x\cdot t),\ 1)$
	root	$(y^2 - x^3,\ \ (x,y)\ ,\ 0)$.

Exercise 2.5.32. Constuct the desingularization tree for the point $P = (0,0)$ on the curve $X : y^4 = x^5 + x^6$.

In general it is not hard to write out expressions for the labels of an immediate offspring of a node.

Proposition 2.5.33. *Let* s *be a node of* Γ_P *which is labeled by* $(P_s(x_s,y_s),\ (X_s,Y_s),\ \Lambda_s)$ *and let* $\Lambda_s = 0$ *(i.e.* Q_s *is a singular point). Then*

a) *The set* I_s *of immediate offsprings of* s *is in a canonical bijection with the set* $M_s \cup N_s$, M_s *being the set of roots of the polynomial* $P_s'(0,t_s) \in k[t_s]$ *where*

$$P_s(x_s,x_s\cdot t_s) = x_s^{a_s}\cdot P_s'(x_s,t_s) \quad \text{and} \quad x_s \nmid P_s'(x_s,t_s)\ .$$

The set N_s *has a single element for* $P_s''(0,0) = 0$ *and is empty for* $P_s''(0,\ 0) \neq 0$ *, where*

$$P_s(u_s\cdot y_s,y_s) = y_s^{b_s}\cdot P_s''(u_s,y_s) \quad \text{and} \quad y_s \nmid P_s''(u_s,\ y_s)\ .$$

We identify the sets I_s *and* $M_s \cup N_s$.

b) *If* $t \in M_s$ *then* $x_t = x_s$, $y_t = t_s - \alpha_t$ *where* α_t *is the root of* $P_s'(0,t_s)$ *corresponding to* t *, and*

$$P_t(x_t,y_t) = P_s'(x_t,\ y_t + \alpha_t) \quad ; \tag{2.5.5}$$

$$X_t(x_t,y_t) = X_s(x_t,\ x_t\cdot(y_t + \alpha_t)) \quad ; \tag{2.5.6}$$

$$Y_t(x_t,y_t) = Y_s(x_t,\ x_t\cdot(y_t + \alpha_t)) \quad ; \tag{2.5.7}$$

$$\Lambda_t = \begin{cases} 0 \ , \ if \ \dfrac{\partial P_t}{\partial x_t}(0,0) = \dfrac{\partial P_t}{\partial y_t}(0,0) = 0 \ ; \\ 1 \ , \ otherwise \ . \end{cases} \tag{2.5.8}$$

c) *If* $t \in N_s$ *then* $x_t = u_s$, $y_t = y_s$, *and*

$$P_t(x_t,y_t) = P''_s(x_t, y_t) \tag{2.5.9}$$

$$X_t(x_t,y_t) = X_s(x_t \cdot y_t, y_t) \tag{2.5.10}$$

$$Y_t(x_t,y_t) = Y_s(x_t \cdot y_t, y_t) \tag{2.5.11}$$

$$\Lambda_t = \begin{cases} 0 \ , \ if \ \dfrac{\partial P_t}{\partial x_t}(0,0) = \dfrac{\partial P_t}{\partial y_t}(0,0) = 0 \ ; \\ 1 \ , \ otherwise \ . \end{cases} \tag{2.5.12}$$

Proof: It follows immediately from the definition of node labels and from formulae for the local monoidal transformation. ■

Proposition 2.5.33 makes it possible to construct Γ_P beginning from its root and thus give a desingularization algorithm for a plane curve. In fact this algorithm is polynomial, i.e. it uses time and space polynomially bounded in terms of input length. In cases we need this will be shown in Part 4.

CHAPTER 2.6

REDUCTIONS AND SCHEMES

The idea of specialization (or reduction) of an algebraic variety is one of the most important for the geometry and especially for the arithmethic of algebraic varieties. By specialization one can obtain for example varieties over finite fields from varieties over algebraic number fields. The study of specialization using the language of quasi-projective varieties has many disadvantages. For these questions the language of schemes which is now the working language of algebraic geometry is much more adequate. One can consider the theory of schemes as the theory of algebraic varieties over arbitrary commutative rings. It should be also remarked that specialization was one of the main sources of the theory of schemes. Moreover the theory of schemes possesses a powerful technique for constructing algebraic varieties based on the notion of a representable functor. Many results of this book can not be obtained without use of moduli schemes. In this chapter we give a brief introduction to this theme.

In Section 2.6.1 we discuss a "naive" approach to the reduction of curves; Section 2.6.2 is devoted to the definition and elementary properties of affine schemes. In Section 2.6.3 we consider sheaves and ringed spaces, in Section 2.6.4 we define schemes and explain their connections with varieties. Section 2.6.5 is devoted to representable functors which play an essential role in the construction of many types of schemes.

In this chapter the word "ring" means "a commutative ring with unity".

2.6.1. Reduction of a curve

Let $X \subset \mathbb{A}^N$ be a smooth affine curve over a field K. Let I_X be the ideal in $K(\mathbb{A}^N) = K[x_1,\ldots,x_N]$ formed by $f \in K(\mathbb{A}^N)$ vanishing on X, and let $F_1,\ldots,F_n$ be its basis; thus $I = (F_1,\ldots,F_n)$ and $K[X] = K[\mathbb{A}^N]/I_X$. Let $\mathcal{O}$ be a valuation subring in K whose fraction field coincides with K, let m be its maximal ideal and let $k = \mathcal{O}/m$ be its residue field. Since the fraction field of $\mathcal{O}$ is K, we can assume that all coefficients of F_i lie in $\mathcal{O}$ and each F_i has a coefficient which does not lie in m. Hence for any $i = 1,\ldots,n$ the polynomial $\overline{F}_i(x) \in k[x_1,\ldots,x_N]$ does not vanish; here $\overline{F}_i(x)$ is obtained from $F_i(x)$ by reduction modulo m of its coefficients. Let us consider an affine closed subset $\overline{X}$ in $\mathbb{A}^N(k)$ which is defined by the system $\overline{F}_1(x) = \ldots = \overline{F}_n(x) = 0$. One can show that there exists a basis $F_1,\ldots,F_n$ of I_X such that this subset is of dimension 1. It is natural to consider $\overline{X}$ as a reduction of X modulo m, but it should be pointed out that $\overline{X}$ depends on the choice of basis in I_X. In general, another choice of basis can give another reduction $\overline{X}'$ which is not isomorphic to $\overline{X}$, moreover $\overline{X}'$ can be irreducible while $\overline{X}$ is reducible. If $\overline{X}$ is an irreducible smooth curve, we say that X has a *good reduction* modulo m. An affine curve X can have non-isomorphic good reductions $\overline{X}$ and $\overline{X}'$. If we assume that X is a smooth projective curve we can define its reduction choosing a basis $\{F_1',\ldots,F_n'\}$ in the ideal I_X of formes vanishing on X. For a smooth projective curve one can prove that its good reduction is unique (up to an isomorphic). Thus good reduction of a smooth projective curve is well-defined.

Let X and Y be smooth projective curves having good reductions modulo m, and let $f : X \longrightarrow Y$ be a regular map. We can define the reduction $\overline{f} : \overline{X} \longrightarrow \overline{Y}$ as follows. Let Γ be the *graph* of f, i.e. the set

$$\Gamma = \{(x,f(x)) \in X \times Y \mid x \in X\} \quad ;$$

Γ is a smooth projective curve isomorphic to X. Hence it has a good reduction $\overline{\Gamma} \subset \overline{X} \times \overline{Y}$, and one can show that there exists a unique regular map $\overline{f} : \overline{X} \longrightarrow \overline{Y}$ whose graph is $\overline{\Gamma}$. Similarly one can define reductions of other objects connected with X such as differential forms, divisors, etc.

The definion of reduction modulo m sketched above has some disadvantages. First of all it gives no idea how to find an appropriate basis $\{F_1,\ldots,F_n\}$ in I_X in the case of good reduction. Further on, if X has no good reduction it is not clear how to define its "true" reduction modulo m. Note also that the above definition of reduction of a morphism is an *ad hoc* definition which is rather difficult to use. The theory of schemes gives an adequate techinique to study these questions. We proceed to describe some elements of this theory.

2.6.2. Spectra of rings

As we have seen in Exercise 2.1.25, an affine variety X over an algebraically closed field k is uniquely determined by its ring of regular functions $k[X]$. In fact the points of X are just maximal ideals of $k[X]$. Recall that for a ring A we denote the set of its maximal ideals by $Max(A)$.

Exercise 2.6.1. Show that the map $X \longrightarrow Max(k[X])$, defined by $P \longmapsto m_P$, where m_P is the ideal of $k[X]$ formed by $f \in k[X]$ vanishing at P, is a bijection.

Note that $f(P) \in k$ equals the image of f in $k[X]/m_P = k$.

We can define a topology on $Max(k[X])$ requiring the bijection $X \longrightarrow Max(k[X])$ to be a homeomorphism in the Zariski topology of X. This topology is also called *Zariski topology.*

Using the language of categories we can express the connection between X and $k[X]$ in the following way:

Exercise 2.6.2. Show that the map $X \longrightarrow Max(k[X])$ defines an equivalence between the category of affine varieties over k and the category Max_k whose objects are sets $Max(A)$ where A is an integral k-algebra of finite type over k, and $Mor_{Max_k}(Max(A), Max(B)) = Hom_k(B, A)$. (*Hint*: If $\varphi : B \longrightarrow A$ is a k-linear homomorphism and $m \in Max(A)$ then $\varphi^{-1}(m) \in Max(B)$).

Spectra. Let A be an arbitrary commutative ring with unity. The set $Max(A)$ can be considered as a generalization of the notion of an affine variety. This generalization has a disadvantage. Let $\varphi : B \longrightarrow A$ be a homomorphism and let $m \in Max(A)$. Then $\varphi^{-1}(m)$ can be a non-maximal ideal of B. For example, if B is a domain which is not a field, A is its fraction field, and φ is the natural embedding then (0) is a maximal ideal of A but not of B. To avoid this difficulty one has to consider the set $Spec\ A$ of all prime ideals of A which is called the *spectrum* of A. Since the inverse image of a prime ideal is also a prime ideal, any homomorphism $\varphi : B \longrightarrow A$ defines the map

$$\varphi^* : Spec\ A \longrightarrow Spec\ B$$

where $\varphi^*(P) = \varphi^{-1}(P)$. Let us define a topology on $Spec\ A$ as follows: for any subset $S \subseteq Spec\ A$ we call the set

$$\overline{S} = \{P' \in Spec\ A \mid P' \supseteq P \text{ for some } P \in S\}$$

its *closure*.

For example, if A is a domain then (0) is a prime ideal whose closure coincides with $Spec\ A$.

Exercise 2.6.3. Show that if A is a k-algebra of a finite type over an algebraically closed field k then the above topology on $Spec\ A$ induces the Zariski topology on $Max(A)$.

We keep the name *Zariski topology* for this topology as well. Note that $P \in Spec\ A$ is closed iff $P \in Max(A)$.

Exercise 2.6.4. Show that $Spec\ A$ is quasi-compact in the Zariski topology, i.e. any open covering has a finite subcovering (the prefix "quasi-" corresponds to non-separability of the Zariski topology).

For $P \in Spec\ A$ let $k(P)$ be the fraction field of A/P. We call elements $f \in A$ *functions* on $Spec\ A$; the *value* of f at P is, by definition, the image of f in $k(P)$. Note that the values of a function at different points lie in general in different fields. Note also that functions with identical values need not to coincide:

Exercise 2.6.5. Show that $f_1(P) = f_2(P)$ for any $P \in Spec\ A$ iff $(f_1 - f_2)$ is *nilpotent*, i.e. if for a positive integer n we have $(f_1 - f_2)^n = 0$.

The set of all nilpotent elements of A is an ideal in A which coincides with the intersection of all its prime ideals; it is called the *nil-radical* of A. If it is prime itself then we call it the *generic point* of $X = Spec\ A$; note that its closure coincides with X. If A has no non-zero nilpotent elements we call X *reduced*.

Exercise 2.6.6. Show that if $X = Spec\ A$ has a generic point then X is irreducible in the Zariski topology.

Exercise 2.6.7. Describe $Spec\ k[T_1,T_2]$ for a field k. To which object do non-maximal non-zero prime ideals correspond ?

Tangent space. Let $P \in Spec\ A$ be a closed point, i.e. $P \in Max\ A$, then P/P^2 is a vector space over $k(P) = A/P$. Let

$$\Theta_P = Hom_{k(P)}(P/P^2, k(P))$$

be the vector space dual to P/P^2 . We call θ_P the *tangent space* to $X = Spec\ A$ at P .

Dimension. The *dimension* of $X = Spec\ A$ (and of A) is defined as

$$dim\ X = dim\ A =$$

$$= sup\ \{n \mid P_0,\ldots,P_n \in Spec\ A,\ P_0 \subset P_1 \subset\ldots\subset P_n\}\ ;$$

$dim\ X$ is either a non-negative integer or ∞ . Note that there exists a Noether ring A with $dim\ A = \infty$ (recall that A is called a *Noether ring* iff there exists no infinitely ascending chain of ideals $I_0 \subset I_1 \subset\ldots$ $\ldots\subset I_j \subset\ldots\subset A$). If A is a local Noether ring then $dim\ A < \infty$. One can show that $dim_{k(P)} \theta_P \geq dim\ X$ for any $P \in Max\ A$.

Singular points. Let A be an integral Noether ring, and let A_P be its *localization at* P , i.e. the subring in the fraction field K of A generated over A by a^{-1} for $a \notin P$. Then A_P is a local Noether ring and hence $dim\ A < \infty$. We call P a *regular* (*smooth, non-singular*) *point* of $X = Spec\ A$ iff $dim\ A_P = dim_{k(P)} \theta_P$ and a *singular point* iff it is not the case (and thus $dim\ A_P < dim\ \theta_P$). Note that if X is an affine variety, $A = k[X]$ and $P \in Max(A)$ then this definition coincides with that from Section 2.1.1. If each $P \in Spec\ A$ is regular we say that $Spec\ A$ and A are *regular*.

Exercise 2.6.8. a) Check that $\mathbb{Z}$ is a one-dimensional regular ring.

b) Show that $Spec\ A$ for $A = \mathbb{Z}[\sqrt{5}]$ is not regular. Find out its singular points.

The pair $(Spec\ A\ ,\ A)$ is called an *affine scheme* (to be more precise, an affine scheme is $Spec\ A$ with an additional structure which is canonically determined by A , see Section 2.6.4 below). The above facts show that affine schemes are adequate generalization of affine varieties. To obtain an adequate generalization of non-affine varieties one has to use sheaves.

2.6.3. Sheaves and ringed spaces

Presheaves. Let X be a topological space. We say that on X there is given an *abelian presheaf* F if for any open $U \subseteq X$ there is given an abelian group $F(U)$ and for any pair $V \subseteq U$ there is given a group homomorphism $\rho_{UV} : F(U) \longrightarrow F(V)$ satisfying the following conditions:

a) $F(\emptyset) = \{0\}$;

b) $\rho_{UU} : F(U) \longrightarrow F(U)$ is the identity map;

c) for $W \subseteq V \subseteq U$ one has $\rho_{UW} = \rho_{VW} \circ \rho_{UV}$.

Elements from $F(U)$ are called *sections* of F over U ; in particular, elements from $F(X)$ are called *global sections*. The maps ρ_{UV} are called *restriction maps*.

Sheaves. We call a presheaf F a *sheaf* if it satisfies the additional requirements which follow:

d) Let U be an open subset of X , let $\{V_i\}$ be its open covering, let $s \in F(U)$, and let $\rho_{U,V_i}(s) = 0$ for any $i \in I$. Then $s = 0$.

e) Let U be an open subset of X , let $\{V_i\}_{i \in I}$ be its open covering, let $s_i \in F(V_i)$, and let

$\rho_{V_i, V_i \cap V_j}(s_i) = \rho_{V_j, V_i \cap V_j}(s_j)$ for any i and $j \in I$. Then there exists $s \in F(U)$ such that $\rho_{U, V_i}(s) = s_i$ for any i .

Exercise 2.6.9. Give an example of a presheaf which is not a sheaf.

Exercise 2.6.10. Let X be a quasi-projective variety and let

$$\mathcal{O}_X(U) = \{\text{the set of functions regular on } U\}$$

for any open subset $U \subseteq X$; the restriction map $\mathcal{O}_X(U) \longrightarrow \mathcal{O}_X(V)$ for $V \subseteq U$ is defined as the usual restriction of regular functions. Show that $\mathcal{O}$ is a sheaf on X in the Zariski topology which is called the *sheaf of regular functions* (or the *structure sheaf*).

Exercise 2.6.11. Let A be an abelian group, let X be a topological space, let $F_A(U) = A$ for any non-empty open $U \subseteq X$, and let ρ_{UV} be the identity map for any non-empty open U and V such that $V \subseteq U$. Check that F_A is a precheaf on X . Find an example of the space X such that F_A is not a sheaf on X , find an example of the space Y such that F_A is a sheaf on Y .

We call F_A the *constant presheaf* (or the *constant sheaf*, if it is the case) on X with the stalk A .

Exercise 2.6.12. Similarly to Exercise 2.6.10 define the *sheaf of rational functions* on a variety X . Show that this sheaf is the constant sheaf with the stalk $A = k(X)$.

Stalks. The *stalk* F_P of a presheaf F at the point $P \in X = Spec\ A$ is an abelian group whose elements are equivalence classes of pairs (U,s) where U is an open neighbourhood of P, and $s \in F(U)$; the pairs (U,s) and (V,t) are equivalent iff there exists an open $W \subseteq U \cap V$ such that $\rho_{UW}(s) = \rho_{VW}(t)$.

Exercise 2.6.13. Show that the stalk of the constant presheaf F_A at any $P \in Spec\ A$ is equal to A.

Exercise 2.6.14. Show that the stalk of the sheaf O_X on a variety X at any $P \in X$ coincides with the local ring $O_{X,P}$.

Morphisms. A *morphism* φ of a presheaf F to a presheaf F' on X is a system of group homomorphisms $\varphi_U : F(U) \longrightarrow F'(U)$ such that for any pair $V \subseteq U$ one has

$$\varphi_V \circ \rho_{UV} = \rho'_{UV} \circ \varphi_U$$

where ρ and ρ' are the restriction maps of F and F', respectively.

If F and F' are sheaves, φ is called a *morphism of sheaves*. We call a morphism $\varphi : F \longrightarrow F'$ an *isomorphism* iff there exists a morphism $\varphi' : F' \longrightarrow F$ such that $\varphi \circ \varphi'$ and $\varphi' \circ \varphi$ are the identity morphisms of F' and F, respectively.

Structure sheaf on $Spec\ A$. Let now $X = Spec\ A$ be an affine scheme.

Let us recall the definition of a *localization* of A. Let S be a *multiplicative system* in A, i.e. a subset in A such that $1 \in S$ and if $s, t \in S$ then $s \cdot t \in S$. By $S^{-1}A$ we denote the ring whose elements are equivalence classes of quotients a/s where $a \in A$, $s \in S$, the quotients a/s and a'/s' being equivalent iff

$t\cdot(a\cdot s' - a'\cdot s) = 0$ for some $t \in S$. In particular if P is a prime ideal then $S_P = A - P$ is a multiplicative system and the corresponding localization $S_P^{-1}A$ is denoted by A_P ; if $S = \{1, f, f^2, \ldots, f^n, \ldots\}$ then $S^{-1}A$ is denoted by A_f .

Exercise 2.6.15. Show that:

a) if A is a domain then the above definitions of A_P coincide;

b) if f is a nilpotent element then $A_f = \{0\}$;

c) if A is a domain then $A \subseteq S^{-1}A$ for any S which does not contain 0.

Let now

$$\mathcal{O}(U) = \left\{ \begin{array}{l} \text{functions } s : U \longrightarrow \prod_{P \in U} A_P \text{ such that for} \\ \text{any } P \in U \text{ there exist an open neighbourhood} \\ V \ni P \text{ , } a \in A \text{ , and } f \in A \text{ such that} \\ s(Q) = a/f \text{ , } f \notin Q \text{ for any } Q \in V \end{array} \right\}$$

(this definition is similar to that of regular functions on a variety, but we consider functions with values in A_P rather than in the ground field). The restriction maps are defined in a natural way.

Exercise 2.6.16. Check that $\mathcal{O}$ is a sheaf and that its stalk $\mathcal{O}_P$ at any $P \in \mathit{Spec}\ A$ coincides with A_P . This sheaf is called the *structure sheaf* on $\mathit{Spec}\ A$.

Note that all the groups $\mathcal{O}(U)$ are rings and all the maps φ_{UV} are ring homomorphisms. A presheaf (respectively, a sheaf) possessing this property is called a *presheaf* (respectively, a *sheaf*) *of rings*.

Ringed spaces. A *ringed space* is a pair $(X,\mathcal{O}_X)$, X being a topological space and $\mathcal{O}_X$ being a sheaf of rings on X . A morphism from $(X,\mathcal{O}_X)$ to $(Y,\mathcal{O}_Y)$ is a pair $(f,\bar{f})$ where $f : X \longrightarrow Y$ is a continuous map, and $\bar{f}$ is a system of ring homomorphisms $\bar{f}_U : \mathcal{O}_Y(U) \longrightarrow \mathcal{O}_X(f^{-1}(U))$ for any open $U \subseteq Y$ such that $\rho_{UV}\circ\bar{f}_V = \bar{f}_U\circ\rho_{f^{-1}(U),f^{-1}(V)}$ for $V \subseteq U \subseteq Y$. A ringed space $(X,\mathcal{O}_X)$ is called a *locally ringed space* iff all the stalks $\mathcal{O}_{X,P}$, $P \in X$ are local rings.

Exercise 2.6.17. Show that a morphism of ringed space $(f,\bar{f}) : (X,\mathcal{O}_X) \longrightarrow (Y,\mathcal{O}_Y)$ induces a ring homomorphism of stalks $\bar{f}_P : \mathcal{O}_{Y,f(P)} \longrightarrow \mathcal{O}_{X,P}$ for any $P \in X$.

A morphism $(f,\bar{f}) : (X,\mathcal{O}_X) \longrightarrow (Y,\mathcal{O}_X)$ is called a *morphism of locally ringed spaces* iff for any $P \in X$ the homomorphism $\bar{f}_P : \mathcal{O}_{Y,f(P)} \longrightarrow \mathcal{O}_{X,P}$ is a *local homomorphism*, i.e. the inverse image of the maximal ideal in $\mathcal{O}_{X,P}$ is the maximal ideal in $\mathcal{O}_{Y,f(P)}$. An *isomorphism* of locally ringed spaces is defined in a natural way.

Exercise 2.6.18. Show that

a) $(Spec\ A\ ,\ \mathcal{O})$ is a locally ringed space for any A ;

b) a ring homomorphism $\varphi : A \longrightarrow B$ induces a morphism of locally ringed spaces

$$(Spec\ B\ ,\ \mathcal{O}) \longrightarrow (Spec\ A\ ,\ \mathcal{O})\ ;$$

conversely, any morphism of locally ringed spaces is induced by a ring homomorphism $\varphi : A \longrightarrow B$;

c) there exists a morphism of ringed spaces $(f,\bar{f}) : (Spec\ B\ ,\ \mathcal{O}) \longrightarrow (Spec\ A\ ,\ \mathcal{O})$ such that there is no ring homomorphism $\varphi : A \longrightarrow B$ which induces $(f,\bar{f})$.

(*Hint*: Let $A = R$ be a valuation ring and let $B = K$ be its fraction field).

Thus using the language of categories one says that the category of affine schemes is a full subcategory of the category of locally ringed spaces but is not that of the category of ringed spaces.

2.6.4. Schemes

Let (X, O_X) be a ringed space and let U be an open subset in X. The sheaf O_X determines the sheaf $O_{X|U}$ on U by

$$O_{X|U}(V) = O_X(V)$$

for an open $V \subseteq U$; $O_{X|U}$ is called the *restriction of* O_X *to* U.

We call a locally ringed space (X, O_X) a *scheme* iff for any $P \in X$ there exists its open neighbourhood U_P such that the locally ringed space $(U_P, O_{X|U_P})$ is isomorphic to an affine scheme $(Spec\ A\ , O)$. Sometimes we write X rather than (X, O_X) ; it means that the sheaf O_X is also given. A *morphism of schemes* is a morphism of locally ringed spaces.

Connection with varieties. In fact the notion of a scheme is a generalization of that of a variety. To explain this we need a definition which follows. Let S be a scheme and let $f_X : X \longrightarrow S$ be a morphism of schemes. We call X a *scheme over* S and write X/S. A *morphism* of X/S to Y/S is a morphism of schemes $\varphi : X \longrightarrow Y$ such that

$f_Y \circ \varphi = f_X$. If $S = Spec\ A$ then we write X/A and call X a *scheme over* A or an A*-scheme*.

Exercise 2.6.19. Let k be a field. Show that $Spec\ A$ is a k-scheme iff A is a k-algebra. A morphism $Spec\ B/k \longrightarrow Spec\ A/k$ corresponds to a k-algebra homomorphism $A \longrightarrow B$.

Let now k be an algebraically closed field, let X be a variety over k , and let $\mathcal{O}_X$ be the sheaf of regular functions on X . Denote by $t(X)$ the set of all the non-empty closed subsets of X ; note that $X \subseteq t(X)$. Define a topology on $t(X)$ as follows: F is closed in $t(X)$ iff $F = t(Y)$ for a closed $Y \subseteq X$.

Exercise 2.6.20. Check that this topology is well-defined and that the natural embedding $X \longrightarrow t(X)$ is continuous and induces a bijection of the set of open subsets of X onto that of $t(X)$.

Exercise 2.6.21. Let $\mathcal{O}_{t(X)}$ be the presheaf defined by $\mathcal{O}_{t(X)}(U) = \mathcal{O}_X(U \cap X)$ for an open $U \subseteq t(X)$. Show that $\mathcal{O}_{t(X)}$ is a sheaf and that $(t(X), \mathcal{O}_{t(X)})$ is a scheme.

Exercise 2.6.22. Show that the set of k-morphisms $t(X)/k \longrightarrow t(Y)/k$ can be identified with the set of regular maps $X \longrightarrow Y$.

Therefore t is a functor which identifies the category of varieties over k with a full subcategory of the category of k-schemes. We can identify a variety with its image $t(X)$ and speak about varieties as about schemes. A scheme of the form $t(X)$ is irreducible (as a topological space), reduced, and has two additional properties: it is a scheme of a *finite type* over k and it is separated over

k . The first property means that $t(X) = \bigcup_{i=1}^{M} Spec\ A_i$ where A_i is a finitely generated k-algebra. To explain the second property we need two more definitions which follow.

Product of schemes. Let $f_X : X \longrightarrow S$ and $f_Y : Y \longrightarrow S$ be schemes over a scheme S . We call a scheme Z the *product* of X and Y over S, and write $Z = X \times_S Y$ iff the following conditions hold:

a) there exist morphisms $p_X : Z \longrightarrow X$ and $p_Y : Z \longrightarrow Y$ such that $f_X \circ p_X = f_Y \circ p_Y$;

b) for any triple (Z', h_X, h_Y) where Z' is a scheme, $h_X : Z' \longrightarrow X$, and $h_Y : Z' \longrightarrow Y$ are morphisms such that $f_X \circ h_X = f_Y \circ h_Y$, there exists a unique morphism $h : Z' \longrightarrow Z$ such that $h_X = p_X \circ h$, and $h_Y = p_Y \circ h$.

Proposition 2.6.23. *For any X/S and Y/S there exists the product $X \times_S Y$.* ■

If $S = Spec\ A$ we write $X \times_A Y$ for $X \times_S Y$; in particular, if $S = Spec\ \mathbb{Z}$ we write just $X \times Y$ for $X \times_S Y$.

Exercise 2.6.24. Prove the existence of the product for affine schemes. Show that if $X = Spec\ A$ and $Y = Spec\ B$ are schemes over $S = Spec\ C$, i.e. A and B are C-algebras then $X \times_S Y = Spec(A \otimes_C B)$.

Exercise 2.6.25. Describe the scheme $Spec\ \mathbb{Q} \times Spec\ \mathbb{Q}$.

Closed embeddings. Let $f : X \longrightarrow Y$ be a morphism of schemes. It is called a *closed embedding* iff any $P \in Y$ has an open heighbourhood such that $f^{-1}(U)$ is an affine scheme and the map $\overline{f}_U : \mathcal{O}_Y(U) \longrightarrow \mathcal{O}_X(f^{-1}(U))$ is surjective. In this case Y is called a *closed subscheme* of X.

Exercise 2.6.26. Let $Y = Spec\ A$ be an affine scheme, and let $f : X \longrightarrow Y$ be a closed embedding. Show that $X = Spec\ B$ for some B and there exists a surjective ring homomorphism $\varphi : A \longrightarrow B$ which induces f.

Exercise 2.6.27. Show that for any X/S there exists a unique morphism (the *diagonal*) $\Delta_X : X \longrightarrow X \times_S X$ such that both $p_1 \circ \Delta_X$ and $p_2 \circ \Delta_X$ coincide with the identity map $X \longrightarrow X$, where p_1 and p_2 are canonical projections of $X \times_S X$ onto X.

A scheme X is called *separated* over S iff Δ_X is a closed embedding.

Exercise 2.6.28. Check that for any B-algebra A the scheme $X = Spec\ A$ is separated over $S = Spec\ B$.

Abstract varieties. Thus to any quasi-projective variety X there corresponds a separated scheme $t(X)$ of a finite type over k; moreover $t(X)$ is an *integral* scheme (i.e. it is irreducible and all its local rings $\mathcal{O}_P$ are domains). One can identify X with $t(X)$ and speak about X as about a scheme. An *abstract variety* is an integral separated scheme of a finite type over a field k. An abstract variety is not in general *quasi-projective*, i.e. is not isomorphic to the intersection of an open and a closed subschemes of $t(\mathbb{P}^n)$.

Exercise 2.6.29. Show that a scheme is of the form $t(X)$ for a quasi-projective variety X iff it is integral separated quasi-projective scheme of a finite type over k.

Let now X be a scheme, let P be its closed point and let $U = Spec\ A$ be an affine neighbourhood of P. We call the tangent space to U at P the *tangent space to* X *at* P ; we call P *non-singular* on X if it is non-singular on U.

Exercise 2.6.30. Show that the above definitions do not depend on the choice of U. (*Hint*: In fact they depend only on $O_{X,P}$).

If X is irreducible then we define $dim\ U$ for any non-empty affine open subscheme $U \subseteq X$.

Exercise 2.6.31. Check that $dim\ X$ is well-defined, i.e. that it does not depend on the choice of U.

Gluing schemes. It is important that one can "glue" a scheme using its open subschemes. To be precise, let $\{X_i\}$, $i \in I$, be a system of schemes, let for any $i, j \in I$, $i \neq j$ there be given an open subscheme $U_{ij} \subseteq X_i$ and let $\varphi_{ij} : U_{ij} \longrightarrow U_{ji}$ be an isomorphism for any i and $j \in I$. Let us assume that the following conditions hold:

a) the morphisms $\varphi_{ij} \circ \varphi_{ji}$ and $\varphi_{ji} \circ \varphi_{ij}$ are identical on U_{ji} and U_{ij}, respectively;

b) $\varphi_{ij}(U_{ij} \cap U_{i\ell}) = U_{ji} \cap U_{j\ell}$ and $\varphi_{i\ell}|_{U_{ij} \cap U_{i\ell}} = \varphi_{j\ell} \circ \varphi_{ij}|_{U_{ij} \cap U_{i\ell}}$ for any triple (i,j,ℓ).

Exercise 2.6.32. Show that under these assumptions there exist a unique scheme X and morphisms $\psi_i : X_i \longrightarrow X$ such that:

a) ψ_i is an isomorphism of X_i onto an open subscheme of X ;

b) $X = \bigcup_i \psi_i(X_i)$;

c) $\psi_i(U_{ij}) = \psi_i(X_i) \cap \psi_j(X_j)$;

d) $\psi_i|_{U_{ij}} = \psi_j \circ \varphi_{ij}|_{U_{ij}}$.

In this situation we say that X is obtained by *gluing* X_i *along* U_{ij} .

Exercise 2.6.33. Let k be a field, let $X_1 = X_2 = \mathbb{A}^1 = \mathit{Spec}\ k[T]$, let $U_{12} = U_{21} = \mathbb{A}^1 - \{0\} = \mathit{Spec}\ k[T, T^{-1}]$, and let φ_{12} and φ_{21} be the identity maps. Show that the gluing of X_1 and X_2 along U_{12} is not separated over k (and hence is not affine).

Exercise 2.6.34. Let A be a ring, let

$$X_1 = \mathit{Spec}\ A[T_1] \ , \ X_2 = \mathit{Spec}\ A[T_2] \ ,$$

let

$$U_{12} = \mathit{Spec}\ A[T_1, T_1^{-1}] \ , \ U_{21} = \mathit{Spec}\ A[T_2, T_2^{-1}]$$

and let $\varphi_{12} : U_{12} \longrightarrow U_{21}$ be induced by an A-algebra isomorphism $\varphi : A[T_2, T_2^{-1}] \longrightarrow A[T_1, T_1^{-1}]$ with $\varphi(T_2) = T_1^{-1}$. Show that for $A = k$ this scheme coincides with $t(\mathbb{P}^1)$.

The scheme $\mathbb{P}^1_A$ obtained by gluing X_1 with X_2 along U_{12} is called the *projective line over* A .

Exercise 2.6.35. Similarly to Exercise 2.6.34 construct $t(\mathbb{P}^n)$ gluing $(n + 1)$ copies of

$$t(\mathbb{A}^n) = \mathit{Spec}\ k[T_1, \ldots, T_n]$$

along suitable open subschemes.

Geometric points. Let $P \in X$ and $K = k(P)$. The point P defines a morphism $f_P : Spec\ K \longrightarrow X$. Indeed, let Q be the unique point of $Spec\ K$, and let $f(Q) = P$. Then let $\bar{f}_U(O(U)) = 0$ if $P \notin U$ and let $\bar{f}_U : A \longrightarrow K$ be the morphism of factorization (over P) for $P \in U = Spec\ A$.

A *geometric point* of X is defined as a morphism $f : Spec\ K \longrightarrow X$, K being a field; we call f also a *K-point* of X . The image of f is called the *centre of* f ; thus any geometric point is centered at a point of X .

Exercise 2.6.36. Let $X = Spec\ \mathbb{R}[T]$. Describe $\mathbb{R}$- and $\mathbb{C}$-points of X .

Fibres. Let $\psi : X \longrightarrow Y$ be a morphism, and let Z be a closed subscheme of Y . We call the scheme $X \times_Y Z$ the *inverse image* of Z .

Exercise 2.6.37. Show that $X \times_Y Z$ is a closed subscheme of X .

In particular, if $Z = P$ is a point of Y then $X_P = X \times_Y Z$ is called the *fibre* of ψ at P . More generally, if $f : Spec\ K \longrightarrow Y$ is a geometric point, we call the K-scheme $X \times_Y Spec\ K$ the *fibre* of ψ over f (or the *geometric fibre*). If all the geometric fibres of ψ are of dimension n then ψ is called a *morphism of relative dimension* n .

Line bundles and invertible sheaves. In the study of schemes an essential role is played by line bundles and invertible sheaves.

Exercise 2.6.38. Define a *line bundle* on a scheme similarly to the case of varieties (cf. the end of Section 2.1.1).

Let now F be a sheaf on a scheme X. The sheaf F is called an *invertible sheaf* iff for any $P \in X$ there exists its open neighbourhood U in X such that the sheaf $F|_U$ is isomorphic to $\mathcal{O}_U$, where $F|_U$ is the restriction of F to U, i.e. a sheaf on U such that $F|_U(V) = F(V)$ for $V \subseteq U$. Note that the notions of a line bundle and of an invertible sheaf are in fact equivalent. Indeed, let $\pi : E \longrightarrow X$ be a line bundle on X, let U be an open subset in X, and let

$$F_E(V) = \left\{ \begin{array}{l} \text{the set of sections } s : U \longrightarrow E \text{, i.e.} \\ \text{of morphisms such that } (\pi|_U) \circ s = id_U \end{array} \right\}$$

Exercise 2.6.39. Show that F_E is an invertible sheaf on X. Check that the rule $E \longmapsto F_E$ gives a canonical bijection of the set of isomorphism classes of line bundles on X onto that of invertible sheaves.

Proper morphisms. Let $F : X \longrightarrow S$ be a separated morphism of a finite type. We call f a *proper morphism* iff for any $S' \longrightarrow S$ the morphism $f' : X \times_S S' \longrightarrow S'$ induced by f is closed, i.e. the image of any closed subset is closed. The scheme X is called a *proper scheme* over S. If $S = Spec\, k$, X is an abstract variety over k, and X is proper over S then X is called a *complete variety*. Completeness in an analogue of projectivity.

Proposition 2.6.40. *Any projective variety is complete.*

■

Exercise 2.6.41. Show that any quasi-projective complete variety is projective.

It should be remarked that there exist complete non-projective varieties (of any dimension ≥ 2), but it is not easy to construct them. On the other hand:

Exercise 2.6.42. Show that any complete curve is projective. (*Hint*: Consider the case of a nonsingular curve, and then deduce the statement from this case, using the normalization of an abstract curve which can be constructed similarly to that of a quasi-projective curve).

Reduction of a curve. Now we can give a definition of a good reduction of a curve which is more "sophisticated" than that of Section 2.6.1. Let $\mathcal{O}$ be a valuation ring with the maximal ideal m and the fraction field K , and let X be a nonsingular projective curve over K .

Let now $\mathcal{X}$ be an integral scheme of relative dimension 1 over a scheme S . We call $\mathcal{X}$ *smooth over* S iff the image of $\mathcal{X}$ is dense in S and all the geometric fibres of $\mathcal{X}$ over S are non-singular. We say that X has a *good reduction modulo* m iff there exists a regular scheme $\mathcal{X}$, proper and smooth over $S = Spec\ \mathcal{O}$, whose geometric fibre over the point $Spec\ K \longrightarrow S$, defined by the natural embeding $\mathcal{O} \hookrightarrow K$, coincides with X .

It is not immediately clear that this "sophisticated" definition is better than the "naive" one of Section 2.6.1. Part 4 of this book gives some evidence of advantages of the sophisticated definition. Note also that the theory of schemes makes it possible to define a "genuine" reduction of a curve which has no good reduction, but we do not need it here.

2.6.5. Representable functors

The techinique of representable functors is a powerful tool for constructing schemes and varieties.

Let (Sch/S) be the category of schemes over a fixed scheme S. If $S = Spec\,\mathbb{Z}$ then (Sch/S) is the category of schemes which is denoted by (Sch). Let F be a *contravariant functor* from (Sch/S) to the category of sets $(Sets)$. Recall that it means that for any scheme X/S there is given a set $F(X/S)$ and for any morphism $f : X/S \longrightarrow Y/S$ of schemes over S there is given a map $F(f) : F(Y/S) \longrightarrow F(X/S)$ (note that $F(f)$ has the opposite direction) such that:

a) $F(id_{X/S}) = id_{F(X/S)}$, i.e. F sends the identity map to the identity map;

b) for morphisms $f : X/S \longrightarrow Y/S$ and $g : Y/S \longrightarrow Z/S$ of schemes over S one has $F(g \circ f) = F(f) \circ F(g)$.

Note that these conditions imply that F maps isomorphisms of schemes over S into bijections.

Let F and G be contravariant functors from (Sch/S) into $(Sets)$. A *morphism of functors* $\theta : F \longrightarrow G$ is a system of maps $\theta_{X/S} : F(X/S) \longrightarrow G(X/S)$ for any $X/S \in Obj(Sch/S)$ such that for any morphism $f : X/S \longrightarrow Y/S$ of schemes over S the diagram

$$\begin{array}{ccc} F(X/S) & \xrightarrow{\theta_{X/S}} & G(X/S) \\ \uparrow{\scriptstyle F(f)} & & \uparrow{\scriptstyle G(f)} \\ F(Y/S) & \xrightarrow{\theta_{Y/S}} & G(Y/S) \end{array}$$

commutes. In particular if $F = G$ and $\theta_{X/S}$ is the identity map for any X/S then θ is a morphism of functors which is called *identical*.

Exercise 2.6.43. Define the *composition* of functor morphisms and *isomorphisms* of functors. Show that for an isomorphism of functors θ all the maps $\theta_{X/S}$ are bijections.

Let now $X/S \in Obj(Sch/S)$. Note that X/S defines a contravariant functor $H_{X/S}$ from (Sch/S) to $(Sets)$. Indeed, let $H_{X/S}(Y/S) = Mor_{(Sch/S)}(Y/S,X/S)$ where $Mor_{(Sch/S)}(Y/S,X/S)$ is the set of morphisms in (Sch/S) from Y/S to X/S ; for $f : Y/S \longrightarrow Z/S$ the map $H_{X/S}(f)$ sends $g : Z/S \longrightarrow X/S$ to $g \circ f : Y/S \longrightarrow X/S$. A functor F which is isomorphic to a functor $H_{X/S}$ is called *representable*. In this case we say that X/S *represents* F .

Exercise 2.6.44. Show that the scheme X/S which represents a functor F is unique up to an isomorphism over S . (*Hint*: if X/S and X'/S both represent F then $Mor(X/S,X/S) = Mor(X/S,X'/S)$).

We are interested mostly in functors F for which the set $F(T/S)$ is the set of isomorphism classes of certain schemes over T (may be, together with some additional structures). Schemes which represent such functors are called *moduli schemes*; they play an essential role in Part 4.

Historical and bibliographic notes to Part 2

The material of this part is mostly classical. We do not try to expose the history of algebraic geometry in these brief notes. Our purpose is rather to sketch the general lines of its developement and to point out presice references for few modern results quoted in this chapter. One who is interested in the history of algebraic geometry should turn to the book [Die] by J.Dieudonné and to the historical sketch in the book [Shf] by I.R.Shafarevitch. The history of elliptic functions is described in Chapter VII of the book [Ab], which is written by C.Houzél. Many valuable information on the history of algebraic geometry can be find in the book [We 1] by A.Weil which is dedicated to number theory.

The origin of the interest to algebraic curves can be traced up to Diophatos who in his "Arithmetic" investigated what in modern terms is the structure of the set of $\mathbb{Q}$-rational points on various algebraic curves. All these curves were of genus 0 or 1; Diophantos worked out a series of original methods for solving such problems. It was not until the seventeenth century that the level achieved in

"Arithmetic" was surpassed in the works of P.Fermat, who investigated a series of problems raised by Diophantos. Fermat obtained a number of properties of elliptic curves (using the language of diophantine equations).

Further development of the subject proceeded in the framework of analysis and was concerned with the theory of elliptic integrals. In the eighteenth century in works of G.C.di Fagnano and especially of L.Euler the principal property of these integrals - the addition theorem - was discovered. In late 1820's N.-H.Abel and C.G.J.Jacobi gave the inversion of elliptic integrals thus creating the theory of elliptic functions. Moreover their study of integrals of the type $\int f(x,y)\cdot dy$, f being a rational function, for x and y related by a polynomial equation $P(x,y) = 0$ led directly to the theory of Jacobians.

The next important step was made in the middle of the nineteenth century by G.F.B.Riemann who proceeded also in the analytical framework. He introduced a new principle for investigations of functions of a complex variable which used the idea of Riemann surface. Riemann proved the inequality $\ell(D) \geq deg\ D - g + 1$; the proof of the Riemann-Roch theorem was completed by his pupil G.Roch.

An algebraic-geometric approach to Abel's and Riemann's results was proposed by R.F.A.Clebsch and was developed in full in the work of his pupil M.Noëther. Their investigations showed the algebraic-geometric nature of Abel's and Riemann's achievements; a rich theory of curves was developed. R.Dedekind and H.Weber proposed a purely algebraic approach to the theory of curves which was based on the consideration of function fields.

In the twenties of our century E.Artin, F.K.Smidt and H.Hasse began to study curves over finite fields. Hasse and his pupils developed a theory of curves over an arbitrary (non-closed) field. A theory of higher-dimensional varieties over an arbitrary fields was developed by B.L.van der Waerden and A.Weil. The latter proved the Riemann hypothesis for curves over finite fields; this proof used higher-dimensional varieties. J.-P.Serre gave a new definition of an algebraic variety based on the notion of a sheaf. A.Grothendieck made a deep reconstruction of algebraic geometry on the basis of schemes. Let us now give some references.

The theory of algebraic curves is contained in the book [Fu]. The analitic viewpoint (the theory of Riemann surfaces) is described in [Spr]. A purely algebraic approach to the theory of curves is presented in the books [Chev], [Cheb]. The book [Se 5] contains a simple proof of the Riemann-Roch theorem, the theory of singular curves and of Jacobians. The book [Shf] contains many facts about algebraic curves; in this part we have often followed it. The theory of elliptic curves and functions is contained in [Lan 1]. A standard textbook on schemes is [Hart]. Many properties of algebraic curves are described also in [Gr/Ha].

The references listed above contain almost all results of this chapter (and many others as well). Here is the list of exeptions with respective references.

Remark 2.2.14 is based on [Roq]. Proposition 2.2.48 and Corollary 2.2.49 were obtained by B.Schoeneberg [Scho]. The theory of the Cartier operator was developed in [Car]; see also Appendix B in [Lan 1].

Theorems 2.3.12 and 2.3.15 were obtained by A.Weil [We 2] and [We 3]. Theorems 2.3.16, 2.3.18, 2.3.19, Remark 2.3.23 and Theorem 2.3.25.a are due to J.-P.Serre [Se 1], [Se 2], [Se 3] and [Se 4]. Theorem 2.3.17 is due to M.Deuring [Deu]; see also [Wat]. A proof of Proposition

2.3.21 can be found in the last section of [We 4]. Theorem 2.3.22 was obtained by V.G.Drinfeld and S.G.Vladuţ [Drf/Vl]. Theorem 2.3.24 is due to Y.Ihara [Ih]; independently this fact (for $q = p^2$ and p^4) was discovered by S.G.Vladuţ and T.Zink [Ts/Vl/Z]. Theorem 2.3.25.b is due to T.Zink [Z]. Theorem 2.3.25.c is due to M.Perret [Pe 2]. Propositions 2.3.26 and 2.3.28 are borrowed from [Vl 1]. Exercise 2.3.27 is due to M.A.Tsfasman [Ts 5], [Ro/Ts]. Formula (2.4.9) is due to P.Deligne and M.Rapoport [De/Ra]. Theorem 2.4.30 is due to W.Waterhaus [Wat]. Theorem 2.4.31 is due to M.A.Tsfasman [Ts 2] and [Ts 3]; see also the papers [Scf] and [Vo] where essentially the same result is independently obtained. The lifting theorem used in the proof is due to M.Deuring [Deu].

Proposition 2.5.33 is due to S.G.Vladuţ [Ma/Vl].

PART 3

AG-CODES

This part is the central one. At last we come to algebraic-geometric constructions of codes. Besides the striking link between two disciplines so far from each other, algebraic-geometric codes (which we abbreviate as AG-codes) have very good parameters, if of course we choose the varieties they come from in a proper way. Their advantages become obvious when we consider asymptotic problems. The best studied AG-codes are those from algebraic curves. The main algebraic-geometric hero is a modular curve, in this part we only list its properties, leaving the detailed analysis for Part 4. Almost all coding corollaries are discussed in this part.

In Chapter 3.1 we give the constructions and study some properties of AG-codes. Chapter 3.2 presents some examples. In particular we discuss codes from curves of small genera; the case studied best of all is that of elliptic curves. Chapter 3.3 is totally devoted to the problem of decoding AG-codes. In Chapter 3.4 we collect the fruit, applying AG-codes to asymptotic problems. Algebraic-geometric codes ameliorate almost every lower bound for the main asymptotics: for linear and for non-linear codes (for q large enough), for polynomially constructable codes (for any q), for polynomially decodable codes, for self-dual codes, for constant-weight codes, etc. As usual we end the part with historical and bibliographic notes.

CHAPTER 3.1

CONSTRUCTIONS AND PROPERTIES

These exist several essentially equivalent ways to construct linear codes starting from algebraic curves (and also from varieties of higher dimensions). For curves, the codes we get can be rather well described: we can bound their parameters and weight spectra, we understand the duality.

In Section 3.1.1 we discuss four types of constructions of AG-codes and estimate the parameters. Then, in Section 3.1.2 we study the possibility to ameliorate the parameters, the automorphism group actions, and some constructions of the second level obtained by applying constructions of Section 1.2.3 to AG-codes. In Section 3.1.3 we study duality for AG-codes, including the question about self-dual AG-codes. Section 3.1.4 is devoted to estimates for the spectra of AG-codes.

All the curves considered in this chapter are projective, smooth, and absolutely irreducible over a finite field $\mathbb{F}_q$. Recall some notation: let X be such a curve then $K = \mathbb{F}_q(X)$ is the field of rational functions on X ; $\Omega(X)$ is the space of rational differential one-forms on X ; $X(\mathbb{F}_{q^r})$ is the set of $\mathbb{F}_{q^r}$-points of X , $N_r = |X(\mathbb{F}_{q^r})|$ is its cardinality, $Div(X)$ is the group of $\mathbb{F}_q$-divisors, $Div^+(X)$ is the semi-group of effective divisors, $Pic\ X$ is the divisor class group (or, equivalently, the group of isomorphism classes of line bundles on X), $J_X = Pic^0 X$ is the Jacobian of X (identified with the algebraic group of divisor classes of degree zero); if $D \in Div(X)$ then

$$L(D) = \{f \in \mathbb{F}_q(X)^* \mid (f) + D \geq 0\} \cup \{0\} ,$$

$$\Omega(D) = \{\omega \in \Omega(X)^* \mid (\omega) + D \geq 0\} \cup \{0\} .$$

3.1.1. AG-constructions

L-construction. Let X be a curve such that $X(\mathbb{F}_q) \neq \emptyset$; let $\mathcal{P} \subseteq X(\mathbb{F}_q)$, $|\mathcal{P}| = n$; $D \in Div(X)$. Let $Supp\ D \cap \mathcal{P} = \emptyset$.

Consider the map

$$Ev_{\mathcal{P}} : L(D) \longrightarrow \mathbb{F}_q^n ,$$

$$Ev_{\mathcal{P}} : f \longmapsto (f(P_1),\ldots,f(P_n)) ,$$

where $\mathcal{P} = \{P_1,\ldots,P_n\}$. We get a code $C = Ev_{\mathcal{P}}(L(D))$. We use the notation

$$C = (X,\mathcal{P},D)_L .$$

Suppose that D is chosen in such a way that any function $f \in L(D)$ has at most b zeroes at $\mathbb{F}_q$-points of the curve X . If $n > b$ then $Ev_{\mathcal{P}}$ is an embedding and

$$k = \ell(D) ,$$

$$d \geq n - b .$$

The Riemann-Roch theorem makes it possible to estimate the parameters of C .

Theorem 3.1.1. *Let X be a curve of genus g and let $0 \leq deg\ D = a < n = |\mathcal{P}|$. Then $C = (X, \mathcal{P}, D)_L$ is an $[n,k,d]_q$-code with*

$$k \geq a - g + 1 ,$$

$$d \geq n - a .$$

Proof: Let $D = D_1 - D_2$, $D_1 \geq 0$, $D_2 \geq 0$. A non-zero function $f \in L(D)$ has at most $a_1 = deg\ D_1$ poles and at least $a_2 = deg\ D_2$ zeroes in $Supp\ D$, since $D + (f)_0 - (f)_\infty \geq 0$. Hence the number of its zeroes out of $Supp\ D$ is at most $a_1 - a_2 = a$. We have supposed that $a < n$, therefore we get $Ev_{\mathcal{P}}(f) \neq 0$ for any function $f \neq 0$, i.e. $Ev_{\mathcal{P}}$ is any embedding. Moreover $Ev_{\mathcal{P}}(f)$ has at least $(n - a)$ non-zero coordinates, thus

$$d \geq n - a .$$

On the other hand $C = Ev_{\mathcal{P}}(L(D)) \simeq L(D)$, i.e.

$$k = \ell(D) \geq a - g + 1$$

according to the Riemann-Roch theorem (Theorem 2.2.17). ∎

Remark 3.1.2. The statement of Theorem 3.1.1 is valid for $0 \leq a < n$ but the first estimate is non-trivial only for $g - 1 < a < n$.

Remark 3.1.3. Later on we shall see that the above inequalities need not be equalities. Let $d_C = n - a$, $k_C = a - g + 1$; the parameters d_C and k_C are called respectively, the *designed distance* and the *designed dimension* of the AG-code C. By the Riemann-Roch theorem, if $n > a \geq 2g - 1$ then $k = k_C$. Using the Riemann-Roch theorem we can also write out the precise formula for the dimension of C for any a:

$$k = \ell(D) - \ell(D - \mathbf{P}) = a - g + 1 + \ell(K - D) - \ell(D - \mathbf{P}),$$

where $\mathbf{P} = \sum_{P_i \in \mathcal{P}} P_i$. In particular $k \geq n - g$ for $a \geq n$, since $\ell(D - \mathbf{P}) \leq deg(D - \mathbf{P}) + 1 = a - n + 1$ (see Exercise 2.1.53). (Note that in some papers on the subject the code $(X,\mathcal{P},D)_L$ is denoted by $C(G,D)$ where G is our D and D is our $\mathbf{P}$).

Example 3.1.4. Let $X = \mathbb{P}^1$, $D = a\cdot\infty$, then $L(D)$ is the space of polynomials of degree at most a. If for $\mathcal{P}$ we take all $\mathbb{F}_q$-points of $\mathbb{P}^1$ except ∞, i.e. $\mathcal{P} = \mathbb{F}_q$, then we get a $[q, a + 1, q - a]_q$-code which is a Reed-Solomon code (cf. Section 1.2.1).

Exercise 3.1.5. Let

$$D \sim D', \quad \mathcal{P} \subseteq X(\mathbb{F}_q) - (Supp\ D \cup Supp\ D').$$

How are the codes $C = (X,\mathcal{P},D)_L$ and $C' = (X,\mathcal{P},D')_L$ related?

Remark 3.1.6. Let X be a curve of genus g, $N = |X(\mathbb{F}_q)|$. If D_1 and D_2 are divisors such that $D_1 - D_2 \geq 0$ then $C_1 = (X,\mathcal{P},D_1)_L \supseteq C_2 = (X,\mathcal{P},D_2)_L$ for any $\mathcal{P} \subseteq X(\mathbb{F}_q) - Supp\ D_1$. Choose a point $P_0 \in X(\mathbb{F}_q)$ and let $D_a = a\cdot P_0$ and $\mathcal{P} \subseteq X(\mathbb{F}_q) - \{P_0\}$. For different values of a we get an embedded family of AG-codes $C_a = (X,\mathcal{P},D_a)_L$, $C_a \subseteq C_{a+1}$, $k_a + d_a \geq n - g + 1$. Note that $n \leq N - 1$.

Let us remark that the L-construction just described can be applied not only to a curve but also to any smooth projective variety X over $\mathbb{F}_q$, to any set $\mathcal{P}$ of its $\mathbb{F}_q$-points, and to any $\mathbb{F}_q$-divisor D on X. Contrary to the case of curves, in general it is difficult to give a satisfactory estimate for the parameters of the code $C = (X,\mathcal{P},D)_L$ although there are some particular cases when such an estimate can be given (cf. the end of Section 3.2.3). Unfortunately on this way there are still no new good families of codes found.

Ω-construction. We still suppose X to be a curve. Let $\mathcal{P} = \{P_1,\dots,P_n\}$ and $\mathbf{P} = P_1 + \dots + P_n \in Div(X)$.

Consider the space of differential forms

$$\Omega(\mathbf{P} - D) = \{\omega \in \Omega(X)^* \mid (\omega) + \mathbf{P} - D \geq 0\} \cup \{0\},$$

i.e. the space of forms having an appropriate zero multiplicities in the support of D and at most simple poles at the points P_i.

Recall that for any point P of X and for any non-zero form $\omega \in \Omega(X)$ in Section 2.2.2 there is defined the residue $Res_P(\omega)$. It is clear that for $P \in X(\mathbb{F}_q)$ and ω defined over $\mathbb{F}_q$ we have $Res_P(\omega) \in \mathbb{F}_q$.

The map

$$Res_{\mathcal{P}} : \Omega(\mathbf{P} - D) \longrightarrow \mathbb{F}_q^n$$

$$Res_{\mathcal{P}} : \omega \longmapsto (Res_{P_1}(\omega),\dots,Res_{P_n}(\omega))$$

defines a code $C = Res_{\mathcal{P}}(\Omega(\mathbf{P} - D))$. We write

$$C = (X,\mathcal{P},D)_\Omega .$$

(In some papers on the subject the code $(X,\mathcal{P},D)_\Omega$ is denoted by $C^*(G,D)$, where G is our D and D is our $\mathbf{P}$).

Let us estimate the parameters.

Theorem 3.1.7. *Let X be a curve of genus g, $2g-2<a$; and let $\mathcal{P}\cap \mathrm{Supp}\, D=\emptyset$. Then the code $C=(X,\mathcal{P},D)_\Omega$ is an $[n,k,d]_q$-code, where*

$$k \geq n - a + g - 1 \quad ,$$

$$d \geq a - 2g + 2 \quad .$$

Proof: Let K be a canonical divisor (i.e. K belongs to the canonical class), $\deg K = 2g-2$. Then

$$\Omega(\mathbf{P} - D) \simeq L(K + \mathbf{P} - D) \quad ,$$

$$\dim \Omega(\mathbf{P} - D) = \ell(K + \mathbf{P} - D) \geq$$

$$\geq (2g - 2 + n - a) - g + 1 = n - a + g - 1 \ ,$$

since

$$\deg\,(K + \mathbf{P} - D) = 2g - 2 + n - a \ .$$

For any differential form ω

$$K \sim (\omega) = (\omega)_0 - (\omega)_\infty \quad ,$$

i.e.

$$2g - 2 = \deg K = \deg\,(\omega)_0 - \deg\,(\omega)_\infty \quad .$$

Let $D = D_1 - D_2$, $D_i \geq 0$, $\mathrm{Supp}\, D_1 \cap \mathrm{Supp}\, D_2 = \emptyset$. If $\omega \in \Omega(\mathbf{P} - D)$ then $(\omega)_0 \geq D_1$, hence

$$\deg(\omega)_\infty = \deg(\omega)_0 - 2g + 2 \geq a - 2g + 2 + \deg D_2 \ ,$$

i.e. the form ω has at least $(a - 2g + 2)$ poles out of $\mathrm{Supp}\, D_2$. Since $\mathcal{P} \cap \mathrm{Supp}\, D = \emptyset$, all these poles are at points $P_i \in \mathcal{P}$, the poles are of order 1, and $Res_{P_i} \omega \neq 0$ iff $P_i \in Supp(\omega)_\infty$. We have supposed that $a > 2g - 2$, therefore $Res_{\mathcal{P}}(\omega) \neq 0$ for any $\omega \neq 0$, i.e. $Res_{\mathcal{P}}$ is an embedding. Moreover, the number of non-zero coordinates is at least $a - 2g + 2$. The dimension $k = dim\,\Omega(\mathbf{P} - D)$ is already estimated above. ∎

Remark 3.1.8. The estimate for k is non-trivial only for $a \leq n + g - 1$. As above we call $d_C = a - 2g + 2$ and $k_C = n - a + g - 1$, respectively the *designed distance* and the *designed dimension* of the AG-code $C = (X,\mathcal{P},D)_\Omega$. For $2g - 2 < a < n$ we have $k = k_C$. The precise value of k for any a is

$$k = \ell(K+\mathbf{P}-D) - \ell(K-D) = n - a + g - 1 - \ell(K-D) + \ell(D-\mathbf{P}).$$

In particular, if $a \leq 2g - 2$ then $k \geq n - g$, since $\ell(K - D) \leq deg(K - D) + 1 = 2g - 1 - a$.

Remark 3.1.9. The designed parameters both for L- and Ω-constructions satisfy

$$d_C + k_C = n - g + 1 .$$

Both constructions have an essential defect, we have to choose points $P_i \notin Supp\, D$. There are several ways out of this problem.

Continuation to the points of $Supp\, D$. For codes $C = (X,\mathcal{P},D)_L$ there exists an elementary construction of lengthening which makes it possible to dispense with the condition $Supp\, D \cap \mathcal{P} = \emptyset$. Let X be a curve, $\mathcal{P} \subseteq X(\mathbb{F}_q)$, $|\mathcal{P}| = n$, $D \in Div(X)$, and let $D = D' + D''$, where $Supp\, D' \cap \mathcal{P} = \emptyset$, $Supp\, D'' \subseteq \mathcal{P}$. For any point $Q_i \in Supp\, D''$

choose a local parameter t_i . If $D'' = \sum_{i=1}^{s} b_i \cdot Q_i$ then for any $f \in L(D)$ the function $t_i^{b_i} \cdot f$ is regular at Q_i . Consider the map

$$Ev'_{\mathcal{P}} : L(D) \longrightarrow \mathbb{F}_q^n ,$$

$$Ev'_{\mathcal{P}} : f \longmapsto (f(P_1),\ldots,f(P_r),t_1^{b_1}\cdot f(Q_1),\ldots,t_s^{b_s}\cdot f(Q_s)) ,$$

where $\{P_1,\ldots,P_r\} = \mathcal{P} - Supp\ D''$. The code $C' = Ev'(L(D))$, which we denote by $(X,\mathcal{P},D)'$, is a lengthening of $(X,\mathcal{P}',D)_L$ (where $\mathcal{P}' = \mathcal{P} - Supp\ D''$) by s positions corresponding to the points of $Supp\ D'' = Supp\ D \cap \mathcal{P}$. The parameters of C' also satisfy the estimates of Theorem 3.1.1.

In fact the described construction of lengthening can be changed for the following much more conceptual one, which is valid for any variety X .

H-construction. Let X be a smooth projective variety over $\mathbb{F}_q$, $\mathcal{L}$ a line bundle on X defined over $\mathbb{F}_q$, and $H^0(\mathcal{L})$ the space of its sections. Let $\mathcal{P} = \{P_1,\ldots,P_n\} \subseteq X(\mathbb{F}_q)$. It is impossible to map $H^0(\mathcal{L})$ to $\mathbb{F}_q^n$ by evaluation at points (as it has been done in the L-construction) since the value of a section at a point is not well defined. But the vanishing of a section at a point is well defined, which is quite near to what we want. There is a natural map (cf. Section 2.1.3)

$$H^0(\mathcal{L}) \longrightarrow \bigoplus_{i=1}^{n} \overline{\mathcal{L}}_{P_i} ,$$

where $\overline{\mathcal{L}}_{P_i}$ is the fibre of $\mathcal{L}$ at P_i , i.e. a one-dimensional vector space over $\mathbb{F}_q$.

Fix an arbitrary *trivialization* at the fibres $\overline{\mathcal{L}}_{P_i}$, i.e. an isomorphism $\overline{\mathcal{L}}_{P_i} \simeq \mathbb{F}_q$ (which is of course equivalent to a choice of a non-zero vector in each $\overline{\mathcal{L}}_{P_i}$). We obtain the map

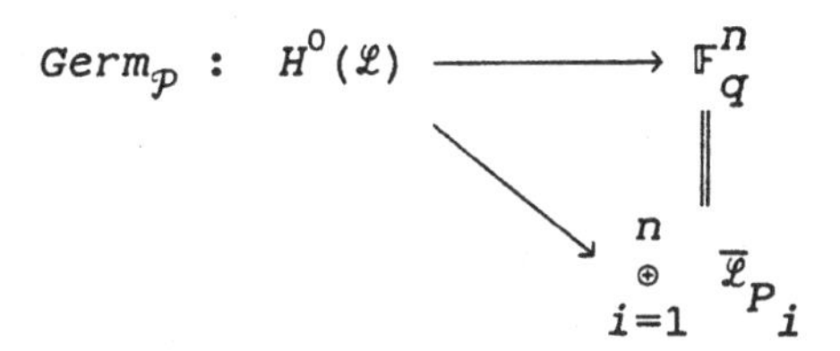

and consider the code $C = Germ_{\mathcal{P}}(H^0(\mathcal{L}))$ which we denote $C = (X,\mathcal{P},\mathcal{L})_H$.

Just as above good estimates for the parameters can be given in the case of curves.

Theorem 3.1.10. *Let* X *be a curve of genus* g, *let* $n = |\mathcal{P}|$, *and* $a = \deg \mathcal{L}$, $0 \le a < n$. *Then the code* $C = (X,\mathcal{P},\mathcal{L})_H$ *is an* $[n,k,d]_q$*-code, where*

$$k \ge a - g + 1 \quad ,$$

$$d \ge n - a \quad .$$

Proof: For any non-zero section $s \in H^0(\mathcal{L})$ the divisor D of its zeroes belongs to the divisor class corresponding to the bundle $\mathcal{L}$ (cf. Section 2.1.3). Therefore the total number of zeroes of s equals a (if counted with proper multiplicities), and the number of zeroes belonging to $\mathcal{P}$ is at most a. Thus $d \ge n - a$. If $a < n$ then the map $Germ_{\mathcal{P}}$ is an embedding and $k = h^0(\mathcal{L}) \ge a - g + 1$ because of the Riemann-Roch theorem. ∎

Corollary 3.1.11. *Let* X *be a curve of genus* g *over* $\mathbb{F}_q$ *, and let* $N = |X(\mathbb{F}_q)| > g - 1$ *. Then for any* $n = g+1,\dots,N$ *and for any* $k = 1,\dots,n-g$ *there exists a linear* $[n,k,d]_q$*-code whose parameters satisfy*

$$k + d = n - g + 1 \ .$$

Proof: Choose a set $\mathcal{P} = \{P_1,\dots,P_n\} \subseteq X(\mathbb{F}_q)$ and a bundle $\mathcal{L} \in Pic\, X$ of degree $a = k + g - 1$, $g - 1 < a < n$ (on X there exist $\mathbb{F}_q$-bundles of any given degree since $X(\mathbb{F}_q)$ is non-empty and we can take a divisor class which is a multiplicity of an $\mathbb{F}_q$-point). By Theorem 3.1.10 the parameters of $C' = (X,\mathcal{P},\mathcal{L})_H$ satisfy

$$k' \geq a - g + 1 = k \ ,$$

$$d' \geq n - a \ .$$

The spoiling lemma (Lemma 1.1.34) shows that then we can construct an $[n,k,d]_q$-code C with

$$d = n - a \ ,$$

i.e.

$$k + d = n - g + 1 \ . \qquad \blacksquare$$

Remark 3.1.12. Strictly speaking the code $C = (X,\mathcal{P},\mathcal{L})_H$ depends on the choice of trivializations, and also (just as the codes $(X,\mathcal{P},D)_L$ and $(X,\mathcal{P},D)_\Omega$) on the ordering of $\mathcal{P}$. However any other choice of trivialization corresponds to the multiplication of i-th basis vector of $\mathbb{F}_q^n$ by a non-zero constant $y_i \in \mathbb{F}_q^*$, and any other choice of ordering of $\mathcal{P}$ to some permutation of the coordinates, i.e. both choices lead to an equivalent code (see Section

1.1.1). Thus the equivalence class of the code $(X,\mathcal{P},\mathcal{L})_H$ is well defined.

Remark 3.1.13. Let D be a divisor corresponding to the bundle $\mathcal{L}$, i.e. D is the divisor of zeroes of some section $s_0 \in H^0(\mathcal{L})$. The section s_0 uniquely defines the following trivialization of $\overline{\mathcal{L}}_{P_i}$ for any $P_i \notin Supp\, D$: any $t \in \overline{\mathcal{L}}_{P_i}$ is mapped to an element $x \in \mathbb{F}_q$ such that $t = x \cdot s_0(P_i)$. Thus D defines trivializations of the fibres $\overline{\mathcal{L}}_{P_i}$ up to a mutual non-zero constant (from $\mathbb{F}_q^*$).

Exercise 3.1.14. Using the last remark prove that if $\mathcal{P} \cap Supp\, D = \varnothing$ then this trivialization yields a code $(X,\mathcal{P},\mathcal{L})_H$ coinciding with $(X,\mathcal{P},D)_L$. What is the relation between the codes $(X,\mathcal{P},\mathcal{L})_H$ and $(X,\mathcal{P},D)'$ for $Supp\, D \cap \mathcal{P} \neq \varnothing$?

Now we are going to translate the *H*-construction into another language.

P-construction. Recall (cf. Section 1.1.2) that the equivalence class of a non-degenerate linear $[n,k,d]_q$-code C is uniquely determined by a projective $[n,k,d]_q$-system $\mathcal{P}$.

If a variety X is given together with its projective embedding $X \subset \mathbb{P}^m$ then any choice of $\mathcal{P} \subseteq X(\mathbb{F}_q)$, such that $|\mathcal{P}| > m$ and $\mathcal{P}$ does not lie in a hyperplane, yields a projective $[n,k,d]_q$-system with $n = |\mathcal{P}|$, $k = m + 1$, $d = n - max\,\{|H \cap \mathcal{P}|\}$, the maximum being taken over all $\mathbb{F}_q$-hyperplanes $H \subset \mathbb{P}^m$.

For a curve X (given with an embedding) its degree $deg\, X$ is defined as the number of $\overline{\mathbb{F}}_q$-points (counted with proper multiplicities) in its intersection with an arbitrary hyperplane (cf. Section 2.1.3). In any case we have $\max_H \{|H \cap \mathcal{P}|\} \le \deg X$, so we have proved the following

Proposition 3.1.15. *Let* $X \subset \mathbb{P}^m$ *be a curve and let* $N = |X(\mathbb{F}_q)|$. *For any* n *such that* $N \ge n > \max\{m, \deg X\}$ *there exists a non-degenerate linear* $[n,k,d]_q$*-code with* $k = m + 1$ *and* $d \ge n - \deg X$. ∎

Let X be a variety and $\mathcal{L}$ a line bundle (a divisor class) on it. The bundle $\mathcal{L}$ corresponds to a map

$$\varphi_{\mathcal{L}} : X \longrightarrow \mathbb{P}(H^0(\mathcal{L})) \simeq \mathbb{P}^{h^0(\mathcal{L})-1}$$

(cf. Section 2.1.3). For a projective $[n,k,d]_q$-system let us take some subset $\mathcal{P} \subseteq X(\mathbb{F}_q)$ mapped to $\mathbb{P}(H^0(\mathcal{L}))$ (if some points are glued together, they are counted with the corresponding multiplicities); here $n = |\mathcal{P}|$, $k = h^0(\mathcal{L})$. According to Section 2.1.3, inverse images of hyperplane sections of $\mathbb{P}(H^0(\mathcal{L}))$ are effective divisors D belonging to the class $\mathcal{L}$. Therefore $d = n - \max_D \{|D \cap \mathcal{P}|\}$ (with the obvious definition of $|D \cap \mathcal{P}|$).

Consider now the case of curves. If X is a curve of genus g and $a = \deg \mathcal{L}$ then $k = h^0(\mathcal{L}) \ge a - g + 1$. Moreover $|D(\mathbb{F}_q)| \le \deg D = a$, i.e. $d \ge n - a$. So we have constructed a linear $[n,k,d]_q$-code $C = (X, \mathcal{P}, \mathcal{L})_P$.

Exercise 3.1.16. Prove that the projective systems $(X, \mathcal{P}, \mathcal{L})_H$ and $(X, \mathcal{P}, \mathcal{L})_P$ are isomorphic, i.e. that the corresponding codes are equivalent.

Remark 3.1.17. If $a \ge 2g + 1$ then according to Corollary 2.2.29 the map $\varphi_{\mathcal{L}}$ is an injection and we get a projective system without multiplicities.

Remark 3.1.18. There is an interesting question: how large is the class of AG-codes? Let C be an arbitrary non-degenerate linear $[n,k,d]_q$-code. It corresponds to a projective system $\mathcal{P}$, i.e. to a set of points in $\mathbb{P}^{k-1}$ with multiplicities. If there are no multiplicities then it is possible to prove that there exists a smooth curve X passing through all these points and through no other $\mathbb{F}_q$-point of $\mathbb{P}^{k-1}$. So, in a sense, any projective system without multiplicities is algebraic-geometric. (The same is also true for systems with multiplicities if we consider maps $X \longrightarrow \mathbb{P}^{k-1}$ with singular images instead of smooth embeddings $X \hookrightarrow \mathbb{P}^{k-1}$). However the embedding $X \hookrightarrow \mathbb{P}^{k-1}$ in general is not given by a complete linear system (i.e. by a line bundle $\mathcal{L}$). We come to the following

Problem 3.1.19. Characterize those linear codes (projective systems) which can be obtained by the H-construction, i.e. by complete linear systems on curves.

3.1.2. Some additional remarks

Parameters. It is fruitful to remember that the parameters of algebraic-geometric codes are not found but only estimated. Some AG-codes have better parameters. To be more precise, for $deg\ D = a < n$ the L-construction gives $k = a - g + 1 + \ell(K - D)$ where $\ell(K - D) = 0$ for $a \geq 2g - 1$, and $0 \leq \ell(K - D) \leq 2g - 1 - a$ for $a \geq 2g - 1$. The problem of finding $\ell(K - D)$ is rather complicated.

On the other hand a function $f \in L(D)$ has at most a poles, but it can however happen that for any $f \in L(D)$ some of its poles are situated out of the set of $\mathbb{F}_q$-points.

This remark is analyzed minutely in Section 3.4.2 below.

It can therefore happen that $k > k_C$ or $d > d_C$ and there are cases when it is possible to prove one of these inequalities (or both). We start considering the simplest case when it is possible to ameliorate the estimate of k for L- and H-constructions. As we have already seen such an amelioration (for L-construction) is possible only for $a = deg\, D < 2g - 1$ and it does take place whenever $\ell(K - D) > 0$, i.e. when D is a special divisor. Unfortunately the information on $\mathbb{F}_q$-rational special divisors we dispose of is rather scarce. However there are still some results in this direction. We start with a bound on how strong such an amelioration can be.

Proposition 3.1.20. *Let D be a special divisor of degree a on a curve X, $\mathcal{L} = \mathcal{O}(D)$, and let $C = (X, \mathcal{P}, \mathcal{L})_H$ be the corresponding $[n,k,d]_q$-code. Then $2k + d_C \le n + 2$.*

Proof: The Clifford theorem (Theorem 2.2.42) gives $k \le \frac{a}{2} + 1$. Since $d_C = n - a$, we are done. ■

Let us expose a case when this bound is attained. Let X be a hyperelliptic curve, $D = m \cdot F$, F being a hyperelliptic divisor (i.e. $F = \pi^*(P)$, where $\pi : X \longrightarrow \mathbb{P}^1$ is a projection of degree 2, and P is an arbitrary $\mathbb{F}_q$-point of $\mathbb{P}^1$), then $a = deg\, D = 2m$, $\ell(D) = m + 1$, and for any $\mathcal{P} \subseteq X(\mathbb{F}_q)$ the code $C = (X, \mathcal{P}, \mathcal{O}(D))_H$ satisfies $2k + d_C = n + 2$. Unfortunately, codes on hyperelliptic curves cannot be longer than $2(q + 1)$.

Remark 3.1.21. The second part of the Clifford theorem (cf. Remark 2.2.43) shows that for an $[n,k,d]_q$-code $C = (X, \mathcal{P}, \mathcal{L})_H$, where $1 \le deg\, \mathcal{L} \le 2g - 3$, the equality $2k + d_C = n + 2$ can take place for hyperelliptic curves only.

The obtained estimate (Proposition 3.1.20) does not minorate the bounds of Section 1.1.4 . Thus the bounds for parameters of codes lead to some bounds for parameters of special $\mathbb{F}_q$-rational divisors on curves. For the most part of these bounds we do not know how to obtain them by purely algebraic-geometric means. Here is the simplest example: the Plotkin bound applied to AG-codes can be written as

$$n - a = d_C \le d \le \frac{n\cdot q^{k-1}\cdot(q - 1)}{q^k - 1} .$$

Thus

$$n \le a\cdot\frac{q^k - 1}{q^{k-1} - 1}$$

for any $\mathbb{F}_q$-rational divisor D of degree a on X , where $k = \ell(D)$, $n = X(\mathbb{F}_q)$. For $k = 2$ we obtain the simplest bound $n \le a\cdot(q + 1)$ which can be easily proved by geometric means.

Exercise 3.1.22. Write out the relations between $n = |X(\mathbb{F}_q)|$, $deg\ D$, and $\ell(D)$ corresponding to other upper bounds for codes.

An important feature of the obtained upper bound $2k + d_C \le n + 2$ is that it does not prevent AG-codes to be very good. In particular, it might even happen that AG-codes from curves over $\mathbb{F}_2$ lie over the asymptotic Gilbert-Varshamov bound.

Exercise 3.1.23. Write out the relations between $|X(\mathbb{F}_2)|$, $deg\ D$, and $\ell(D)$ which yield the parameters of $C = (X, X(\mathbb{F}_q), \mathcal{O}(D))_H$ to be over the Gilbert-Varshamov bound for $q = 2$.

Our present knowledge of special divisors is, alas, too poor to give any idea how to construct divisors satisfying this relation (and whether they do exist or not).

One of the few classes of special divisors which we know a bit better are multiplicities of Weierstrass points (cf. Section 2.2.3). Let $P \in X(\mathbb{F}_q)$ be a Weierstrass point on a curve of genus $g \geq 2$ and let $(a_1, \ldots, a_{g-1})$ be the sequence of its non-gaps ($2 \leq a_1 < \ldots < a_{g-1} \leq 2g - 1$) i.e. $\ell(a_i \cdot P) > \ell((a_i - 1)\cdot P)$ for any $i = 1, \ldots, g - 1$. Then $\ell(a_i) = i + 1$ and choosing $D = a_i \cdot P$ we get a code $C = (X, \mathcal{P}, \mathcal{O}(D))$ with $k = \ell(D) = i + 1$, $d_C = n - a_i$. The obtained value of k is better than k_C for any i satisfying $i + 1 > a_i - g + 1$. Since P is a Weierstrass point this inequality holds at least for one i. If P is a hyperelliptic point we get the example discussed after Proposition 3.1.21.

Weierstrass points and some more subtle technique using the Cartier operator sometimes lead to sharper estimates for the parameters of AG-codes obtained by the Ω-construction. Let $C = (X, \mathcal{P}, D)_\Omega$ and $D = \sum_{i=1}^{m} a_i \cdot P_i$ where P_i are prime divisors, i.e. points of X rational over some extension of the ground field $\mathbb{F}_q$ (cf. Section 2.3.1). Let $a_i = \beta_i \cdot q + r_i$ with $\beta_i, r_i \in \mathbb{Z}$, $0 \leq r_i \leq q - 1$ and let $(s_{i1} \leq \ldots \leq s_{ig})$ be the sequence of gaps at the point P_i. Let

$$w_i = \begin{cases} \sup\{h \mid s_{ih} \leq \beta_i\} & \text{if } \beta_i > 0, \\ 0 & \text{if } \beta_i = 0, \end{cases}$$

$$\varepsilon_i = \begin{cases} 1 & \text{if } a_i \equiv -1 \ (\mathrm{mod}\ q) \text{ and } \\ & (a_i+1)/q \text{ is not a gap at } P_i, \\ 0 & \text{else}. \end{cases}$$

As usual,

$$a = deg\ D = \sum a_i \cdot deg\ P_i\ , \quad n = |\mathcal{P}|\ .$$

Proposition 3.1.24. *For* $n \le a \le n + 2g - 2$ *the parameters of the code* $C = (X,\mathcal{P},D)_\Omega$ *satisfy the following inequalities*

$$k \ge n - \sum_{i=1}^{m} (a_i - \beta_i + w_i) \cdot deg\ P_i + g - 1\ ,$$

$$d \ge a + \sum_{i=1}^{m} \varepsilon_i \cdot deg\ P_i - 2g + 2\ . \blacksquare$$

We cut the proof of the proposition into a series of exercises.

Exercise 3.1.25. Prove that

$$k \ge n + g - 1 - a - \beta - w$$

whenever $D = a \cdot Q$, $Q \in X(\mathbb{F}_q)$. (*Hint*: Consider the filtration

$$\Omega(P) \supseteq \Omega(\mathbf{P} - Q) \supseteq \ldots \supseteq \Omega(\mathbf{P} - a \cdot Q)$$

Then use Exercise 2.2.52 and the equality $\Omega(-a \cdot Q) = \Omega(-(a + 1) \cdot Q)$ whenever $(a + 1)$ is a non-gap at Q to prove that in the filtration there are at least $\beta - w$ equalities).

Exercise 3.1.26. Prove that in the above situation $d \ge a + 1 - 2g + 2$. (*Hint*: If $(a + 1)/q$ is a non-gap at Q then $\Omega(\mathbf{P} - a \cdot Q) = \Omega(\mathbf{P} - (a + 1) \cdot Q)$).

Exercise 3.1.27. Investigate the case $D = a \cdot \sum_{i=1}^{m} \sigma^i(Q)$, where Q is a point of X of degree m , and σ is a generator of $Gal(\mathbb{F}_{q^m}/\mathbb{F}_q)$. (*Hint*: Consider the filtration

$$\Omega(P) \geq \Omega(P - R) \geq \Omega(P - 2\cdot R) \geq \Omega(P - a\cdot R) ,$$

where $R = \sum_{i=1}^{m} \sigma^i(Q)$).

Exercise 3.1.28. Prove the proposition in the general case. (*Hint*: Consider the filtration

$$\Omega(P) \geq \Omega(P - R_1) \geq \ldots \geq \Omega(P - a_1\cdot R_1) \geq$$

$$\geq \Omega(P - a_1\cdot R_1 - R_2) \geq \ldots \geq \Omega(P - D + R_m) \geq \Omega(P - D) ,$$

where $R_i = \sum_{j=1}^{b_i} \sigma_i^j(P_i)$ for a point P_i of degree b_i , and σ_i is a generation of $Gal(\mathbb{F}_{q_i}/\mathbb{F}_q)$, $q_i = q^{b_i}$).

The amelioration of parameters given by Proposition 3.1.24 is also most significant in the case of hyperelliptic points.

Exercise 3.1.29. Let X be a hyperelliptic curve such that no hyperelliptic point of it lie in $X(\mathbb{F}_q)$, and let $D = \sum_{i=1}^{2g+2} a_i\cdot Q_i$ be an $\mathbb{F}_q$-divisor. Write out the estimates for k and d given by Proposition 3.1.24.

Automorphisms. Let $G \leq Aut_{\mathbb{F}_q}(X)$ be a subgroup in the group of $\mathbb{F}_q$-automorphisms of X . In general there is no G-action on the AG-code $(X,\mathcal{P},D)_L$. However if the set $\mathcal{P}$ and the divisor D are G-invariant then G acts naturally on $(X,\mathcal{P},D)_L$. Indeed in this case for any $g \in G$ and any $f \in L(D)$ the function $g^*(f)$ lies in $L(D)$ since $gD = D$.

Therefore a code vector $(f(P_1),\ldots,f(P_n))$ is mapped to $(g^*(f)(P_1),\ldots,g^*(f)(P_n)) = (f(g(P_1)),\ldots,f(g(P_n)))$, which is also a code vector. Thus G is naturally mapped to $Aut(C) \cap S_n$ and, according to the properties of group codes (cf. Section 1.2.2), we have $C \subseteq \mathbb{F}_q[G/H_1]\oplus\ldots\oplus\mathbb{F}_q[G/H_m]$ where H_i is the stabilizer of a point $Q_i \in \mathcal{P}$. We obtain

Proposition 3.1.30. *Let* $G \subseteq Aut_{\mathbb{F}_q}(X)$, *let* $\mathcal{P}$ *be a* G*-invariant subset of* $X(\mathbb{F}_q)$, *and* D *be a* G*-invariant* $\mathbb{F}_q$*-divisor on* X, $Supp\, D \cap \mathcal{P} = \varnothing$. *Then* $C = (X,\mathcal{P},D)_L$ *is a group code:* $C \subseteq \mathbb{F}_q[G/H_1]\oplus\ldots\oplus\mathbb{F}_q[G/H_m]$, *where* H_i *is a stabilizer of* $Q_i \in \mathcal{P}$, $\{Q_1,\ldots,Q_m\}$ *being the set of orbit representatives of the action of* G *on* $\mathcal{P}$. ■

Later on, in Part 4, we shall show that such codes are sometimes obtained from modular curves.

Example 3.1.31. Let

$$X = \mathbb{P}^1 ,\ D = a\cdot\infty ,\ \mathcal{P} = X(\mathbb{F}_q) - \{\infty\} ,\ Aut(X) = PGL(2,\mathbb{F}_q) .$$

The subgroup G of affine automorphisms $x \longmapsto a\cdot x + b$, $a \in \mathbb{F}_q^*$, $b \in \mathbb{F}_q$, stabilizes $\mathcal{P}$ and D. Hence G acts on the Reed-Solomon code $C = (X,\mathcal{P},D)_L$; moreover $C \subseteq \mathbb{F}_q[G/H]$ since G acts transitively on $\mathcal{P}$, H being the stabilizer of $P_1 = 0 \in X(\mathbb{F}_q)$, $H = \{x \longmapsto a\cdot x \mid a \in \mathbb{F}_q^*\} \simeq \mathbb{F}_q^*$. Let us remark that $G/H \simeq \mathbb{F}_q$.

The example $X = \mathbb{P}^1$, $D = a\cdot\infty$, $\mathcal{P} = X(\mathbb{F}_q) - \{0,\infty\}$ is quite similar.

Concatenation. Let us apply concatenation (cf. Section 1.2.3) to AG-codes we have just constructed.

Recall the definition $N_q(g) = max\ |X(\mathbb{F}_q)|$, the maximum being taken over all curves over $\mathbb{F}_q$ of genus g. It is known that $N_q(g) \le q + 1 + \lceil 2\sqrt{q} \rceil \cdot g$ (cf. Section 2.3.2, some other estimates for $N_q(g)$ are also given there).

Corollary 3.1.11 tells us that there exist $[N,K,D]_q$-codes with $N \le N_q(g)$ and $K + D = N - g + 1$ for any sensible K (i.e. $K \ge 0$ and $D = N - g + 1 - K \ge 1$).

Proposition 3.1.32. *For any linear* $[n,k,d]_q$*-code, any* $g \ge 0$ *and any* K *such that* $0 \le K \le N_{q^k}(g) - g$ *there exists a linear code with parameters*

$$[n \cdot N_{q^k}(g)\ ,\ k \cdot K\ ,\ d \cdot (N_{q^k}(g) - K - g + 1)]_q\ .$$

■

Exercise 3.1.33. Prove the proposition. (*Hint*: Use concatenation and Corollary 3.1.11).

In particular, the field descent (i.e. $[n,n,1]_q$-code) yields

Corollary 3.1.34. *For any* $g \ge 0$, *any* $n \ge 1$, *and any* K *such that* $0 \le K \le N_{q^n}(g) - g$ *there exists a linear code with parameters*

$$[n \cdot N_{q^n}(g)\ ,\ n \cdot K\ ,\ N_{q^n}(g) - K - g + 1]_q\ .$$

Remark 3.1.35. To use Proposition 3.1.32 and Corollary 3.1.34 we surely need some lower estimates for $N_q(g)$. In Chapters 3.2 and 3.4 we shall give some results of this kind.

Exercise 3.1.36. What codes can be obtained from AG-codes by using other constructions of Section 1.2.3 ?

Subfield restriction. The restriction of AG-codes also gives many interesting codes. Recall that restricting codes of genus zero we obtain Goppa codes and BCH-codes (see Section 1.2.2).

Proposition 3.1.37. *For any* $g \geq 0$, $m \geq 1$ *and any* K *such that*

$$\left\lceil \frac{(m-1)\cdot N_{q^m}(g)}{m} \right\rceil \leq K \leq N_{q^m}(g) - g$$

there exists a linear

$$[N_{q^m}(g)\ ,\ m\cdot K - (m-1)\cdot N_{q^m}(g),\ N_{q^m}(g) - K - g + 1]_q\text{-code.} \blacksquare$$

Exercise 3.1.38. Prove the proposition.

Remark 3.1.39. Let $n = N_{q^m}(g)$, $K = n - a + g - 1$. (cf. the Ω-construction in Section 3.1.1). Then Proposition 3.1.37 yields an

$$[n,\ n - m\cdot(a - g + 1)\ ,\ a - 2g + 2]_q\text{-code;}$$

for any $n \leq N_{q^m}(g)$ the parameters are of course the same. For the BCH case $(g = 0\ ,\ n = q^m - 1)$ we get

$$[q^m - 1\ ,\ q^m - m\cdot(a + 1)\ ,\ a + 2]_q\ .$$

Sometimes (cf. Section 1.2.2) the parameters are in fact better (the codimension is about $(q - 1)/q$ times less).

Amelioration of parameters. It turns out that this phenomenon takes place in the general situation as well.

Theorem 3.1.40. *Let* $m \geq 1$ *and* $2g - 2 < a < n$. *Consider an* $[n, n - a + g - 1, \geq a - 2g + 2]_{q^m}$*-code* $C_0 = (X, \mathcal{P}, D)_\Omega$ *where* $\mathcal{P} \subseteq X(\mathbb{F}_{q^m})$, $|\mathcal{P}| = n$, $D \geq 0$, $\deg D = a$. *If* $D = q \cdot D_1$ *and* $D_1 = \sum a_i \cdot Q_i$, $0 \leq a_i \leq q - 1$ *then the field restriction gives a code* $C = C_0 \cap \mathbb{F}_q$ *with the parameters*

$$[n, \geq n - m \cdot a \cdot (q - 1)/q - 1, \geq a - 2g + 2]_q .$$

Proof: Later on (in Section 3.1.3) we shall see that C_0 is dual to $C_1 = (X, \mathcal{P}, D)_L$. Let $b = a/q = \deg D_1$, and suppose that $b > g$. By the Riemann-Roch theorem $\ell(D_1) \geq b - g + 1$. Let $1, f_1, \ldots, f_{b-g}$ be independent vectors in $L(D_1)$. Let us show that the vectors $1, f_1, \ldots, f_{b-g}, f_1^q, \ldots, f_{b-g}^q$ are also independent. Indeed we know that $x_0 + \sum x_i \cdot f_i \neq \sum y_i \cdot f_i^q$ since the left hand side has a pole of order at most $q - 1$ (because of the condition $a_i \leq q - 1$) and the order of any pole of the right hand side is divisible by q (if there is no pole then $\sum y_i \cdot f_i^q = y_0$, let $z_i^q = y_i$ then $(\sum z_i \cdot f_i)^q = z_0^q$, i.e. $z_0 = \sum z_i \cdot f_i$ contradicting the independence of $1, f_1, \ldots, f_{b-g}$).

Extend the set $M = \{1, f_1, \ldots, f_{b-g}, f_1^q, \ldots, f_{b-g}^q\}$ to a basis of $L(D)$. Since $C = C_1^\perp \cap \mathbb{F}_q^n$, an element $v \in \mathbb{F}_q^n$ lies in C iff $(v, f) = 0$ for any element f of this basis. In general $(v, f) \in \mathbb{F}_{q^m}$, i.e. each equation $(v, f) = 0$ gives m $\mathbb{F}_q$-linear conditions. But in fact for $f = 1$ it is just one condition and the condition $(v, f_i) = 0$ is equivalent to $(v, f_i^q) = 0$. Thus we see that the number of $\mathbb{F}_q$-conditions is at least by

$$(b - g)\cdot m + (m - 1)$$

less that the *a priori* estimate $m\cdot\ell(D) = m\cdot(a - g + 1)$, i.e. $dim\, C \geq n - m\cdot(q - 1)\cdot b - 1$.

In the case $b \leq g$ we set $M = \{1\}$ and proceed as before. The *a priori* estimate $m\cdot(a - g + 1)$ is ameliorated by $(m - 1)$ and we get $dim\, C \geq n - m\cdot(a - g) - 1$ which (for $b \leq g$) is even better than the estimate of the theorem. ∎

Later on, in Section 3.4.3, we shall see that for large m this amelioration is rather strong. However it makes sense even for $m = 1$, though in a very restricted zone for $a = deg\, D$.

Corollary 3.1.41. *Let* $q\cdot g \leq a < n$. *If* $D = q\cdot D_1$ *and* $D_1 = \sum a_i\cdot Q_i$, $0 \leq a_i \leq q - 1$ *then the parameters of the code* $C = (X,\mathcal{P},D)_\Omega$ *are*

$$[n\ ,\ \geq n - a\cdot(q - 1)/q - 1\ ,\ \geq a - 2g + 2]_q$$

where $n = |\mathcal{P}|$, $a = deg\, D$. ∎

Remark 3.1.42. Using a subtler technique of linear algebra it is possible to show that the statement of Theorem 3.1.40 holds in the following more general situation. For an arbitrary divisor D let D_1 be such that $D \geq q\cdot D_1$, then

$$k \geq \begin{cases} n - 1 - m\cdot(\ell(D) - \ell(D_1)) & \text{if } D \text{ is effective,} \\ n - m\cdot(\ell(D) - \ell(D_1)) & \text{if } D \text{ is not effective;} \end{cases}$$

the same technique also leads to some estimates for any

linear code obtained by field restriction in terms of the filtration corresponding to the morphism $x \longmapsto x^q$.

3.1.3. Duality and spectra

(L-Ω)-duality. Recall the residue formula: for any differential form $\omega \in \Omega_X$ we have

$$\sum_{Q \in Supp(\omega)_\infty} Res_Q \omega = 0 .$$

Theorem 3.1.43. *The codes* $C_L = (X,\mathcal{P},D)_L$ *and* $C_\Omega = (X,\mathcal{P},D)_\Omega$ *are dual to each other.*

Proof: Consider first the case $2g - 2 < a < n$ when both maps $Ev_{\mathcal{P}}$ and $Res_{\mathcal{P}}$ are embeddings. Let $f \in L(D)$, $\omega \in \Omega(\mathrm{P} - D)$ where $\mathrm{P} = \sum_{P_i \in \mathcal{P}} P_i$, then the residue formula yields

$$\sum_{i=1}^{n} f(P_i) \cdot Res_{P_i}(\omega) = \sum_{i=1}^{n} Res_{P_i}(f \cdot \omega) = 0$$

since all the poles of the form $f \cdot \omega$ lie in $\mathcal{P}$ (possible poles of f are killed by zeros of ω). Thus any code vector of C_L is orthogonal to any code vector of C_Ω , i.e. $C_\Omega \le C_L^\perp$. On the other hand,

$$dim\, C_L^\perp = n - dim\, C_L \le n - (a - g + 1) \le dim\, C_\Omega ,$$

hence $C_\Omega = C_L^\perp$ and the estimates for the dimensions of C_L and C_Ω are in fact equalities.

For an arbitrary a we have $Ker(Ev_{\mathcal{P}}) = L(D - \mathrm{P})$, $Ker(Res_{\mathcal{P}}) = \Omega(-D)$. The Riemann-Roch theorem shows that the dimensions of C_L and C_Ω are complementary. Since the proof of orthogonality did not use any condition on a, the theorem is proved. ∎

H-**duality.** The line bundle $\tilde{\Omega}(\mathbf{P})$ where $\mathbf{P} = \sum_{P_i \in \mathcal{P}} P_i$ has the canonical trivialization

$$Res_{P_i} : \overline{\Omega}(\mathbf{P})_{P_i} \xrightarrow{\sim} \mathbb{F}_q$$

(recall that elements of the fiber $\overline{\Omega}(\mathbf{P})_{P_i}$ are classes of differential forms, whose pole at P_i is at most of order one, modulo forms regular at P_i).

Let $\mathcal{L}$ be a line bundle on X and $\mathcal{M} = \tilde{\Omega}(\mathbf{P}) \otimes \mathcal{L}^{-1}$. Fix a trivialization $t_i : \overline{\mathcal{L}}_{P_i} \xrightarrow{\sim} \mathbb{F}_q$, $\ell_0 = t_i^{-1}(1)$. Let us define trivialization $t_i' : \overline{\mathcal{M}}_{P_i} \xrightarrow{\sim} \mathbb{F}_q$ by the condition $t_i'(m) = Res_{P_i}(\ell_0 \cdot m)$. We say that t_i' is *concordant* with t_i.

Theorem 3.1.44. *Suppose that* $\mathcal{L}$ *and* $\tilde{\Omega}(\mathbf{P}) \otimes \mathcal{L}^{-1}$ *have concordant trivializations. Then the codes* $C = (X, \mathcal{P}, \mathcal{L})_H$ *and* $C' = (X, \mathcal{P}, \tilde{\Omega}(\mathbf{P}) \otimes \mathcal{L}^{-1})_H$ *are dual.*

Proof: By the Riemann-Roch theorem the codes C and C' have complementary dimensions. Let $v \in C$ and $v' \in C'$ correspond to the sections $s \in H^0(\mathcal{L})$ and $s' \in H^0(\mathcal{M})$ and let $s_i \in \overline{\mathcal{L}}_{P_i}$ and $s_i' \in \overline{\mathcal{M}}_{P_i}$ be the images of s and s'. Then

$$(v, v') = \sum_i t_i(s_i) \cdot t_i'(s_i') =$$

$$= \sum_i Res_{P_i}(s_i \cdot s_i') = \sum_i Res_{P_i}(s \otimes s') = 0$$

because of the residue formula (since all the poles of a section of $\tilde{\Omega}(\mathbf{P})$ lie in $\mathcal{P}$). ■

Exercise 3.1.45. Show that for any a such that $0 \le a \le n - g + 1$, the parameters of the code dual to C satisfy

$$k^{\perp} + d^{\perp} \ge n - g + 1 .$$

Self-dual AG-codes. Let us look at self-duality for AG-codes.

Let $y = (y_1, \dots, y_n) \in \mathbb{F}_q^n$, $y_i \ne 0$ for any i. Recall that a code C is called *quasi-self-dual with respect to* $y \in (\mathbb{F}_q^*)^n$ if $n = 2k$ and for any $u, v \in C$

$$\sum_{i=1}^{n} u_i \cdot v_i \cdot y_i = 0$$

(i.e. if u and v are *y-orthogonal*). A code is called *quasi-self-dual* if there exists such y and *self-dual* if $y = (1, \dots, 1)$.

Theorem 3.1.46. *Let* $n > 2g - 2$ *be even, and* $a = \frac{n}{2} + g - 1$. *If* $K + \mathbf{P} \sim 2D$, *where* $\mathbf{P} = \sum_{P_i \in \mathcal{P}} P_i$, *then an* $[n, \frac{n}{2}, \ge \frac{n}{2} - g + 1]_q$*-code* $C = (X, \mathcal{P}, D)_L$ *is quasi-self-dual. Moreover, there exists a unique (up to a multiplicative constant) form* $\omega_0 \in \Omega(\mathbf{P} - 2D)$ *such that the code* C *is quasi-self-dual with respect to* $y = (y_1, \dots, y_n)$, *where* $y_i = Res_{P_i}(\omega_0) \ne 0$. *In particular, if* $Res_{P_i}(\omega_0) = \dots = Res_{P_n}(\omega_0)$ *then* C *is self-dual.*

Proof: Since $n > 2g - 2$ we have $n > a > 2g - 2$ and hence $k = a - g + 1 = \frac{n}{2}$, $d \ge n - a = \frac{n}{2} - g + 1$. Let

$2D \sim K + \mathbf{P}$, then $\dim \Omega(\mathbf{P} - 2D) = \dim \Omega(-K) = 1$. Let $\omega_0 \in \Omega(\mathbf{P} - 2D)$, $\omega_0 \neq 0$. If we suppose that $Res_{P_i}(\omega_i) = 0$ for some i then $\omega_0 \in \Omega(\mathbf{P} - 2D - P_i)$ which is impossible since $\deg(\mathbf{P} - 2D - P_i) < 2 - 2g$. Therefore all residues $y_i = Res_{P_i}(\omega_0)$ are non-zero. For any two functions $f, g \in L(D)$ the form $\omega = f \cdot g \cdot \omega_0 \in \Omega(\mathbf{P})$, i.e. it has no poles out of $\mathcal{P}$. Hence

$$\sum_{i=1}^{n} f(P_i) \cdot g(P_i) \cdot y_i = \sum_{i=1}^{n} Res_{P_i} \omega = 0 ,$$

i.e. any two elements of the code are y-orthogonal. ■

Corollary 3.1.47. *If the image of the divisor* $\mathbf{P} = \sum_{P_i \in \mathcal{P}} P_i$ *in the group* $(Pic\ X)(\mathbb{F}_q)$ *is divisible by 2 then there exists a quasi-self-dual AG-code with parameters* $[n, \frac{n}{2}, \geq \frac{n}{2} - g + 1]_q$. *If, moreover,* q *is even then there exists a self-dual code with these parameters.*

Proof: The class field theory shows that the canonical divisor K is always divisible by 2 in the group $(Pic\ X)(\mathbb{F}_q)$ (cf. Theorem 2.3.21). Let $D = (K + \mathbf{P})/2$ and use Theorem 3.1.46. For an even q one can also apply Exercise 1.1.33. ■

It would be desirable to construct sets $\mathcal{P} \subseteq X(\mathbb{F}_q)$ with an even sum $\mathbf{P}$ and even $n = |\mathcal{P}|$ as large as possible.

Theorem 3.1.48. *Let* $N = |X(\mathbb{F}_q)| > 2g$. *Then there exists* $\mathcal{P} \subseteq X(\mathbb{F}_q)$ *with an even* $n = |\mathcal{P}| \geq N - 2g - 1$ *and an even sum.*

Proof: Let $J = (Pic\ X)(\mathbb{F}_q)$ and let $J_0 \subset J$ be the kernel of the map $deg : J \longrightarrow \mathbb{Z}$. Since $\mathbb{Z}$ has no torsion, we have $J/2 = \mathbb{Z}/2 \times J_0/2$. On the other hand J_0 is finite, hence the order of $J_0/2$ equals the order of the kernel of multiplication by 2. The latter group is embedded into $(\mathbb{Z}/2)^{2g}$, therefore $J/2 \subseteq (\mathbb{Z}/2)^{2g+1}$.

Order $\mathbb{F}_q$-points of X in such a way that the images of $P_{r+1},\dots,P_N$ form a basis of $J/2$ over $\mathbb{F}_2$. We have just proved that $r \geq N - 2g - 1$. Let us write out the image of $P_1 + \dots + P_r$ in $J/2$ in this basis:

$$P_1 + \dots + P_r = \sum_{i=r+1}^{N} \varepsilon_i \cdot P_i \ (mod\ 2)\ ,$$

where $\varepsilon_i = 0$ or 1. Reorder $P_{r+1},\dots,P_N$ so that $\varepsilon_i = 1$ for $r+1 \leq i \leq s$ (if all $\varepsilon_i = 0$ set $s = r$) and $\varepsilon_i = 0$ for $i > s$. Then $P_1 + \dots + P_s \equiv 0\ (mod\ 2)$, $s \geq r \geq N - 2g - 1$, and we can take $\mathcal{P} = \{P_1,\dots,P_s\}$. Of course s is always even since a divisor of an odd degree is never zero modulo 2. ■

Corollary 3.1.49. *In the notation of Theorem* 3.1.48 *there exists a quasi-self-dual (self-dual if* q *is even)* $[n, \frac{n}{2}, \geq \frac{n}{2} - g + 1]_q$*-code with some even* $n \geq N - 2g - 1$. ■

Remark 3.1.50. The sufficient condition of Theorem 3.1.46 is not in general necessary. Consider an elliptic curve $y^2 + y = x^3$. This curve has 3 points over $\mathbb{F}_2$, i.e. $|X(\mathbb{F}_2)| = 2 + 1 - (\omega + \bar{\omega}) = 3$, hence $\omega = \pm\sqrt{2}\cdot i$, and therefore $|X(\mathbb{F}_4)| = 4 + 1 - (\omega^2 + \bar{\omega}^2) = 9$. Theorem 2.4.31 shows that $X(\mathbb{F}_4) \simeq (\mathbb{Z}/3)^2$. Let P_0 be the infinite point, $P_1 = (0,0)$, $P_2 = (\varepsilon,1)$, $P_3 = (\varepsilon,\varepsilon)$, $P_4 = (\varepsilon,\varepsilon^2)$, where ε is a generator of $\mathbb{F}_4^*$. The code $C_L = (X,\{P_1,\dots,P_4\},2P_0)_L$ is a $[4,2,3]_4$-code, and the dual code C_Ω has the same parameters ($L(2P_0)$ is generated by 1 and x , $\Omega(P_1 + P_2 + P_3 + P_4 - 2P_0)$ is generated by $\omega_1 = \frac{x^2 \cdot dx}{y(y-\varepsilon)}$ and $\omega_2 = \frac{dx}{y-\varepsilon}$). The calculation of residues shows that $C_L = C_\Omega$. On the other hand $P_1 + P_2 + P_3 + P_4 \sim P_1 + 3P_0$ is not equivalent to $4P_0$.

Exercise 3.1.51. Check all the statements of Remark 3.1.50.

There are however situations when the condition of Theorem 3.1.46 is not only sufficient but also necessary.

Theorem 3.1.52. *Let* $g \geq 1$, *let* $D = \sum a_j \cdot Q_j$ *be an effective divisor of degree* $a = \frac{n}{2} + g - 1 \geq 4g - 1$ *such that* $a_j \leq a - (2g - 1)$ *for any* j ; $\mathcal{P} \subseteq X(\mathbb{F}_q) - \mathrm{Supp}\, D$. *The code* $C = (X,\mathcal{P},D)_L$ *is self-dual iff there exists a form* $\omega \in \Omega(\mathbf{P} - 2D)$ *with* $Res_{P_i}(\omega) = 1$ *for any* $i = 1,\dots,n$.

Proof: Let $C = C^\perp = (X,\mathcal{P},D)_\Omega$. Since $1 \in L(D)$, there exists a form $\eta \in \Omega(\mathbf{P} - D)$ with $Res_{P_i}(\eta) = 1$. Let us show that $\eta \in \Omega(\mathbf{P} - 2D)$. This condition is local, i.e. it is enough to prove that $(\eta) \geq 2a_j \cdot Q_j$ for any j . Let Q be one of the points Q_j and let $D = b \cdot Q + D_0$, $Q \notin \mathrm{Supp}\, D_0$, by the supposition of the theorem $\deg D_0 \geq 2g - 1$. We can suppose that Q is defined over $\mathbb{F}_q$ (otherwise we can take an extension of the ground field, the condition $(\eta) \geq 2D$ does not change). By the supposition of the theorem there exists a divisor $D' = b \cdot Q + D'_0$, $0 \leq D' \leq D$ such that $\deg(D - D') = 2g - 1$, and $\deg D' \geq 2g$. The Riemann-Roch theorem then yields $L(D') \neq L(D' - Q)$. Let $f \in L(D') - L(D' - Q)$, i.e. f has a pole of order exactly b at Q . By self-duality there exists a form $\omega \in \Omega(\mathbf{P} - D)$ with $Res_{P_i}(\omega) = f(P_i)$ (since $f \in L(D') \subseteq L(D)$). We have $f\cdot\eta \in \Omega(\mathbf{P} - (D - D'))$ and $Res_{P_i}(f\cdot\eta) = Res_{P_i}(\omega)$, i.e. $\omega - f\cdot\eta \in \Omega(-(D - D'))$, on the other hand $deg(D - D') = 2g - 1$ and a non-zero regular differential form cannot have $(2g - 1)$ zero, therefore $\omega = f\cdot\eta$. Hence the order of zero of η at Q satisfies

$$\nu_Q(\eta) = \nu_Q(\omega) - \nu_Q(f) = \nu_Q(\omega) + b \geq 2b ,$$

i.e. $(\eta) \geq 2b\cdot Q$. This is valid for any point $Q \in Supp\, D$, therefore $(\eta) \geq 2D$, i.e. $\eta \in \Omega(P - 2D)$. ■

Remark 3.1.53. It can be shown that for $g = 0$ the criterion of Theorem 3.1.52 is valid for any effective divisor D , and for $g = 1$ for any effective divisor of degree $a \geq 3$.

Spectra. What is it possible to say about the weight spectrum of an algebraic-geometric code C ? Let

$$W_C(x:y) = \sum_{v\in C} x^{n-w(v)}\cdot y^{w(v)} = x^n + \sum_{r=d}^{n} A_r\cdot x^{n-r}\cdot y^r$$

be the enumerator of C , and $W_C(x)$ its non-homogeneous form. In Section 1.1.3 we have described it in terms of some values B_ℓ . Let $d \geq n - a$ for some integer $a \geq 0$, then

$$W_C(x) = x^n + \sum_{\ell=0}^{a} B_\ell\cdot(x - 1)^\ell ,$$

$$B_\ell = \sum_{j=n-a}^{n-\ell} \binom{n-j}{\ell}\cdot A_j ,$$

where we put $A_1 = \ldots = A_{d-1} = 0$.

In Exercise 1.1.27 the value B_ℓ is interpreted as a sum of values $(q^{\ell(i_1,\ldots,i_\ell)} - 1)$, where $\ell(i_1,\ldots,i_\ell)$ is the dimension of the subcode consisting of vectors with zeros at the positions $i_1,\ldots,i_\ell$, the sum being taken over all sets $\{i_1,\ldots,i_\ell\} \subseteq \{1,\ldots,n\}$ of cardinality ℓ.

Theorem 3.1.54. *Let* X *be a curve of genus* g, $\mathcal{L} \in Pic\ X$, $deg\ \mathcal{L} = a$, $\mathcal{P} \subseteq X(\mathbb{F}_q)$, $|\mathcal{P}| = n$; and let $C = (X,\mathcal{P},\mathcal{L})_H$. *Then*

$$W_C(x) = x^n + \sum_{\ell=0}^{a} B_\ell \cdot (x - 1)^\ell ,$$

where for $0 \le \ell \le a - 2g + 1$

$$B_\ell = \binom{n}{\ell} \cdot (q^{a-\ell-g+1} - 1) ,$$

and for $a \ge \ell \ge a - 2g + 2$

$$\binom{n}{\ell} \cdot (q^{\lceil (a-\ell)/2 \rceil + 1} - 1) \ge B_\ell \ge max\left\{0, \binom{n}{\ell} \cdot (q^{a-\ell-g+1} - 1)\right\}.$$

Proof: Theorem 3.1.10 and Exercise 3.1.45 show that the parameters of C satisfy

$$k + d \ge n - g + 1 ,$$

$$(n - k) + d^\perp \ge n - g + 1 ,$$

i.e. C is a code of genus at most g (in the sense of definition given in Section 1.1.3).

For such a code, all the statements of the theorem except the left inequality follow from Theorem 1.1.16. Example 1.1.19 shows that for an arbitrary linear code of genus at most g the left inequality can fail, but its weaker version with the change of $(\lceil (a - \ell)/2 \rceil + 1)$ by

$(a - \ell + 1)$ is always valid. Let us now prove this inequality for AG-codes. If fact the subcode consisting of code vectors of C with zeros at the positions $i_1, \ldots, i_\ell$ is the image at $\mathbb{F}_q^n$ of the subspace

$$H^0(\mathcal{L} \otimes \mathcal{O}(-P_{i_1} - \ldots - P_{i_\ell})) \subseteq H^0(\mathcal{L})$$

If the line bundle $\mathcal{L}' = \mathcal{L} \otimes \mathcal{O}(-P_{i_1} - \ldots - P_{i_\ell})$ is non-special, i.e. if $h^0(\Omega \otimes (\mathcal{L}')^{-1}) = 0$, then the Riemann-Roch theorem yields

$$h^0(\mathcal{L}') = deg\ \mathcal{L}' - g + 1 = a - \ell - g + 1 ,$$

but $a - \ell \leq 2g - 2$, hence

$$h^0(\mathcal{L}') = a - \ell - g + 1 \leq (a - \ell)/2 + 1 .$$

On the other hand, if $\mathcal{L}'$ is special then we can apply the Clifford theorem (Theorem 2.4.42) showing that $h^0(\mathcal{L}') \leq (a - \ell)/2 + 1$. ■

Exercise 3.1.55. Prove Theorem 3.1.54 without using Theorem 3.1.16. (*Hint*: Use the Riemann-Roch and Clifford theorems).

Thus for an algebraic-geometric code on a curve of genus g in the spectrum there are $(2g - 1)$ unknown parameters B_ℓ, $a - 2g + 2 \leq \ell \leq a$. For $g = 0$ Theorem 3.1.54 completely determines the spectrum of a Reed-Solomon code. Later on, in Section 3.2.2, we shall write out the spectrum for the simplest non-trivial case $g = 1$ (i.e. for elliptic codes).

CHAPTER 3.2

EXAMPLES

The general discussion of the previous chapter leaves us a bit in the air without examples of AG-codes for which it is possible to calculate the parameters and to compare them with codes obtained by non-algebraic-geometric constructions.

In Section 3.2.1 we consider AG-codes on curves of small genera and calculate parameters of some binary codes obtained from them. Section 3.2.2 is devoted to a detailed analysis of the first non-trivial case (codes on curves of genus one); we calculate their spectra, answer the question whether they can be MDS or not, study when they are self-dual. In Section 3.2.3 we analyse some other families of AG-codes, in particular those obtained from curves with maximum possible number of $\mathbb{F}_q$-points, and from simplest higher-dimensional varieties. Section 3.2.4 is devoted to minute analysis of some particular examples (up to writing out the generator matrix explicitly).

3.2.1. Codes of small genera

The less the genus the simpler is mostly the analysis of curves and of corresponding codes.

Codes of genus zero. Any smooth curve of genus zero over $\mathbb{F}_q$ is $\mathbb{F}_q$-isomorphic to the projective line $\mathbb{P}^1$; the number of $\mathbb{F}_q$-points on $\mathbb{P}^1$ equals $q + 1$. We get therefore a family of $[n,k,d]_q$-codes with parameters

$$n \le q + 1 \ ,$$

$$d = 1 \ , \ \dots \ , n \ ,$$

$$k = n + 1 - d \ .$$

Exercise 3.2.1. Check that in this way for $n \le q$ it is possible to construct a family of embedded codes with $k = 1,\dots,n$. (*Hint*: Use L-construction with $D = k\cdot P_0$).

Remark 3.2.2. For $n = q + 1$ it is already impossible to construct an embedded family. The reason is that if such a family existed it would be possible to construct $[q + 2, k, q + 3 - k]_q$-codes (using the construction of a code from an embedded pair, cf. Section 1.2.3), and for $k = 4$ such codes do not exist (cf. the properties of n-sets in Section 1.1.2).

As it has been already explained (cf. Proposition

1.2.4) for a code C of genus zero the spectrum is uniquely determined:

$$W_C(x) = x^n + \sum_{i=0}^{k-1} \binom{n}{i} \cdot (q^{k-i} - 1) \cdot (x - 1)^i .$$

The spectrum of the dual code is of the same form. For $n = 2k$ there is an equality $W_{C^\perp}(x) = W_C(x)$, i.e. the code is formally self-dual. Moreover,

Exercise 3.2.3. Show that an algebraic-geometric $[n, n/2, n/2 + 1]_q$-code C of genus zero is always quasi-self-dual. For an even q there exist self-dual AG-codes with these parameters. (*Hint*: If $g = 0$ then any divisor of even degree is divisible by 2 and all the divisors of a given degree are equivalent).

The codes on curves of genus zero (i.e. on $X = \mathbb{P}^1$) have been essentially described in Section 1.2.1: they are just Reed-Solomon codes for $\mathcal{P} = \mathbb{A}^1(\mathbb{F}_q)$ and their natural extensions for $\mathcal{P} = \mathbb{P}^1(\mathbb{F}_q)$.

Codes of genus one. *Elliptic codes* are the codes obtained from elliptic curves (i.e. curves of genus $g = 1$). They are codes of genus at most one (in the sense of definition of Section 1.1.3). Elliptic codes exist for any

$$n \le N_q(1) ,$$

$$k = 1, \ldots , n - 1 ,$$

where $N_q(1)$ is the maximum possible number of $\mathbb{F}_q$-points on a curve of genus one. For $q = p^m$ Theorem 2.3.17 gives

$$N_q(1) = \begin{cases} q + \lceil 2\sqrt{q} \rceil , & \text{for } p \mid \lceil 2\sqrt{q} \rceil , \text{ and odd } m \ge 3 , \\ q + \lceil 2\sqrt{q} \rceil + 1 , & \text{else} . \end{cases}$$

The spectrum of an elliptic code C has (in terms of B_i, cf. Theorem 3.1.54) only one unknown coefficient B_k. Later on, in Section 3.2.2, we shall calculate B_k; it can differ for different curves and divisors.

Codes of genera two and three. Consider now codes on curves of genus $g = 2$. They exist for any

$$n \le N_q(2) \quad ,$$

$$k = 1, \ldots, n-2 \quad ,$$

moreover (cf. Theorem 2.3.18) for $q = p^m$ and $m \equiv 0 (\mathrm{mod}\ 2)$

$$N_q(2) = \begin{cases} q + 4\sqrt{q} + 1 & \text{for} \quad q \ne 4, 9 \\ 20 & \text{for} \quad q = 9 \\ 10 & \text{for} \quad q = 4 \end{cases}$$

and for $m \equiv 1 (\mathrm{mod}\ 2)$

$$N_q(2) = \begin{cases} q + 2 \cdot \lceil 2\sqrt{q} \rceil + 1 & \text{if } q \text{ is non-special,} \\ q + 2 \cdot \lceil 2\sqrt{q} \rceil & \text{if } q \text{ is special and} \\ & 2\sqrt{q} - \lceil 2\sqrt{q} \rceil > (\sqrt{5} - 1)/2 \,, \\ q + 2 \cdot \lceil 2\sqrt{q} \rceil - 1 & \text{if } q \text{ is special and} \\ & 2\sqrt{q} - \lceil 2\sqrt{q} \rceil < (\sqrt{5} - 1)/2 \,. \end{cases}$$

Recall that q is special iff either $p | \lceil 2\sqrt{q} \rceil$, or q is of the form $q = \ell^2 + 1$ or $q = \ell^2 + \ell + 1$ or $q = \ell^2 + \ell + 2$ for some integer ℓ.

Codes on curves of genus $g = 3$ satisfy

$$k + d \ge n - 2 \quad ,$$

$$n \le N_q(3) \,, \quad k = 1, \ldots, n-3 \quad .$$

For $N_q(3)$ the following values are known (cf. Theorem 2.3.19)

q	2	3	4	5	7	8	9	11	13	16	17	19	25
$N_q(3)$	7	10	14	16	20	24	28	28	32	38	40	44	56

Binary codes. Let us briefly consider some binary codes obtained from AG-codes of small genera by the help of constructions of Section 1.2.3 (cf. also Section 3.1.2).

First of all note that restricting the field we get BCH codes from codes of genus zero. These codes have many interesting properties, we have already discussed them in Section 1.2.2.

Is is also possible to construct binary analogues of BCH codes from curves of genus $g \geq 1$ over $\mathbb{F}_q$, $q = 2^m$.

Example 3.2.4. Consider an elliptic curve X over $\mathbb{F}_{16}$ with the maximum possible number of points. We have already pointed out that on X there are 25 points, a bit later we shall give an equation for such a curve (see Example 3.2.53). Let us construct AG-codes starting with this curve X, a divisor D of degree 8, and some 21 points of these 25. The parameters of C are at least $[21,8,13]_{16}$. Concatenation of the outer code C with the inner parity-check $[5,4,2]_2$-code gives a $[105,32,26]_2$-code. Its shortening is a $[104,32,25]_2$-code, which is rather good (the best $[104,k,25]_2$-code known before AG-codes was non-linear and had $k = 31.585$).

Example 3.2.5. Consider a curve X of genus 3 over $\mathbb{F}_8$ with 24 points (cf. Example 3.2.58) and a bundle of degree 11. We obtain a $[24,9,13]_8$-code C. Concatenating it with the parity-check $[4,3,2]_2$-code we get a

$[96,27,26]_2$-code. Shortening yields a $[95,27,25]_2$-code. These parameters are quite good (the best one known before AG-codes was $[95,24,25]_2$).

Different considerations applied to curves of genera 1, 2, or 3 (using different linear and non-linear combinatorical constructions like concatenation, generalized concatenation, shortening, etc.) give a number of binary codes with good parameters. Table A.1.4 of Appendix containes some cases when their parameters are better than those "before AG-codes".

Of course, such tables are not very interesting. However we know no other ways to compare codes produced by some new method with those known before, and such tables are to some extend good to characterize new methods.

3.2.2. Elliptic codes

An elliptic curve X (i.e. a curve of genus 1) is an algebraic group (cf. Section 2.4.1). This determins many properties of the corresponding AG-codes.

Spectra of elliptic codes. An elliptic code is a code of genus at most 1. Its weight enumerator according to Theorem 1.1.16 equals

$$W_C(x) = x^n + \sum_{i=0}^{k-1} \binom{n}{i} \cdot (q^{k-i} - 1) \cdot (x - 1)^i + B_k \cdot (x - 1)^k ,$$

$$k + d \geq n \quad .$$

As we have already mentioned in Remark 1.1.17 $B'_{n-k} = B_k$ and for $n = 2k$ these codes are formally self-dual, i.e. $W_C(x) = W_{C^\perp}(x)$.

There are two cases:

either $B_k = B'_{n-k} = 0$, $k + d = n + 1$, i.e. the code is a code of genus zero (later on, in Theorems 3.2.19 and 3.2.21, we shall see that this case is quite rare),

or $0 < B_k = B'_{n-k} \le \binom{n}{k}\cdot(q - 1)$, $k + d = n$.

Now we are going to find out B_k . If it is positive then $B_k = A_d$ is the number of vectors of the minimum weight.

Let X be an elliptic curve. Recall (cf. Section 2.4.4) that $X(\mathbb{F}_q)$ is equipped with a structure of an abelian group which depends on the choice of the zero element $P_0 \in X(\mathbb{F}_q)$. This group is a subgroup of $X(\overline{\mathbb{F}}_q)$. Each divisor $D = \sum a_i\cdot Q_i$, where $Q_i \in X(\overline{\mathbb{F}}_q)$ corresponds to some point P_D equal to the sum (in the sense of the group law) $\sum a_i\cdot Q_i$. If D is defined over $\mathbb{F}_q$, then $P_D \in X(\mathbb{F}_q)$; for a principle divisor (f) this sum is zero: $P_{(f)} = P_0$, hence P_D depends only on the divisor class, i.e. we have well defined $P_{\mathcal{L}}$ for a line bundle $\mathcal{L}$.

Theorem 3.2.6. *For an elliptic $[n,k,d]_q$-code $C = (X,\mathcal{P},\mathcal{L})_H$*

$$B_k = (q - 1)\cdot M(X(\mathbb{F}_q),\mathcal{P},P_{\mathcal{L}},k) ,$$

where for a finite abelian group H , its element $h \in H$, subset $\mathfrak{h} \subseteq H$, and an integer $a > 0$, by $M(H, \mathfrak{h}, h, a)$ we denote the number of representation of h as a sum of a different elements from $\mathfrak{h}$.

Proof: The number $B_k = A_{n-k}$ equals the number of non-zero sections $s \in H^0(\mathcal{L})$ having exactly k zeroes. Exercise 1.1.27 tells us that

$$B_k = \sum_{Q \subset \mathcal{P}, |Q|=k} (q^{\ell(Q)} - 1) .$$

It is easy to see that

$$\ell(Q) = h^0(\mathcal{L}_Q) ,$$

where $\mathcal{L}_Q = \mathcal{L} \otimes (\bigotimes_{P \in Q} \mathcal{O}(-P_\ell))$ is a bundle of degree zero.

If $\mathcal{L}_Q \simeq \mathcal{O}$ then $h^0(\mathcal{L}_Q) = 1$, otherwise $h^0(\mathcal{L}_Q) = 0$ (since any section $s \in H^0(\mathcal{L}_Q)$ yields a homomorphism $\mathcal{O} \longrightarrow \mathcal{L}_Q$, which is an isomorphism since $deg\, \mathcal{L}_Q = 0$, cf. Section 2.1.3). Hence

$$B_k = (q - 1) \cdot M$$

where M is the number of sets Q such that $\mathcal{L}_Q \simeq \mathcal{O}$, i.e. $\mathcal{L} \simeq \bigotimes_{P_\ell \in Q} \mathcal{O}(P_\ell)$. We know that the group law on X corresponds to the group law on $Pic\, X$, i.e.

$$\bigotimes_{P_\ell \in Q} \mathcal{O}(P_\ell) \simeq \mathcal{O}(P_Q) \otimes (\mathcal{O}(P_0))^{k-1} ,$$

where $P_Q = \sum_{P_\ell \in Q} P_\ell \in X(\mathbb{F}_q)$. On the other hand

$$\mathcal{L} \simeq \mathcal{O}(P_{\mathcal{L}}) \otimes (\mathcal{O}(P_0))^{k-1} .$$

Therefore M is the number of sets Q such that $P_Q = P_{\mathcal{L}}$, and the theorem is proved. ■

Further on we consider the case of codes of maximum

length, i.e. $\mathcal{P} = X(\mathbb{F}_q)$. For an arbitrary abelian group H let

$$M(H,h,a) = M(H,H,h,a)$$

Proposition 3.2.7. *The number $M(H,h,a)$ depends only on the class of h in the factor-group H/aH .*

Proof: Let $h = \sum_{i=1}^{a} h_i$ be a representation of h as a sum of different elements, then for any $h_0 \in H$ there is a representation $h + a\cdot h_0 = \sum_{i=1}^{a} (h + h_0)$ and all the elements $(h + h_0)$ are also different. Therefore the numbers of representations of h and $(h + a\cdot h_0)$ are equal. ■

Corollary 3.2.8. *If a and the order N of H are co-prime, then for any $h \in H$*

$$M(H,h,a) = \binom{N}{a}/N$$

Proof: in this case $aH = H$. ■

Corollary 3.2.9. *Let the degree k of a bundle $\mathcal{L}$ be co-prime to the order n of $X(\mathbb{F}_q)$. Then the code $C = (X,X(\mathbb{F}_q),\mathcal{L})_H$ is an $[n, k, n - k]_q$-code and its enumerator equals*

$$W_C(x) = x^n + \sum_{i=0}^{k-1} \binom{n}{i}\cdot(q^{k-i} - 1)\cdot(x - 1)^i +$$

$$+ \frac{\binom{n}{k}\cdot(q - 1)}{n}\cdot(x - 1)^k \quad .$$

■

If a and N are not co-prime then $M(H,h,a)$ can be still calculated, but it is much more dificult. Here is the answer.

Let H be an abelian group of order N, $H = \sum_{j=1}^{G} H_j$, $H_j \simeq \mathbb{Z}/p_j^{\alpha_j}$ being its primary decomposition. Each element $h \in H$ is represented as $h = (h_1,\ldots,h_G)$, where $h_j \in \mathbb{Z}$, $0 \le h_j \le p_j^{\alpha_j} - 1$.

Define $\varepsilon_j(h) \in \mathbb{Z} \cup \{\infty\}$ as the maximum power of p_j dividing h_j (and $\varepsilon_j(h) = \infty$ if $h_j = 0$). For each prime $p|N$ set $G_p = \{j \mid p_j = p\}$. Set

$$e_p(h) = \begin{cases} \min\limits_{j\in G_p} \varepsilon_j(h) & \text{if there exists } j \in G_p \text{ such that } \varepsilon_j(h)\neq\infty \\ \sum\limits_{j\in G_p} \alpha_j & \text{if } \varepsilon_j(h) = \infty \text{ for any } j \in G_p . \end{cases}$$

Let

$$\Delta(h) = \prod_{p|N} p^{e_p(h)} .$$

Let $t|N$ and $t = \prod_p p^{\delta_p}$. Define an integer

$$R_H(t) = \prod_{p|N} p^{\sum_{j\in G_p} \min(\delta_p,\alpha_j)} .$$

In particular, $R_H(1) = 1$. Let $\mu(m)$ be the Möbius function.

Theorem 3.2.10. *The number $M(H,h,a)$ of representations of an element h of an abelian group H as the sum of a* *different summands is given by the formula*

$$M(H,h,a) = \frac{1}{N} \cdot \sum_{\ell \mid (N,a)} \binom{N/\ell}{a/\ell} \cdot (-1)^{a-a/\ell} \cdot \sum_{t \mid (\ell,\Delta(h))} \mu(\ell/t) \cdot R_H(t) \quad .$$

■

Exercise 3.2.11. Prove the theorem.

Now let us ask another question: what is the minimal possible value $B_{k,min}(q,N,a)$ of vectors of the minimum weight in an elliptic code (for some fixed q, N, and a). To answer this question one shouls find out the value

$$M(H,a) = \min_{h \in H} M(H,h,a) \quad .$$

The formula for $M(H,a)$ appeares to be simpler than that for $M(H,h,a)$.

For an abelian group H define r_H as the maximum integer such that $(\mathbb{Z}/2)^{r_H} \le H$.

Theorem 3.2.12. a) *If* $N \equiv 0 (mod\ 4)$, $a \equiv 2 (mod\ 4)$, *then*

$$M(H,a) = \frac{1}{N} \cdot \left[\sum_{m \mid (a,N)} \binom{N/m}{a/m} \cdot \mu(m) \cdot (-1)^{a/m} - \right.$$

$$\left. - 2^{r_H} \cdot \sum_{m \mid (a/2,N/2)} \binom{N/2m}{a/2m} \cdot \mu(m) \right] .$$

b) *For any other pair* (N,a)

$$M(H,a) = \frac{1}{N} \cdot \sum_{m \mid (a,N)} \binom{N/m}{a/m} \cdot \mu(m) \cdot (-1)^{a-a/m} \quad .$$

■

Exercise 3.2.13. Prove the theorem.

To find out the minimum possible number of vectors of the minimum weight one should also minimize $M(H,a)$ over all $H = X(\mathbb{F}_q)$ of order N. This is not difficult since we know the possible structure of the group $X(\mathbb{F}_q)$ (see Theorem 2.4.31). In particular $r_H = 1$ or 2.

Theorem 3.2.14. *If* N *is such that there exists an elliptic curve over* $\mathbb{F}_q$ $(q = p^e)$ *having exactly* N *points, then*

a) *if* $N \equiv 0 \pmod 4$, $a \equiv 2 \pmod 4$, q *is odd, and moreover either* $N \not\equiv 1 \pmod p$, *or* q *is a square and* $N = (\sqrt{q} \pm 1)^2$, *or* q *is not a square,* $p \equiv 3 \pmod 4$ *and* $N = q + 1$ *then*

$$B_{k,min}(N,a) = \frac{q-1}{N} \cdot \left[\sum_{m|(a,N)} \binom{N/m}{a/m} \cdot \mu(m) \cdot (-1)^{a-a/m} - \right.$$

$$\left. - 4 \cdot \sum_{m|(a/2,N/2)} \binom{N/2m}{a/2m} \cdot \mu(m) \right],$$

b) *if* $N \equiv 0 \pmod 4$, $a \equiv 2 \pmod 4$, *and either* q *is even, or* q *is odd and no auxilary condition of* **a)** *holds then*

$$B_{k,min}(N,a) = \frac{q-1}{N} \cdot \left[\sum_{m|(a,N)} \binom{N/m}{a/m} \cdot \mu(m) \cdot (-1)^{a-a/m} - \right.$$

$$\left. - 2 \cdot \sum_{m|(a/2,N/2)} \binom{N/2m}{a/2m} \cdot \mu(m) \right],$$

c) *if either* $N \not\equiv 0 \pmod 4$, *or* $a \not\equiv 2 \pmod 4$, *then*

$$B_{k,min}(N,a) = \frac{q-1}{N} \cdot \sum_{m|(a,N)} \binom{N/m}{a/m} \cdot \mu(m) .$$

■

Exercise 3.2.15. Prove the theorem.

Remark 3.2.16. For given q and N Theorem 2.4.30 tells us whether there exists an elliptic curve over $\mathbb{F}_q$ with exactly N points, or not.

The formulae of Theorems 3.2.12 and 3.2.14 are not very interesting. What is interesting is that it appears possible to calculate $B_{k,min}$ explicitly.

Remark 3.2.17. For curves of genus $g \geq 2$ it is much more difficult to calculate $B_k, B_{k-1}, \dots, B_{k-2g+2}$. For example, for $g = 2$ the enumerator equals

$$W_C(x) = x^n + \sum_{i=0}^{k-1} \binom{n}{i} \cdot (q^{k-i} - 1) \cdot (x - 1)^i +$$

$$+ B_k \cdot (x - 1)^k + B_{k-1} \cdot (x - 1)^{k-1} + B_{k-2} \cdot (x - 2)^{k-2} \quad ;$$

$$k + d \geq n - 1 \quad .$$

It is related to the enumerator of the dual code as follows:

$$B'_{n-k+1} = \frac{B_{k-1} - \binom{n}{k-1} \cdot (q - 1)}{q} \quad ,$$

$$B'_{n-k} = B_k \quad ,$$

$$B'_{n-k-1} = q \cdot B_{k+1} + \binom{n}{k+1} \cdot (q - 1) \quad .$$

These codes (unlike codes of genus zero and elliptic codes) are not in general formally self-dual.

Problem 3.2.18. Calculate B_{k-1}, B_k, B_{k+1} for a curve of genus 2 in terms of geometry of the curve and of the divisor. Calculate $B_{k-g+1}, \dots, B_{k+1}$ for a curve of an arbitrary genus.

Elliptic MDS-codes. Recall that an $[n,k,d]_q$-code is called an MDS-code (or a code of genus zero) iff $k + d = n + 1$. Such are AG-codes on the projective line, their length $n \le q + 1$. A code of genus at most one is an MDS-code iff $B_k = 0$. Therefore a question: is it possible to construct an elliptic MDS-code whose length is more than $q + 1$, or not?

We start with elliptic codes with $\mathcal{P} = X(\mathbb{F}_q)$.

Theorem 3.2.19. *Let* $C = (X, X(\mathbb{F}_q), \mathcal{L})_H$ *be an* $[n,k,d]_q$-*code constructed from an elliptic curve and all its* $\mathbb{F}_q$-*points. If* $k > 0$ *and* $B_k = 0$, *then either* $X(\mathbb{F}_q) = \mathbb{Z}/2$, *or* q *is odd and* $X(\mathbb{F}_q) \simeq (\mathbb{Z}/2)^2$. ■

Exercise 3.2.20. Prove the theorem. (*Hint*: Either derive it from Theorem 3.2.12, or prove it directly using only Theorem 3.2.6).

For elliptic codes constructed from proper subsets $\mathcal{P} \subset X(\mathbb{F}_q)$ the answer is not so nice. However it is still possible to prove the negative answer to the above question.

First of all note that for $q \le 11$ there is no non-trivial MDS-code longer than $q + 1$ (cf. Section 1.2.1, the remark after Problem 1.1.12).

For the rest values of q we have

Theorem 3.2.21. *For* $q \ge 13$ *there is no elliptic code of length* $n > q + 1$ *such that* $B_k = 0$.

Proof: Suppose that $B_k = 0$ for an elliptic code $C = (X, \mathcal{P}, \mathcal{L})_H$, i.e. $M(H, \mathcal{P}, h_0, a) = 0$ where $H = X(\mathbb{F}_q)$, $|H| = N$, $|\mathcal{P}| = n$, $h_0 = P_{\mathcal{L}} \in X(\mathbb{F}_q)$ is the sum of the divisor points, $a = deg\ \mathcal{L}$.

Let $a = s + 2t$, define an $(a,s,\mathcal{P})$-representation of h_0, as the decomposition

$$h_0 = (h_1 + \ldots + h_s) +$$

$$+ (h_{s+1} + (-h_{s+1}) + \ldots + h_{s+t} + (-h_{s+t})) \quad ,$$

where all the elements $h_1,\ldots,h_{s+t};-h_{s+1},\ldots,-h_{s+t}$ are different and belong to $\mathcal{P}$.

Let us prove several auxiliary lemmas.

Lemma 3.2.22. *If* h_0 *has an* $(a,s,\mathcal{P})$*-representation and* $s + a \le 2n - 5 - N$ *then* h_0 *also has an* $(a + 2,s,\mathcal{P})$*-representation.*

Proof: Let

$$\mathcal{P}^* = \{h \in \mathcal{P} \mid -h \in \mathcal{P},\ h \ne -h\} .$$

Since $H = X(\mathbb{F}_q) \subseteq (\mathbb{Z}/N)^2$ (cf. Theorem 2.4.31), there are at most four elements $h \in H$ such that $h = -h$; denote the set of such elements by H_0. An element $h \in \mathcal{P}$ does not lie in $\mathcal{P}^*$ iff either $-h \notin \mathcal{P}$ (there are at most $|H - \mathcal{P}| = N - n$ such elements), or $h \in H_0$. Hence

$$|\mathcal{P}^*| \ge n - 4 - (N - n) = = 2n - 4 - N .$$

For a given $(a,s,\mathcal{P})$-representation of h_0 let

$$R = \mathcal{P}^* - \{\pm h_i\}_{i=1,\ldots,s+t} ;$$

$$|R| \ge 2n - 4 - N - 2s - 2t = 2n - 4 - s - a .$$

If $h^* \in R$ then $-h^* \in R$ and $\pm h^*$ are out of the given $(a,s,\mathcal{P})$-representation of h_0 ; adding $h^* + (-h^*)$ to it we get an $(a + 2,s,\mathcal{P})$-representation. The set R is clearly non-empty if $s + a \le 2n - N - 5$. ■

Remark 3.2.23. It is not difficult to see that the statement of Lemma 3.2.20 is also valid if either $(\mathbb{Z}/2)^2$ does not lie in H and $s + a \le 2n - 3 - N$, or $\mathbb{Z}/2$ does not lie in H and $s + a \le 2n - 2 - N$.

Lemma 3.2.24. *If*

$$N - n < \left]\frac{N-2}{3} - \frac{2}{3(N-1)}\right[$$

then for any $h_0 \in H$ *there is a* $(3,3,\mathcal{P})$*-representation.*

Proof: By Theorem 3.2.12 the number of representations of h_0 as a sum of three different elements from H equals

$$M(H,h_0,3) \ge M(H,3) \ge \frac{1}{N}\cdot\left(\binom{N}{3} - \frac{N}{3} \right) .$$

Let us estimate the number of representations $h_0 = h + h_1 + h_2$ such that one of the elements does not belong to $\mathcal{P}$ (we suppose that $h \notin \mathcal{P}$). For a fixed h there are at most $(N-1)/2$ possibilities for a pair $\{h_1,h_2\}$ (of course, $\{h_1,h_2\} = \{h_2,h_1\}$). Therefore there are at most $(N-n)\cdot(N-1)/2$ possibilities for $h_0 = h + h_1 + h_2$, $h \notin \mathcal{P}$, and the lemma follows. ■

Lemma 3.2.25. *If* $N - n < \left]\frac{N-4}{2}\right[$ *then for any* h_0 *there exists a* $(2,2,\mathcal{P})$*-representation.* ■

Lemma 3.2.26. *If* $a \le 2n - 6 - N$ *and*

$$N - n < A(N) = \min\left\{ \left]\frac{N-4}{2}\right[, \left]\frac{N-2}{3} - \frac{2}{3(N-1)}\right[\right\}$$

then for any $h_0 \in H$ *we have* $M(H,\mathcal{P},h_0,a) > 0$. ■

Exercise 3.2.27. Prove Lemma 3.2.25. (*Hint*: It is similar to Lemma 3.2.24), and Lemma 3.2.26. (*Hint*: Use induction over a applying Lemma 3.2.22).

Now we continue the proof of Theorem 3.2.21. Recall that we are interested in the case $N > n > q + 1$ for $q \geq 13$ (for $q \leq 11$ we know that there are no non-trivial MDS-code with $n > q + 1$, and the case $N = n$ is considered in Theorem 3.2.19). In particular $n \geq 15$, $N \geq q + 3 \geq 16$.

Let us find out when the conditions of Lemma 3.2.26 can be violated. We start with the inequality for $(N - n)$. It is clear that for $N \geq 16$

$$A(N) = \left]\frac{N - 2}{3} - \frac{2}{3(N - 1)}\right[\geq \left]\frac{N - 2}{3}\right[.$$

Since $q + \lceil 2\sqrt{q}\rceil + 1 \geq N > n \geq q + 2$ we have

$$N - n \leq \lceil 2\sqrt{q}\rceil - 1$$

The inequality for $(N - n)$ can be violated only for

$$2\sqrt{q} - 1 \geq \lceil 2\sqrt{q} - 1\rceil \geq N - n \geq A(N) \geq$$

$$\geq \left]\frac{N - 2}{3}\right[\geq \frac{N - 2}{3} \geq \frac{q + 1}{3} ,$$

i.e. if $q - 6\sqrt{q} + 4 \leq 0$, which is possible only for $q \leq 27$. For $q = 27$ we have $\lceil 2\sqrt{q}\rceil - 1 = 9 < \left]\frac{q + 1}{3}\right[$. So only the values $q = 13, 16, 17, 19, 23, 25$ are left, and are to be considered separately.

Exercise 3.2.28. Check that for all these values of

q the inequalities

$$n \geq q + 2 \ , \ N \leq q + \lceil 2\sqrt{q} \rceil + 1 \ , \ N - n \geq \left] \frac{N - 2}{3} \right[$$

are contradictory, i.e. the condition of the lemma for $(N - n)$ is always satisfied (for $q \geq 13$).

Then we pass to the next condition.

For $q \geq 13$ the other condition of the lemma $a \leq 2n - 6 - N$ can be in general violated. Let us use the following trick. If an elliptic code constructed from a bundle of degree a has $B_k = 0$ then the dual code has $B'_{k'} = 0$ for $a' = n - a$. Therefore Lemma 3.2.26 cannot be applied only if $a \geq 2n - N - 5$ and $n - a \geq 2n - N - 5$. Adding these inequalities we get $2(N - n) + 10 \geq n$, i.e.

$$2 \cdot \lceil 2\sqrt{q} \rceil + 8 \geq 2 \cdot (N - n) + 10 \geq n \geq q + 2 \ ,$$

$$q - 2\lceil 2\sqrt{q} \rceil - 6 \leq 0$$

and this is impossible for $q \geq 27$.

Exercise 3.2.29. Give the detailed proof for the cases $q = 13, 16, 17, 19, 23, 25$. (*Hint*: Use the description of the group $X(\mathbb{F}_q)$ given by Theorem 2.4.31). ∎

Self-duality. The formula for the spectrum yields immediately the following result.

Exercise 3.2.30. Check that any elliptic code with $n = 2k$ is formally self-dual.

The question about longest possible quasi-self-dual elliptic codes can be analyzed nearly up to the end. We start with a simple auxiliary statement.

Proposition 3.2.31. *Let* H *be an abelian group of order* N *. If* N *is odd then the equality* $\sum_{h\in\mathcal{P}} h = 0$ *holds for* $\mathcal{P} = \mathcal{P}_0 = H$ *and* $\mathcal{P} = \mathcal{P}_1 = H - \{0\}$ *. If* N *is even then either it holds for* $\mathcal{P} = \mathcal{P}_0$ *and* $\mathcal{P} = \mathcal{P}_1$ *, or there exists* $h_0 \in H$ *such that it holds for* $\mathcal{P} = \mathcal{P}_2 = H - \{h_0\}$ *and* $\mathcal{P} = \mathcal{P}_3 = H - \{0,h_0\}$ *.*

Proof: Let H_0 be the subgroup of elements of order two. If $h \notin H_0$ then $-h \neq h$, hence $\sum_{h\notin H_0} h = 0$. In a group of odd order $H_0 = \{0\}$. In a group of even order $H_0 \simeq (\mathbb{Z}/2)^r$. If $r \geq 2$ then $\sum_{h\in H_0} h = 0$ (this is proved easily using induction over r). In the case $r = 1$ we have to exclude the only non-trivial element of order two. ■

Theorem 3.2.32. *Let* N *be the number of* $\mathbb{F}_q$*-points on an elliptic curve. If* N *is odd then there exists a quasi-self-dual* $\left[N - 1,\frac{N - 1}{2},\frac{N - 1}{2}\right]_q$*-code; if* N *is even then there exists either a quasi-self-dual* $\left[N,\frac{N}{2},\frac{N}{2}\right]_q$*-code or a quasi-self-dual* $\left[N - 2,\frac{N - 2}{2},\frac{N - 2}{2}\right]_q$*-code. If* q *is even, there exist corresponding self-dual codes.* ■

Exercise 3.2.33. Prove the theorem.

Remark 3.2.34. Because of Theorem 2.3.17, the maximum number of $\mathbb{F}_q$-points on an eliptic curve is at least $q + \lceil 2\sqrt{q}\,\rceil$, Theorem 3.2.32 yields the existence of a quasi-self-dual $\left[n,\frac{n}{2},\frac{n}{2}\right]_q$-code for some $n \geq q + \lceil 2\sqrt{q}\,\rceil - 2$ (self-dual for $q = 2^m$).

Remark 3.2.35. Theorem 3.2.32 is a bit more precise than Theorem 3.1.46. A similar result can be also obtained for hyperlliptic curves of an arbitrary genus g . Such a curve embedded into its Jacobian J_X has the same property: if $P \in X$ then $-P \in X$.

Summing up, we know almost everything about elliptic codes. In Chapter 3.3 we shall also see that there exists a decoding algorithm up to $\left\lceil \frac{d-1}{2} \right\rceil$. Comparing elliptic codes with MDS-codes we see that their parameters $k + d = n$ are worse by 1, but that they are approximately by $2\sqrt{q}$ longer.

3.2.3. Other families of AG-codes

Up to now we have considered curves of small genera ($g = 0, 1, 2, 3$). Now we pass to the "middle" case, when the genus is not so small but is still not very large with respect to q .

Maximum codes. The parameters of AG-code (to be precise the sum of their designed parameters) are determined by the genus of a curve: $k_c + d_c = n - g + 1$. Hence, the parameters are the better, the more is the number of $\mathbb{F}_q$-points on a curve of the given genus. Recall that in Section 2.3.2 we have defined

$$N_q(g) = \max_X |X(\mathbb{F}_q)| \quad ,$$

the maximum being taken over all $\mathbb{F}_q$-curves X of genus g .

It is known (cf. Section 2.3.2) that

$$N_q(g) \le q + 1 + \lceil 2\sqrt{q} \rceil \cdot g \quad .$$

A curve attaining the equality is called a *maximum curve.* For a fixed q maximum curves exist only for a limited number of values of g.

Hermitian codes. There exists a class of maximum curves for which the properties of corresponding codes are rather well known and interesting. We mean *Hermitian curves* X_r with the affine equations

$$x^{r+1} + y^{r+1} + 1 = 0$$

over $\mathbb{F}_q$, where $q = r^2$, $r = p^m$ being an arbitrary power of a prime.

Exercise 3.2.36. Check that the Hermitian curve is a smooth plane curve, and that its genus equals $(r^2 - r)/2$.

To describe $\mathbb{F}_q$-points of X_r it is rather convenient to use the following change of variables:

$$u = b/(y - b\cdot x) \quad , \quad v = x\cdot u - a \quad ,$$

where $a,b \in \mathbb{F}_q$ satisfy $a^r + a = b^{r+1} = -1$. (The existence of such b follows from the fact that $\mathbb{F}_q^*$ is a cyclic group of order $r^2 - 1$. The existence of such a is also easy to prove).

Exercise 3.2.37. Show that this change of variables gives an isomorphism between X_r and the curve X'_r given by

$$v^r + v = u^{r+1} \quad .$$

Exercise 3.2.38. Show that the cardinality of $X'_r(\mathbb{F}_q)$ is at least $r^3 + 1$. (*Hint:* $X'_r(\mathbb{F}_q)$ consists of the unique infinite point of X'_r, i.e. the mutual pole of u and v, and the points (α,β), where $\alpha \in \mathbb{F}_q$ is arbitrary and β is any solution of $\beta^r + \beta = \alpha^{r+1}$, all these solutions lying in $\mathbb{F}_q$).

Since

$$r^3 + 1 = r^2 + 1 + 2\cdot g_r\cdot r \quad ,$$

g_r being the genus of X'_r, on it there are exactly $(r^3 + 1)$ points over $\mathbb{F}_q$ and the curves X'_r and X_r are maximum curves over $\mathbb{F}_q$.

We denote the mutual pole of u and v by Q and the point (α,β) by $P_{\alpha,\beta}$.

Exercise 3.2.39. Check that the principle divisors of functions $(u - \alpha)$ and $(v - \beta)$ are given by the following formulae:

$$(u - \alpha) = \sum_{\beta^r+\beta=\alpha^{r+1}} P_{\alpha,\beta} - r\cdot Q \quad ;$$

$$(v - \beta) = \begin{cases} (r + 1)\cdot P_{0,\beta} - (r + 1)\cdot Q & \text{if } \beta^r + \beta = 0 \\ \sum\limits_{\alpha^{r+1}=\beta^r+\beta} P_{\alpha,\beta} - (r + 1)\cdot Q & \text{if } \beta^r + \beta \neq 0. \end{cases}$$

Exercise 3.2.40. Check that the principle divisor of $z = u^q - u$ equals $(z) = \mathbf{P} - r^3\cdot Q$, where $\mathbf{P} = \sum_{\alpha,\beta} P_{\alpha,\beta}$. Derive that the residue of the differential form $\omega = dz/z$ at any points $P_{\alpha,\beta}$ equals 1 and its divisor is given by

$$(\omega) = (r^3 + r^2 - r - 2)\cdot Q - D \quad .$$

For any elements $\gamma, \delta \in \mathbb{F}_q$ such that $\delta^r + \delta = \gamma^{r+1}$ there is an $\mathbb{F}_q$-automorphism $\sigma = \sigma(\gamma,\delta)$ of X'_r given by

$$\sigma u = u + \gamma \quad , \quad \sigma v = v + \gamma^r \cdot u + \delta \quad .$$

Exercise 3.2.41. Show that automorphisms $\sigma(\gamma,\delta)$ form a subgroup G_r of order r^3 which acts transively on the set $\{P_{\alpha,\beta}\}$ and stabilizes the point Q .

Now let $n = r^3$ and $m \in \mathbb{N}$. By C_m we denote the q-ary code

$$C_m = (X'_r, \mathcal{P}, m\cdot Q)_L \quad ,$$

$\mathcal{P} = Supp\ \mathbf{P} = \{P_{\alpha,\beta}\} = X'_r(\mathbb{F}_q) - Q$ being the set of affine $\mathbb{F}_q$-points of X'_r .

Exercise 3.2.42. Prove that for any m the codes C_m and $C_{r^3+r^2-r-2-m}$ are dual. (*Hint*: Use Theorem 3.1.44 and Exercise 3.2.40).

In particular if q is even and $m = (r^3+r^2-r-2)/2$, the code C_m is self-dual.

To calculate $k_m = dim\ C_m$ we need the following

Exercise 3.2.43. Check that for any $m \geq 0$ the space $L(m\cdot Q)$ has the following basis:

$$\{u^i \cdot v^j \mid 0 \leq i \ ; \ 0 \leq j \leq r - 1 \ ; \ i\cdot r + j\cdot(r + 1) \leq m\} \quad .$$

For $m \in N$ define $\nu(m) \in \mathbb{N}$ as follows. Let $\bar{m} \leq m$ be the largest integer of the form $i\cdot r + j\cdot(r + 1)$,

$i \geq 0$, $0 \leq j \leq r - 1$, and let

$$\overline{m} = t \cdot r + s \ , \ 0 \leq s \leq r - 1 \ .$$

Set

$$\nu(m) = 1 + \frac{t \cdot (t + 1)}{2} + min\{t, s\} \ .$$

Exercise 3.2.44. Prove that $k_m = dim\ C_m$ is given by

$$k_m = \begin{cases} \nu(m) & \text{if } m \leq r^2 - r - 2 \\ m + 1 - \dfrac{r^2 - r}{2} & \text{if } r^2 - r - 2 < m < r^3 \\ r^3 - \nu(r^3+r^2-r-2-m) & \text{if } m \geq r^3 \end{cases}$$

(*Hint*: Use the results of Exercises 3.2.43 and 3.2.42).

We are also able to write out a generator (or a parity-check) matrix of the Hermitian code C_m . Indeed let M_m be the $(\nu(m) \times n)$-matrix whose rows are vectors

$$u(\alpha,\beta) = (\alpha^i \cdot \beta^j) \ ,$$

$$i \cdot r + j \cdot (r + 1) \leq m \ , \ 0 \leq i \ , \ 0 \leq j \leq r - 1 \ ,$$

for all $\alpha, \beta \in \mathbb{F}_q$ such that $\alpha^{r+1} = \beta^r + \beta$.

Proposition 3.2.45. a) *For* $0 \leq m < n = r^3$ *the matrix* M_m *is a generator matrix of the code* C_m .

b) *For* $r^2 - r - 2 < m \leq r^3 + r^2 - r - 2$ *the matrix* $M_{r^3+r^2-r-2-m}$ *is a parity check matrix of* C_m . ■

For the minimum distance d_m of C_m we have the usual estimate $d_m \ge r^3 - m$. Sometimes one can show this estimate to be tight.

Proposition 3.2.46. *Let* $m = i\cdot r + j\cdot(r+1) \le r^3 - 1$ *for some* $i \ge 0$ and $0 \le j \le r-1$. If *either* $j = 0$ *or* $i \le r^3 - r^2$ *(in particular, if* $m \le r^3 - r^2$ *) then* $d_m = r^3 - m$.

Proof: Let $j = 0$ and let $\alpha_1, \ldots, \alpha_i$ be different elements of $\mathbb{F}_q$. Then the function $f = \prod_{e=1}^{i}(u - \alpha_e)$ has exactly $i\cdot r$ different zeroes in $\mathcal{P}$. Now let $i \le r^2 - r - 1$. Consider the set

$$A = \{\alpha \in \mathbb{F}_q \mid \alpha^{r+1} \neq 1\}.$$

Its cardinality is $r^2 - r - 1$, and we can choose i different elements $\alpha_1, \ldots, \alpha_i \in A$. Let us put $t_1 = \prod_{\ell=1}^{i}(u - \alpha_\ell)$. Choose j different elements $\varepsilon_1, \ldots, \varepsilon_j$ such that $\varepsilon_\ell^r + \varepsilon = 1$ and let $t_2 = \prod_{\ell=1}^{i}(v - \varepsilon_\ell)$. It is easy to check that $t = t_1\cdot t_2 \in L(m\cdot Q)$ has exactly m different zeroes in $\mathcal{P}$. If $m \le r^3 - r^2$ then either $j = 0$ or $j \ge 1$ and $i \le r^2 - r - 1$. ∎

Since the group G_r acts transitively on $\mathcal{P}$ and leaves Q invariant, the code C_m is represented by an ideal I_m in the group algebra $\mathbb{F}_q[G_r]$.

Exercise 3.2.47. Write out the ideal I_m explicitly and estimate the parameters of C_m without the help of the Riemann-Roch theorem. (*Hint*: Use the above representation of elements of G_r in the form (γ, δ). Introduce the following elements of the group algebra:

$$u : G_r \longrightarrow \mathbb{F}_q \quad , \qquad v : G_r \longrightarrow \mathbb{F}_q \quad ,$$

$$u : (\gamma,\delta) \longmapsto \gamma \quad , \quad v : (\gamma,\delta) \longmapsto \delta \quad .$$

Show that I_m is generated by $u^i \cdot v^j$ for $0 \le i$, $0 \le j \le r - 1$, $i \cdot r + j \cdot (r + 1) \le m$; this gives a formula for k. To estimate d check that a function $F : G_r \longrightarrow \mathbb{F}_q$ lying in I_m and having more than m zeroes vanishes identically; doing this use the properties of Vandermonde determinants).

Hyperelliptic codes. A curve X is called hyperelliptic if it has a map $f : X \longrightarrow \mathbb{P}^1$ of degree 2. Such a curve always has a singular plane model given by the affine equation $y^2 + A(x) \cdot y + B(x) = 0$, the map f being given by the projection $f(x,y) = x$. If $g \ge 2$, the map f is unique up to an automorphism of $\mathbb{P}^1$ (i.e. for any $f' : X \longrightarrow \mathbb{P}^1$, $deg\, f' = 2$, there exists $\sigma \in Aut(\mathbb{P}^1)$ such that $\sigma \circ f' = f$), and is therefore defined over $\mathbb{F}_q$. For any g there exist hyperelliptic curves of genus g. Obviously $|X(\mathbb{F}_q)| \le 2g + 2$. Hence these curves can be maximum curves only for $g \le (q + 1)/\lceil 2\sqrt{q}\,\rceil$ being in this respect worse than the Hermitian curves, for which $g = (q - \sqrt{q})/2$.

One of advantages of hyperelliptic codes is the existence of a decoding algorithm up to $\left\lceil \frac{d - 1}{2} \right\rceil$ (cf. Section 3.3.3).

Multi-dimensional constructions. We are mostly unable to calculate (or to estimate reasonably) the parameters of codes constructed with the help of multi-dimensional varieties. There are however some exceptions.

Consider the projective space $\mathbb{P}^m$ over $\mathbb{F}_q$ and the bundle $\mathcal{L} = \mathcal{O}(1)$ on it. Its sections are linear forms in $(m + 1)$ variables, $dim\, H^0(\mathbb{P}^m, \mathcal{O}(1)) = m + 1$. Zeroes of such a section form a hyperplane $H \subset \mathbb{P}^m$; if the form f is defined over $\mathbb{F}_q$ then H is also defined over $\mathbb{F}_q$,

$$|H(\mathbb{F}_q)| = \frac{q^m - 1}{q - 1}, \quad |(\mathbb{P}^m - H)(\mathbb{F}_q)| = q^m .$$

We obtain a code C with parameters

$$\left[\frac{q^m - 1}{q - 1}, m + 1, q^m\right]_q .$$

This code is equidistant, i.e. weights of its non-zero code vectors are equal (and equal q^m); for this code

$$d/n = \frac{q^k}{q^k - 1} \cdot \frac{q - 1}{q},$$

i.e. the Plotkin bound (Theorem 1.1.39) is tight for it. We have in fact already considered these codes in Section 1.2.2, these are just first order Reed-Muller codes. Here is a natural generalization:

Again consider $X = \mathbb{P}^m$, and let $\mathcal{L} = \mathcal{O}(r)$. The sections of $\mathcal{O}(r)$ are forms of degree r, the zeroes of such a section form a hypersurface $S \subset \mathbb{P}^m$ of degree r (which can be reducible and non-reduced).

Exercise 3.2.48. Show that for $deg\, S = r \le q$ there is an $\mathbb{F}_q$-point in $\mathbb{P}^m$ out of S. (*Hint*: Use induction over the dimension).

Exercise 3.2.49. Show that for any hypersurface $S \subset \mathbb{P}^n$ of degree $r \le q$ there is an estimate

$$|S(\mathbb{F}_q)| \le r \cdot \frac{q^m - 1}{q - 1} .$$

(*Hint*: Consider the pencil of lines passing through a point out of S).

This estimate can be strengthened. If fact it is natural to suppose that the maximum number of points is attained when the hypersurface is just a union of r hyperplanes passing through a fixed plane of codimension two.

Exercise 3.2.50. Prove that

$$\max_S \{ |S(\mathbb{F}_q)| \} = r \cdot q^{m-1} + \frac{q^{m-1} - 1}{q - 1} \quad ,$$

the maximum being taken over all hypersurfaces $S \subset \mathbb{P}^m$ of degree $r \le q$. (*Hint*: Consider separately the case when all $\mathbb{F}_q$-points of S lie on its hyperplane components. In the other case fix an $\mathbb{F}_q$-point P_0 of S which does not lie on such components and consider the set of pairs (H,P) , P being an $\mathbb{F}_q$-point of S , $P \ne P_0$ and $H \supset \{P, P_0\}$. Calculate the number of such pairs in two different ways and compare the results).

Thus the H-construction yields a code C whose parameters (up to a spoiling) are

$$\left[\frac{q^{m+1} - 1}{q - 1} , \binom{m+r}{r} , q^m - (r - 1) \cdot q^{m-1} \right]_q .$$

Hypersurface codes. Consider a smooth hypersurface $X \subset \mathbb{P}^{m+1}$ of degree r and the bundle $\mathcal{L} = \mathcal{O}(1)$. Linear forms $s \in H^0(X,\mathcal{O}(1))$ vanish on hyperplane sections $S = X \cap H$. Suppose that the set of $\mathbb{F}_q$-points of X does not lie in a hyperplane. Then the H-construction gives us a code C whose parameters (up to a spoiling) are

$$\left[n,\ m+2,\ n - r\cdot q^{m-1} - \frac{q^m - 1}{q - 1} \right]_q .$$

Problem 3.2.51. Estimate the possible length n of this code, i.e. the maximum of $|X(\mathbb{F}_q)|$ over all smooth m-dimensional hypersurfaces X of degree r .

In general high-dimensional AG-codes are quite poorly studied. We are not even sure that the H-construction is good, may be one needs something else. Let us ask a question in this direction.

Problem 3.2.52. Discover a construction of codes starting from an abelian variety A such that for the Jacobian $A = J_X$ this construction gives an AG-code on the curve X .

3.2.4. Some particular examples

Now we are going to discuss some examples of AG-codes without any particular system. All these codes are on maximum curves.

Example 3.2.53. Consider the plane curve E given

over $\mathbb{F}_2$ by the affine equation $y^2 + y = x^3 + x + 1$, i.e. by the homogeneous equation $y^2 \cdot z + y \cdot z^2 = x^3 + x \cdot z^2 + z^3$.

Exercise 3.2.54. Check that the curve $E \subset \mathbb{P}^2$ is absolutely irreducible and smooth (therefore it is an elliptic curve). Check that $|E(\mathbb{F}_2)| = 1$, $|E(\mathbb{F}_4)| = 5$, $|E(\mathbb{F}_{16})| = 25$. (*Hint*: The only $\mathbb{F}_2$-point is the infinity; $|X(\mathbb{F}_q)| = 2^r + 1 - \omega^r - \bar{\omega}^r$, where $q = 2^r$, and $|\omega| = \sqrt{2}$; since $|X(\mathbb{F}_2)| = 1$, we get $\omega = 1 + i$). Write out the coordinates of $\mathbb{F}_4$- and $\mathbb{F}_{16}$-points of E.

Note that since $25 = 16 + 1 + 2 \cdot 4$, E is a maximum curve over $\mathbb{F}_{16}$.

Exercise 3.2.55. Let $Q = (0:1:0)$ be the only infinite point of E. Prove that for any $m \geq 0$ the space $L(m \cdot Q)$ has a basis $\{x^i \cdot y^j \mid 2i + 3j \leq m\}$. Write out generator matrices of codes $(E, \mathcal{P}, m \cdot Q)_L$ and $(E, \mathcal{P}', m \cdot Q)_L$ over $\mathbb{F}_4$ and $\mathbb{F}_{16}$, respectively, where $\mathcal{P} = E(\mathbb{F}_4) - \{Q\}$, $\mathcal{P}' = E(\mathbb{F}_{16}) - \{Q\}$ for $m = 2, 8, 12, 16$. Calculate their spectra.

Example 3.2.56. Consider the curve E_1 over $\mathbb{F}_{25}$ given by $y^2 + y = x^3$.

Exercise 3.2.57. Show that $E_1 \subset \mathbb{P}^2$ is a maximum (over $\mathbb{F}_{25}$) elliptic curve, i.e. $|E_1(\mathbb{F}_{25})| = 36$ (one point at infinity and 35 points in $\mathbb{A}^2$). Write out their coordinates and generator matrices of codes $(E_1, \mathcal{P}_1, m \cdot Q)_L$ for $m = 2, 8, 16$, where $\mathcal{P}_1 = E(\mathbb{F}_{25}) - \{Q\}$. (*Hint*: Use the same method as in Exercise 3.2.55).

Example 3.2.58 (*the Klein quartic*). Consider the curve $X \subset \mathbb{P}^2$ given by $x^3 \cdot y + y^3 \cdot z + z^3 \cdot x = 0$.

Exercise 3.2.59. Show that X is a smooth curve of genus 3 , and that it is a maximum curve over $\mathbb{F}_8$. (*Hint*: $24 = 8 + 1 + 3\cdot\lceil 2\sqrt{8}\rceil$). Write out the coordinates of $\mathbb{F}_8$-points.

Exercise 3.2.60. Prove that the divisor of the linear form z equals $(z) = Q_1 + 3\cdot Q_2$, where $Q_1 = (1:0:0)$, $Q_2 = (0:1:0)$. Show that the space $L(m\cdot(z))$ is generated by $\{x^i\cdot y^j \mid 2i + 3j \le 3m\ ,\ \ i - 2j \le m\}$. Restrict this generating system to a basis of $L(m\cdot(z))$ and write out generator matrices of codes $(X,\mathcal{P},m\cdot(z))_L$ for $m = 2,\ 3,\ 5,$ where $\mathcal{P} = X(\mathbb{F}_8) - \{Q_1,Q_2\}$.

Example 3.2.61. Consider a smooth model Y of the curve over $\mathbb{F}_{25}$ given by the equation $y^2 = x^6 + 1$. The projection $(x,y) \longmapsto x$ defines a map $f : Y \longrightarrow \mathbb{P}^1$ of degree two, i.e. Y is hyperelliptic.

Exercise 3.2.62. Show that the genus of Y equals 2. (*Hint*: Use the Hurwitz formula).

Exercise 3.2.63. Show that $|Y(\mathbb{F}_{25})| = 46$ and Y is a maximum curve. Find out the coordinates of $\mathbb{F}_{25}$-points. (*Attention*: There are two infinite points $Q_1,\ Q_2 \in Y(\mathbb{F}_{25})$ corresponding to $x = y = \infty$).

Exercise 3.2.64. Find out a basis of $L(D)$ for $D = 10\cdot(Q_1 + Q_2)$ and write out the generator matrix of the code $(Y,\mathcal{P},D)_L$ where $\mathcal{P} = Y(\mathbb{F}_{25}) - \{Q_1,Q_2\}$. (*Hint*: Consider Y as a singular plane curve given by $y^2\cdot z^4 = x^6 + z^6$ with one singularity $Q = (0:1:0)$. Construct the desingularization tree of Q , according to Section 2.5.4 and express the local parameters $x_1 = x/y$ and $z_1 = z/y$ at the points Q_1 and Q_2 respectively, in terms of local parameters (u_1,t_1) and (u_2,t_2) . Using these expressions rewrite the condition $(P(x,y)) + D \geq 0$ as a system of linear conditions for coefficients of the polynomial $P(x,y)$. Solving these equations find out the basis).

Example 3.2.65. Consider the Hermitian curve X_5 (given by $x^5 + y^5 = 1$) over $\mathbb{F}_{16}$ and the code $C = (X_5, \mathcal{P}, 37\cdot Q)_L$ where (in the notation of the previous section) $\mathcal{P} = X_5(\mathbb{F}_{16}) - \{Q\}$.

Exercise 3.2.66. Show that C is a self-dual code of length 64 over $\mathbb{F}_{16}$, its minimum distance being equal to 27 . (*Hint*: See Proposition 3.2.46). Write out a generator matrix of C . Show that in the spectrum of C we have $A_r > 0$ for $r \geq 27$. (*Hint*: For all r except 63, 62, 61, 58, 57, 53 this follows from the argument we used to prove Proposition 3.2.46. Similar functions t can be constructed for other values of r . For example, for $r = 53$ one can take

$$t = (u - \alpha_1)\cdot(u - \alpha_2)\cdot(v - \gamma_1)\cdot(v - \gamma_2)\cdot(v - \gamma_3) ,$$

where $\alpha_1, \alpha_2 \neq 0$, $\gamma_i^3 + 1 = 0$ for $i = 1,2,3;$ $\gamma_i \neq \gamma_j$ for $i \neq j$).

All the curves we have considered up to this moment were plane. Here is a space curve.

Example 3.2.67. Consider a curve $Z \subset \mathbb{P}^3$ given over $\mathbb{F}_3$ by the equations

$$\begin{cases} x\cdot t - z^2 = 0 \\ y^3 - y\cdot t^2 - x^2\cdot z = 0 \end{cases} .$$

Exercise 3.2.68. Show that Z is a smooth absolutely irreducible curve of genus 4 cannonically embedded into $\mathbb{P}^3$. (*Hint*: Consider the divisor of the form dx and show that $(dx) = 6\cdot Q$, where $Q = (1:0:0:0)$).

Exercise 3.2.69. Show that Z is a maximum curve over $\mathbb{F}_{81}$; $|Z(\mathbb{F}_{81})| = 154$. Find out the coordinates of $\mathbb{F}_{81}$-points.

Exercise 3.2.70. Show that $L(m\cdot Q)$ has the basis $\{y^i\cdot z^j \mid 5i + 3j \le m, \quad 0 \le j \le 2, \ i \ge 0\}$. Write out a generator matrix of $(Z,\mathcal{P},m\cdot Q)_L$ for $m = 3, 6, 20, 50,$ where $\mathcal{P} = Z(\mathbb{F}_{81}) - \{Q\}$.

To end the chapter devoted to examples we express our regret that such remarkable codes as the Golay codes are left out of it. The following question looks appropriate:

Problem 3.2.71. Express the Golay codes as AG-codes of the form $(X,\mathcal{P},D)_L$, $(X,\mathcal{P},D)_\Omega$, or $(X,\mathcal{P},\mathcal{L})_H$ for some curves X over $\mathbb{F}_2$ and $\mathbb{F}_3$, some sets $\mathcal{P} \subseteq X(\mathbb{F}_q)$, and some divisors D (bundles $\mathcal{L}$).

CHAPTER 3.3

DECODING

In this chapter we shall see that algebraic-geometric codes can be effectively decoded. In Section 3.3.1 we present a generalization of the decoding algorithm for Reed-Solomon codes (given in Section 1.2.1) to the case of an arbitrary curve; we call it the basic algorithm. Unfortunately the basic algorithm corrects $g/2$ errors less than one would like. The reason is that the Riemann-Roch theorem does not answer the question about the exact value of the dimension $\ell(D)$ for $0 \le deg\ D \le 2g - 2$. Section 3.3.2 is devoted to a modification of the basic algorithm leading in particular in Section 3.3.3 to algorithms decoding elliptic codes and many hyperelliptic codes (for some proper choice of the divisor) up to the half of the designed minimum distance. In Section 3.3.3 we also consider plane curves, for which it is possible to correct about $g/4$ errors more than in the general case, and also codes obtained from AG-codes by concatenation and field restriction.

3.3.1. Basic algorithm

Consider an AG-code $C = (X, \mathcal{P}, D)_\Omega$ over $\mathbb{F}_q$. If $deg\ D = a$, $|\mathcal{P}| = n$, $2g - 2 < a \le n + g - 1$, then its designed parameters are

$$k_C = n - a + g - 1 \quad ,$$

$$d_C = a - 2g + 2 \quad .$$

Theorem 3.1.43 shows that a parity-check matrix of C is given by $(f_i(P_j))$, where $f_1, \ldots, f_m$ is a basis of $L(D)$.

Basic systems. For a vector $v \in \mathbb{F}_q^n$ and a function $f \in L(D)$ we define the syndrome

$$s(v,f) = \sum_{P_i \in \mathcal{P}} v_i \cdot f(P_i) \quad .$$

Note that if $v = u + e$, where $u \in C$ and e is the error vector, $I = \{i \mid e_i \ne 0\}$ being the set of "error coordinates" (or *error locators*), then

$$s(v,f) = \sum_{i \in I} e_i \cdot f(P_i) \ .$$

Let $t = |I|$.

Consider an auxiliary divisor D' of degree b such that $Supp\ D' \cap \mathcal{P} = \emptyset$. By $g_1, \ldots, g_\ell$ we denote a basis of $L(D')$, and by $h_1, \ldots, h_r$ a basis of $L(D - D')$. Of course $g_i \cdot h_j \in L(D)$. Let

$$s_{ij} = s_{ij}(v) = s(v, g_i h_j) \quad .$$

The crucial role for decoding of C is played by the system

$$\sum_{i=1}^{\ell} s_{ij} \cdot x_i = 0 \quad , \; j = 1,\ldots,r \quad . \tag{3.3.1}$$

Proposition 3.3.1. *If*

$$\ell(D') > t$$

then the system (3.3.1) *has a non-trivial solution. If*

$$a - b > t + 2g - 2$$

then for any solution $y = (y_1,\ldots,y_\ell)$ *of this system, the function*

$$g_y = \sum_{i=1}^{\ell} y_i \cdot g_i$$

vanishes at the point P_i *for any* $i \in I$.

Proof: If $\ell(D') > t$ then

$$\ell(D' - \sum_{i \in I} P_i) \geq \ell(D') - t > 0 \ .$$

Let $g \in L(D' - \sum_{i \in I} P_i)$, $g \neq 0$. Express g over the basis of $L(D')$: $g = \sum y_i \cdot g_i$. The vector $y = (y_1,\ldots,y_\ell)$ is a solution of (3.3.1) since $g(P_k) = 0$ for $k \in I$ and therefore

$$\sum_i s_{ij}(v) \cdot y_i = \sum_i s_{ij}(e) \cdot y_i = \sum_i \sum_{k \in I} e_k \cdot g_i(P_k) \cdot h_j(P_k) \cdot y_i =$$

$$= \sum_{k \in I} e_k \cdot g(P_k) \cdot h_j(P_k) = 0 \quad ,$$

If $a - b > t + 2g - 2$ then the Riemann-Roch theorem yields

$$\ell(D - D' - \sum_{i \in I} P_i) = a - b - t - g + 1$$

since

$$deg\,(D - D' - \sum_{i \in I} P_i) \geq 2g - 1 \,.$$

Because of the same reason,

$$\ell(D - D') = a - b - g + 1 \,.$$

Let $\mathcal{P}_I = \{P_i\}_{i \in I}$. The kernel of the map $Ev_{\mathcal{P}_I} : L(D - D') \longrightarrow \mathbb{F}_q^t$ equals $L(D - D' - \sum_{i \in I} P_i)$, however

$$\ell(D - D') - \ell(D - D' - \sum_{i \in I} P_i) = t \,,$$

hence $Ev_{\mathcal{P}_I}$ is surjective, and therefore for any $i \in I$ there exists $F_i \in L(D - D')$ such that $F_i(P_i) = 1$ and $F_i(P_j) = 0$ for $j \in I - \{i\}$.

Since the functions $F_i \in L(D - D')$ are linear combinations of the functions h_j, the equations $\sum_i s(v, g_i \cdot F_j) \cdot x_i = 0$ follow from the system (3.3.1). For any solution $y = (y_1, \ldots, y_\ell)$ we have

$$0 = \sum_i s(v, g_i \cdot F_j) \cdot y_i = \sum_i \sum_{k \in I} e_k \cdot g_i(P_k) \cdot F_j(P_k) \cdot y_i =$$

$$= \sum_{k \in I} e_k \cdot g_y(P_k) \cdot F_j(P_k) = e_j \cdot g_y(P_j) \quad ,$$

i.e. $g_y(P_j) = 0$ for any $j \in I$. ∎

Thus, the above properties of the divisor D' are sufficient to find a function g_y vanishing at all points

P_i which correspond to error locators (and may be also vanishing at some "extra" points). Denote the set of points $P_i \in \mathcal{P}$ such that $g_y(P_i) = 0$ by I_y ; we have just proved that $I_y \supseteq I$.

Consider the system

$$\sum_{i \in I_y} f_j(P_i) \cdot z_i = s(v, f_j) \quad , \tag{3.3.2}$$

$f_1, \ldots, f_m$ being a basis of $L(D)$. The error vector e is a solution of this system, since $s(v, f_i) = s(e, f_i)$.

Proposition 3.3.2. *If*

$$a - b > 2g - 2$$

then the system (3.3.2) *has at most one solution.*

Proof: Let z and z' be two different solutions of this system. Then $(z - z')$ is a solution of

$$\sum_{i \in I_y} f_j(P_i) \cdot x_i = 0 \ ,$$

i.e. the vector $x = (x_1, \ldots, x_n)$, where $x_i = z_i - z'_i$ for $i \in I_y$ and $x_i = 0$ for $i \notin I_y$ is a non-zero code vector: $x \in C - \{0\}$. Moreover $\|x\| \le |I_y| \le deg\ D' = b$ since $g_y \in L(D')$. On the other hand by supposition $b < a - 2g + 2 = d_c \le d$. The weight of a non-zero vector cannot be less than the minimum distance. The contradiction proves the proposition. ■

Algorithm. Thus a choice of an auxiliary divisor D' yields the following algorithm of decoding the code $C = (X, \mathcal{P}, D)_\Omega$, which we call the *basic algorithm*:

1. Calculate a basis $\{f_i\}$ of $L(D)$, a basis $\{g_j\}$ of $L(D')$, and a basis $\{h_k\}$ of $L(D - D')$.

2. For a given vector $v \in \mathbb{F}_q^n$ calculate its syndromes $s(v, g_j \cdot h_k)$ and $s(v, f_i)$.

3. Find out a solution y of the linear system (3.3.1).

4. Find out (sorting out the points P_i) those i for which $g_y(P_i) = 0$.

5. Solve the linear system (3.3.2).

Propositions 3.3.1 and 3.3.2 immediately yield

Theorem 3.3.3. *Let* $C = (X, \mathcal{P}, D)_\Omega$, $2g - 2 < a = \deg D \le n - g + 1$. *If for a positive integer* t *there exists a divisor* D' *of some degree* b *such that* $\mathrm{Supp}\, D' \cap \mathcal{P} = \emptyset$ *and*

$$\ell(D') > t \quad ,$$

$$a - b > t + 2g - 2$$

then the basic algorithm corrects t *errors.* ■

Corollary 3.3.4. *The basic algorithm corrects any*

$$t \le t_0 = \max_{D'} \{\min \{\ell(D') \, , \, a - \deg D' - 2g + 2\}\} - 1$$

errors, the maximum being taken over all divisors D' *such that* $\mathrm{Supp}\, D' \cap \mathcal{P} = \emptyset$. ■

Note that to use the basic algorithm we must know D' explicitly. Using the H-construction or choosing $D = a \cdot P_0$ for some $P_0 \in X(\mathbb{F}_q)$ we easily solve the problem of finding D'.

Exercise 3.3.5. Extend the results of this section to codes $C = (X,\mathcal{P},\mathcal{L})_H$ (to dispense with the conditions $Supp\ D \cap \mathcal{P} = \varnothing$ and $Supp\ D' \cap \mathcal{P} = \varnothing$).

Let us estimate t_0 . Knowing the result of Exercise 3.3.5 we just forget the condition $Supp\ D' \cap \mathcal{P} = \varnothing$.

Proposition 3.3.6. $t_0 \geq \left\lceil \frac{d_C - 1 - g}{2} \right\rceil$.

Proof: Choose a point P_0 and let $D' = b\cdot P_0$, where $b = t_0 + g$. By the Riemann-Roch theorem $\ell(D') \geq t_0 + 1$. Hence

$$t_0 \geq a - b - 2g + 1 = a - t_0 - 3g + 1 ,$$

i.e.

$$2t_0 \geq a - 3g + 1 = d_C - 1 - g .$$

■

Exercise 3.3.7. Check that if the bases of $L(D)$, $L(D')$, and $L(D - D')$ are already given then the basic algorithm uses at most $c\cdot n^3$ operations in the field $\mathbb{F}_q$, where the constant c does not depend on q . (*Hint*: In fact it needs at most $c((d^2 + g^2)\cdot n)$ operations, which is better than $c\cdot n^3$ when q grows. To prove this just note that a system of m linear equations in n variables needs $O(m^2\cdot n)$ operations in $\mathbb{F}_q$).

3.3.2. Modified algorithm

In this section we modify the basic algorithm using it iteratively for $b = 1, 2, \ldots$. The *modified algorithm* is especially effective for $D = a \cdot P$, P being a Weierstrass point. However it may be of use in many other cases as well, for example it is possible to apply it to decode elliptic codes up to $\lceil (d_C - 1)/2 \rceil$.

For a fixed D' and a given $v \in \mathbb{F}_q^n$ (which we want to decode) let V denote the space of solutions of the system (3.3.1).

Exercise 3.3.8. Consider the embedding

$$L\,(D' - \sum_{i \in I} P_i) \hookrightarrow V \quad .$$

mapping each function $g \in L(D' - \sum_{i \in I} P_i) \subseteq L(D')$ to the set of coefficients of its expansion over the basis $\{g_i\}$. Check that this embedding is well defined and if $D'' = D - D' - \sum_{i \in I} P_i$ is non-special (i.e. if $\ell(K - D'') = 0$) then this map is an isomorphism. (*Hint*: Cf. the proof of Proposition 3.3.1) .

Now we are going to consider the case $D = a_1 \cdot h$ where H is an effective divisor of degree h ; $a = a_1 \cdot h$. Let

$$S(H) = \max_{i \in \mathbb{Z}} \left\{ \left\lceil \frac{i \cdot h + h + 1}{2} \right\rceil - \ell(i \cdot H) \right\} ,$$

then for any $i \in \mathbb{Z}$

$$\ell(i \cdot H) \geq \left\lceil \frac{i \cdot h + h + 1}{2} \right\rceil - S(H) \quad .$$

The value of $S(H)$ influences greatly the correcting ability of the following method (which we call the *modified algorithm*).

Algorithm. We shall solve system (3.3.1) for divisors $D' = H, 2H, 3H, \ldots$. Let b_1 be the smallest integer for which there exists a solution (in particular, because of Exercise 3.3.8, we have

$$\ell((b_1 - 1)\cdot H - \sum_{i\in I} P_i) = 0).$$

Proposition 3.3.9. *Let*

$$t \le \left\lceil \frac{d_C - 1}{2} \right\rceil - S(H) .$$

Then the divisor $D'' = (a_1 - b_1)\cdot H - \sum_{i\in I} P_i$ *is non-special.*

Proof: Suppose it is special, i.e. $\ell(K - D'') > 0$. Then there exists an effective divisor E equivalent to $K - D''$;

$$deg\ E = 2g - 2 - a_1\cdot h + b_1\cdot h + t .$$

Then the divisor $(b_1 - 1)\cdot H - \sum_{i\in I} P_i$ is equivalent to $K - E - (a_1 - 2b_1 + 1)\cdot H$, hence

$$\ell(K - E - (a_1 - 2b_1 + 1)\cdot H) = 0 ,$$

and therefore

$$\ell(K - (a_1 - 2b_1 + 1)\cdot H) \le deg\ E$$

(adding a point to a divisor we add at most 1 to the dimension, and adding an effective divisor at most its degree). By the Riemann-Roch theorem

$$\ell(K - (a_1 - 2b_1 + 1)\cdot H) =$$

$$= \ell((a_1 - 2b_1 + 1)\cdot H) - a_1\cdot h + 2b_1\cdot h - h + g - 1 .$$

On the other hand, the definition of $S(H)$ yields

$$\ell((a_1 - 2b_1 + 1)\cdot H) \geq$$

$$\geq \left\lceil \frac{(a_1 - 2b_1 + 1)\cdot h + h + 1}{2} \right\rceil - S(H) =$$

$$= \left\lceil \frac{a_1\cdot h - 2g + 1}{2} \right\rceil + g - b_1\cdot h + h - S(H) \quad .$$

We get

$$deg\ E \geq \ell(K - (a_1 - 2b_1 + 1)\cdot H) \geq$$

$$\geq \left\lceil \frac{d_C - 1}{2} \right\rceil + 2g - 1 - a_1\cdot h + b_1\cdot h - S(H) \geq$$

$$\geq t + 2g - 1 - a_1\cdot h + b_1\cdot h = deg\ E + 1 \ .$$

which is contradictory. ■

This proposition and Exercise 3.3.8 show that when t satisfies the condition of Proposition 3.3.9 we have

$$V = L(b_1\cdot H - \sum_{i\in I} P_i) \ ,$$

i.e. each solution of (3.3.1) yields a function g vanishing at error locators. Now we have to find out the errors, which is similar to Proposition 3.3.2.

Exercise 3.3.10. Prove that under the condition on t the system (3.3.2) has a unique solution. (*Hint*: Use the definition of $S(H)$ and the fact that $\ell((b_1 - 1)\cdot H) \le t$, since $\ell((b_1 - 1)\cdot H - \sum_{i\in I} P_i) = 0$).

Thus we have proved

Theorem 3.3.11. *For a positive divisor H let $D = a_1\cdot H$ be such that $2g - 2 < a = \deg D \le n - g + 1$. Then the modified algorithm described above decodes any*

$$t \le \left\lceil \frac{d_C - 1}{2} \right\rceil - S(H)$$

errors for the code $C = (X,\mathcal{P},D)_\Omega$. ■

Exercise 3.3.12. The modified algorithm obviously needs at most $c\cdot n^4$ operations in $\mathbb{F}_q$. Find a method to solve systems (3.3.1) for $D'_i = i\cdot H$, $i = 1,\dots,b_1$ which lowers this estimate to $c\cdot n^3$. (*Hint*: The system (3.3.1) for D'_i is obtained from those for D'_{i-1} by crossing out some equations and adding some new variables. If one of the systems is triangular then the other is "rather close" to a triangular form).

Remark 3.3.13. Contrary to the situation with the basic algorithm, we are unable to make the modified algorithm applicable to the H-construction, or to the case of an arbitrary D.

3.3.3. Some remarks

Estimation of $S(H)$. The question about the value of $S(H)$ is now crucial.

Exercise 3.3.14. Prove that

$$S(H) \geq \left\lceil \frac{h-1}{2} \right\rceil ,$$

and that if either h or $\left\lceil \frac{g-1}{h} \right\rceil$ is even then

$$S(H) \leq \left\lceil \frac{g+h}{2} \right\rceil - 1 ,$$

and if neither of them is even then

$$S(H) \leq \left\lceil \frac{g+h-1}{2} \right\rceil .$$

(*Hint*: To prove the second inequality use the fact that $S(H)$ is the maximum of the difference of values at integers of two functions. One of them is almost linear, and the other function $\ell(i \cdot H)$ has a polygon lower bound consisting of the horizontal line $\ell(i \cdot H) \geq 1$ and the line given by the Riemann-Roch theorem. Maximum is attained at their meeting point).

If $S(H) = 0$, the modified algorithm corrects $\left\lceil \frac{d_C - 1}{2} \right\rceil$ errors, just as one would like. However this is a rare case.

Hyperelliptic curves. From the definition it is clear that $S(H)$ is the smaller, the larger is the speciality index of $i \cdot H$. Consider the "most favorable" case.

Let a curve X of genus $g \geq 1$ have a map $f : X \longrightarrow \mathbb{P}^1$ of degree 2, i.e. either $g = 1$ and X is elliptic, or X is hyperelliptic. For H we choose a hyperelliptic divisor, i.e. $H = f^*(P)$, $P \in \mathbb{P}^1(\mathbb{F}_q)$; $h = deg\ H = 2$.

Exercise 3.3.15. Prove that $S(H) = 0$. (*Hint*: in this case the Clifford theorem yields $\ell(i \cdot H) = i + 1$ for $i \leq g$).

Thus for a hyperelliptic (or an elliptic) curve X and a multiplicity of a hyperplane section $D = a_1 \cdot H$, the code $C = (X, \mathcal{P}, a_1 \cdot H)_\Omega$ can be decoded up to $\left\lceil \frac{d_C - 1}{2} \right\rceil$. In particular decoding exists for any curve X of genus $g \leq 2$ (for an appropriate choice of the divisor), since any curve of genus two is hyperelliptic.

Unfortunately this is the only case when $S(H) = 0$.

Exercise 3.3.16. Prove that if $S(H) = 0$ then either $g \leq 1$ or H is a hyperelliptic divisor (and X is a hyperelliptic curve). (*Hint*: Set $i = 1$ and 2 in the definition of $S(H)$).

Plane curves. For a smooth plane curve $X \subset \mathbb{P}^2$ and a hyperplane section divisor H it is possible to calculate $S(H)$ precisely. It this case $h = deg\ H = deg\ X$ equals the degree of an irreducible equation defining X.

Exercise 3.3.17. Prove that in this case

$$S(H) = \left\lceil \frac{(h - 1)^2}{8} + \frac{1}{2} \right\rceil .$$

(*Hint*: Check that

$$\ell(i \cdot H) = \frac{(i + 2) \cdot (i + 1)}{2}$$

for $i < h$, and that the value of i defining $S(H)$ lies in this zone).

The result of Exercise 3.3.17 means that for large g it is possible to decode a plane curve code up to about $\frac{d_c - 1}{2} - \frac{g}{4}$ instead of the general estimate $\frac{d_c - 1}{2} - \frac{g}{2}$.

Exercise 3.3.18. Estimate $S(H)$ for a smooth irreducible curve $X \subset \mathbb{P}^3$ and also for $X \subset \mathbb{P}^m$, m being fixed. Show that for a fixed m and for large g we get a bit better estimate than in the general case (which however tends to $\frac{d_c - 1}{2} - \frac{g}{2}$ when $m \longrightarrow \infty$).

Fermat curves. Consider the curve

$$x^m + y^m = z^m ,$$

m and q being co-prime. Let $P = (1:0:1) \in X(\mathbb{F}_q)$.

Exercise 3.3.19. Show that if $m \equiv 0 (mod\ 4)$ then

$$S(P) = \frac{m \cdot (m - 4)}{8} + 1 .$$

(*Hint*: Consider the map $\pi : X \longrightarrow \mathbb{P}^1$, $\pi(x:y:z) = (y:x - z)$ and use the method of Proposition 2.2.48. Show that the obtained values of non-gaps determine the gap structure at P). Consider the case of other m.

Tightening. In a number of cases Theorem 3.3.11 can be tightened.

Exercise 3.3.20. Prove that if $D = a_1 \cdot H$, a_1 being even and $a_1 > \frac{2g - 2}{h} + 1$, then the modified algorithm corrects any $\left\lceil \frac{d_c - 1}{2} \right\rceil - s(H)$ errors, where

$$s(H) = \max_{i \geq 1} \left\{ \left\lceil \frac{i \cdot h + h + 1}{2} \right\rceil - \ell(i \cdot H) \right\} .$$

(*Hint*: Read the proof of Theorem 3.3.11 attentively). Derive that if X is elliptic and $D = 2a_2 \cdot H$, it is possible to correct $\left\lceil \frac{d_c - 1}{2} \right\rceil$ errors.

Exercise 3.3.21. For $H = P \in X(\mathbb{F}_q)$ let

$$s_0(P) = \max_j \{j + 1 - \ell((2j + 1) \cdot P)\}$$

Prove that if $D = a \cdot P$, a being even, then the modified algorithm corrects $\left\lceil \frac{d_c - 1}{2} \right\rceil - s_0(P)$ errors. Prove that if $s_0(P) = 0$ then either X is hyperelliptic, or $g(X) \leq 4$. Give examples of non-hyperelliptic curves of genus 3 and 4 and points on them such that $s_0(P) = 0$. (*Hint*: The gap sequence should be $\{1,2,4\}$ or $\{1,2,5\}$ for a curve of genus 3, and $\{1,2,4,7\}$ for a curve of genus 4).

To end this discussion let us point out that the main question on decoding of AG-codes is still open:

Problem 3.3.22. Find out an algorithm decoding any AG-code up to $\left\lceil \frac{d_c - 1}{2} \right\rceil$.

Remark 3.3.23. Problem 3.3.22 being open, there are however some results which can be summed up as follows.

Let X be a curve of genus g over $\mathbb{F}_q$. Consider the map

$$\psi_s : (Div^+_{g-1}(\mathbb{F}_q))^s \longrightarrow (J_X(\mathbb{F}_q))^{s-1} ,$$

$$(D_1, \ldots, D_s) \longmapsto ([D_1 - D_2],[D_2 - D_3],\ldots,[D_{s-1} - D_s])$$

$Div^+_{g-1}(\mathbb{F}_q)$ being the set of effective $\mathbb{F}_q$-divisors of degree $g - 1$, $(Div^+_{g-1}(\mathbb{F}_q))^s$ being its cartesian s-power, $J_X(\mathbb{F}_q)$ being the group of $\mathbb{F}_q$-points of the Jacobian, and $[D]$ denoting the image in J_X of a divisor of degree zero. If for some s the map ψ_s is not surjective then, applying the basic algorithm parallelly to some divisors $D' = D'_1 , \ldots , D'_s$, we decode $C = (X,\mathcal{P},D)_\Omega$ up to $\left\lceil\frac{d_C - 1}{2}\right\rceil$. Moreover, it can be proves that ψ_s is not surjective for $s = 2g$ at least if either

a) X is a maximum curve, or

b) $q \geq 37$, or $q \geq 16$ and g is large enough.

Thus in these cases there exists an algorithm of decoding. Unfortunately this is an existence theorem, we are still unable to find out necessary divisors $D'_1,\ldots,D'_s$. In particular, even for q large enough, this algorithm does not lead to polynomial decoding of asymptotically good codes (cf. Section 3.4.4) up to $\left\lceil\frac{d_C - 1}{2}\right\rceil$.

Concatenated and restricted codes. In Section 3.1.3 we have constructed some families of codes, applying concatenation and field restriction to AG-codes. It is not difficult to give decoding algorithms for such codes.

Exercise 3.3.24. Write out a decoding algorithm up to $\left\lceil \frac{d\cdot(D_C - g) - 1}{2} \right\rceil$ for concatenated codes with a fixed inner $[n,k,d]_q$-code and outer algebraic-geometric $[N, K, \geq D_C]_{q^b}$-codes, which needs at most $c_1(n\cdot N)^3$ operations in $\mathbb{F}_q$, where c_1 is a constant independent of q, and $n\cdot N$ is the length of codes obtained. (*Hint*: Use Proposition 1.2.31 and Exercise 1.2.30).

Exercise 3.3.25. Write out a decoding algorithm up to $\left\lceil \frac{d_C - g - 1}{2} \right\rceil$ for q-ary restriction of q^m-ary AG-codes (m being fixed), which needs at most $c\cdot N^3$ operations in $\mathbb{F}_q$.

Our final remark concerns any linear code.

Remark 3.3.26. The basic algorithm can be in fact applied to any linear code, having a so-called *correcting pair*. Let A, B, C be linear codes in $\mathbb{F}_q^n$ such that

1) $A*B \subseteq C^\perp$, where $A*B = \{a*b \mid a \in A,\ b \in B\}$, and $a*b = (a_1\cdot b_1,\ldots,a_n\cdot b_n)$;

2) $k(A) \geq t$;

3) $d(B^\perp) > t$;

4) $n < d(A) + d(C)$.

Then (A,B) is called a t-correcting pair for C. If there exists such a pair with $t \leq (d(C) - 1)/2$ then there is an algorithm of complexity $O(n^3)$ correcting t errors. For $C = (X,\mathcal{P},D)_\Omega$ we can set $A = (X,\mathcal{P},D')_L$, $B = (X,\mathcal{P},D - D')_L$ and obtain the basic algorithm.

CHAPTER 3.4

ASYMPTOTIC RESULTS

The advantages of algebraic-geometric codes are most illustrious when we consider asymptotic problems. It turns out that the parameters of AG-codes constructed from families of algebraic curves are the better, the higher is (asymptotically) the ratio of the number of $\mathbb{F}_q$-points on them to their genus. In Section 3.4.1 we establish the basic algebraic-geometric asymptotic bound, it is a line intersecting the Gilbert-Varshamov bound if q is large enough (thus ameliorating it on a segment). This result shows that highly un-random codes given by subtle algebraic geometry (modular curves are used) can be asymptotically better than random codes (recall that with probability one any linear code lies on the Gilbert-Varshamov bound). Then, in Section 3.4.2, we see that a probabilistic argument can be also applied to AG-codes themselves. An expurgation method leads (varying the divisor) to the expurgation bound. For small q this bound coincides with the Gilbert-Varshamov bound, and if q is large enough, it is a bit better than the maximum of the Gilbert-Varshamov bound and the basic AG-bound, smoothing the angles at points of their intersection. The basic AG-bound is constructive, applying to it various constructions of Section 1.2.3 we (in Section 3.4.3) ameliorate the polynomial Bloch-Ziablov bound for any q and δ. The progress due to AG-codes leads also to amelioration of some other asymptotic bounds (in Section 3.4.4). For non-linear codes (q being large enough) it can be shown that the expurgation bound is not the best possible; moreover, the amelioration of the Gilbert bound can be extended to arbitrary alphabets. The bound for codes with polynomial construction and polynomial decoding is now also sometimes better than the Gilbert-Varshamov bound. The corresponding bounds for self-dual codes and for constant-weight codes can be also ameliorated by the use of AG-codes.

3.4.1. Basic AG-bound

Let us apply Corollary 3.1.11 to families of curves of growing genus.

Theorem 3.4.1. *If over* $\mathbb{F}_q$ *there exists a family of curves* X_i *of genus* $g_i \longrightarrow \infty$ *such that*

$$\gamma = \lim\inf \frac{g_i}{N_i} < 1$$

where $N_i = |X_i(\mathbb{F}_q)|$ *, then*

$$\alpha_q^{lin}(\delta) \geq 1 - \gamma - \delta \quad .$$

Proof: According to Corollary 3.1.11 the codes constructed from X_i have parameters $[n_i, k_i, d_i]_q$, where

$n_i = 1,2,\ldots,N_i$ and $k_i = 0,1,\ldots,n_i - g_i$, and

$$k_i + d_i = n_i - g_i + 1 \ ,$$

i.e. $R_i = 1 - \dfrac{g_i - 1}{N_i} - \delta_i$. Passing to the limit over i, we see that the points (δ_i, R_i) are dense on the line segment

$$R = 1 - \gamma - \delta, \quad 0 \le \delta \le 1 - \gamma \quad .$$

■

Consider the quantity

$$A(q) = \lim\sup \frac{N}{g} \quad ,$$

the limit being taken over all $\mathbb{F}_q$-curves, and let

$$\gamma_q = \frac{1}{A(q)} \quad .$$

Corollary 3.4.2. $\alpha_q^{lin} \ge 1 - \gamma_q - \delta$. ■

Recall (cf. Theorem 2.3.22) that $\gamma_q \ge (\sqrt{q} - 1)^{-1}$.

Corollary 3.4.3. *Let q be an even power of a prime. Then*

$$\alpha_q^{lin}(\delta) \ge R_{TVZ}(\delta) = 1 - (\sqrt{q} - 1)^{-1} - \delta \quad .$$

Proof: In Part 4 we shall prove that in this case $A(q) = \sqrt{q} - 1$ (cf. Theorem 4.2.38). ■

The bound $R_{TVZ}(\delta)$ is called the *basic algebraic-geometric bound.*

Now we are going to compare the obtained bound (right now we restrict ourselves to the case $q = p^{2m}$) with the Gilbert-Varshamov bound (cf. Section 1.3.2)

$$R_{GV}(\delta) = 1 - H_q(\delta) \quad .$$

The Gilbert-Varshamov bound is concave, its ends are $(0,1)$ and $((q-1)/q, 0)$. Therefore the basic AG-bound which is a line segment connecting points $(0, 1 - (\sqrt{q} - 1)^{-1})$ and $(1 - (\sqrt{q} - 1)^{-1}, 0))$, either totally lies below it, or intersects it in two points, or is tangent to it.

Theorem 3.4.4. *The basic AG-bound* R_{TVZ} *lies totally lower than the Gilbert-Varshamov bound* R_{GV} *for* $q = p^{2m} < 49$. *For* $q = p^{2m} \geq 49$ *these bounds intersect, and* R_{TVZ} *ameliorates* R_{GV} *on the segment* (δ_1, δ_2) , *where* δ_1 *and* δ_2 *are the roots of the equation*

$$H_q(\delta) - \delta = (\sqrt{q} - 1)^{-1} \quad .$$

Proof: Consider the tangent to R_{GV} parallel to the basic AG-bound; if it is higher than the basic AG-bound then the bounds do not intersect, otherwise they do intersect and the intersection equation is as above.

The tangent is given by

$$R = 1 - (log_q(2q - 1) - 1) - \delta$$

(cf. Exercise 1.3.14).

The equation $H_q(\delta) - \delta = \gamma_q$ has two roots iff

$$\gamma_q < log_q(2q - 1) - 1 \quad .$$

Theorem 2.3.22 states that $\gamma_q \geq (\sqrt{q} - 1)^{-1}$, and if $q = p^{2m}$ then (cf. Theorem 4.2.38) there is the equality. It is easy to see that for $q = p^{2m}$ the above inequality holds iff $q \geq 49$. ■

Exercise 3.4.5. Check the last statement.

Exercise 3.4.6. Check that for $q = p^{2m+1} \leq 41$ the basic AG-bound lies lower than R_{GV} .

Remark 3.4.7. The basic AG-bound does not depend on the construction of AG-codes we use. Using the L-construction (or the Ω-construction) we can put $D = a \cdot P_0$, where $P_0 \in X(\mathbb{F}_q)$, and construct codes of length $n = N - 1$ evaluating at the other $\mathbb{F}_q$-points of X . The loss of 1 in the length does not influence asymptotic properties. On the contrary the next section will need the H-construction, notwithstanding the asymptotic character of the result.

3.4.2. Expurgation bound

The Gilbert-Varshamov bound (cf. Sections 1.1.4 and 1.3.2) is proved by an expurgation method, i.e. we construct the object we need (linear codes in our case) step by step (say choosing the basis vectors one by one) using the random choice till it is for sure possible. The basic AG-bound, on the contrary, is proved by direct algebraic-geometric constructions.

Here we are going to unite these two ideas. An AG-code is constructed from a curve X and a line bundle $\mathcal{L}$ on it. If the curve is fixed, then the estimate for the parameters (cf. Theorem 3.1.10) depends only on $a = deg\ \mathcal{L}$. However it turns out that among bundles of given degree one can often choose such that the minimum distance is much better than the *a priori* estimate given by Theorem 3.1.10. The reason is that the number of bundles giving codes with at most given distance can be bound from above, and the total number of bundles of given degree can be bound from below using Proposition 2.3.26 (the expurgation method!).

In this section q is an even power of a prime. The codes are constructed by the H-construction using all $\mathbb{F}_q$-points of X.

Let $\mathcal{P} \subseteq X(\mathbb{F}_q)$. Let

$$C_a(\mathcal{P}) = \{ D \in Div^+(X) \mid deg\, D = a\ ,\ Supp\, D \cap X(\mathbb{F}_q) \subseteq \mathcal{P} \}\ ;$$

$$C_{a,r} = \{ D \in Div^+(X) \mid deg\, D = a\ ,\ |Supp\, D \cap X(\mathbb{F}_q)| = r \}\ .$$

By $J_{a,d}$ we denote the set of line bundles $\mathcal{L} \in Pic(X)$ such that $deg\, \mathcal{L} = a$ and the minimum distance of $C = (X, X(\mathbb{F}_q), \mathcal{L})_H$ equals exactly d.

Lemma 3.4.8. *There exists an embedding of sets*

$$\alpha : J_{a,d} \hookrightarrow C_{a,n-d}\ .$$

In particular

$$|J_{a,d}| \le |C_{a,n-d}|\ .$$

Proof: Let $\mathcal{L} \in J_{a,d}$. Then there exists a section $s \in H^0(\mathcal{L})$ such that the divisor of its zeros satisfies

$$|Supp\, D_s \cap X(\mathbb{F}_q)| = n - d\ ,$$

i.e. $D_s \in C_{a,n-d}$. Choose any s with this property and let $\alpha(\mathcal{L}) = D_s$. The divisor class D_s corresponds to the bundle $\mathcal{L}$, hence α never glues non-isomorphic bundles together. ∎

By $J_a = J_{X,a}$ we denote the set of $\mathbb{F}_q$-divisors of degree a on X.

Lemma 3.4.9. *Consider a family of curves X_i with $g_i \longrightarrow \infty$ and $\lim(N_i/g_i) = A$. Then for any a*

$$\log_q |J_{X_i,a}| \geq g_i \cdot \left(1 + A \cdot \log_q\left(\frac{q}{q-1}\right)\right) - o(g_i) \quad .$$

Proof: On any curve X having an $\mathbb{F}_q$-point there exist $\mathbb{F}_q$-bundles of any degree a. Fix one such bundle $\mathcal{L}_0$ then any bundle of degree a equals $\mathcal{L}_0 \otimes \mathcal{L}$, where $\deg \mathcal{L} = 0$, i.e. $\mathcal{L} \in J_X(\mathbb{F}_q)$. Therefore we have an isomorphism $J_{X,a} \simeq J_X(\mathbb{F}_q)$. Proposition 2.3.26 now yields the lemma. ∎

Lemma 3.4.10. *Consider a family of curves X_i with $g_i \longrightarrow \infty$ and let sequences of integers $\{a_i\}$ and $\{\ell_i\}$ be such that*

$$\liminf \frac{a_i - \ell_i}{g_i} > 0$$

and there exists the limit

$$\lambda = \lim_{g_i \longrightarrow \infty} \frac{\ell_i}{g_i} > 0 \quad .$$

Then

$$\log_2 |C_{a_i,\ell_i}| \leq N_i \cdot H_2(\ell_i/N_i) + f_{a_i,\ell_i} + o(g_i) \quad ,$$

where

$$f_{a_i,\ell_i} = \begin{cases} a_i \cdot H_2(\ell_i/a_i) & \text{for } \ell_i/a_i > (q-1)/q \,, \\ a_i \cdot \log_2 q - \ell_i \cdot \log_2(q-1) & \text{for } \ell_i/a_i \leq (q-1)/q \,. \end{cases}$$

Proof: A divisor from $C_{a,\ell}$ is determined by a choice of a set of ℓ points $\{P_1,\ldots,P_\ell\}$ such that $Supp\, D \cap X(\mathbb{F}_q) = \{P_1,\ldots,P_\ell\}$ and of an effective divisor $D' = D - (P_1 + \ldots + P_\ell)$ of degree $(a - \ell)$; the divisor D' can be any one from $C_{a-\ell}(\{P_1,\ldots,P_\ell\})$. Therefore

$$|C_{a,\ell}| = \binom{N}{\ell}\cdot d_{a,\ell} \quad , \tag{3.4.1}$$

where $d_{a,\ell} = |C_{a,\ell}(\{P_1,\ldots,P_\ell\})|$ depends only on a and ℓ , and does not depend on the set $\{P_1,\ldots,P_\ell\}$.

A divisor of degree $(a - \ell)$ is a sum of b_1 points of degree 1, b_2 points of degree 2, etc.;

$$b_1 + 2b_2 + 3b_3 + \ldots + (a - \ell)\cdot b_{a-\ell} = a - \ell$$

(cf. the end of Section 2.3.1). The points of degree 1 are taken from the set $\{P_1,\ldots,P_\ell\}$, and the rest points are arbitrary.

Thus for a fixed sequence $(b_1,b_2,b_3,\ldots,b_{a-\ell})$, the number of divisors from $C_{a-\ell}(\{P_1,\ldots,P_\ell\})$ with given $(b_1,b_2,\ldots,b_{a-\ell})$ equals

$$\prod_{s=1}^{a-\ell}\binom{B_s + b_s - 1}{b_s} \quad , \tag{3.4.2}$$

where for $s \geq 2$ by B_s we denote (as in Section 2.3.2) the number of points of degree s on the curve X , and $B_1 = \ell$. To give an upper estimate for the product (3.4.2) we can change B_s by any $t_s \geq B_s$. Let $t_1 = \ell$, and for $s = 2,\ldots,c = \lceil 2\cdot log_q g\rceil$ let

$$t_s = \frac{9g\cdot q^{s/2}}{s\cdot(c + 1 - s)}$$

($t_s \geq B_s$ because of Proposition 2.3.28), and for $s \geq c + 1$ let

$$t_s = \frac{4 \cdot q^s}{s}$$

(we have

$$s \cdot B_s \leq N_s = q^s + 1 - \sum_{i=1}^{2g} \omega_i^s \leq q^s + 1 + 2g \cdot q^{s/2} < 4q^s = s \cdot t_s \, ,$$

since $2 \cdot \log_q g \leq s$).

Thus

$$d_{a,\ell} \leq \sum_{(b_1, b_2, \ldots)} \prod_{s=1}^{a-\ell} \binom{t_s + b_s - 1}{b_s}$$

the sum being taken over the sequences $(b_1, b_2, \ldots)$ with $b_1 + 2b_2 + 3b_3 + \ldots = a - \ell$; hence

$$\log_2(d_{a,\ell}) \leq \log_2(p(a - \ell)) + \\ + \max \left\{ \sum_{s=1}^{a-\ell} \log_2 \binom{t_s + b_s - 1}{b_s} \right\} \tag{3.4.3}$$

the maximum being taken over the same sequences $(b_1, b_2, \ldots)$, and $p(n)$ being the *number of partitions*, the well known number theory function satisfying the classical inequality

$$p(n) < e^{\pi\sqrt{2n/3}} \quad .$$

Since $a \leq const \cdot g$ we see that asymptotically

$$\log_2(p(a - \ell)) = o(g) \quad .$$

Let us now bound the maximum value of the sum in (3.4.3).

By the Stirling formula (for $b_s \neq 0$) we have

$$\log_2\binom{t_s + b_s - 1}{b_s} \le \log_2\binom{t_s + b_s}{b_s} \le$$

$$\le (t_s + b_s)\cdot H_2\left(\frac{b_s}{t_s + b_s}\right) =$$

$$= b_s\cdot\log_2\left(1 + \frac{t_s}{b_s}\right) + t_s\cdot\log_2\left(1 + \frac{b_s}{t_s}\right).$$

Since those b_s which are equal to zero do not add anything, we get

$$\log_2 d_{a,\ell} \le \max\left\{\sum_{s=1}^{a-\ell}\left(t_s\cdot\log_2\left(1 + \frac{b_s}{t_s}\right) + b_s\cdot\log_2\left(1 + \frac{t_s}{b_s}\right)\right\} + \right.$$

$$+ o(g) .$$

Dispensing with the condition that b_s are integers, we get

$$\log_2 d_{a,\ell} \le \max\left\{\sum_{s=1}^{a-\ell}\left(t_s\cdot\log_2\left(1 + \frac{x_s}{t_s}\right) + \right.\right.$$

$$\left.\left.+ x_s\cdot\log_2\left(1 + \frac{t_s}{x_s}\right)\right) \;\middle|\; \sum s\cdot x_s = a - \ell\right\} + o(g) ,$$

x_s being continuous variables.

This expression can be estimated by the Lagrange method. The derivative over x_s of the function we maximize being equal to $\log_2\left(1 + \frac{t_s}{x_s}\right)$, we see that the maximum is attained for $x_s = \frac{t_s}{\mu^s - 1}$, where μ is given by the equation

$$\sum_{s=1}^{a-\ell} \frac{s \cdot t_s}{\mu^s - 1} = a - \ell .$$

Note that $\mu > 1$, otherwise the sum would be negative. The maximum equals

$$f_{a,\ell} = \sum_{s=1}^{a-\ell} t_s \cdot \log_2\left(\frac{\mu^s}{\mu^s - 1}\right) + \sum_{s=1}^{a-\ell} x_s \cdot \log_2(\mu^s) =$$

$$= \sum_{s=1}^{a-\ell} t_s \cdot \log_2\left(\frac{\mu^s}{\mu^s - 1}\right) + (a - \ell) \cdot \log_2 \mu .$$

On the other hand

$$\frac{(a - \ell) \cdot t_{a-\ell}}{\mu^{a-\ell} - 1} = (a - \ell) \cdot x_{a-\ell} \leq a - \ell ,$$

i.e. $t_{a-\ell} \leq \mu^{a-\ell} - 1$, hence

$$\log_2 \mu \geq \log_2 q - o(1)$$

($t_{a-\ell} = 4q^{a-\ell}/(a-\ell)$ since $a - \ell \geq const \cdot g$ by a supposition of the lemma).

Let us estimate the sum

$$\sum_{s=2}^{c} t_s \cdot log_2\left(\frac{\mu^s}{\mu^s - 1}\right) \le \sum_{s=2}^{c} \frac{t_s}{(ln\ 2)\cdot(\mu^s - 1)} =$$

$$= \frac{1}{ln\ 2} \cdot \sum_{s=2}^{c} \frac{9g\cdot q^{s/2}}{s\cdot(c + 1 - s)\cdot(\mu^s - 1)} \le$$

$$\le \frac{9g}{ln\ 2} \cdot \sum_{s=2}^{c} \frac{q^{s/2}}{s\cdot(c + 1 - s)\cdot q^s} + o(g) \le$$

$$\le \frac{9g}{c\cdot ln\ 2} \cdot \sum_{s=2}^{c} q^{-s/2} + o(g) = o(g) \quad ,$$

and the sum

$$\sum_{s=c+1}^{a-\ell} t_s \cdot log_2\left(\frac{\mu^s}{\mu^s - 1}\right) \le \sum_{s=c+1}^{a-\ell} \frac{4\cdot q^s}{s\cdot(ln\ 2)\cdot(\mu^s - 1)} \le$$

$$\le \frac{4}{c\cdot ln\ 2} \cdot \sum_{s=c+1}^{a-\ell} (q/\mu)^s + o(g) \le const\cdot\frac{a - \ell}{c} + o(g) = o(g) \ .$$

Substituting these estimates into the formula for $f_{a,\ell}$ we get

$$f_{a,\ell} = (a - \ell)\cdot log_2\mu + \ell\cdot log_2 \frac{\mu}{\mu - 1} + o(g) \quad .$$

It is clear that $log_2\mu \ge log_2(q) - o(1)$. However to calculate $f_{a,\ell}$ we need more precise information on the value of μ . Consider two cases.

a) Suppose that $\mu \ge q\cdot(1 + \varepsilon)$ for some $\varepsilon > 0$. Then

$$\sum_{s=2}^{c} \frac{s\cdot t_s}{\mu^s - 1} \le \sum_{s=2}^{c} \frac{9g\cdot q^{s/2}\cdot s}{s\cdot(c + 1 - s)} \le$$

$$\le \frac{9g}{c} \cdot\sum_{s=2}^{c} s\ (\sqrt{q}/\mu)^s + o(g) = o(g) \quad ,$$

since the sum $\sum_{s=1}^{\infty} (\sqrt{q}/\mu)^s$ converges. We also have

$$\sum_{s=c+1}^{a-\ell} \frac{s \cdot t_s}{\mu^s - 1} \le \sum_{s=c+1}^{a-\ell} \frac{4q^s}{\mu^s - 1} = o(g) \quad .$$

Therefore

$$\frac{\ell}{\mu - 1} = (a - \ell) - \sum_{s=2}^{a-\ell} \frac{s \cdot t_s}{\mu^s - 1} = a - \ell - o(g) \quad ,$$

i.e.

$$\mu = \frac{a}{a - \ell} + o(g) \quad ,$$

$$f_{a,\ell} = -(a - \ell) \cdot log_2 \frac{a - \ell}{a} - \ell \cdot log_2 (1 - \frac{a - \ell}{a}) + o(g) =$$

$$= a \cdot H_2(\ell/a) + o(g).$$

Note that in this case $\frac{a}{a - \ell} = \mu > q$, i.e.

$$\frac{\ell}{a} > 1 - \frac{1}{q} \quad .$$

b) Now let $\mu \le q$. Since $log_2 \mu \ge log_2 q - o(1)$, we have

$$f_{a,\ell} = (a - \ell) \cdot log_2 q + \ell \cdot log_2 \frac{q}{q - 1} + o(g) =$$

$$= a \cdot log_2 q - \ell \cdot log_2 (q - 1) + o(g) \ .$$

In this case (since $a - \ell \geq \frac{\ell}{\mu - 1} = \frac{\ell}{q - 1} + o(g)$) we get

$$\ell/a \leq 1 - \frac{1}{q} \quad .$$

Note also that for $\ell/a = 1 - \frac{1}{q}$ the formula for a) and for b) coincide.

The Stirling formula and (3.4.1) yield

$$log_2|C_{a,\ell}| = log_2\binom{N}{\ell} + log_2 d_{a,\ell} \leq N\cdot H_2(\ell/N) + f_{a,\ell} + o(g),$$

and we obtain the desired result. ■

Based on the estimate of the total number of $\mathbb{F}_q$-divisors (Lemma 3.4.9) and on the estimate of the number of divisors whose support includes a given number of $\mathbb{F}_q$-points (Lemmas 3.4.8 and 3.4.10) we are now able to prove the lower bound.

Theorem 3.4.11. *Let* q *be an even power of a prime. Fix a family* $\{X_i\}$ *of curves of genus* $g_i \longrightarrow \infty$ *over* $\mathbb{F}_q$ *such that* $\lim \frac{|X_i(\mathbb{F}_q)|}{g_i} = \sqrt{q} - 1$ *and let* $\beta_q(\delta)$ *be the lower asymptotic bound for linear codes of the form* $C = (X_i, X_i(\mathbb{F}_q), \mathcal{L})_H$. *Then*

a) If $q < 49$, *then*

$$\alpha_q^{lin}(\delta) \geq \beta_q(\delta) \geq R_{GV}(\delta) \quad .$$

b) If $q \geq 49$, *let* $\gamma = (\sqrt{q} - 1)^{-1}$, *and let* δ_1' *and*

δ'_4 *be the roots of*

$$H_q(\delta) + \frac{q}{q-1}\cdot(1-\delta) = 1 + \gamma \quad ,$$

and δ'_2 *and* δ'_3 *be the roots of*

$$H_q(\delta) + (1-\delta)\cdot log_q(q-1) = 1 + \gamma \quad ,$$

$$0 < \delta'_1 < \delta'_2 < \delta'_3 < \delta'_4 < \frac{q-1}{q} \quad .$$

Then

$$\alpha_q^{lin}(\delta) \geq \beta_q(\delta) \geq R_V(\delta) \quad ,$$

where

$$R_V(\delta) = R_{GV}(\delta) \quad for \quad \delta \in [0,\delta'_1] \cup [\delta'_4, \frac{q-1}{q}] \quad ;$$

$$R_V(\delta) = R_{TVZ}(\delta) = 1 - \gamma - \delta \quad for \quad \delta \in [\delta'_2, \delta'_3] \; ;$$

and for $\delta \in [\delta'_1,\delta'_2] \cup [\delta'_3,\delta'_4]$ *the function* $R_V(\delta) = R_V^0(\delta)$ *is given by the implicit equation*

$$(R_V^0(\delta) + \gamma)\cdot H_q\left(\frac{1-\delta}{R_V^0(\delta) + \gamma} \right) + H_q(\delta) = 1 + \gamma \quad . \qquad (3.4.4)$$

Exercise 3.4.12. Check that $R_V(\delta)$ is everywhere continuously differentiable, and that the second derivative is not continuous at δ'_1, δ'_2, δ'_3, δ'_4 .

The behaviour of the *expurgation bound* R_V for $q = p^{2m} \geq 49$ is shown on Fig.3.1 on the next page. Cf. also Diagrams A.2.2 and Tables A.2.4 of Appendix.

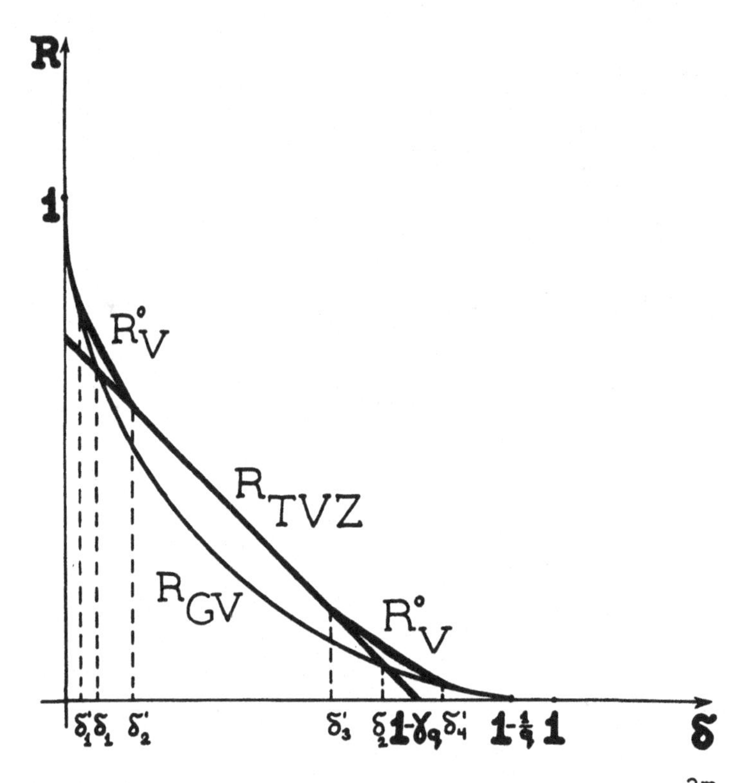

Behaviour of lower asymptotic bounds for $q = p^{2m} \geq 49$:

R_{GV} *is the Gilbert-Varshamov bound;*

R_{TVZ} *is the basic AG-bound;*

R_V^0 *is the expurgation bound (where it differs from* R_{GV} and R_{TVZ}).

(*For* $q = 49$ *we have* $\delta'_1 \approx 0.28$, $\delta_1 \approx 0.37$, $\delta'_2 \approx 0.39$, $\delta'_3 \approx 0.61$, $\delta_2 \approx 0.62$, $\delta'_4 \approx 0.67$, $1 - \gamma_q \approx 0.83$, and $1 - \frac{1}{q} \approx 0.98$).

Fig.3.1

Proof of the theorem: Fix some $R \in (0,1)$ and a curve X of genus g from the considered family. Let

$$a = \lceil (\gamma^{-1} \cdot R + 1) \cdot g \rceil .$$

Then for an AG-code $C = (X, X(\mathbb{F}_q), \mathcal{L})_H$, where $\mathcal{L}$ is any line bundle of degree a, its parameters satisfy

$$k \geq a - g + 1 \geq \gamma^{-1} \cdot R \cdot g \quad ,$$

$$d \geq n - a \geq n - \gamma^{-1} \cdot R \cdot g - g \quad .$$

Since (for $g \longrightarrow \infty$) the length n of C, which equals the number of $\mathbb{F}_q$-points on X, behaves like $g \cdot (\sqrt{q} - 1)$, these inequalities become

$$k/n \geq R \quad ,$$

$$\delta = d/n \geq 1 - R - \gamma \quad .$$

Using spoiling (Lemma 1.1.34) we can put $k/n \sim R$. Further on we consider only the relative distance $\delta = d/n$. For any $\mathcal{L}$ we can of course obtain (as in Corollary 3.4.3) $\delta \sim 1 - R - \gamma$. We are going to prove that it is possible to choose a line bundle $\mathcal{L}_0$ of degree a such that δ is asymptotically on the bound of the theorem. For $\delta \in (\delta_2', \delta_3')$ it is obvious.

For other values of δ we shall prove that for any $\varepsilon > 0$ and any large enough g we have

$$\sum_{d \leq d_\varepsilon} |C_{a,n-d}| < |J_a|$$

where $d_\varepsilon = n \cdot (\delta - \varepsilon)$, $\delta = R_V^{-1}(R)$. Indeed from this inequality and Lemma 3.4.8 one can see that the number of line bundles of degree a such that the minimum distance is at most d_ε is less than the total number of $\mathbb{F}_q$-bundles of degree a:

$$\sum_{d \leq d_\varepsilon} |J_{a,d}| \leq \sum_{d \leq d_\varepsilon} |C_{a,n-d}| < |J_a| \quad , \tag{3.4.5}$$

hence there exists a bundle $\mathcal{L}_0$ such that the minimum distance is more than d_ε.

Note that $d_\varepsilon \le const\cdot g$, and therefore

$$log_q\left(\sum_{d\le d_\varepsilon} |C_{a,n-d}|\right) \le log_q\left(\max_{d\le d_\varepsilon}\left\{|C_{a,n-d}|\right\}\right) + o(g) .$$

To prove (3.4.5) it suffices to prove (use Lemma 3.4.9 and the logarithm of (3.4.5)) that

$$lim\ sup\ \left(\frac{1}{g}\cdot log_q\left(max\left\{|C_{a,n-d}|\right\}\right)\right) <$$

$$< 1 + \gamma^{-1}\cdot log_q\left(\frac{q}{q-1}\right) . \tag{3.4.6}$$

According to Lemma 3.4.10 the left hand side of (3.4.6) is at most

$$\max_{d\le d_\varepsilon}\left\{(\sqrt{q} - 1)\cdot H_2\left(1 - \frac{d}{n}\right) + f_{a,n-d}/g\right\} .$$

Let

$$y = d/n \quad , \quad f(y) = (\sqrt{q} - 1)\cdot H_2(y) + f_{a,n-d}/g .$$

Since we are interested in asymptotics, we can suppose that $a/g = \gamma^{-1}\cdot R + 1$ (which is in fact true up to $o(1)$). Then we see that

$$f(y) = \gamma^{-1}\cdot H_2(y) + (\gamma^{-1}\cdot R + 1)\cdot H_2\left(\frac{1-y}{R+\gamma}\right) ,$$

for

$$1 \ge \frac{n-d}{a} = \frac{1-y}{R+\gamma} \ge 1 - \frac{1}{q} ,$$

and that

$$f(y) = \gamma^{-1}\cdot H_2(y) + (\gamma^{-1}\cdot R + 1)\cdot log_2 q - \gamma^{-1}\cdot(1-y)\cdot log_2(q-1) ,$$

for

$$\frac{1 - y}{R + x} \le 1 - \frac{1}{q} .$$

Passing to derivatives (recall that $H_2'(x) = log_2 \frac{1 - x}{x}$), we get

$$\gamma\cdot f'(y) = log_2 \frac{1 - y}{y} - log_2 \frac{R + \gamma - 1 + y}{1 - y}$$

for

$$1 \ge \frac{1 - y}{R + \gamma} \ge 1 - \frac{1}{q} ,$$

and

$$\gamma\cdot f'(y) = log_2 \frac{1 - y}{y} + log_2(q - 1)$$

for

$$\frac{1 - y}{R + \gamma} \le 1 - \frac{1}{q} .$$

Note that in the first case $f'(y) > 0$ for

$$(1 - y)^2 > y\cdot(R + \gamma - 1 + y),$$

i.e. for $y < \frac{1}{1 + R + \gamma}$; however in this case

$$y \le 1 - (R + \gamma)\cdot\frac{q - 1}{q} < \frac{1}{1 + R + \gamma}$$

(since $R + \gamma \ge \gamma > \frac{1}{q - 1}$). In the second case $f'(y) > 0$ for $y \le \frac{q - 1}{q}$. Hence $f(y)$ is increasing on the interval $(0,(q - 1)/q)$. Therefore

$$\max_{d \le d_\varepsilon} f(d/n) = f(d_\varepsilon/n) < f(\delta) .$$

Thus the inequality (3.4.6) holds for

$$f(\delta) \le 1 + \gamma^{-1}\cdot(1 - log_2(q - 1)) \quad . \tag{3.4.7}$$

For $1 \ge \frac{1-\delta}{R+\gamma} \ge 1 - \frac{1}{q}$ we have

$$f(\delta) - (1 + \gamma^{-1}\cdot(1 - log_2(q - 1))) =$$

$$= \gamma^{-1}\cdot(H_2(\delta) + (R + \gamma)\ log_2 q - (1 - \delta)\cdot log_2(q - 1) -$$

$$- \gamma - 1 + log_2(q - 1)) =$$

$$= \gamma^{-1}\cdot log_2 q\ \left(H_q(\delta) + (R + \gamma)\cdot H_q\left(\frac{1-\delta}{R+\gamma}\right) - (1 + \gamma)\right) \quad .$$

Thus, since $f(\delta)$ is increasing, the inequality (3.4.7) (and hence (3.4.5)) holds for any δ less or equal to the solution of

$$H_q(\delta) + (R + \gamma)\cdot H_q\left(\frac{1-\delta}{R+\gamma}\right) = 1 + \gamma \tag{3.4.8}$$

subject to the auxiliary condition

$$1 - (R + \gamma) \le \delta \le 1 - \frac{q-1}{q}\cdot(R + \gamma) \quad . \tag{3.4.9}$$

For $\frac{1-\delta}{R+\gamma} \le 1 - \frac{1}{q}$ we get

$$f(\delta) - (1 + \gamma^{-1}\cdot(1 - log_2(q - 1))) =$$

$$= \gamma^{-1}\cdot(H_2(\delta) + (R + \gamma)\cdot log_2 q - (1 - \delta)\cdot log_2(q - 1) -$$

$$- \gamma - 1 + log_2(q - 1)) = \gamma^{-1}\cdot(log_2 q)\cdot(H_q(\delta) + R - 1) \quad .$$

We have proved that (3.4.7) holds for any δ less or equal than the solution of

$$R = 1 - H_q(\delta) \tag{3.4.10}$$

subject to the condition

$$\delta \geq 1 - \frac{q-1}{q}\cdot(R + \gamma) \quad . \tag{3.4.11}$$

The equation (3.4.10) is that of the Gilbert-Varshamov bound.

Summing up, we have first shown that there exist line bundles of the given degree on curves of the given family such that the corresponding codes have the given R and $\delta \geq 1 - (R + \gamma)$. We have also shown that if the minimum distance is less than the bound given by (3.4.8) and (3.4.9), (3.4.10) and (3.4.11), then we can choose a bundle such that the distance of the code is better than δ .

The curve (3.4.8) intersects the line $\delta = 1 - (R + \gamma)$ at δ'_2 and δ'_3 , and the line $\delta = 1 - \frac{q-1}{q}\cdot(R + \gamma)$ at δ'_1 and δ'_4 (where δ'_i are defined in the statement of the theorem). The bound (3.4.4) takes place between these lines (then the condition (3.4.9) is satisfied), i.e. for $\delta \in [\delta'_1, \delta'_2] \cup [\delta'_3, \delta'_4]$. Over the second line the Gilbert-Varshamov bound (3.4.10) takes place (since (3.4.11) holds there). It is easy to check that at δ'_1 and δ'_4 the bounds (3.4.8) and (3.4.10) meet.

If $q < 49$ (i.e. for $q = 4, 9, 16, 25$) the line

$$\delta = 1 - \frac{q-1}{q}\cdot(R + \gamma)$$

is everywhere lower than (3.4.4) (cf. Theorem 3.4.4) and δ can be "pulled up" to the Gilbert-Varshamov bound for any R . For $q \geq 49$ both equations giving the values of δ'_2, δ'_3

and δ'_1, δ'_4 are solvable, and the bound is assembled of five pieces. ∎

Corollary 3.4.13. *Let* q *be an even power of a prime. Then*

$$\alpha_q(\delta) \geq \alpha_q^{lin}(\delta) \geq R_V(\delta) \quad .$$

∎

The bound R_V is the best estimate known for α_q^{lin} (but not for α_q , cf. Section 3.4.4).

Exercise 3.4.14. Prove that (in the situation of Theorem 3.4.11) the parameters of $C = (X_i, X_i(\mathbb{F}_q), \mathcal{L})_H$ lie on the bound $R_V(\delta)$ with probability one. (It is useful to compare this result with Remark 1.3.17).

3.4.3. Constructive bounds

AG-codes also make it possible to ameliorate the lower bounds for α_q^{pol} and $\alpha_q^{pol,lin}$ described in Section 1.3.4.

Theorem 3.4.15. *If* q *is an even power of a prime then*

$$\alpha^{pol,lin}(\delta) \geq R_{TVZ}(\delta) = 1 - (\sqrt{q} - 1)^{-1} - \delta \quad .$$

∎

To prove this theorem it suffices to give a polynomial algorithm of construction for AG-codes on some family of curves X_i of genus $g_i \longrightarrow \infty$ over $\mathbb{F}_q$ such that

$$\lim\sup \frac{|X_i(\mathbb{F}_q)|}{g_i} = \sqrt{q} - 1 \quad .$$

This is rather difficult and will be done later in Section 4.3.2. Let us remark that since (for q large enough) the bound R_{TVZ} is partly higher than R_{GV}, Theorem 3.4.15 obviously ameliorates the polynomial bound R_{BZ} for such q and some δ .

Now we are going to use some constructions of Section 1.2.3 (cf. also Sections 1.3.4 and 3.1.2).

Concatenation bound. Applying Corollary 1.3.36 to the bound R_{TVZ} we obtain the following result valid for any q .

Theorem 3.4.16. *Let*

$$R_{KTV(lin)}(\delta) = \max\left\{(1 - (q^{k/2} - 1)^{-1})\cdot\frac{k}{n} - \frac{k}{d}\cdot\delta\right\} \quad ,$$

the maximum being taken over all linear $[n,k,d]_q$*-codes such that if* q *is an odd power of a prime then* k *is even. Then*

$$\alpha_q^{pol,lin}(\delta) \geq R_{KTV(lin)}(\delta) \quad .$$

■

Exercise 3.4.17. Prove the theorem.

Unfortunately we do not know the precise value of $R_{KTV(lin)}(\delta)$. The reason is, of course, that we do not know the parameters of all q-ary codes. However, each code from the set described in the theorem gives a line segment

$$R = (1 - (q^{k/2} - 1)^{-1})\cdot\frac{k}{n} - \frac{k}{d}\cdot\delta \qquad (3.4.12)$$

which is a lower bound for $R_{KTV(lin)}$ and for $\alpha_q^{pol,lin}$.

If we have some subset of such codes then the corresponding bound is assembled from parts of such segments. Consider first the behavior of $R_{KTV(lin)}$ at the ends of the interval.

Proposition 3.4.18. *For* $\delta \longrightarrow 0$ *the bound* $R_{KTV(lin)}(\delta)$ *behaves at worst as*

$$1 - R \sim -2\delta\cdot log_q\delta \quad .$$

For $x = \left(\frac{q-1}{q} - \delta\right) \longrightarrow 0$ *it behaves at worst as*

$$R \sim \begin{cases} -2\cdot(\sqrt{q} - 1)\cdot q^4\cdot(\sqrt{q}^3 - 1)^{-3}\cdot x^3\cdot log_q x & \text{for } q = p^{2m} \ , \\ -2q^6\cdot(q+1)^2\cdot(q^3 - 1)^{-3}\cdot x^3\cdot log_q x & \text{for } q = p^{2m+1} . \end{cases}$$

■

Exercise 3.4.19. Prove the proposition. (*Hint*: Estimate the asymptotic behavior of the nodes of the polygon defined as the maximum of segments (3.4.12) corresponding to trivial $[\ell,\ell,1]_q$- or $[2\ell,2\ell,1]_q$-codes and first order affine Reed-Muller $[q^m , m+1 , q^m - q^{m-1}]_q$- or $[q^{2m-1} , 2m , q^{2m-1} - q^{2m-2}]_q$-codes).

Remark 3.4.20. To compare recall the behavior of the Gilbert-Varshamov bound at the upper end

$$1 - R_{GV}(\delta) \sim -\delta\cdot log_q\delta \quad \text{for} \quad \delta \longrightarrow 0$$

and at the lower end

$$R_{GV}\left(\frac{q-1}{q} - x\right) \sim \frac{q^2}{2(q-1)\cdot ln\ q}\cdot x^2 \quad \text{for} \quad x \longrightarrow 0 \ ;$$

and that of the Bloch-Ziablov bound

$$1 - R_{BZ}(\delta) \sim \frac{ln\ q}{2}\cdot\delta\cdot(log_q\delta)^2$$

for $\delta \longrightarrow 0$ and

$$R_{BZ}\left(\frac{q-1}{q} - x\right) \sim \frac{q^3}{6(q-1)^2\cdot ln\ q}\cdot x^3$$

for $x \longrightarrow 0$ (cf. Exercises 1.3.13 and 1.3.34, and also Tables A.2.4 of Appendix).

Theorem 3.4.21. *For any q (equal to a power of a prime) and for any $\delta \in \left(0, \frac{q-1}{q}\right)$ we have*

$$R_{KTV(lin)}(\delta) > R_{BZ}(\delta) \quad .$$

Sketch of proof: We do not give the details here because the proof uses a lot of computation. The idea is as follows.

Consider the polygon of Exercise 3.4.19. Accurate evaluation of the difference between this polygon and the asymptotics of Proposition 3.4.18, as well as of the difference between R_{BZ} and its asymptotics (cf. Remark 3.4.20), proves the theorem for δ ranging over intervals $(0,\alpha)$ and $(\beta,\frac{q-1}{q})$ for some explicitly calculated α and β . For large enough values of q we get $\beta < \alpha$ and we are done. For small q one adds some segments (3.4.12) for few other codes. To prove that the obtained polygon is higher than R_{BZ} it is enough to show this for its nodes

(R_{BZ} being concave). Far from the ends of $(0,\frac{q-1}{q})$ this can be done by straight-forward calculation using the decreasing of R_{BZ} and tables of its values. Since the constructed polygon is "much higher" than R_{BZ} (cf. Tables A.2.4), the estimates can be rather rough. ■

Exercise 3.4.22. Give a complete proof.

For $\delta \longrightarrow 0$ the bound $R_{KTV(lin)}$ behaves twice worse than R_{GV}. Now we are going to give another bound, good for small values of δ.

Restriction bound. Let us now apply the field restriction.

Theorem 3.4.23. *Let*

$$R_{KT}(\delta) = \max_{m} \left\{ 1 - \frac{2m\cdot(q-1)}{q\cdot(q^{m/2}-1)} - \frac{m\cdot(q-1)}{q}\cdot\delta \right\} ,$$

the maximum being taken over integers $m \geq 1$ *such that* q^m *is a square. Then*

$$\alpha_q^{pol,lin}(\delta) \geq R_{KT}(\delta) .$$

Sketch of proof: Let q^m be an even power of a prime. Consider a family of curves X of growing genus g over $\mathbb{F}_{q^m}$ such that the ratio of the number of $\mathbb{F}_{q^m}$-points to the genus is asymptotically maximal, i.e.

$$\lim\sup \frac{N}{g} = q^{m/2} - 1 .$$

We shall also need a point over an extension $\mathbb{F}_{q^{mb}}$ of $\mathbb{F}_{q^m}$ which is not defined over any smaller field:

Exercise 3.4.24. Prove that for $b \geq g$ there always exist an $\mathbb{F}_{q^{mb}}$-point on X which is not defined over any smaller field. (*Hint*: Use the fact that

$$\big|\, |X(\mathbb{F}_{q^{mb}})| - q^{mb} - 1 \big| \leq 2g\cdot q^{mb/2} ,$$

cf. Exercise 2.3.14).

Let $b \geq g$. Fix such a point Q and let $Q = Q_1, \ldots, Q_b$ be its conjugates over $\mathbb{F}_{q^m}$. Let $D_1 = \sum Q_i$, $\mathcal{P} = X(\mathbb{F}_{q^m})$. Up to a spoiling Theorem 3.1.40 yields an $[N,\ N - m\cdot(q - 1)\cdot b - 1,\ q\cdot b - 2g + 2]_q$-code. Asymptotically we get a family of codes with

$$R \sim 1 - m\cdot(q - 1)\cdot\beta ,$$

$$\delta \sim q\cdot\beta - 2\gamma ,$$

where $\gamma = (q^{m/2} - 1)^{-1}$, $\beta = \lim\inf b/N \geq \gamma$. The latter inequality is not essential, since for $b < g$ the *a priori* estimate $k \geq N - m\cdot(q\cdot b - g + 1)$ is better than that of Theorem 3.1.40.

A proper choice of β (depending on δ) and the remark that field restriction is a polynomial operation and that we restrict polynomially constructable codes, yield

$$\alpha_q^{pol,lin}(\delta) \geq 1 - \frac{2m\cdot(q - 1)}{q\cdot(q^{m/2} - 1)} - \frac{m\cdot(q - 1)}{q}\cdot\delta .$$

■

Theorem 3.4.25. *For* $\delta \longrightarrow 0$ *the bound* $R_{KT}(\delta)$ *behaves as*

$$R \sim 1 + 2\cdot\frac{(q - 1)}{q}\cdot\delta\cdot\log_q\delta ;$$

in particular, for $q = 2$ *this bound behaves for* $\delta \longrightarrow 0$ *exactly as the Gilbert-Varshamov bound (in the sense that the main asymptotic terms of* $1 - R$ *coincide).* ∎

Exercise 3.4.26. Give a detailed proof of Theorem 3.4.23. Check the statement of Theorem 3.4.25.

Let us remark that for $q = p^{2m}$ we have

$$R_{KT}(\delta) \le \max \{R_{GV}(\delta)\ ,\ R_{TVZ}(\delta)\}\ .$$

On the other hand R_{KT} always intersects R_{TVZ} and for $q \ge 361$ it also intersects R_{GV} .

3.4.4. Other bounds

AG-codes lead to ameliorations of a number of other asymptotic bounds.

Non-linear codes. Since concatenation can be applied to non-linear codes as well, Theorem 3.4.16 has the following obvious analogue.

Theorem 3.4.27. *Let*

$$R_{KTV}(\delta) = \max \left\{(1 - (q^{k/2} - 1)^{-1})\cdot\frac{k}{n} - \frac{k}{d}\cdot\delta\right\}\ ,$$

the maximum being taken over all (linear and non-linear) $[n,k,d]_q$*-codes such that* $k \in \mathbb{N}$ *and* k *is even if* q *is an odd power of a prime. Then*

$$\alpha_q^{pol} \ge R_{KTV}(\delta)\ .$$

∎

This theorem makes sense since (for a fixed finite length) there do exist non-linear codes whose parameters are better than those of linear codes. Numerical values of $R_{KTV}(\delta)$ are given in Tables A.2.4.

Now we are going to apply non-linear constructions of Section 1.2.3. Up to the end of the discussion of non-linear codes the alphabet cardinality q is arbitrary (not necessarily a power of a prime). Recall that for a non-linear $[n,k,d]_q$-code C, in general, $k = log_q|C|$ is not an integer.

Theorem 3.4.28. *Let*

$$R^0_{LT}(\delta) = max \left\{ (log_M p^{2m}) \cdot \left(\frac{(p^m - 2) \cdot 2m}{(p^m - 1) \cdot n} - \frac{2m}{d} \cdot \delta \right) \right\} ,$$

here the maximum is taken over all $[n,k,d]_{p^{2m}}$*-codes, and a prime* p *and an integer* m *are such that* $M = q^k \geq p^{2m}$. *Then*

$$\alpha_q(\delta) \geq \alpha_q^{pol}(\delta) \geq R^0_{LT}(\delta) .$$

■

Exercise 3.4.29. Prove the theorem. (*Hint*: Apply Exercise 1.3.38 on the alphabet extension to the bound R_{KTV} for $\alpha^{pol}_{p^{2m}}$).

Theorem 3.4.30. *Let* $q \geq 46$. *Then for* δ *from some interval we have*

$$\alpha_q^{pol}(\delta) > R_{GV}(\delta) .$$

Proof: Applying the part of Theorem 1.3.19 concerning alphabet restriction to the basic AG-bound

$$\alpha_{q'}(\delta) \geq 1 - \frac{1}{\sqrt{q'} - 1} - \delta ,$$

where $q' \geq q$, $q' = p^{2m}$ is an even power of a prime, we get

$$\alpha_q^{pol}(\delta) \geq R_{LZ}(\delta) = 1 - ((\sqrt{q'} - 1)^{-1} + \delta) \cdot log_q(q') \quad .$$

The obtained bound is a line segment, and we can compare it with the Gilbert-Varshamov bound just as in the proof of Theorem 3.4.4.

According to Exercise 1.3.14 the tangent to $R_{GV}(\delta)$ with the angle $-log(q')$ touches $R_{GV}(\delta)$ at the point

$$\delta_0 = (q - 1)/(q' + q - 1) \ .$$

At this point we have

$$R_{GV}(\delta_0) = 1 + \frac{q'}{q' + q - 1} \cdot log_q(q') - log_q(q' + q - 1) \ ,$$

$$R_{LZ}(\delta_0) = 1 - \left(\frac{1}{\sqrt{q'} - 1} + \frac{q - 1}{q' + q - 1} \right) \cdot log_q(q') \ .$$

The curve $R_{GV}(\delta)$ and the line $R_{LZ}(\delta)$ intersect in two points iff $R_{LZ}(\delta_0) > R_{GV}(\delta_0)$, which is clearly equivalent to

$$p^{2m} + q - 1 > p^{2m \cdot p^m/(p^m-1)} \quad ,$$

i.e.

$$q' = p^{2m} \geq q > p^{2m \cdot p^m/(p^m-1)} - p^{2m} + 1 \quad . \qquad (3.4.13)$$

To prove the theorem for a given q it is enough to show that q belongs to an interval of the form (3.4.13).

Let $p = 2$ then (3.4.13) becomes

$$2^{2m} \geq q > 2^{2m \cdot 2^m/(2^m-1)} - 2^{2m} + 1 = 2^{2m} \cdot (2^{2m/(2^m-1)} - 1) + 1$$

and the intervals corresponding to m and $(m + 1)$ intersect iff

$$2^{2(m+1)/(2^{m+1}-1)} < \frac{5}{4} ,$$

i.e. for $m \geq 5$. The left end of the interval corresponding to $m = 5$ is less than 258, i.e. for $q \geq 258$ the theorem is proved.

Now let successively $p = 17, 13, 11$. For $m = 1$ we get the intervals [289,124], [169,91], [121,76]. For $p = 3,\ m = 2;\ p = 2,\ m = 3;\ p = 7,\ m = 1$ we get [81,61], [64,53], [49,46], respectively.

It is left to prove the theorem for $q = 50, 51, 52$. We apply the alphabet extension (cf. Theorem 1.3.19) to the basic AG-bound, getting (for $p^{2m} \leq q$)

$$\alpha_q^{pol}(\delta) > \bar{R}_{LZ}(\delta) = \left(1 - \frac{1}{p^m - 1} - \delta\right) \cdot log_q(p^{2m}) .$$

The comparison between R_{GV} and $\bar{R}_{LZ}$ is similar with the previous argument. Exercise 1.3.14 shows that the tangent with the angle $-log_q p^{2m}$ is tangent to R_{GV} at the point $\delta_0 = \dfrac{q - 1}{p^{2m} + q - 1}$, and

$$R_{GV}(\delta_0) = 1 + \frac{p^{2m} \cdot log_q(p^{2m})}{p^{2m} + q - 1} - log_q(p^{2m} + q - 1) ,$$

$$\bar{R}_{LZ}(\delta_0) = \left(1 - \frac{1}{p^m - 1} - \frac{q - 1}{p^{2m} + q - 1} \right) \cdot log_q(p^{2m}) .$$

The condition $\bar{R}_{LZ}(\delta_0) > R_{GV}(\delta_0)$ is equivalent to

$$\frac{p^{2m} - 1}{p^{2m/(p^m - 1)} - 1} > q \geq p^{2m} .$$

Putting $p = 7$, $m = 1$ we get it for $q = 50$, 51 and 52. ∎

Here is one simple theorem more, bounding $\alpha_q(\delta)$ in the non-linear case. This is the best lower bound known today.

Theorem 3.4.31. *For any* q

$$\alpha_q(\delta) \geq R_{LT}(\delta) = max\ \{R'_{LT}(\delta), R''_{LT}(\delta)\}\ ,$$

where

$$R'_{LT}(\delta) = max\ \{1 - (1 - R_V^{(q')}(\delta))\cdot log_q(q')\}\ ,$$

the maximum being taken over all even powers of primes $q' = p^{2m} \geq q$ ($R_V^{(q')}(\delta)$ *being the expurgation bound for* q'*-ary codes*), *and*

$$R''_{LT}(\delta) = max\ \{R_V^{(q')}(\delta)\cdot log_q(q')\}\ ,$$

the maximum being taken over $q' = p^{2m} \leq q$.

Proof: Apply Theorem 1.3.19 to the expurgation bound. ∎

This result is reasonable even when q is itself an even power of a prime (this is by no means obvious, since if in the statement we change the expurgation bound by the Gilbert-Varshamov bound, we get the same Gilbert-Varshamov bound).

Exercise 3.4.32. Check that for $q = 256^2$ there exist δ_0 such that $R_{LT}(\delta_0) \geq R_V(\delta_0)$. (*Hint*: Put $\delta_0 = 0.0146$

and calculate $R_{TVZ}(\delta_0)$. Then estimate $R_{LT}(\delta_0)$, noting that

$$R_{LT}(\delta_0) \geq 1 - (1 - R_{TVZ}(\delta_0))\cdot log_{256}263$$

and show that $\delta_0 > \delta_2$, where δ_2 is the smallest root of

$$H_q(\delta) + (1 - \delta)\cdot log_q(q - 1) = 1 + \gamma_q$$

for $q = 256^2$).

It is likely that such δ_0 exists whenever q is large enough. On the other hand for small q (say, for $q \leq 256$) the bounds $R_{LT}(\delta)$ and $R_V(\delta)$ most likely coincide.

Polynomially decodable codes. In Section 3.4.3 we have studied asymptotic bounds for codes with polynomial construction complexity. However we have not considered decoding problem for them. Now let us look for codes that are both polynomially constructable and polynomially decodable.

Consider a polynomially constructable family of $[n_i,k_i,d_i]_q$-codes C_i and a polynomial (in n) decoding algorithm for these codes correcting any t_i errors (we do not restrict ourselves to the case $t_i = \lceil(d_i - 1)/2\rceil$, t_i can be smaller). Let $\tau = lim\ inf\ (t_i/n_i)$. The family C_i corresponds to the point $(\delta = 2\tau\ ,\ R)$; here δ can be smaller than $lim\ inf\ (d_i/n_i)$; δ is the *relative decoding distance*.

Exercise 3.4.33. Define

$$\alpha_q^{pol\ dec}(\delta) \quad \text{and} \quad \alpha_q^{pol\ dec,lin}(\delta)$$

similarly to the definition of α_q^{pol} (cf. Theorem 1.3.25). Prove that

$$\alpha_q^{pol\ dec,lin}(\delta) \geq R_Z(\delta) \quad \text{and} \quad \alpha_q^{pol\ dec,lin}(\delta) \geq R_{BZ}(\delta) .$$

Just as for α_q^{pol}, application of AG-codes ameliorates this bound.

Theorem 3.4.34. *Let* $q = p^{2m}$. *Then*

$$\alpha_q^{pol\ dec,lin}(\delta) \geq R_{SV}(\delta) = 1 - \frac{2}{\sqrt{q} - 1} - \delta .$$ ■

Exercise 3.4.35. Prove the theorem. (*Hint*: It follows from Corollary 3.3.4 and Proposition 3.3.6).

Exercise 3.4.36. Check that for $q = p^{2m} \geq 361$ the bound R_{SV} is (on an interval of δ) higher than the Gilbert-Varshamov bound.

Theorem 3.4.37. *Let*

$$R'_{SV(lin)}(\delta) = max \left\{ \left(1 - \frac{2}{q^{k/2} - 1}\right) \cdot \frac{k}{n} - \frac{k}{d} \cdot \delta \right\} ,$$

the maximum being taken over all linear $[n,k,d]_q$*-codes such that* q^k *is a square. Then*

$$\alpha_q^{pol\ dec,lin}(\delta) \geq R'_{SV(lin)}(\delta) .$$

If we dispense with the linearity condition, we get the bound $R'_{SV}(\delta)$ *for which*

$$\alpha_q^{pol\ dec}(\delta) \geq R'_{SV}(\delta) .$$ ■

Theorem 3.4.38. *Let*

$$R''_{SV(lin)}(\delta) = \max_{m} \left\{ 1 - \frac{3m\cdot(q-1)}{q\cdot(q^{m/2}-1)} - \frac{m\cdot(q-1)}{q}\cdot\delta \right\} ,$$

the maximum being taken over integers $m \geq 1$ *such that* q^m *is a square. Then*

$$\alpha_q^{pol\ dec,lin}(\delta) \geq R''_{SV}(\delta) .$$

■

Exercise 3.4.39. Prove Theorems 3.4.37 and 3.4.38. (*Hint*: Similar to the proofs of Theorems 3.4.16 and 3.4.23, taking into account Proposition 3.3.6).

Exercise 3.4.40. For $q = 2$ prove that

$$R_{SV(lin)}(\delta) > R_{BZ}(\delta)$$

for any δ in the interval $(0,1/2)$. (*Hint*: Cf. Theorem 3.4.21).

Self-dual codes. Consider the asymptotic problem for self-dual (and quasi-self-dual) codes. For such codes $R = 1/2$ and the question is about the value of

$$\delta_q^{sd} = \lim\sup \delta ,$$

lim sup being taken over all self-dual codes (respectively, δ_q^{qsd} for quasi-self-dual, and also $\delta_q^{sd,pol}$ *and* $\delta_q^{qsd,pol}$ for polynomial families).

It is known that there exist self-dual codes on the Gilbert-Varshamov bound (Theorem 1.3.24). There are therefore questions about an asymptotic amelioration of this bound and about its polynomiality. For $q \geq 49$ we would like to have an answer to the following

Problem 3.4.41. Do there exist quasi-self-dual AG-codes lying on the basic AG-bound ? (If yes, then $\delta_q^{qsd} \geq \frac{1}{2} - \frac{1}{\sqrt{q} - 1}$).

Here is a weaker bound:

Theorem 3.4.42. *If q is an even power of a prime then*

$$\delta_q^{qsd} \geq \delta_{Sch} = \frac{1}{2} - \frac{1}{\sqrt{q} - 3} .$$

If q is an even power of 2 then

$$\delta_q^{sd} \geq \delta_{Sch} . \blacksquare$$

Exercise 3.4.43. Prove the theorem. (*Hint*: Use Corollary 3.1.49 for a family of curves with $N/g \sim \sqrt{q} - 1$).

Exercise 3.4.44. Show that for $q = p^{2m} \geq 121$ (respectively, for $q = 2^{2m} \geq 256$) the obtained bound for δ_q^{qsd} (respectively, for δ_q^{sd}) is better than the Gilbert-Varshamov bound.

Note that the method of Theorem 3.1.48 and Corollary 3.1.49 says nothing about $\delta^{qsd,pol}$ and $\delta^{sd,pol}$.

Constant-weight codes. A non-linear code $C \subset \mathbb{F}_q^n$ is called a constant-weight code of weight w iff $\|v\| = w$ for any $v \in C$.

Exercise 3.4.45. Let $n \longrightarrow \infty$, $W/n \longrightarrow \omega$. Define $\alpha_q(\omega,\delta)$. Let

$$R_G(\omega,\delta) = H_2(\omega) - \omega\cdot H_2(\delta/2\omega) - (1 - \omega)\cdot H_2(\delta/(2 - 2\omega)) .$$

Prove that

$$\alpha_2(\omega,\delta) \geq R_G(\omega,\delta)$$

and generalize this inequality ("the Gilbert bound") to the case of $\alpha_q(\omega,\delta)$.

Here is a simple construction of constant-weight codes. Let C be an $[n,k,d]_q$-code. Enumerate the elements of $\mathbb{F}_q$ by integers $1,2,\ldots,q$ and map the i-th element to the binary vector with 1 on place i and 0 everywhere else. Map each $x \in \mathbb{F}_q^n$ to $\varphi(x) \in \mathbb{F}_2^{qn}$ obtained from x by the above change of each coordinate.

Exercise 3.4.46. Prove that the image $\varphi(C) \in \mathbb{F}_2^{qn}$ has the same cardinality q^k , the minimum distance $2d$, and that it is a constant-weight binary code of weight n .

Theorem 3.4.47. *Let* $\omega = p^{-2m}$. *Then*

$$\alpha_2(\omega,\delta) \geq R_{EZ}(\omega,\delta) = -\left(\omega - \frac{\delta}{2} - \frac{\omega\cdot\sqrt{\omega}}{1 - \sqrt{\omega}}\right)\cdot \log_2\omega . \quad \blacksquare$$

Exercise 3.4.48. Prove the theorem. (*Hint*: Apply Exercise 3.4.46 to AG-codes). Show that for small enough ω the bound of Theorem 3.4.47 is better (for δ in some interval) than that of Exercise 3.4.45.

Problem 3.4.49. Generalize the bound of Theorem 3.4.47 to the case of $\alpha_q(\omega,\delta)$ for an arbitrary q .

Historical and bibliographic notes to Part 3

Algebraic geometric codes were discovered by V.D.Goppa [Go 1] in 1980. His discovery was a reward for his more than ten years research on the possible generalizations of Reed-Solomon codes, BCH codes, and classical Goppa codes (cf.[Go 1], [Go 2], [Go 3]). In his paper [Go 1] he used the Ω-construction (generalizing that of classical Goppa codes) and proved Theorem 3.1.7.

After this wonderful discovery it was not difficult to find the L-construction. It was known to V.D.Goppa, to Yu.I.Manin, and to other mathematicians. The H-construction is due to Yu.I.Manin and first appeared in [Ma/Vl]. Its interpretation in terms of projective systems has been never published before.

Proposition 3.1.24 is due to Y.Driencourt and J.F.Michon [Dr/Mi 2]. Proposition 3.1.30 was independently proved by S.G.Vladuţ [Vl 4] and J.P.Hansen [Ha 1], [Ha 2]. Proposition 3.1.32 and Corollary 3.1.36 are due to G.L.Katsman, M.A.Tsfasman, and S.G.Vladuţ [Ka/Ts/Vl 1], [Ka/Ts/Vl 2]. Theorem 3.1.40 was proved by G.L.Katsman and M.A.Tsfasman [Ka/Ts 2]. Remark 3.1.42 is based on the paper [St 4] by H.Stichtenoth.

Duality for AG-codes is first mentioned by V.D.Goppa [Go 2]. Self-dual algebraic-geometric codes were studied by Y.Driencourt and J.F.Michon [Dr/Mi 3], G.L.Katsman and M.A.Tsfasman (unpublished), and H.Stichtenoth [St 2]. Theorem 3.1.48 and Corollary 3.1.49 are due to W.Scharlau [Sch]; Remark 3.1.50, Theorem 3.1.52, and Remark 3.1.53 to Y.Driencourt and H.Stichtenoth [Dr/St]. Theorem 3.1.54 was obtained by G.L.Katsman and M.A.Tsfasman [Ka/Ts 1].

Codes of genera 1, 2, and 3 and their binary posterity were studies by A.M.Barg, G.L.Katsman, S.N.Litsyn, and M.A.Tsfasman [Ba/Ka/Ts]. This paper is based on the papers by J.-P.Serre [Se 1]-[Se 4]. Elliptic codes were studied by Y.Driencourt and J.F.Michon [Dr/Mi 1]. The calculation of B_k for such codes and the result on the non-existence of elliptic MSD-codes of length more than $(q + 1)$ is due to G.L.Katsman and M.A.Tsfasman [Ka/Ts 1].

Hermitian codes were introduced by V.D.Goppa and studied by G.Tiersma [Ti] and H.Stichtenoth [St 3]. In our exposition we follow [St 3]. The result of Exercise 3.2.47 is contained in [Vl 4].

The interpretation of Reed-Muller codes as multi-dimensional AG-codes is due to Yu.I.Manin [Ma/Vl], who many times pointed out the significance of multi-dimensional constructions. Unfortunately here the progress is quite moderate. The upper estimate in Exercise 3.2.50 was proved by J.-P.Serre [Se 6].

The examples of Section 3.2.4 are mostly new, with the exception of Example 3.2.65 and Exercise 3.2.66 taken from [St 3].

The idea of the decoding algorithm (Chapter 3.3) is due to J.Justesen [Ju/al], the precise statements and proofs to A.N.Skorobogatov and S.G.Vladuţ [Sk/Vl]. The basic algorithm was later independently discovered by V.Yu.Krachkovskii

[Kr]; S.Porter [Por] discovered another algorithm for decoding AG-codes, generalizing the Euclid algorithm of decoding Reed-Solomon codes. Remark 3.3.23 (except for part b) is due to R.Pellikaan [Pel 2], and part b to S.G.Vladuţ [Vl 5]. Remark 3.3.26 is based on the paper [Pel 1] by R.Pellikaan.

The results of Section 3.4.1 (the basic AG-bound) are due to M.A.Tsfasman, S.G.Vladuţ, and T.Zink [Ts/Vl/Z]; the results of Section 3.4.2 (the expurgation bound) to S.G.Vladuţ [Vl 1]. Theorem 3.4.15 is due to S.G.Vladuţ [Vl 2], [Ma/Vl]; Theorem 3.4.16 and 3.4.21 to G.L.Katsman, M.A.Tsfasman, and S.G.Vladuţ [Ka/Ts/Vl 1], [Ka/Ts/Vl 2]; Theorems 3.4.23 and 3.4.25 to G.L.Katsman and M.A.Tsfasman [Ka/Ts 2]. Theorem 3.4.30 is due to S.N.Litsyn and V.A.Zinoviev [Li/Zi]; Theorems 3.4.28, 3.4.31 and Remark 3.4.32 to S.N.Litsyn and M.A.Tsfasman [Li/Ts 1] (and unpublished notes). Theorems 3.4.37 and 3.4.38 are proved by A.N.Skorobogatov and S.G.Vladuţ [Sk/Vl]. Theorem 3.4.42 is due to W.Scharlau [Sch]; Theorem 3.4.47 to T.Ericson and V.A.Zinoviev [Er/Zi 1], [Er/Zi 2].

PART 4

MODULAR CODES

This part is devoted to codes arising from modular curves. We consider two types of curves: classical modular curves (their reductions in fact) and Drinfeld modular curves. Classical modular curves yield codes over $\mathbb{F}_{q322}q = p^2 ,_2p$ being a prime, while Drinfeld curves yield codes over $\mathbb{F}_q$, where $q = p^{2a}$ is an arbitrary even power of a prime.

We are going to show that on both types of modular curves there is a lot of rational points, and therefore the obtained codes have good asymptotic parameters. Moreover for codes of both classes there exist polynomial algorithms of construction. For Drinfeld curves these algorithms are much more conceptual and simple. One can put a natural question: why should we consider codes on classical modular curves, those on Drinfeld modular curves being in many aspects more conceptual and easier to study, and q being any even power of a prime (and not only p^2 as in the classical case). The answer is that though having the same asymptotic parameters, in many other aspects the two classes of codes differ greatly. For example one can rather easily obtain codes of moderate length on classical modular curves which is quite a problem for the case of Drinfeld curves, so that the question of a practical use nowadays makes sense only for the first class. Besides the two classes have different group-theoretic properties, and so on. Thus parallel consideration of classical modular curves and Drinfeld curves is fully justified. Let us also remark that reductions of some Shimura curves (which generalize classical modular curves) also provide asymptotically maximal ratio of the number of $\mathbb{F}_q$-points to the genus for any $q = p^{2a}$. Therefore codes on such curves also have very good parameters. No other property of these codes is yet investigated.

In Chapter 4.1, basing on the study of moduli schemes of elliptic curves, we consider codes on reductions of classical modular curves. Chapter 4.2 is devoted to codes on Drinfeld curves representing moduli varieties of elliptic modules. In Chapter 4.3 we show that both considered families of codes are polynomially constructable (unfortunately the proof of this result is rather cumbersome and technical; to understand it the reader will need a fair amount of patience and diligence; however having done this, one will convince oneself in polynomial constructibility of a large class of AG-codes, not only of modular ones, see Remark 4.3.33).

CHAPTER 4.1

CODES ON CLASSICAL MODULAR CURVES

This chapter is devoted to codes on classical modular curves or, to put it more precisely, on those obtained from modular curves by reduction to a finite characteristic. As we have shown in Part 3, the properties of an AG-code constructed from a curve are determined by its geometry and arithmetic. Here we give some necessary information on the structure of modular curves and of their reductions, and some corollaries for codes on such curves.

Section 4.1.1 contains an analytical description of modular curves, of their morphisms, and of their fields of rational functions. In Section 4.1.2 we present the theory of moduli schemes of elliptic curves, which makes it possible to prove non-singularity of reductions of modular curves and existence of a great number of $\mathbb{F}_{p^2}$-rational points on these reductions. Section 4.1.3 is devoted to codes on reductions of modular curves: we estimate their parameters taking into account the improvements due to

Weierstrass points, we also describe the group properties of a subclass of such codes.

4.1.1. Classical modular curves

The basic method to study classical modular curves is to consider their analytical description as factors of the upper half-plane over discrete subgroups of automorphisms.

Riemann surface $\Gamma\backslash H^*$. By H we denote the upper half-plane of the complex plane $\mathbb{C}$:

$$H = \{ z \in \mathbb{C} \mid Im\ z > 0 \} .$$

The *special linear group* over $\mathbb{Z}$

$$SL(2,\mathbb{Z}) = \left\{ \begin{pmatrix} a & c \\ b & d \end{pmatrix} \mid a,b,c,d \in \mathbb{Z} \ , \ a\cdot d - b\cdot c = 1 \right\}$$

acts naturally on H (on the left). This action is defined by the rule

$$\gamma : z \longmapsto \frac{a\cdot z + b}{c\cdot z + d} ,$$

where $\gamma = \begin{pmatrix} a & b \\ c & d \end{pmatrix} \in SL(2,\mathbb{Z})$. Here the element $\begin{pmatrix} -1 & 0 \\ 0 & -1 \end{pmatrix}$ acts trivially, so that the action of $SL(2,\mathbb{Z})$ induces an action of the *modular group*

$$\Gamma(1) = PSL(2,\mathbb{Z}) = SL(2,\mathbb{Z})/\{\pm 1\}$$

which is the factor group of $SL(2,\mathbb{Z})$ over its centre. Further on we write $\gamma = \begin{pmatrix} a & b \\ c & d \end{pmatrix} \in \Gamma(1)$ meaning that γ is the image in $\Gamma(1)$ of $\begin{pmatrix} a & b \\ c & d \end{pmatrix} \in SL(2,\mathbb{Z})$.

We are interested in the following standard subgroups of $\Gamma(1)$. Let us fix a positive integer N and put

$$\Gamma(N) = \left\{ \begin{pmatrix} a & b \\ c & d \end{pmatrix} \in \Gamma(1) \mid a,d \equiv 1 (mod\ N),\ b,c \equiv 0 (mod\ N) \right\} ,$$

$$\Gamma_0(N) = \left\{ \begin{pmatrix} a & b \\ c & d \end{pmatrix} \in \Gamma(1) \mid c \equiv 0 (mod\ N) \right\} .$$

Sometimes to define $\Gamma(N)$ and $\Gamma_0(N)$ one writes them in the following form:

$$\Gamma(N) = \left\{ \begin{pmatrix} a & b \\ c & d \end{pmatrix} \mid \begin{pmatrix} a & b \\ c & d \end{pmatrix} \equiv \begin{pmatrix} 1 & 0 \\ 0 & 1 \end{pmatrix} (mod\ N) \right\} ,$$

$$\Gamma_0(N) = \left\{ \begin{pmatrix} a & b \\ c & d \end{pmatrix} \mid \begin{pmatrix} a & b \\ c & d \end{pmatrix} \equiv \begin{pmatrix} * & * \\ 0 & * \end{pmatrix} (mod\ N) \right\} ,$$

asterisks denoting arbitrary elements from $\mathbb{Z}$. Clearly, $\Gamma(N) \subset \Gamma_0(N)$ for $N \neq 1$. Note that for $N \geq 3$,

$$\begin{pmatrix} -1 & 0 \\ 0 & -1 \end{pmatrix} \notin \Gamma(N) ;$$

$\Gamma(N)$ is called the *principal congruence subgroup of level N*.

Exercise 4.1.1. Show the existence

a) of a natural group isomorphism

$$\Gamma(1)/\Gamma(N) \xrightarrow{\sim} PSL(2,\mathbb{Z}/N) ,$$

$PSL(2,\mathbb{Z}/N)$ being the factor of a special linear group $SL(2,\mathbb{Z}/N)$ over its center; in particular, $\Gamma(N)$ is normal in $\Gamma(1)$;

b) of a natural bijection of sets

$$\Gamma(1)/\Gamma_0(N) \xrightarrow{\sim} \mathbb{P}^1(\mathbb{Z}/N) ,$$

$\mathbb{P}^1(\mathbb{Z}/N)$ being the projective line over the ring $\mathbb{Z}/N$, i.e. the set of one-dimensional free $\mathbb{Z}/N$-submodules of $(\mathbb{Z}/N)^2$.

Exercise 4.1.2. Deduce from the previous exercise that for the indices of subgroups $\Gamma(N)$ and $\Gamma_0(N)$ in the group $\Gamma(1)$ we have

$$\text{a)} \quad [\Gamma(1):\Gamma(N)] = \begin{cases} \frac{N^3}{2}\cdot\prod_{p|N}(1-p^{-2}) & \text{for } N \geq 3 \\ 6 & \text{for } N = 2 \end{cases}$$

$$\text{b)} \quad [\Gamma(1):\Gamma_0(N)] = N\cdot\prod_{p|N}(1+p^{-1}) \ .$$

(*Hint*: Use complete induction over N).

Exercise 4.1.3. Show that for $N \geq 3$ the group $\Gamma(N)$ contains no element of finite order.

A congruence subgroup Γ of $\Gamma(1)$ is a subgroup containing $\Gamma(N)$ for some N. Further on by Γ we denote an arbitrary congruence subgroup of $\Gamma(1)$ (in fact we are mostly interested in the cases $\Gamma = \Gamma(N)$ and $\Gamma = \Gamma_0(N)$).

Suppose that Γ is equipped with a discrete topology while H has a usual complex topology (hence H is homeomorphic to an open disc).

Exercise 4.1.4. Check that the topological space $\Gamma\backslash H$ is Hausdorf in the factor topology.

The space $\Gamma\backslash H$ is canonically equipped by a structure of a Riemann surface (non-compact!), i.e. by an analytical structure. Indeed, let $N \geq 3$ and let $\Gamma_N = \Gamma \cap \Gamma(N)$. According to Exercise 4.1.3, the group Γ_N contains no elements of finite order.

Exercise 4.1.5. Show that the action of Γ_N on H is *free*, i.e. has no fixed points. Moreover, for any $v \in H$ there exists an open neighbourhood U_v homeomorphic to its image $V_v \subset \Gamma_N \backslash H$.

Exercise 4.1.6. Show that on $\Gamma_N \backslash H$ there exists a unique Riemann surface structure such that homeomorphisms between U_v and V_v are complex-analytic isomorphisms.

Since $\Gamma'_N = \Gamma/\Gamma_N$ is a finite group, we can define the Riemann surface $\Gamma \backslash H$ as a factor of $\Gamma_N \backslash H$ over Γ'_N .

Exercise 4.1.7. Show that the Riemann surface structure on $\Gamma \backslash N$ does not depend on the choice of N .

An essential disadvantage of the Riemann surface $\Gamma \backslash H$ is that it is not compact. A canonical compactification $\Gamma \backslash H^*$ of the surface $\Gamma \backslash H$ is constructed as follows. Consider the set $\mathbb{P}^1(\mathbb{Q}) = \mathbb{Q} \cup \{\infty\}$ consisting of rational numbers and of the symbol ∞ . Each element of $\mathbb{P}^1(\mathbb{Q})$ can be written in the form a/b , where $a, b \in \mathbb{Z}$, and for $b = 0$ we put $a/b = \infty$.

The group $\Gamma(1)$ acts naturally on $\mathbb{P}^1(\mathbb{Q})$: if $\gamma = \begin{pmatrix} a & b \\ c & d \end{pmatrix}$, then $\gamma(r) = \dfrac{a \cdot r + b}{c \cdot r + d}$, where $\gamma(\infty) = a/c$ and $\gamma(r) = \infty$ if $c \cdot r + d = 0$.

Let $\Gamma \backslash H^* = (\Gamma \backslash H) \cup (\Gamma \backslash \mathbb{P}^1(\mathbb{Q}))$.

Exercise 4.1.8. Show that the factor set $\Gamma \backslash \mathbb{P}^1(\mathbb{Q})$ is finite. (*Hint*: Apply Exercise 4.1.1.b).

The finite set $\Gamma \backslash \mathbb{P}^1(\mathbb{Q})$ is called the set of cusps of the Riemann surface $\Gamma \backslash H^*$ (or of the group Γ).

Let us define a complex structure on $\Gamma\backslash H^*$ such that its restriction to the open subset $\Gamma\backslash H$ coincides with one defined above, complex-analytic neighbourhoods of $r \in \mathbb{Q}$ being open discs tangent to the line $Im(z) = 0$ at r and neighbourhoods of ∞ being open half-planes of the form $Im(z) > d$.

Theorem 4.1.9. *The set $\Gamma\backslash H^*$ with the above complex structure is a connected compact Riemann surface.* ■

Exercise 4.1.10. Show that if $\Gamma = \Gamma(1)$, then $\Gamma\backslash H$ can be identified with $\mathbb{C}$ and $\Gamma\backslash H^*$ with the Riemann sphere $\mathbb{P}^1(\mathbb{C})$.

Let us consider holomorphic functions $f : H \longrightarrow \mathbb{C}$ obtained by the composition

$$H \xrightarrow{\pi_{\Gamma(1)}} \Gamma(1)\backslash H \xrightarrow[\sim]{\theta} \mathbb{C}$$

for some complex analytic isomorphisms θ ,

$$\pi_{\Gamma(1)} : H \longrightarrow \Gamma(1)\backslash H$$

being the natural projection.

Exercise 4.1.11. Show that the automorphism group of $\mathbb{C}$ as of a Riemann surface consists of affine maps of the form $\{z \longmapsto a\cdot z + b\}$, $a \in \mathbb{C}^*$, $b \in \mathbb{C}$.

Hence there exists a unique function f of this form such that $f(i) = 1728$ and $f(\rho) = 0$, where $i = \sqrt{-1}$ and $\rho = \dfrac{1 + \sqrt{-3}}{2}$. We denote such a function by j .

Exercise 4.1.12. Let $z \in H$, let $\Lambda_z = \mathbb{Z} + \mathbb{Z}\cdot z$ be a lattice in $\mathbb{C}$ and let E_z be a complex elliptic curve such that $E_z(\mathbb{C}) = \mathbb{C}/\Lambda_z$. Prove that then $j(z) = j(E_z)$ is the absolute invariant of E_z.

The curves $X(N)$ **and** $X_0(N)$. If Γ is an arbitrary congruence subgroup of $\Gamma(1)$, then there exists a unique (up to an isomorphism) smooth projective curve X_Γ over $\mathbb{C}$ such that X_Γ considered as a Riemann surface is isomorphic to $\Gamma\backslash H^*$, and the Riemann surface $\Gamma\backslash H$ is naturally isomorphic to the smooth affine curve Y_Γ over $\mathbb{C}$. We call X_Γ and Y_Γ *modular curves.*

We are mostly interested in curves corresponding to the groups $\Gamma = \Gamma(N)$ and $\Gamma = \Gamma_0(N)$, which are denoted by

$$X(N) = X_{\Gamma(N)}, \quad X_0(N) = X_{\Gamma_0(N)};$$

$$Y(N) = Y_{\Gamma(N)}, \quad Y_0(N) = Y_{\Gamma_0(N)}.$$

The subgroup $\Gamma(N)$ being normal in $\Gamma(1)$, the group $\Gamma(1)/\Gamma(N) \simeq PSL(2,\mathbb{Z}/N)$ acts on the curve $X(N)$ (and on $Y(N)$) according to the usual formula: if $\gamma = \begin{pmatrix} a & b \\ c & d \end{pmatrix} \in \Gamma(1)$ and $\bar{z}$ is the image of $z \in H \cup \mathbb{P}^1(\mathbb{Q})$ in $X(N)$, then

$$\gamma : \bar{z} \longmapsto \overline{\left(\frac{a\cdot z+b}{c\cdot z+d}\right)}.$$

Since $\Gamma(N)$ is normal in $\Gamma(1)$, the action is well-defined.

Exercise 4.1.13. Show that if $n|N$, then $X(N)/\overline{\Gamma(n)} = X(n)$, $\overline{\Gamma(n)}$ being the image of $\Gamma(n)$ in $\Gamma(1)/\Gamma(N)$.

The subgroup $\Gamma_0(N)$ is not normal in $\Gamma(1)$. Nevertheless one has

Exercise 4.1.14. Show that for $N = m\cdot n$ the element

$$\tau_m = \frac{1}{\sqrt{m}}\cdot\begin{pmatrix} 0 & 1 \\ -m & 0 \end{pmatrix} \in SL(2,\mathbb{R})$$

lies in the normalizer of $\Gamma_0(N)$ in $SL(2,\mathbb{R})$.

Exercise 4.1.15. Deduce from the above exercise that τ_m defines an automorphism of the curve $X_0(N)$. We denote it by w_m . Show that w_m are involutions (i.e. elements of order 2) and that they commute with each other.

Thus,

$$Aut(X_0(N)) \supseteq (\mathbb{Z}/2)^{\sigma_0},$$

where $\sigma_0 = \sigma_0(N)$ is the number of divisors of N .

The tower of modular curves. Riemann surfaces of the form $\Gamma\backslash H^*$ form a natural tower: for two congruence subgroups $\Gamma' \leq \Gamma \leq \Gamma(1)$, there exists a natural complex-analytic projection

$$\pi_{\Gamma,\Gamma'} : \Gamma'\backslash H^* \longrightarrow \Gamma\backslash H^* .$$

In particular, for any Γ there exists a natural projection

$$\pi_{\Gamma(1),\Gamma} = \pi_\Gamma : \Gamma\backslash H^* \longrightarrow \Gamma(1)\backslash H^* = \mathbb{P}^1_{\mathbb{C}} .$$

The ramification structure of the projection

$$\pi_{\Gamma,\Gamma'} : \Gamma'\backslash H^* \longrightarrow \Gamma\backslash H^*$$

can be easily found. Indeed, let $a \in H \cup \mathbb{P}^1(\mathbb{Q}) = H^*$. By $\Gamma_a \subset \Gamma(1)$ we denote the stabilizer of a .

Exercise 4.1.16. Let $a \in H^*$, and let $\bar{a}$ be the image of a in $\Gamma'\backslash H^*$. Show that the ramification index of the projection $\pi_{\Gamma,\Gamma'}$ at $\bar{a}$ equals $[\Gamma_a : \Gamma'_a]$.

Hence ramification points of the projection $\pi_{\Gamma,\Gamma'}$ are exactly the images in $\Gamma'\backslash H^*$ of points $a \in H^*$ such that $\Gamma_a \neq \Gamma_a'$.

Let $S = \begin{pmatrix} 0 & -1 \\ 1 & 0 \end{pmatrix} \in \Gamma(1)$ and $T = \begin{pmatrix} 1 & 1 \\ 0 & 1 \end{pmatrix} \in \Gamma(1)$.

One can calculate the groups Γ_a for a subgroup $\Gamma \subset \Gamma(1)$ starting from the following result describing the groups $\Gamma(1)_a$.

Exercise 4.1.17. Show that S is an element of order 2 and $U = S \cdot T = \begin{pmatrix} 0 & -1 \\ 1 & 1 \end{pmatrix}$ is of order 3 in $\Gamma(1)$.

Remark 4.1.18. In fact $\Gamma(1)$ is the *free product* of its cyclic subgroups $\langle S \rangle$ and $\langle U \rangle$. This means that any element $\gamma \in \Gamma(1)$ can be uniquely written in the form $\gamma = S^a \cdot U^{b_1} \cdot S \cdot U^{b_2} \cdot \ldots \cdot S \cdot U^{b_k}$, where $a \in \{0,1\}$, $b_k \in \{0,1,2\}$, and $b_j \in \{1,2\}$ for $j \neq k$.

Exercise 4.1.19. Prove the following facts:

a) Let $z \in H$, $\rho = (-1 + \sqrt{-3})/2$, $i = \sqrt{-1}$. If $z \notin \Gamma(1) \cdot i \cup \Gamma(1) \cdot \rho$, then $\Gamma(1)_z = \{1\}$. If $z = \gamma(i)$, $\gamma \in \Gamma(1)$, then $\Gamma(1)_z = \gamma \cdot \langle S \rangle \cdot \gamma^{-1}$ is a subgroup of order 2. If $z = \gamma(\rho)$, $\gamma \in \Gamma(1)$, then $\Gamma(1)_z = \gamma \cdot \langle U \rangle \cdot \gamma^{-1}$ is a subgroup of order 3.

b) The action of $\Gamma(1)$ on $\mathbb{P}^1(\mathbb{Q})$ is transitive. If $a = \gamma(\infty)$, $a \in \mathbb{P}^1(\mathbb{Q})$, $\gamma \in \Gamma(1)$, then $\Gamma(1)_a = \gamma \cdot \langle T \rangle \cdot \gamma^{-1}$, $\langle T \rangle$ being the infinite cyclic subgroup generated by T.

Applying the results of the above exercises together with the Hurwitz formula to the projection $\pi_\Gamma : X_\Gamma \longrightarrow \mathbb{P}^1_{\mathbb{C}}$, one can calculate the genus of X_Γ. We shall write out the answer for curves $X(N)$ and $X_0(N)$, which are most interesting for us.

Proposition 4.1.20. a) *Let* $N \geq 3$. *Then the genus of the curve* $X(N)$ *equals*

$$g(N) = 1 + \frac{(N - 6)\cdot\mu_N}{12\cdot N} ,$$

where $\mu_N = [\Gamma(1):\Gamma(N)] = \frac{N^3}{2}\cdot\prod_{p|N} (1 - p^{-2})$.

b) For the genus $g_0(N)$ *of the curves* $X_0(N)$ *we have*

$$g_0(N) = 1 + \frac{\mu}{12} - \frac{\nu_2}{4} - \frac{\nu_3}{3} - \frac{\nu_\infty}{2}$$

where

$$\mu = [\Gamma(1):\Gamma_0(N)] = N\cdot\prod_{p|N}(1 + p^{-1}) ,$$

and ν_2 *is the number of non-*$\Gamma_0(N)$*-equivalent points of* H *which are* $\Gamma(1)$*-equivalent to* $z = i$:

$$\nu_2 = \begin{cases} 0 & \text{for } 4|N \\ \prod_{p|N} \left(1 + \left(\frac{-1}{p}\right)\right) & \text{for } 4\nmid N , \end{cases}$$

where

$$\left(\frac{-1}{p}\right) = \begin{cases} 0 & \text{for } p = 2 \\ 1 & \text{for } p \equiv 1 (\text{mod } 4) \\ -1 & \text{for } p \equiv 3 (\text{mod } 4) , \end{cases}$$

ν_3 *is the number of non-$\Gamma_0(N)$-equivalent points of* H, *which are* $\Gamma(1)$*-equivalent to* ρ :

$$\nu_3 = \begin{cases} 0 & \text{for } 9\mid N \\ \prod_{p\mid n} \left(1 + \left(\frac{-3}{p}\right)\right) & \text{for } 9\nmid N \end{cases},$$

where

$$\left(\frac{-3}{p}\right) = \begin{cases} 0 & \text{for } p = 3 \\ 1 & \text{for } p \equiv 1 (\mathrm{mod}\ 3) \\ -1 & \text{for } p \equiv 2 (\mathrm{mod}\ 3) \end{cases},$$

and ν_∞ *is the number of cusps of* $X_0(N)$:

$$\nu_\infty = \sum_{\substack{d\mid N \\ d>0}} \varphi((d,N/d)) ,$$

φ *being the Euler function.* ∎

Corollary 4.1.21. *Let* $\ell \neq 11$ *either be a prime or* $\ell = 1$. *Then* $g_0(11\cdot\ell) = \ell$. ∎

Further on we shall need the description of ramification in the tower

$$X_0(11\cdot\ell) \xrightarrow{\varphi} X_0(11) \xrightarrow{\psi} \mathbb{P}^1 ,$$

where

$$\varphi = \pi_{\Gamma_0(11),\Gamma_0(11\cdot\ell)} , \quad \psi = \pi_{\Gamma_0(11)} , \quad \theta = \psi\circ\varphi = \pi_{\Gamma_0(11\cdot\ell)} .$$

Exercise 4.1.22. Check the following facts:

a) The degrees of projections φ , ψ , and θ equal, respectively:

$$deg\ \varphi = \ell + 1\ , \quad deg\ \psi = 12\ , \quad deg\ \theta = 12\cdot(\ell + 1)\ .$$

b) The projection θ is unramified outside the sets $\theta^{-1}(0)$, $\theta^{-1}(\infty)$, $\theta^{-1}(1728)$. At the points lying over 0 , 1728, and $\infty \in \mathbb{P}^1_{\mathbb{C}}$, the ramification of θ is described as follows:

the set $\theta^{-1}(0)$ consists of $4\cdot(\ell + 1)$ points $R_1,\ldots,R_{4(\ell+1)}$ each of them having the ramification index 3;

the set $\theta^{-1}(1728)$ consists of $6\cdot(\ell + 1)$ points $P_1,\ldots,P_{6(\ell+1)}$ each of them having the ramification index 2 ;

the set $\theta^{-1}(\infty)$ consists of four points Q_1,Q_2,Q_3,Q_4, with ramification indices, respectively,

$$e_1 = e_{Q_1} = 1,\ e_2 = e_{Q_2} = \ell,$$

$$e_3 = e_{Q_3} = 11, \quad e_4 = e_{Q_4} = 11\cdot\ell\ .$$

The ramification of the projection ψ is described as follows:

the set $\psi^{-1}(0)$ consists of points R'_1,R'_2,R'_3,R'_4, each of them having the ramification index 3 ;

the set $\psi^{-1}(1728)$ consists of points $P'_1,\ldots,P'_6$, each of them having the ramification index 2 ;

the set $\psi^{-1}(\infty)$ consists of points Q'_1 and Q'_2 with ramification indices $e'_1 = e_{Q'_1} = 1,\ e'_2 = e_{Q'_2} = 11$.

The projection φ is unramified outside $\theta^{-1}(0)$.

Moreover, $\varphi^{-1}(Q_1') = \{Q_1, Q_2\}$, and the corresponding ramification indices of φ are equal to $e_1'' = e_{Q_1} = 1$, $e_2'' = e_{Q_2} = \ell$; and $\varphi^{-1}(Q_2') = \{Q_3, Q_4\}$, $e_3'' = 1$, $e_4'' = \ell$.

Thus, for the ramification divisor B_θ of the projection θ we have:

$$B_\theta = \sum_{i=1}^{6\cdot(\ell+1)} P_i + 2\cdot \sum_{m=1}^{4\cdot(\ell+1)} R_m +$$

$$+ (\ell - 1)\cdot Q_2 + 10\cdot Q_3 + (11\cdot\ell - 1)\cdot Q_4 .$$

Exercise 4.1.23. Find out the ramification structure in the towers

$$X_0(32\cdot\ell) \longrightarrow X_0(32) \longrightarrow \mathbb{P}^1 ,$$

$$X_0(2^n) \longrightarrow X_0(32) \longrightarrow \mathbb{P}^1 ,$$

and

$$X_0(3^m) \longrightarrow X_0(27) \longrightarrow \mathbb{P}^1 ,$$

where ℓ is a prime, $n \geq 5$, $m \geq 3$ being integers. (*Hint*: The curves $X_0(32)$ and $X_0(27)$ are elliptic).

Weierstrass points. Since Weierstrass points play a considerable role in the study of AG-codes, we are rather interested in Weierstrass points on curves X_Γ. A complete description of Weierstrass points on such curves is obtained only for several small values of N; however using Proposition 2.2.48 and Corollary 2.2.49 one can easily find a rather representative set of Weierstrass points on each of these curves. Cusps are particularly convenient since they quite often happen to be Weierstrass points.

Recall that the canonical involution w_N permutes 0 and ∞; it follows that 0 is a Weierstrass point iff ∞

is. Here we restrict ourselves to the case of the cusp $P = \infty$ (or to 0, which is equivalent), since it is this point that can be most readily shown to be Weierstrass, its weight being sufficiently high. Moreover we are interested mostly in the case of curves $X_0(p^2 \cdot n)$, p being a prime and $n \geq 2$.

Exercise 4.1.24. Show that the natural projection

$$X_0(p^2 \cdot n) \longrightarrow X_0(p \cdot n)$$

is of degree p, and that the point $0 \in X_0(p^2 \cdot n)$ has the ramification index equal to p (i.e. 0 is *totally ramified*). (*Hint*: Write out the set of representatives of right cosets of $\Gamma_0(p \cdot n)$ over the subgroup $\Gamma_0(p^2 \cdot n)$).

Exercise 4.1.25. Show that if

$$g(X_0(p^2 \cdot n)) \geq p \cdot g(X_0(p \cdot n)) + 1$$

then the points 0 and ∞ are Weierstrass points on $X_0(p^2 \cdot n)$. (*Hint*: Proceed as in Proposition 2.2.48).

Proposition 4.1.26. *The point* $\infty \in X_0(p^2 \cdot n)$ *is a Weierstrass point at least in the following cases:*

a) $p \geq 5$ *and* $p | n$, *or*
$p = 3$ *and* $3 | n$, $n \neq 3$, $n \neq 9$, *or*
$p = 2$ *and* $2 | n$, $n \neq 2$, $n \neq 2 \cdot \ell$, $n \neq 4 \cdot \ell$, ℓ *being a prime;*

b) n *is divisible by three different primes;*

c) n *is divisible by* m^2, *where* $p \nmid m$ *and* $p^2 \cdot n \neq 36$;

d) n *is divisible by two different primes and at least one of two congruences* $x^2 + 1 \equiv 0 \pmod{p \cdot n}$, *or* $x^2 + x + 1 \equiv 0 \pmod{p \cdot n}$ *has a solution;*

e) $p \equiv n \equiv 1 \pmod{12}$. ∎

Exercise 4.1.27. Prove the above proposition, for each case pointing out a non-gap at ∞ not exceeding $g_0(p^2 \cdot n)$. (*Warning*: This exercise is intended only for an assiduous reader).

Thus for the curves $X_0(N)$, where N is divisible by a square, ∞ is quite often a Weierstrass point.

This effect is most important for curves of the form $X_0(2^n)$, $n \geq 7$ (note that for $n \leq 5$ the genus of these curves is at most 1, and for $n = 6$ the point ∞ is in fact a Weierstrass point, but Exercise 4.1.25 does not yield this fact). Hence for the curves $X_0(2^n)$ one can point out many non-gaps which do not exceed the genus of the curve.

Exercise 4.1.28. Considering projections

$$X_0(2^n) \longrightarrow X_0(2^m), \quad n \geq 7, \ m < n - 1,$$

write out non-gaps not exceeding the genus of $X_0(2^n)$ (existing due to the method used in Proposition 2.2.48). Give a lower bound for the weight of ∞ on $X_0(2^n)$.

Hyperellipticity of the curves $X_0(n)$ is the other problem which is closely connected to Weierstrass points. One can easily write out a series of small values of n for which $X_0(n)$ is hyperelliptic, for example $n = 37$.

However for large n the curves $X_0(n)$ are never hyperelliptic.

Proposition 4.1.29. *For* $n \geq 72$ *the curves* $X_0(n)$ *are not hyperelliptic.* ∎

Automorphic functions. The elements of the field of rational functions on the curve X_Γ, or in other terms meromorphic functions on the Riemann surface $\Gamma\backslash H^*$, can be considered as functions on H invariant under Γ such that their only singularities (both on $\mathbb{Q}\cup\{\infty\}$ and on H) are poles. These functions are called *automorphic* under Γ.

Let us start with the case $\Gamma = \Gamma(1)$. Here $\Gamma\backslash H^* \simeq \mathbb{P}^1_{\mathbb{C}}$. Hence, the field of functions automorphic under $\Gamma(1)$ coincides with $\mathbb{C}(j)$. Any such function has a canonical expansion into a Laurent series in the neighbourhood of infinity which is called its *q-expansion*. Since $T \in \Gamma(1)$, for any $f \in \mathbb{C}(j)$ and $z \in H$ we have $f(Tz) = f(z+1) = f(z)$.

Exercise 4.1.30. Check that the function $q = e^{2\pi iz}$ can be considered as a local parameter in a neighbourhood of $\infty \in \mathbb{P}^1_{\mathbb{C}}$.

A local expansion over q of the function $\tilde{f}$ induced on $\mathbb{P}^1_{\mathbb{C}}$ by f, is called the q-expansion of f:

$$f = \sum_{m=-N}^{\infty} a_m \cdot q^m .$$

The number of negative powers of q in this expansion is finite since f is meromorphic.

The theory of elliptic functions yields the following

Proposition 4.1.31. *For the q-expansion of j we have*

$$j(z) = q^{-1} \cdot (1 + \sum_{m=1}^{\infty} c_m \cdot q^m) ,$$

c_m being integers. ■

In Section 4.3.1 we give more precise information on the coefficients c_m.

Remark that j is normalized (multiplied by 1728) just in order to satisfy Proposition 4.1.31.

Now let Γ be a congruence subgroup such that $\Gamma \supseteq \Gamma(N)$. Then

$$T^N = \begin{pmatrix} 1 & N \\ 0 & 1 \end{pmatrix} \in \Gamma$$

so that for any automorphic function f on the subgroup Γ we have $f(T^N \cdot z) = f(z + N) = f(z)$ for $z \in H$.

Therefore any such function in a neighbourhood of infinity has an expansion into a Laurent series of the form

$$f(z) = \sum_{k=-m}^{\infty} a_k \cdot q^{k/N} , \quad a_k \in \mathbb{C} .$$

This expansion is called the q-expansion of f at infinity and is in fact an expansion of the rational function $\tilde{f}$ on X_Γ defined by the function f in powers of a local parameter at the cusp of X_Γ corresponding to ∞ . In a similar way one can define q-expansions of automorphic functions at other cusps, but they are of no use for us.

Note that $T \in \Gamma$ for $\Gamma = \Gamma_0(N)$, and therefore the functions automorphic under $\Gamma_0(N)$ have expansions in integral powers of q .

The field of automorphic functions under $\Gamma_0(N)$ has the following explicit description:

Theorem 4.1.32. *The field of functions automorphic under* $\Gamma_0(N)$ *coincides with* $\mathbb{C}(j(z), j(N \cdot z))$. ■

In particular the function $j_N(z) = j(N \cdot z)$ is invariant under $\Gamma_0(N)$. There exists an involution of the field $\mathbb{C}(j,j_N)$ which corresponds to the involution w_N on the curve $X_0(N)$; it permutes j and j_N .

Modular equation. Thus the functions $j(z)$ and $j(N\cdot z)$ satisfy a relation of the form $F(j(z),j(N\cdot z)) = 0$, F being a polynomial in two variables. Moreover, since all the coefficients of the q-expansions of $j(z)$ and $j(N\cdot z)$ are integers, one can choose F having integral coefficients. The minimum relation of the form $F(j,x) = 0$ satisfied by $j = j(z)$ and $x = j_N = j(N\cdot z)$ is called the *modular equation*. The basic properties of the modular equation can be resumed as follows:

Theorem 4.1.33. *Let*

$$\Phi_N(x,j) = \prod_{\alpha\in A} (x - j(\alpha(z))) ,$$

α running over the set A of all matrices of the form $\alpha = \begin{pmatrix} a & b \\ 0 & d \end{pmatrix}$, where $a,b,d \in \mathbb{Z}$, $a > 0$, $d > 0$, $a\cdot d = N$, $0 \le b < d$, $(a,b,d) = 1$, and $\alpha(z) = \dfrac{a\cdot z + b}{d}$. Then:

a) $\Phi_N(x,j) \in \mathbb{Z}[x,j]$, *and*

$$\Phi_N(x,j) = x^{\psi(N)} + j^{\psi(N)} + \sum_{a,b<\psi(N)} c_{ab}\cdot x^a \cdot j^b ,$$

where $\psi(N) = [\Gamma(1):\Gamma_0(N)] = N\cdot\prod_{p|N}(1 + p^{-1})$ *; moreover*

$$\Phi_N(x,j) = \Phi_N(j,x) ,$$

so that

$$c_{ab} = c_{ba} \in \mathbb{Z}.$$

b) The polynomial $\Phi_N(x,j)$ *is absolutely irreducible and satisfies the relation*

$$\Phi_N(j, j_N) = 0 \ . \blacksquare$$

Thus, the modular equation $\Phi_N(x,j) = 0$ defines an affine model of the curve $X_0(N)$. Moreover, since Φ_N has integral coefficients, the curve $X_0(N)$ is defined over $\mathbb{Q}$. Further on we shall be interested mostly in the curve $X_0(N)$ over $\mathbb{Q}$, however to study its properties one has to use the Riemann surface $\Gamma_0(N)\backslash H^*$ as well.

4.1.2. Reductions of modular curves

To study reductions of modular curves, one has to consider moduli schemes of elliptic curves.

Modular curves as moduli varieties. Above we have already mentioned that the function $j : H \longrightarrow \mathbb{P}^1_{\mathbb{C}}$ defining the isomorphism

$$j : \Gamma(1)\backslash H^* \xrightarrow{\sim} \mathbb{P}^1_{\mathbb{C}}$$

is in fact the absolute invariant of a variable elliptic curve $\mathbb{C}/(\mathbb{Z} + z\cdot\mathbb{Z})$. Since the absolute invariants are equal iff complex elliptic curves are isomorphic, we see that the set $\Gamma(1)\backslash H = Y(1)$ is in a bijection with the set of isomorphism classes of complex elliptic curves, the bijection being defined by the absolute invariant

$$j : (\text{elliptic curve } E) \longmapsto j(E) \ .$$

This is the basic fact for the further argument, leading to an interpretation of modular curves as moduli varieties of elliptic curves with additional structures. Before we pass to the general theory let us give two examples more.

Consider the curve $Y_0(N) = \Gamma_0(N)\backslash H$ and its natural projection onto $\mathbb{A}^1 = Y(1)$. The fibre of

$$\pi_{\Gamma_0(N)} : Y_0(N) \longrightarrow Y(1)$$

over the point $j(z)$ coincides (according to Exercise 4.1.1) with the set $\mathbb{P}^1(\mathbb{Z}/N)$ of cyclic subgroups of order N in the group $E_N \simeq (\mathbb{Z}/N)^2$ of points of order N on the curve $E = \mathbb{C}/(\mathbb{Z}+z\cdot\mathbb{Z})$. This shows that the set $Y_0(N)$ is in a natural bijection with the set of isomorphism classes of pairs (E,C_N), E being a complex elliptic curve and C_N being its cyclic subgroup of order N (clearly, $C_N \subset E_N$).

To state a similar result for $Y(N)$ we need the following fact.

Exercise 4.1.34. Show that the following construction defines a non-degenerate skew-symmetric pairing (the *Weil pairing*)

$$E_N \times E_N \longrightarrow \mu_N ,$$

$$(s , t) \longmapsto e(s,t) ,$$

which defines a canonical identification $\Lambda^2 E_N = \mu_N$. The construction is as follows: let $\mathbb{C}/(\mathbb{Z}+z\cdot\mathbb{Z}) \xrightarrow{\sim} E(\mathbb{C})$ be an analytic isomorphism; $E_N = \frac{1}{N}\Lambda/\Lambda$, where $\Lambda = \mathbb{Z} + z\cdot\mathbb{Z}$. Then $s \in E_N$ is defined by a pair (a,b) of residues modulo N. Let $e(s,t) = e^{2\pi i\cdot(a\cdot b'-a'\cdot b)/N}$ for $s = (a,b)$ and $t = (a',b')$.

Considering the curve $Y(N)$ and its projection onto $Y(1)$, we see that the points $Y(N)$ are in a natural bijection with the set of pairs (E,α_N) , E being a complex elliptic curve and α_N a so-called "structure of level N with $det\ \alpha_N = 1$ ", i.e. an isomorphism

$$\alpha_N : E_N \xrightarrow{\sim} (\mathbb{Z}/N)^2$$

such that the isomorphism

$$\Lambda^2\alpha_N : \Lambda^2 E_N = \mu_N \xrightarrow{\sim} \Lambda^2((\mathbb{Z}/N)^2) = \mathbb{Z}/N$$

maps $e^{2\pi i/N}$ to 1 .

Exercise 4.1.35. Verify the assertion on the interpretation of points of $Y(N)$ as pairs (E,α_N) .

The above examples motivate the choice of additional structures necessary to interpret modular curves as moduli varieties.

The general theory of moduli schemes requires relative elliptic curves over schemes.

Relative elliptic curves. Let S be a scheme. A *relative elliptic curve* E/S is a proper smooth morphism $p : E \longrightarrow S$ such that all its geometrical fibres are elliptic curves and the following requirements are satisfied:

a) there is a morphism $e : S \longrightarrow E$ called the *zero section* of E over S such that the map $p \circ e : S \longrightarrow S$ is identical;

b) there is a morphism (of *addition*) $m : E \times_S E \longrightarrow E$

such that the following diagrams commute

$$\begin{array}{ccc} E \times_S E \times_S E & \xrightarrow{m \times id_E} & E \times_S E \\ {\scriptstyle id_E \times m}\downarrow & & \downarrow{\scriptstyle m} \\ E \times_S E & \xrightarrow{m} & E \end{array}$$

(*associativity*);

$$\begin{array}{ccc} E \times_S E & \xrightarrow{i} & E \times_S E \\ {\scriptstyle m}\downarrow & & \downarrow{\scriptstyle m} \\ E & = & E \end{array}$$

(*commutativity*),

where $i : E \times_S E \longrightarrow E \times_S E$ is the morphism of permutation (it is defined on coordinate rings of affine open subvarieties in E by $a \otimes b \longmapsto b \otimes a$), and id_E is the identity morphism;

$$\begin{array}{ccc} E = E \times_S S & \xrightarrow{id_E \times e} & E \times_S E \\ {\scriptstyle id_E}\downarrow & & \downarrow{\scriptstyle m} \\ E & = & E \end{array}$$

(*the existence of a neutral element*);

c) There exists a morphism (*inversion*) $\psi : E \longrightarrow E$ such that the following diagram commutes:

$$\begin{array}{ccccc} E \times_S E & & \xrightarrow{\psi \times id_E} & & E \times_S E \\ {\scriptstyle \Delta}\uparrow & & & & \downarrow{\scriptstyle m} \\ E & \xrightarrow{p} & S & \xrightarrow{e} & E \end{array}$$

(*the existence of an inverse*).

Remark 4.1.36. If a scheme G/S is equipped with morphisms e , m , ψ satisfying the requirements a) - c) , G is called a *commutative group scheme*. A *morphism* of commutative group schemes over S is a morphism of schemes over S commuting with morphisms e , m , and ψ . One can define an *isomorphism* of commutative group schemes over S in a natural way.

Exercise 4.1.37. Show that for $S = Spec\, k$, k being an algebraically closed field, the notions of a relative elliptic curve E/S and of an elliptic curve over k coincide. What are the maps e , m , ψ in this case?

Exercise 4.1.38. Define a morphism of relative elliptic curves $E/S \longrightarrow E'/S$, in particular define isomorphic relative elliptic curves.

The category of elliptic curves over S is denoted by (Ell/S) .

Exercise 4.1.39. Show that if E/S is a relative elliptic curve then the set $E(S)$ of sections $\varphi : S \longrightarrow E$ of the morphism p (i.e. of φ such that $p \circ \varphi = id$) is an abelian group. (*Hint*: The group structure on $E(S)$ is determined by e , m , and ψ).

The subgroup of $E(S)$ consisting of elements of order dividing N is denoted by $E_N(S)$.

An isomorphism

$$\alpha_N : E_N(S) \xrightarrow{\sim} (\mathbb{Z}/N)^2$$

is called a *structure of level* N *on* E/S (or just a *level structure*).

Exercise 4.1.40. Show that if E/S has a structure of level N, then N is invertible on S, i.e. N is invertible in all local rings O_P, $P \in S$.

Moduli schemes of elliptic curves. To construct moduli schemes, one has to use the technique of representable functors. Let $F(N)$ be the following contravariant functor on the category of schemes:

$$F(N)(S) = \left\{ \begin{array}{l} \text{the set of isomorphism classes of pairs} \\ (E,\alpha_N),\ E/S \text{ being a relative elliptic curve} \\ \text{and } \alpha_N \text{ being a structure of level } N \text{ on } E/S \end{array} \right\}.$$

Note that for any ring A the category (Sch/A) is a sub-category in the category (Sch), and the notion of restriction $F(N)|_A$ of the functor $F(N)$ to the category (Sch/A) makes sense. The following fact is basic for the theory of moduli schemes of elliptic curves.

Theorem 4.1.41. *Let* $N \geq 3$. *The functor* $F(N)|_{\mathbb{Z}[1/N]}$ *is representable by an affine smooth over* $\mathbb{Z}[1/N]$ *scheme* $Y(N)/_{\mathbb{Z}[1/N]}$. ■

The scheme $Y(N)/_{\mathbb{Z}[1/N]}$ is called the *moduli scheme of elliptic curves with structure of level* N.

The Weil pairing makes it clear that $Y(N)/_{\mathbb{Z}[1/N]}$ is in fact a scheme over the ring $A_N = \mathbb{Z}[\zeta_N, 1/N]$, $\zeta_N = e^{2\pi i/N}$ being a primitive root of unity.

Proposition 4.1.42. *Geometric fibres of the morphism* $Y(N)/_{\mathbb{Z}[1/N]} \longrightarrow \operatorname{Spec} A_N$ *are smooth affine irreducible curves.* ■

Exercise 4.1.43. Show that

$$(Y(N)/_{\mathbb{Z}[1/N]}) \otimes_{A_N} \mathbb{C} = Y(N) \ .$$

Thus, for the curve $Y(N)$ there exists a form defined over the field $\mathbb{Q}(\zeta_N)$, which has good reduction over primes not dividing N .

The following functors $P(N)$ and $F_0(N)$ are of primary importance for us:

$$P(N)(S) = \left\{\begin{array}{l}\text{the set of isomorphism classes of pairs} \\ (E,\beta_N),\ E/S \text{ being a relative elliptic curve} \\ \text{and } \beta_N \text{ being a projective structure of level } N\end{array}\right\},$$

where the *projective structure* β_N is a class of structures α_N of level N up to a multiplication by an element $\gamma \in (\mathbb{Z}/N)^*$;

$$F(N)(S) = \left\{\begin{array}{l}\text{the set of isomorphism classes of pairs} \\ (E,C_N),\ E/S \text{ being a relative elliptic curve} \\ \text{and } C_N \text{ being a cyclic subgroup } E \text{ of order } N\end{array}\right\}.$$

Unfortunately, the functors $P(N)$ and $F_0(N)$ are never representable, even when restricted to the category (Sch/k) , k being an algebraically closed field. Thus the corresponding moduli schemes do not exist. Let us introduce the following definition.

Let $F : (Sch/A)^0 \longrightarrow (Sets)$ be a contravariant functor on the category of schemes over A . A scheme S_F is called a *coarse moduli scheme* for F iff the following statements hold:

a) There exists a morphism of functors

$$\tau : F \longrightarrow Hom_{Sch/A}(\cdot, S_F) \ .$$

b) For any algebraically closed field k, which is an A-algebra, the morphism τ induces an isomorphism

$$\tau(k) : F(Spec\ k) \xrightarrow{\sim} S_F(k) .$$

c) The scheme S_F is *universal* for morphisms of F to representable functors, i.e. for any morphism

$$\lambda : F \longrightarrow Hom_{Sch/A}(\cdot, B)$$

to a functor representable by an A-scheme B, there exists an A-morphism $f : B \longrightarrow S_F$ such that the diagram

$$\begin{array}{ccc} & \overset{\lambda}{\nearrow} & Hom_{Sch/A}(\cdot, B) \\ F & & \uparrow \theta_f \\ & \underset{\tau}{\searrow} & Hom_{Sch/A}(\cdot, S_F) \end{array}$$

commutes, the functor morphism θ_f being induced by f.

The definition yields the uniqueness of the coarse moduli scheme for F, once it exists.

Theorem 4.1.44. *The restrictions of functors* $P(N)$ *and* $F_0(N)$ *to a subcategory of* $\mathbb{Z}[1/N]$*-schemes have coarse moduli schemes* $Y_P(N)/_{\mathbb{Z}[1/N]}$ *and* $Y_0(N)/_{\mathbb{Z}[1/N]}$ *.* ∎

Proposition 4.1.45. *The scheme* $Y_P(N)/_{\mathbb{Z}[1/N]}$ *is a form of the scheme* $Y(N)$ *over the ring* $B_N = K_N \cap \mathbb{Z}[\zeta_N, 1/N]$ *,* K_N *being the unique quadratic subfield in* $\mathbb{Q}(\zeta_N)$ *, i.e.*

$$(Y_P(N)/_{\mathbb{Z}[1/N]}) \otimes_{B_N} \mathbb{Z}[\zeta_N, 1/N] = Y(N)/_{\mathbb{Z}[1/N]} . \blacksquare$$

Corollary 4.1.46. *The fibres of the morphism* $Y_P(N)/_{\mathbb{Z}[1/N]} \longrightarrow Spec\ B_N$ *are smooth absolutely irreducible affine curves, and*

$$(Y_P(N)/_{\mathbb{Z}[1/N]}) \otimes_{B_N} \mathbb{C} = Y(N) \ . \blacksquare$$

Proposition 4.1.47. *The fibres of the projection* $Y_0(N)/_{\mathbb{Z}[1/N]} \longrightarrow Spec\ \mathbb{Z}[1/N]$ *are smooth absolutely irreducible affine curves, and*

$$(Y_0(N)/_{\mathbb{Z}[1/N]}) \otimes_{\mathbb{Z}[1/N]} \mathbb{C} \simeq Y_0(N) \ . \blacksquare$$

Thus $Y_P(N)/_{\mathbb{Z}[1/N]}$ is a form of the complex curve $Y(N)$ over the ring B_N, and $Y_0(N)/_{\mathbb{Z}[1/N]}$ is a form of $Y_0(N)$ over $\mathbb{Z}[1/N]$.

To clarify the situation described by Proposition 4.1.45 and Corollary 4.1.46, let us describe the case of $Y(N)$ and $Y_P(N)$ considered over $\mathbb{Q}$. To put it more exactly, let

$$Y(N)_{\mathbb{Q}} = (Y(N)/_{\mathbb{Z}[1/N]}) \otimes \mathbb{Q}$$

and

$$Y_P(N)_{\mathbb{Q}} = (Y_P(N)/_{\mathbb{Z}[1/N]}) \otimes \mathbb{Q} \ .$$

The curve $Y(N)_{\mathbb{Q}}$ is a curve over $\mathbb{Q}$ which becomes reducible when considered over $\mathbb{Q}(\zeta_N)$. This curve has $\varphi(N)$ irreducible components (note that $\varphi(N) = [\mathbb{Q}(\zeta_N):\mathbb{Q}]$) and the action of the Galois group $Gal(\mathbb{Q}(\zeta_N)/\mathbb{Q})$ permutes these components transitively. In terms of function fields this means that the field $\mathbb{Q}(Y(N)_{\mathbb{Q}})$ contains $\mathbb{Q}(\zeta_N)$ and $\mathbb{Q}(Y(N)_{\mathbb{Q}}) = \mathbb{Q}(\zeta_N)(Y)$ for some curve Y. Considering the subfield in $\mathbb{Q}(Y(N)_{\mathbb{Q}})$ fixed under the subgroup $Gal(\mathbb{Q}(\zeta_N)/K_N)$, which is of index 2 in $Gal(\mathbb{Q}(\zeta_N)/\mathbb{Q})$, we obtain the field $K_N(Y_P(N)_{\mathbb{Q}})$ such that $Y_P(N)_{\mathbb{Q}}$ is an absolutely irreducible curve defined over the field K_N (when considering over $\mathbb{Q}$ it has two irreducible components).

Schemes $X(N)/_{\mathbb{Z}[1/N]}$ **and** $X_0(N)/_{\mathbb{Z}[1/N]}$. The schemes

$$Y(N)/_{\mathbb{Z}[1/N]}\ ,\quad Y_P(N)/_{\mathbb{Z}[1/N]}\ ,\quad Y_0(N)/_{\mathbb{Z}[1/N]}$$

give a satisfactory description of reductions of affine curves $Y(N)$ and $Y_0(N)$. The study of reductions of projective curves $X(N)$ and $X_0(N)$ is also important. A more thorough investigation of moduli schemes of elliptic curves makes it possible to prove the following fundamental result:

Theorem 4.1.48. a) *There exists a regular proper over* $\mathbb{Z}[1/N]$ *scheme* $X(N)/_{\mathbb{Z}[1/N]}$ *containing* $Y(N)/_{\mathbb{Z}[1/N]}$ *as an open set and such that all fibres of the projecton* $X(N)/_{\mathbb{Z}[1/N]} \longrightarrow \mathrm{Spec}\ \mathbb{Z}[1/N]$ *are smooth proper curves and the closed subscheme* $X(N)/_{\mathbb{Z}[1/N]} - Y(N)/_{\mathbb{Z}[1/N]}$ *is a disjoint union of several copies of* $\mathrm{Spec}\ A_N$ *; moreover* $X(N)/_{\mathbb{Z}[1/N]}$ *is a scheme over* A_N *, fibres of the projection* $X(N)/_{\mathbb{Z}[1/N]} \longrightarrow \mathrm{Spec}\ A_N$ *are smooth absolutely irreducible curves, and*

$$(X(N)/_{\mathbb{Z}[1/N]})\otimes_{A_N}\mathbb{C} \simeq X(N)\ .$$

b) *There exists a regular proper over* $\mathbb{Z}[1/N]$ *scheme* $X_P(N)/_{\mathbb{Z}[1/N]}$ *containing the scheme* $Y_P(N)/_{\mathbb{Z}[1/N]}$ *as an open dense subset. The scheme* $X_P(N)/_{\mathbb{Z}[1/N]}$ *is a scheme over* B_N *and*

$$(X_P(N)/_{\mathbb{Z}[1/N]})\otimes_{B_N}A_N = X(N)/_{\mathbb{Z}[1/N]}\ .$$

c) *There exists a regular proper over* $\mathbb{Z}[1/N]$ *scheme* $X_0(N)/_{\mathbb{Z}[1/N]}$ *containing* $Y_0(N)/_{\mathbb{Z}[1/N]}$ *as an open dense subscheme; moreover, fibres of* $X_0(N)/_{\mathbb{Z}[1/N]} \longrightarrow \mathrm{Spec}\, \mathbb{Z}[1/N]$ *are smooth absolutely irreducible curves, and the closed subscheme* $X_0(N)/_{\mathbb{Z}[1/N]} - Y_0(N)/_{\mathbb{Z}[1/N]}$ *is isomorphic to a disjoint union of several copies of* $\mathrm{Spec}\, \mathbb{Z}[1/N]$ *, and*

$$(X_0(N)/_{\mathbb{Z}[1/N]}) \otimes_{\mathbb{Z}[1/N]} \mathbb{C} \simeq X_0(N) \; . \blacksquare$$

Thus, the curves $X(N)$ and $X_0(N)$ have good reductions over any prime ideal not dividing N . Since the curve $X_0(N)$ can be defined over $\mathbb{Q}$, for any $p \nmid N$ there exists a good reduction $X_0(N)/p$ of the curve $X_0(N)$ modulo p , $X_0(N)/p = (X_0(N)/_{\mathbb{Z}[1/N]}) \otimes_{\mathbb{Z}[1/N]} \mathbb{F}_p$.

Note that by Theorem 4.1.48 the genus of curves $X(N)$ and $X_0(N)$ is preserved under reduction. Moreover, the reductions inherit a number of structures of $X(N)$ and $X_0(N)$. In particular $X_0(N)/p$ form towers similar to the towers described in Exercises 4.1.22 and 4.1.23, with the same degrees of maps and the same ramification structures. Moreover, the curve $X_0(N)/p$ inherits from $X_0(N)$ the set of involutions w_d , $d|N$. Remark as well that the hyperellepticicity of $X_0(N)$ is equivalent to that of $X_0(N)/p$.

Exercise 4.1.49. Prove the last statement. (*Hint*: The reduction of a map defined by the canonical class gives a map defined by the canonical class of the reduction).

Proposition 4.1.29 together with Exercise 4.1.49 yield non-hyperellepticity of curves $X_0(N)/p$ for $N \geq 72$.

The reductions of $X(N)$ also inherit a number of its properties; later on we shall need the action of the group

$PSL(2,\mathbb{Z}/N)$ on these reductions. To be more precise recall that $PSL(2,\mathbb{Z}/N)$ acts on $X(N)$. By Theorem 4.1.48 $X(N)$ can be defined over the field K_N . The action of $PSL(2,\mathbb{Z}/N)$ is also defined over K_N and can be extended to the action of $PSL(2,\mathbb{Z}/N)$ on $X_P(N)/_{\mathbb{Z}[1/N]}$, preserving fibres of the projection $X_P(N) \longrightarrow Spec\ B_N$. Therefore, if $\mathfrak{p}$ is a prime ideal dividing p in K_N (i.e. either $\mathfrak{p} = (p)$ or $(p) = \mathfrak{p}\cdot\mathfrak{p}'$ for the complex conjugate ideal $\mathfrak{p}'$), then the reduction $X(N)/\mathfrak{p}$ of the curve $X(N)$ (to be more precise of its form over K_N) over the ideal $\mathfrak{p}$ is good, i.e. it is a smooth projective absolutely irreducible curve over the residue field $k(\mathfrak{p})$ of $\mathfrak{p}$ and $PSL(2,\mathbb{Z}/N)$ acts on this curve (note that $k(\mathfrak{p})$ is isomorphic to $\mathbb{F}_{p^2}$ for $\mathfrak{p} = (p)$ and to $\mathbb{F}_p$ for $(p) = \mathfrak{p}\cdot\mathfrak{p}'$).

Remark 4.1.50. It is possible to describe the action of $PSL(2,\mathbb{Z}/N)$ on $X(N)/\mathfrak{p}$ in more invariant terms. Indeed, if $\theta : F \longrightarrow F$ is a morphism of representable contravariant functor $F : (Sch/A) \longrightarrow (Sets)$ to itself which is an isomorphism (i.e. it has an inverse morphism), then θ induces an automorphism θ^* of the scheme M_F representing F . This follows immediately from the definition of a representable functor. In particular an action of a group G on the functor F induces an action of G on M_F .

The functor $F(N)$ has a natural action of the group $GL(2,\mathbb{Z}/N)$ defined by

$$\gamma : (E,\alpha_N) \longmapsto (E,\alpha'_N) ,$$

α'_N being a structure of level N obtained by composing α_N and $\gamma \in GL(2,\mathbb{Z}/N)$:

$$\alpha'_N : E_N(S) \xrightarrow{\alpha_N} (\mathbb{Z}/N)^2 \xrightarrow{\gamma} (\mathbb{Z}/N)^2 .$$

Hence, the scheme $Y(N)/_{\mathbb{Z}[1/N]}$ is equipped with the action of the group $GL(2,\mathbb{Z}/N)$ which can be extended to the action of $GL(2,\mathbb{Z}/N)$ on $X(N)/_{\mathbb{Z}[1/N]}$. This action preserves the fibration over $Spec\ A_N$ and defines the action of $GL(2,\mathbb{Z}/N)$ on the fibres. In fact the action of $GL(2,\mathbb{Z}/N)$ on a fibre over $\mathfrak{p} \in Spec\ A_N$ permutes its components, on the contrary the action of $SL(2,\mathbb{Z}/N)$ preserves them. When passing to the scheme $X_P(N)/_{\mathbb{Z}[1/N]}$, we get the described action of $PSL(2,\mathbb{Z}/N)$.

Exercise 4.1.51. Show that the factor of the curve $X_P(N)/\mathfrak{p}$ over $PSL(2,\mathbb{Z}/N)$ coincides with $\mathbb{P}^1$, and the projection $X_P(N)/\mathfrak{p} \longrightarrow \mathbb{P}^1_{k(\mathfrak{p})}$ is given by j-invariant (i.e. a pair (E,β_N) representing a point of $X_P(N)/\mathfrak{p}$ is mapped to $j(E)$).

$\mathbb{F}_{p^2}$**-rational points.** As we have already seen, the theory of moduli schemes of elliptic curves shows that $X(N)$ and $X_0(N)$ have good reductions outside the divisors of N. The same theory shows that on such curves the upper asymptotic bound for the ratio of the number of $\mathbb{F}_{p^2}$-rational points to the genus is tight. To be more precise let us fix a prime p coprime to N. Then there exists a smooth projective absolutely irreducible curve $X_N = X_0(N)/p$ over $\mathbb{F}_p$; moreover, over the field $k(\mathfrak{p}) \subseteq \mathbb{F}_{p^2}$, $\mathfrak{p}$ being a prime ideal dividing p in the field K_N, there exists a smooth projective absolutely irreducible curve $X'_N = X_P(N)/\mathfrak{p}$. Let $g_N = g(X_N) = g_0(N)$, $g'_N = g(X'_N) = g(N)$ and let R_N be the number of $\mathbb{F}_{p^2}$-rational points on X_N, and R'_N the number of $\mathbb{F}_{p^2}$-points on X'_N. We have

Theorem 4.1.52.

$$\lim_{n \longrightarrow \infty} R_N/g_N = p - 1 \,, \tag{4.1.1}$$

$$\lim_{n \longrightarrow \infty} R'_N/g'_N = p - 1 \,. \tag{4.1.1'}$$

Proof: To begin with, we remark that there exists a tower of curves

$$X'_N \xrightarrow{\lambda_N} X_N \xrightarrow{\theta_N} \mathbb{P}^1 \,,$$

θ_N being a map of degree

$$\psi(N) = N \cdot \prod_{\ell | N} (1 + \ell^{-1})$$

defined by j-invariant, and the composition

$$\pi_N = \theta_N \circ \lambda_N \; : \; X'_N \longrightarrow \mathbb{P}^1$$

is also defined by j-invariant and is of degree equal to the order μ_N of the group $PSL(2,\mathbb{Z}/N)$, i.e. to $\frac{N^3}{2} \cdot \prod_{\ell | N} (1 - \ell^{-2})$, so that λ is of degree

$$deg\ \lambda_N = deg\ \pi_N / deg\ \theta_N = \frac{N^2}{2} \cdot \prod_{\ell | N} (1 - \ell^{-1}) \,.$$

Hence it suffices to prove the formula (4.1.1') since it is clear that $R_N \geq R'_N / deg\ \lambda_N$, and by the Hurwitz formula $g_N \leq g'_N / deg\ \lambda_N$, so that $R_N/g_N \geq R'_N/g'_N$. The inequality $lim\ sup\ R_N/g_N \leq p - 1$ follows from Theorem 2.3.22. To prove (4.1.1') if suffices to give an adequate lower estimate for R'_N. To do this let us consider all points of X'_N lying over supersingular values of j-invariant. The estimate for R'_N is implied by the following facts:

a) the number of such points equals $\mu_N \cdot \frac{p-1}{12}$;

b) all such points are $\mathbb{F}_{p^2}$-rational.

Thus

$$R'_N \geq \frac{\mu_N \cdot (p-1)}{12}$$

which yields (4.1.1′), since (cf. Proposition 4.1.20)

$$g'_N = g(N) = 1 + \frac{\mu_N \cdot (N-6)}{12 \cdot N} .$$

We start with a). The point $P \in X_N$ is represented by a pair (E, β_N) , E being an elliptic curve and β_N being a projective structure of level N . The group $PSL(2, \mathbb{Z}/N)$ of order μ_N acts on the fibre over the point $j(E) \in \mathbb{P}^1$ in such a way that the order of the isotropy group of this point equals $|Aut\ E|/2$ (since isomorphic curves give the same point of $X(N)$). Hence the fibre over $j(E)$ contains $2 \cdot \mu_N / |Aut\ E|$ points, and the total number of points lying over supersingular values of j (by (2.4.9)) is equal to

$$\sum_{j(E)} 2 \cdot \mu_N \cdot |Aut\ E|^{-1} = \mu_N \cdot (p-1)/12 ,$$

the sum being taken over all supersingular values $j(E)$.

Now we are going to prove b) . Let (E, β_N) be a supersingular elliptic curve over $\overline{\mathbb{F}}_p$ with a projective structure of level N . It suffices to prove that the pair (E, β_N) is preserved by the action of the Frobenius endomorphism F over $\mathbb{F}_{p^2}$ which is a topological generator of the group $Gal(\overline{\mathbb{F}}_p / \mathbb{F}_{p^2})$. Since the degrees of F and of the endomorphism of multiplication by p coincide, and both morphisms are purely inseparable, $F = \varepsilon \circ p$ in the ring

$End_{\overline{\mathbb{F}}_p}(E)$, ε being the automorphism of E . If $|Aut\ E| = 2$, then $\varepsilon = \pm 1$, $F = \pm p$, and F preserves β_N . If $|Aut\ E| \neq 2$, then one has to consider the representative of the point P on X'_N defined by the pair (E,β'_N) obtained from (E,β_N) by the action of ε . For this representative F coincide with $\pm p$. ∎

Exercise 4.1.53. Check the assertion on the order of the isotropy subgroup of a point and the last assertion in the proof of the theorem. Check that for a supersingular curve E there exists $\varepsilon \in Aut_{\overline{\mathbb{F}}_p}(E)$ such that $F = \varepsilon \circ p$. (*Hint:* Since p annihilates the kernel of the $\mathbb{F}_p$-Frobenius morphism $F_0 : E \longrightarrow E'$, we see that $p = f \circ F_0$, $f : E' \longrightarrow E$ being a morphism of degree p ; now check that any morphism of degree p is of the form $\varepsilon \circ F_0$).

Remark 4.1.54. In fact

$$R'_N = (p - 1) \cdot g_N + c \cdot g_N^{2/3} + o(g_N^{2/3})$$

for some $c > 0$. From the proof of Theorem 2.3.22 one can conclude that

$$N_1 \le (p - 1) \cdot g + O(g/log\ g) ,$$

N_1 being the number of $\mathbb{F}_q$-points on a curve of genus g . We do not know whether the difference $N_1 - (p - 1) \cdot g$ can grow faster than $O(g_N^{2/3})$, or not.

Remark 4.1.55. From the proof of Theorem 4.1.52 it follows that the points of X_N lying over supersingular values of j , are $\mathbb{F}_{p^2}$-rational and that their number is at least $(p - 1) \cdot g_0(N)$.

Exercise 4.1.56. Calculate the number S_N of "supersingular" points on reductions $X_0(N)/p$ for $N = \ell$, $11\cdot\ell$, 2^n, 3^m, where ℓ is a prime, $\ell \neq 11$, $\ell \neq p$, and n and m are positive integers. Calculate the genus of corresponding curves.

4.1.3. Codes

Let us now consider some properties of codes obtained by the help of reductions of modular curves.

Parameters. We use the notation introduced in the proof of Theorem 4.1.52. Thus $X_N = X_0(N)/p$ is an absolutely irreducible smooth projective curve over $\mathbb{F}_p$ of genus $g_N = g_0(N)$ and $X'_N = X_P(N)/p$ is an absolutely irreducible smooth projective curve of genus $g'_N = g(N)$ over $\mathbb{F}_{p^2}$ or $\mathbb{F}_p$. By $\mathcal{P}$ we denote the set of supersingular points of X_N, $\mathcal{P} \subseteq X_N(\mathbb{F}_{p^2})$, by $\mathcal{P}'$ the set of supersingular points of X'_N, we know that $\mathcal{P}' \subseteq X'_N(\mathbb{F}_{p^2})$. Let $N \geq 3$.

Proposition 4.1.57. a) *Let D be a divisor of degree a on X_N defined over $\mathbb{F}_{p^2}$, and let $\operatorname{Supp} D \cap \mathcal{P} = \emptyset$. The $[n,k,d]_{p^2}$-code $C = (X_N, \mathcal{P}, D)_L$ has the following parameters: $n = |\mathcal{P}| = 1,\ldots,n_0$, where*

$$n_0 \geq \frac{N\cdot(p-1)}{12}\cdot\prod_{\ell\mid N}(1+\ell^{-1}) ,$$

$$k \geq a - g_0(N) + 1 ,$$

$$d \geq n - a ,$$

here the product is taken over all prime divisors of N *and* $g_0(N)$ *is given by the formula of Proposition* 4.1.20.*b* .

b) Let D' *be a divisor of degree* a' *on* X'_N *defined over* $\mathbb{F}_{p^2}$ *and let* $\mathrm{Supp}\, D' \cap \mathcal{P}' = \emptyset$. *The* $[n',k',d']_{p^2}$*-code* $C' = (X'_N, \mathcal{P}', D')_L$ *has the following parameters:* $n' = |\mathcal{P}| = 1,\ldots,n'_0$ *, where*

$$n'_0 \geq \frac{N^3\cdot(p-1)}{24}\cdot\prod_{\ell|N}(1-\ell^{-2}) ,$$

$$k' \geq a' - \frac{N^2\cdot(N-6)}{24}\cdot\prod_{\ell|N}(1-\ell^{-2}) ,$$

$$d' \geq n' - a' ,$$

the product being taken over all prime divisors of N .

Proof: The proposition follows immediately from Proposition 4.1.20 and Theorem 4.1.52. ■

Remark 4.1.58. Since the length of C can be at least

$$\frac{N\cdot(p-1)}{12}\cdot\prod_{\ell|N}(1+\ell^{-1}) ,$$

shortening it (if necessary), we can assume that

$$n = \rceil\frac{N\cdot(p-1)}{12}\cdot\prod_{\ell|N}(1+\ell^{-1})\lceil ,$$

The set of numbers of such form is rather dense:

Exercise 4.1.59. a) Show that the number $\rho(n)$ of the positive integers of the form

$$m = \rceil\frac{N\cdot(p-1)}{12}\cdot\prod_{\ell|N}(1+\ell^{-1})\lceil ,\ m \leq n ,$$

is asymptotically at least $\frac{p-1}{12}\cdot\frac{n}{\ln n}$, i.e. $\rho(n) \geq f(n)$, where $\lim_{n\to\infty} \frac{12\cdot f(n)\cdot \ln n}{(p-1)\cdot n} = 1$. (*Hint:* Consider the case of a prime N and apply the prime distribution theorem).

b) For $p = 7$ (i.e. for codes over $\mathbb{F}_{49}$) check the following table giving the code length n as a function of N:

N	4	5	6	7	8	9	10	11	12	13	14	15	16	17	18	19	20	21	22
n	3	3	6	4	6	6	9	6	12	7	12	12	12	9	18	10	18	16	18

N	23	24	25	26	27	28	29	30	31	32	33	34	35	36	37	38	39	40
n	12	24	15	21	18	24	15	36	16	24	24	27	24	36	19	30	28	36

Using Weierstrass points, we can as usual ameliorate the code parameters for small k. We shall now consider only the point ∞ on $X_0(N)$ and restrict ourselves to the values $N = 2^m$, in this case the effect obtained by using Weierstrass points is maximal. Recall (cf. Proposition 4.1.20) that the genus $g_N = g_0(N)$ of the curve $X_N = X_0(N)/p$ is given by:

$$g_0(2^m) = 2^{m-3} + 1 - \delta_m ,$$

where

$$\delta_m = \begin{cases} 3\cdot 2^{m/2-2} & \text{for an even } m ; \\ 2^{(m-1)/2} & \text{for an odd } m . \end{cases}$$

Proposition 4.1.60. *Let* $D = a\cdot\infty$ *be a divisor of degree* a *on* X_{2^m} *and let* $C = (X_{2^m}, \mathcal{P}, D)_L$ *be an*

$[n,k,d]_{p^2}$-code. *Then for the parameters of* C *we have:* $n = |\mathcal{P}| = 1,\ldots,n_0$, *where*

$$n_0 \geq 2^{m-3}\cdot(p-1)\ ,$$

$$k \geq k(m,a)\ ,$$

$$d \geq n - a\ ,$$

and $k(m,a)$ *is defined as follows:*

$$k(m,a) = \ell \quad iff \quad 2^{m-3} - 2^{m-2-\ell} \leq a < 2^{m-3} - 2^{m-3-\ell}\ ,$$

$$\ell = 1,2,\ldots,\lceil \tfrac{m}{2} \rceil - 1\ .$$

(Note that for such a the standard estimate $a - g + 1$ is not positive).

Proof: The only estimate which really needs a proof is that for k. Considering projections $X_{2^m} \longrightarrow X_{2^{2j}}$ and $X_{2^m} \longrightarrow X_{2^{2j+1}}$ for $2j + 1 \leq m$, we get the following estimates (using Exercise 4.1.25):

$$\ell(2^{m-2j}\cdot(2^{2j-3} - 3\cdot 2^{j-2} + 1 + i)\cdot\infty) \geq i + 1\ ,$$

$$\ell(2^{m-2j-1}\cdot(2^{2j-2} - 2^{j} + 1 + i)\cdot\infty) \geq i + 1\ .$$

The maximum of these estimates for $j = 1,2,\ldots,\lceil \frac{m}{2} \rceil$ coincides with $k(m,a)$. ∎

Exercise 4.1.61. Check the formula for $k(m,a)$. Deduce an analogue of Proposition 4.1.60 for X_N, $N = 3^m$.

Group properties. Since the group $PGL(2,\mathbb{Z}/N)$ acts on the curve X'_N preserving fibres of the projection $X'_N \longrightarrow \mathbb{P}^1$, in particular preserving the set $\mathcal{P}'$ and the set of cusps of X'_N, it is natural to expect the codes on X'_N constructed from $\mathcal{P}'$ and cusp divisors to be group codes.

To be precise, let S_∞ be the set of points on X'_N which are mapped to $\infty \in \mathbb{P}^1$ by the natural projection $X'_N \longrightarrow \mathbb{P}^1$. Every point $Q \in S_\infty$ is a ramification point of this projection with index N, hence

$$\nu_\infty = |S_\infty| = \mu_N/N = \frac{N^2}{2} \cdot \prod_{\ell | N}(1-\ell^{-2}) \ .$$

Consider a divisor $F_\infty = \sum_{Q\in S_\infty} Q$ of degree $deg\, F_\infty = \nu_\infty = |S_\infty|$. Since the point $\infty \in \mathbb{P}^1$ is $\mathbb{F}_p$-rational, the divisor F_∞ is defined over $\mathbb{F}_p$ (and hence over $\mathbb{F}_{p^2}$). Let $D' = b \cdot F_\infty$ be a divisor supported in S_∞. Consider the code $C' = (X'_N, \mathcal{P}', D')_L$ over $\mathbb{F}_{p^2}$.

The definition of D', the invariance of $\mathcal{P}'$ under $PGL(2,\mathbb{Z}/N)$, and Proposition 3.1.30 yield the following

Proposition 4.1.62. *The code C' has a natural action of the group $PGL(2,\mathbb{Z}/N)$. The number of orbits of this action equals the number $s(p)$ of supersingular values of j-invariant in characteristic p.* ■

Corollary 4.1.63. *The code C' can be realized as a $PGL(2,\mathbb{Z}/N)$-submodule in the algebra $\mathbb{F}_{p^2}[PGL(2,\mathbb{Z}/N)]^{s(p)}$.* ■

Remark 4.1.64. If we restrict ourselves to the values of N such that $p \nmid (\ell^2 - 1)$ for all prime divisors $\ell | N$ (recall that we have also assumed $p \nmid N$) then the algebra

$\mathbb{F}_{p^2}[PGL(2,\mathbb{Z}/N)]$ is semi-simple (since the order is not divisible by p) which essentially simplifies the study of the group code C' . Note that for $p \geq 5$ such N can be as great as you like.

Problem 4.1.65. Give an explicit description of C' as of a submodule in $\mathbb{F}_{p^2}[PGL(2,\mathbb{Z}/N)]$ and estimate its parameters without any use of the Riemann-Roch theorem.

Remark 4.1.66. Although we cannot solve the above problem, the Riemann-Roch theorem gives us some information on the structure of C' as of a $PGL(2,\mathbb{Z}/N)$-module. Let $a_0 = \lceil N/6 \rceil$. Then by the Riemann-Roch theorem, for $m \geq a_0$ we have

$$\ell(m \cdot F_\infty) = \mu_N \cdot (12 \cdot m \cdot N + 6)/12$$

since $deg\ (m \cdot F_\infty) > 2 \cdot g'_N - 2$. Thus

$$L((m + 1) \cdot F_\infty)/L(m \cdot F_\infty)$$

is a $PGL(2,\mathbb{Z}/N)$-module of dimension

$$\mu_N/N = \frac{N^2}{2} \cdot \prod_{\ell | N} (1 - \ell^{-2}) \ .$$

Hence C' considered as a $PGL(2,\mathbb{Z}/N)$-module has a filtration of the form: $C' = C_b \supset C_{b-1} \supset \ldots \supset C_{a_0}$, where C_m for $b \geq m \geq a_0$ is the image of $L(m \cdot F_\infty)$ in C' , and all the factors C_{m+1}/C_m for $m = a_0, \ldots, b - 1$ are $PGL(2,\mathbb{Z}/N)$-modules of dimension μ_N/N . In fact, applying a more subtle algebraic-geometric and group theory technique, one can determine the isomorphism classes of $PGL(2,\mathbb{Z}/N)$-modules of the form C_{m+1}/C_m . However this procedure is far beyond the methods of this book.

It looks likely that a solution of Problem 4.1.65 lies on this lines.

CHAPTER 4.2

CODES ON DRINFELD CURVES

Considering codes on reductions of classical modular curves we obtain codes over $\mathbb{F}_{p^2}$ with good asymptotic properties. To obtain codes with similar properties over $\mathbb{F}_{q^2}$, q being an arbitrary power of a prime p, one has to consider Drinfeld modular curves. The latter have a number of advantages as compared with the classical modular curves; they are in many aspects simpler. Unfortunately this does not regard all the aspects of the theory of modular curves. In some cases classical modular curves are more convenient since they come from characteristic zero, which gives a number of essential results unknown for Drinfeld modular curves (e.g. a characterization of their Weierstrass points).

In this chapter we give the necessary information on Drinfeld curves and codes on them. In Section 4.2.1 we discuss the basic properties of elliptic modules which in the Drinfeld theory play the role of elliptic curves. Section 4.2.2 is devoted to the construction and basic properties of Drinfeld modular curves. In Section 4.2.3 we consider codes on these curves.

Throughout this section we use the following notation: $A = \mathbb{F}_q[C']$, where $C' = C - \{\infty\}$ is an affine absolutely irreducible smooth curve obtained from a complete curve C over $\mathbb{F}_q$ by deleting an $\mathbb{F}_q$-rational point ∞. In fact we are interested here only in the case $C = \mathbb{P}^1$, $C' = \mathbb{A}^1$, $A = \mathbb{F}_q[T]$, and the reader loses almost nothing by restricting himself to this case. For $a \in A$ we write $|a| = |a|_\infty$ for the value of normalized absolute value corresponding to the point $\infty \in C$. For $A = \mathbb{F}_q[T]$ we have $|a| = q^{-deg\, a}$. By $k = \mathbb{F}_q(C)$ we denote the fraction field of A; the extension of $|\cdot|$ to k is denoted by the same symbol: $|\frac{a}{b}| = \frac{|a|}{|b|}$ for $a,b \in A$, $b \neq 0$. By k_∞ we denote the completion of k by this absolute value, by $\bar{k}_\infty$ the algebraic closure of k_∞. Reading this section it is useful to keep in mind the following list of analogies between the objects of classical theory and those of Drinfeld theory.

Ring $\mathbb{Z}$	Ring A
Field $\mathbb{Q}$	Field k
Absolute value $\vert\cdot\vert$	Absolute value $\vert\cdot\vert_\infty$
Field $\mathbb{R}$	Field k_∞
Field $\mathbb{C}$	Field $\bar{k}_\infty$
Elliptic curve E	Elliptic module φ
Abelian group $Hom(E,E')$	A-module $Hom(\varphi,\psi)$
Ring $End_{\mathbb{Q}}(E)$	Ring $End_k(\varphi)$
Algebra $End_{\mathbb{Q}}(E)\otimes\mathbb{Q}$	Algebra $End_k(\varphi)\otimes_A k$

Lattice $\Lambda \subset \mathbb{C}$	Lattice $\Gamma \subset \bar{k}_\infty$
Positive integer N (structure level)	Ideal $I \subset A$ (structure level)
Group $PSL(2,\mathbb{Z})$	Group $GL(2,A)$
Complex modular curve $X(N)$	Drinfeld curve $\overline{M(I)\otimes\bar{k}_\infty}$
Field $\mathbb{Q}(\zeta_N)$, $\zeta_N = e^{2\pi i/N}$	Field $k(I)$
Groups $\Gamma_0(N)$ and $\Gamma_1(N)$	Groups $\Gamma_0(I)$ and $\Gamma_1(I)$
Reductions of modular curves $X_0(N)/p$ and $X_p(N)/\mathfrak{p}$	Curves $X_0(I)$ and $C(I)$ over $\mathbb{F}_q$
Supersingular elliptic curve E in characteristic p	Supersingular elliptic module φ of "characteristic" (T)

4.2.1. Elliptic modules

Elliptic modules over a field. Let K be a field of characteristic p which is an A-algebra, and let $i : A \longrightarrow K$ be the structure morphism, i.e. $i(a) = a\cdot 1$. The ideal $Ker\ i \in Spec\ A$ is called the *"characteristic"* of the field K and is denoted by "*char*"K . If $Ker\ i = (0)$, we say that K has a *"general characteristic"*, and if $Ker\ i \neq (0)$, then K is a field of *"finite characteristic"*.

By $K\{\tau\}$ we denote the *ring of non-commutative polynomials* in τ , i.e. of expressions of the form

$$\sum_{i=0}^{n} a_i\cdot\tau^i \quad , \; a_i \in K$$

with multiplication defined by $\tau\cdot a = a^p\cdot\tau$ for $a \in K$. By

D we denote the natural projection

$$D : K\{\tau\} \longrightarrow K \quad ,$$

$$D : \sum_{i=0}^{n} a_i \cdot \tau^i \longmapsto a_0 \quad .$$

The degree map $deg : K\{\tau\} \longrightarrow \mathbb{Z}$ is defined by

$$deg(\sum_{i=0}^{n} a_i \cdot \tau^i) = p^n \; ,$$

for $a_n \neq 0$.

An elliptic A-module over K is a ring homomorphism

$$\varphi : A \longrightarrow K\{\tau\}$$

such that $\varphi(A)$ is not contained in K and the composition $D \circ \varphi$ coincides with i .

Proposition 4.2.1. **a)** *The homomorphism φ is injective.*

b) *There exists a positive integer d such that for any $a \in A$*

$$deg\, \varphi(a) = |a|^d \; .$$

Proof: a) If $Ker\, \varphi \neq (0)$, then $\varphi(A)$ is a subfield in $K\{\tau\}$, and hence $\varphi(A) \subseteq K$ which contradicts the definition of an elliptic module.

b) Let us consider the map $a \longmapsto deg\, \varphi(a)$ from A to $\mathbb{Z}$. It is easy to check that this map can be extended to an absolute value on k . Since for $a \in A$, $a \neq 0$ we have $deg\, \varphi(a) \geq 1$, this absolute value cannot correspond to a point $v \in Spec\, A$. Hence $deg\, \varphi(a) = |a|^d$ for some positive rational d ; later on (in Corollary 4.2.8) we shall show that d is an integer. ∎

A positive integer d defined in Proposition 4.2.1 is called the *rank* of the elliptic A-module φ . Later on we shall be mostly interested in the case $d = 2$ and $A = \mathbb{F}_q[T]$. In this case the elliptic A-module over K is defined by one element $\varphi(T) \in A$.

Exercise 4.2.2. Show that

$$\varphi(T) = i(T) + c_1 \cdot F + c_2 \cdot F^2 \ , \ c_i \in K \ , \ c_2 \neq 0 \ ,$$

where $F = \tau^m$ denotes the Frobenius element over $\mathbb{F}_q$, $q = p^m$. (*Hint*: This follows from the commutativity of the ring $\varphi(A)$).

Homomorphisms. A *homomorphism* from an elliptic A-module $\varphi : A \longrightarrow K\{\tau\}$ to an elliptic A-module $\psi : A \longrightarrow K\{\tau\}$ is an element $u \in K\{\tau\}$ such that $u \cdot \varphi(a) = \psi(a) \cdot u$ for any $a \in A$. The set of homomorphisms from the module φ to the module ψ is denoted by $Hom(\varphi,\psi)$. It has a natural structure of an A-module. A non-zero element from $Hom(\varphi,\psi)$ is called an *isogeny*; if $Hom(\varphi,\psi) \neq (0)$, the modules φ and ψ are called *isogenous*.

Exercise 4.2.3. Show that isogenous modules are of the same rank.

An element $u \in K\{\tau\}$ defines an isomorphism of elliptic A-modules φ and ψ over K iff $u \in K^*$ and $\psi(a) = u \cdot \varphi(a) \cdot u^{-1}$ for any $a \in A$.

In the case when $d = 2$ and $A = \mathbb{F}_q[T]$ one can easily give a criterion for elliptic A-modules to be isomorphic and obtain a complete description of the group $Aut_K(\varphi)$ of automorphisms of an elliptic A-module φ over K.

Exercise 4.2.4. Let

$$\varphi(T) = i(T) + c_1 \cdot F + c_2 \cdot F^2 ,$$

$$\psi(T) = i(T) + d_1 \cdot F + d_2 \cdot F^2 ,$$

be elliptic A-modules over K. Show that φ is isomorphic to ψ iff for some $\lambda \in K^*$ we have

$$d_1 = \lambda^{q-1} \cdot c_1 \quad \text{and} \quad d_2 = \lambda^{q^2-1} \cdot c_2 .$$

Exercise 4.2.5. Show that

$$Aut_K(\varphi) = \begin{cases} \mathbb{F}_q^* & \text{for } c_1 \neq 0 , \\ \mathbb{F}_{q^2}^* \cap K & \text{for } c_1 = 0 . \end{cases}$$

Note that the value $j(\varphi) = c_1^{q+1}/c_2$ depends only on the isomorphism class of φ; it is an analogue of the classical j-invariant of an elliptic curve.

Exercise 4.2.6. Let $\varphi : A \longrightarrow K\{\tau\}$ be an elliptic A-module and let B be a K-algebra. Let us define an action of A on B by $a \cdot b = \varphi(a) \cdot b$. Show that this equips B with a structure of A-module $G_\varphi(B)$. Show that for $B = K$ the structure of A-module $G_\varphi(K)$ does not coincide with the initial one defined by the morphism i.

Points of finite order. The points of finite order of an elliptic A-module φ are elements of A-torsion in the module $G_\varphi(\bar{K})$, $\bar{K}$ being the algebraic closure of K . The submodule of A-torsion in $G_\varphi(\bar{K})$ is denoted by $T(\varphi)$.

Proposition 4.2.7. *There is an isomorphism*

$$T(\varphi) = \oplus_v (k_v/A_v)^{d_v} ,$$

where A_v *is the completion of* A *defined by* v , k_v *is its fraction field, the sum is taken over all closed points* $v \in \mathrm{Spec}\, A$, d_v *are integers,* $d_v = d$ *for* $v \neq$ *"char"* K *and* $d_v < d$ *for* $v =$ *"char"* K . *In particular, if* K *has a "general characteristic", then* $T(\varphi) = (k/A)^d$.

Proof: Since the A-module $T(\varphi)$ is divisible and is a module of A-torsion, it is of the form $\oplus_v (k_v/A_v)^{d_v}$. Let $m_v \in A$ be the maximal ideal in A corresponding to the point v . Since the class number of A is finite, there exists n such that $m_v^n = (a)$ for some $a \in A$. Let

$$M = \{x \in G_\varphi(\bar{K}) \mid a \cdot x = 0\} ,$$

then $M = (A/(a))^{d_v}$, hence

$$|M| = |A/(a)|^{d_v} = |a|^{d_v} .$$

On the other hand it is clear that $|M| \leq deg\, \varphi(a) = |a|^d$, the condition $|M| = |a|^d$ being equivalent to the fact that the constant term of $\varphi(a)$ does not vanish, i.e. that $v \neq$ "char"K . ∎

Corollary 4.2.8. *The rank* d *is an integer.* ∎

The group $Hom(\varphi,\psi)$ (which is an A-module) closely resembles the group $Hom(X,Y)$, X and Y being elliptic curves.

Proposition 4.2.9. *Let* φ *and* ψ *be elliptic* A*-modules over* K *. Then* $Hom(\varphi,\psi)$ *is a projective* A*-module of rank at most* d^2 *, and the map*

$$Hom(\varphi,\psi)\otimes_A A_v \longrightarrow Hom(T_v(\varphi),T_v(\psi))$$

is injective for any $v\neq\infty$, $v\neq$ *"char"*K *, where* $T_v(\varphi)$ *is a submodule of* $T(\varphi)$ *consisting of elements annihilated by some power of the ideal* m_v . ■

Exercise 4.2.10. Prove Proposition 4.2.9. (*Hint*: Consider the action of homomorphisms on points of finite order).

Corollary 4.2.11. *The ring* $End(\varphi)\otimes_A k$ *is a skew field. If* K *has a general "characteristic", then* $End(\varphi)\otimes_A k$ *is a field of degree at most* d *over* k *.*

Propositions 4.2.7, 4.2.9 and Corollary 4.2.11 show that for $d=2$ the points of finite order and the endomorphism rings of elliptic modules have structures entirely parallel to the case of elliptic curves.

Lattices. Just as in the case of elliptic curves over $\mathbb{C}$ which can be described in terms of lattices, there is a similar description for elliptic A-modules over a finite extension L of k_∞ . The field L is complete with respect to a non-archimedian absolute value which extends that on k_∞ .

A *lattice* over L is a finitely generated discrete A-submodule Γ in the separable closure L_s of L which is invariant under the action of the Galois group $Gal(L_s/L)$. Such a module is projective over A ; the *rank* of Γ is the rank of the corresponding projective module. A *morphism* from the lattice Γ_1 to a lattice Γ_2 is an element $\alpha \in L$ such that $\alpha \cdot \Gamma_1 \subseteq \Gamma_2$. A morphism composition is just multiplication in L.

Proposition 4.2.12. *The category of elliptic A-modules of rank d over L is equivalent to that of lattices of rank d over L .* ■

Recall that it means that to any lattice Γ there corresponds an elliptic module φ_Γ , and to each morphism of lattices $\alpha : \Gamma_1 \longrightarrow \Gamma_2$ there corresponds a morphism of elliptic modules $u_\alpha : \varphi_{\Gamma_1} \longrightarrow \varphi_{\Gamma_2}$, such that any elliptic A-module ψ is of the form φ_Γ for a suitable lattice Γ and the map $Hom_L(\Gamma_1,\Gamma_2) \longrightarrow Hom_L(\varphi_{\Gamma_1},\varphi_{\Gamma_2})$ is bijective.

We are not going to prove this assertion, we just give a method to construct the elliptic A-module φ_Γ corresponding to a lattice Γ .

Exercise 4.2.13. Show that the following product converges and defines an *integral function* $f_\Gamma : L \longrightarrow \bar{L}$, i.e. a function of the form $\sum_{i=0}^{\infty} a_i \cdot z^i$, where $a_i \in L$, $z \in \bar{L}$:

$$f_\Gamma(z) = z\cdot \prod_{\gamma\in\Gamma-\{0\}} (1 - z\cdot\gamma^{-1}) .$$

(*Hint*: Use the discreteness of Γ).

Exercise 4.2.14. Show that the function f_Γ is additive, i.e. $f_\Gamma(z + w) = f_\Gamma(z) + f_\Gamma(w)$.

Exercise 4.2.15. Let us consider the A-module structure on $\bar{L}$ defined by the restriction of $f_\Gamma : L \longrightarrow \bar{L}$ to A. Show that there exists a unique (up to an isomorphism) elliptic A-module $\varphi = \varphi_\Gamma$ over L such that the obtained A-module is isomorphic to the A-module $G_\varphi(\bar{L})$.

4.2.2. Drinfeld curves

Now we proceed to the construction and basic properties of *Drinfeld modular curves*. Here we assume that $A = \mathbb{F}_q[T]$ and $d = 2$. Note however that a great deal of results of this section is valid *mutatis mutandis* for any A and d.

Moduli schemes and their compactifications. To define moduli schemes of elliptic modules, one has to consider elliptic modules over an arbitrary A-scheme S. An *elliptic module over* S is a pair (L,φ) consisting of a line bundle L on S and an $\mathbb{F}_q$-linear homomorphism

$$\varphi : A \longrightarrow End_{\mathbb{F}_q}(L)$$

such that the following conditions hold:

a) For any $a \in A$ we have

$$\varphi(a) = i(a) + \sum_{i=1}^{2\cdot|a|} \varphi_i(a)\cdot F^i ,$$

$i(a)$ being the image of a under the natural action of A on L, and $F \in End_{\mathbb{F}_q}(L)$ being the morphism of raising to the q-th power, here

$$\varphi_i \in H^0\left(S, L^{\otimes(1-q^i)}\right) .$$

b) For any $a \in A$, $a \neq 0$, the section

$$\varphi_{2\cdot|a|}(a) \in H^0\left(S, L^{\otimes(1-q^{2\cdot|a|})}\right)$$

never vanishes on S .

Exercise 4.2.15. Check that for $S = Spec\, K$ the definition of an elliptic module over S coincides with the definition of that over K given above.

To construct moduli schemes of elliptic modules one has (just as in the classical case of elliptic curves) to fix some additional data called "level structure".

Let $(0) \neq I \subset A$ be a proper ideal, p_I a monic polynomial generating this ideal, $I = p_I \cdot A$, and $G = (L, \varphi)$ an elliptic module over S .

Exercise 4.2.16. Consider the maximal subscheme G_I in $L \times S$ such that the element $\varphi(p_I)$ vanishes on it. Show that G_I is a commutative group scheme. In particular if $S = Spec\, K$ then G_I is a subgroup in $G_\varphi(K)$ annihilated by multiplication by p_I .

A *structure of level* I on the module φ is an isomorphism of commutative group schemes over S :

$$\lambda : (I^{-1}/A)^2 \times S \longrightarrow G_I ,$$

which is compatible with the action of A .

Proposition 4.2.7 describing division points of elliptic A-modules shows that a structure of level I on G can exist only in the case when for any point S its "characteristic" does not belong to the set

$$V(I) = \{P \in \mathit{Spec}\ A \mid P \supseteq I\}\ .$$

By

$$F(I) : (\mathit{Sch})^{\circ} \longrightarrow (\mathit{Sets})$$

we denote the following functor:

$$F(I)(S) = \left\{\begin{matrix}\text{the set of isomorphism classes of pairs } (G,\lambda), \\ \text{where } G = (L,\varphi) \text{ is an elliptic } A\text{-module} \\ \text{over } S \text{ and } \lambda \text{ is a structure of level } I\end{matrix}\right\}.$$

There is the following important

Theorem 4.2.17. *The functor* $F(I)$ *is representable by an affine scheme* $M(I)$ *of a finite type over* A . *Fibres of the morphism* $M(I) \longrightarrow \mathit{Spec}\ A$ *over* $\mathit{Spec}\ A[I^{-1}] = \mathit{Spec}\ A - V(I)$ *are smooth curves, and over* $V(I)$ *they are empty.* ■

The scheme $M(I)$ is equipped by a natural action of $GL(2,A/I)$ since this group acts on the functor $F(I)$ in the following way: if G is an elliptic module over S , $\lambda : (I^{-1}/A)^2 \times S \xrightarrow{\sim} G_I$ being a structure of level I and $h \in GL(2,A/I)$, then the action of h is given by $h : (G,\lambda) \longmapsto (G,\lambda')$, λ' being a structure of level I on G defined by the composition

$$\lambda' : (I^{-1}/A)^2 \times S \xrightarrow{h^{-1}\times id_S} (I^{-1}/A)^2 \times S \xrightarrow{\lambda} G_I\ .$$

Note that here the subgroup $\mathbb{F}_q^* \subset GL(2,A/I)$ acts on $M(I)$ trivially and thus in fact it is the action of $GL(2,A/I)/\mathbb{F}_q^*$.

The scheme $M(I)$ has a canonical compactification. Let $k[I]$ be the maximal abelian extension of k of conductor I in which the point ∞ splits completely, and let $B(I)$ be an integral closure of $A[I^{-1}]$ in the field $k[I]$. The splitting condition means that the ring $k(I)\otimes_k k_\infty$ is isomorphic to a direct sum of copies of k_∞, and the prime ideals completely splitting in $k(I)$ are just prime ideals $\mathfrak{p} \in Spec\, A$ of the form (α), where $\alpha \in A$, $\alpha \equiv 1 (mod\, I)$ (condition of complete splitting of $\mathfrak{p}$ means that the ring $k(I)\otimes_k k_\mathfrak{p}$ is isomorphic to the direct sum of $k_\mathfrak{p}$, where $k_\mathfrak{p}$ is the completion of k by the norm defined by $\mathfrak{p}$). According to the class field theory, $k(I)$ does exist, it is unique, and there is a canonical isomorphism of $A(I)^*/\mathbb{F}_q^*$ onto the group $Gal(k(I)/k)$ (this isomorphism is defined by so-called Artin symbol). We write $M^1(I)$ for the scheme $Spec\, B(I)$.

Theorem 4.2.18. *There exists a unique scheme $\overline{M(I)}$ containing $M(I)$ as an open dense subscheme, whose fibres over $Spec\, A[I^{-1}]$ are smooth complete curves, and the scheme $Cusps(I) = \overline{M(I)} - M(I)$ is isomorphic to a disjoint union of finite number of copies of $M^1(I)$.* ■

Irreducible components of the smooth complete curve $\overline{M(I)}\otimes_A k_\infty$ are analogues of the Riemann surface $\Gamma(N)\backslash H^*$ and can be similarly described in terms of non-archimedian function theory. We are unable to discuss this description now; here are two corollaries. The first concerns cusps and the second the genus of irreducible components of $\overline{M(I)}\otimes_A k_\infty$.

Proposition 4.2.19. *On any irreducible component of the curve $\overline{M(I)}\otimes_A k_\infty$ the part of the finite set $Cusps(I)\otimes_A k_\infty = \overline{M(I)}\otimes_A k_\infty - M(I)\otimes_A k_\infty$ lying on this component* *is canonically bijective to the factor-set* $\Gamma(I)\backslash\mathbb{P}^1(k)$ *under the natural action of* $\Gamma(I)$ *on* $\mathbb{P}^1(k)$:

$$\begin{pmatrix} a & b \\ c & d \end{pmatrix} : z \longmapsto \frac{a\cdot z + b}{c\cdot z + d} \quad ; \quad z \in \mathbb{P}^1(k) \ ,$$

where

$$\Gamma(I) = \left\{ \begin{pmatrix} a & b \\ c & d \end{pmatrix} \in GL(2,A) \ \middle| \ \begin{pmatrix} a & b \\ c & d \end{pmatrix} \equiv \begin{pmatrix} 1 & 0 \\ 0 & 1 \end{pmatrix} (mod\ I) \right\} \subset GL(2,A)$$

is the principle congruence subgroup of level I *in* $GL(2,A)$. ■

Proposition 4.2.20. *The genus of any irreducible component of the curve* $\overline{M(I)}\otimes_A k_\infty$ *equals*

$$1 + (q^2 - 1)^{-1}\cdot q^{2m}\cdot(q^m - q - 1)\cdot\prod_{r\mid p_I}(1 - q^{-2\cdot deg\ r}) \ ,$$

where $m = deg\ p_I$ *, and the product is taken over the set of irreducible monic polynomials* r *dividing* p_I . ■

The reducibility of $\overline{M(I)}\otimes_A k_\infty$ is closely connected with the fact that the above field $k(I)$ is the natural field of definition for this curve. This is parallel to the fact that classical modular curves $X(N)$ are defined over $\mathbb{Q}(\zeta_n)$.

Proposition 4.2.21. *There exists a morphism* $\pi(I) : \overline{M(I)} \longrightarrow M^1(I)$ *whose fibres are complete smooth absolutely irreducible curves.* ■

The action of $GL(2,A/I)$ on $M(I)$ extends to its action on $\overline{M(I)}$. On the other hand the scheme $M^1(I)$ has the action of the finite Galois group $Gal(k(I)/k)$, which is canonically isomorphic (by the class field theory) to the group $(A/I)^*/\mathbb{F}_q^*$. Note that the morphism $\pi(I)$ is compatible with these actions. To be more precise:

Proposition 4.2.22. *For every* $g \in GL(2,A/I)$ *the following diagram:*

$$\begin{array}{ccc} \overline{M(I)} & \xrightarrow{\ g\ } & \overline{M(I)} \\ {\scriptstyle \pi(I)}\downarrow & & \downarrow{\scriptstyle \pi(I)} \\ M^1(I) & \xrightarrow{\ \overline{\det g}\ } & M^1(I) \end{array}$$

commutes. Here we write $\overline{\det g}$ *for the image of* $\det g \in (A/I)^*$ *in* $(A/I)^*/\mathbb{F}_q^*$. ■

Modular curves over $\mathbb{F}_q$. Further on we suppose the ideal I to be prime so that the polynomial p_I is irreducible. We also assume the degree m of the polynomial p_I to be odd and co-prime with $(q-1)$. We are interested in modular curves over $\mathbb{F}_q$, i.e. curves obtained from the smooth complete (reducible) curve $X(I) = \overline{M(I)} \otimes_A \mathbb{F}_q$ over $\mathbb{F}_q$. Here (and further on) the expressions of the form $\otimes_A \mathbb{F}_q$ mean that the structure of A-algebra on $\mathbb{F}_q$ is determined by the factorization morphism over the ideal generated by T, $\nu : A \longrightarrow A/(T) = \mathbb{F}_q$. Thus all the elliptic A-modules considered below are of "characteristic" (T). From Theorems 4.2.17 and 4.2.18 it follows that Propositions 4.2.20 and 4.2.22 remain valid for the curve $X(I)$.

Let us consider the following subgroups of $GL(2,A)$ which are analogous to the subgroups of the classical modular group $PSL(2,\mathbb{Z})$ introduced above:

$$\Gamma_0(I) = \left\{ \begin{pmatrix} a & b \\ c & d \end{pmatrix} \in GL(2,A) \mid c \in I \right\},$$

$$\Gamma_1(I) = \left\{ \begin{pmatrix} a & b \\ c & d \end{pmatrix} \in GL(2,A) \mid c \in I \;,\; a - 1 \in I \right\},$$

$$H(I) = \left\{ \begin{pmatrix} a & b \\ c & d \end{pmatrix} \in GL(2,A) \mid b \in I \;,\; c \in I \;,\; a - d \in I \right\}.$$

Let $\overline{\Gamma_0(I)}$, $\overline{\Gamma_1(I)}$, and $\overline{H(I)}$ be the images of these subgroups in $GL(2,A/I)$. Thus $\overline{\Gamma_0(I)}$ is the subgroup of upper triangular matrices, i.e. of matrices of the form $\begin{pmatrix} * & * \\ 0 & * \end{pmatrix}$, $\overline{\Gamma_1(I)}$ is the subgroup of matrices of the form $\begin{pmatrix} 1 & * \\ 0 & * \end{pmatrix}$, and $\overline{H(I)}$ coincides with the centre of the group $GL(2,A/I)$ consisting of matrices of the form $\begin{pmatrix} r & 0 \\ 0 & r \end{pmatrix}$ for $r \in (A/I)^*$.

Let us define smooth complete curves $X_0(I)$, $X_1(I)$, and $C(I)$ as follows:

$$X_0(I) = X(I)/\overline{\Gamma_0(I)},$$

$$X_1(I) = X(I)/\overline{\Gamma_1(I)},$$

$$C(I) = X(I)/\overline{H(I)}.$$

In our assumptions on the ideal I these curves possess a number of nice properties:

Proposition 4.2.23. a) *The curves* $X_0(I)$, $X_1(I)$, *and* $C(I)$ *are absolutely irreducible.*

b) *The curve* $C(I)$ *is a form of a component of* $X(I)$.

Proof: a) By Proposition 4.2.21 the fibres of the morphism $X(I) \longrightarrow M^1(I)\otimes_A \mathbb{F}_q$ are absolutely irreducible. Using Proposition 4.2.2 and surjectivity of the map $det : G \longrightarrow (A/I)^*$, where G is any of the groups $\overline{\Gamma_0(I)}$, $\overline{\Gamma_1(I)}$, or $\overline{H(I)}$, we get the assertion.

b) This assertion follows from the fact that in our assumptions about I the determinant defines an isomorphism

$$\overline{det} : \overline{H(I)}/\mathbb{F}_q^* \xrightarrow{\sim} (A/I)^*/\mathbb{F}_q^* . \blacksquare$$

Exercise 4.2.24. Give a detailed proof of Proposition 4.2.23.

Embeddings $\overline{H(I)} \subset \overline{\Gamma_0(I)}$, $\overline{\Gamma_1(I)} \subset \overline{\Gamma_0(I)}$ define natural projections $C(I) \longrightarrow X_0(I)$, $X_1(I) \longrightarrow X_0(I)$.

Note that there exists a canonical isomorphism $\overline{\Gamma_0(I)}/\overline{\Gamma_1(I)}\cdot\mathbb{F}_q^* = (A/I)^*/\mathbb{F}_q^*$.

Proposition 4.2.25. *The action of the group* $\overline{\Gamma_0(I)}/\overline{\Gamma_1(I)}\cdot\mathbb{F}_q^* = (A/I)^*/\mathbb{F}_q^*$ *on* $X_1(I)$ *is free.*

Proof: The fact follows from the description of automorphism groups of elliptic modules of rank 2 given at the end of Section 4.2.1, since our assumptions imply that the group $(A/I)^*/\mathbb{F}_q^*$ does not contain $\mathbb{F}_{q^2}^*/\mathbb{F}_q^*$. Indeed, since m is odd we have

$$|(A/I)^*/\mathbb{F}_q^*| = \frac{q^m - 1}{q - 1} \equiv 1(mod(q + 1)) . \blacksquare$$

Since $X_0(I)$ is identified with the factor of $X_1(I)$ over the action of $(A/I)^*/\mathbb{F}_q^*$, we obtain

Corollary 4.2.26. *The morphism* $X_1(I) \longrightarrow X_0(I)$ *is unramified. Its degree equals* $(q^m - 1)/(q - 1)$. ■

Proposition 4.2.19 gives the description of the set $Cusps(I)$ which is the image of $\overline{M(I)} - M(I)$ in $X(I)$. Hence we can deduce the following description of the sets $Cusps_0(I)$ and $Cusps_1(I)$, i.e. of the images of $\overline{M(I)} - M(I)$ in $X_0(I)$ and $X_1(I)$, respectively.

Proposition 4.2.27. **a)** *The set* $Cusps_0(I)$ *consists of* 2 *points.*

b) *The set* $Cusps_1(I)$ *consists of* $2\cdot(q^m - 1)/(q - 1)$ *points.* ■

Exercise 4.2.28. Prove Proposition 4.2.27, using Proposition 4.2.19. (*Hint*: b) follows from a) and Corollary 4.2.26; to prove a) consider the set $\Gamma_0(I)\backslash\mathbb{P}^1(k)$).

A more detailed analysis considering the action of Galois groups makes it possible to prove

Proposition 4.2.29. **a)** *Any point of* $Cusps_0(I)$ *is* $\mathbb{F}_q$*-rational.*

b) *There exists a unique point of* $Cusps_0(I)$ *such that all points of* $Cusps_1(I)$ *lying over it are* $\mathbb{F}_q$*-rational.* ■

Further on we shall write Q_1 for the point from $Cusps_0(I)$ defined in Proposition 4.2.29.b and Q_2 for the other point from $Cusps_0(I)$. Let us remark that no point of $Cusps_1(I)$ lying over Q_2 is $\mathbb{F}_q$-rational.

Modular equation. The classical modular equation defines a plane (singular) model of the curve $X_0(N)$. In the case of Drinfeld curves the curve $X_1(I)$ rather than $X_0(I)$ has a convenient plane model, this explains our interest to $X_1(I)$.

To write out the corresponding equation explicitly, one has to appeal to an interpretation of $X_1(I)$ as a moduli variety over $\mathbb{F}_q$. To be more precise, let

$$Y_1(I) = X_1(I) - Cusps_1(I)$$

be the corresponding affine curve.

Proposition 4.2.30. *For any extension L of $\mathbb{F}_q$ there exists a canonical bijection of sets*

$$Y_1(I)(L) \simeq \left\{\begin{matrix}\text{the set of isomorphism classes of pairs} \\ (\varphi,c),\ \text{where } \varphi \text{ is an elliptic } A\text{-module} \\ \text{over } L \text{ and } c \neq 0 \text{ is a point of} \\ \text{order } I \text{ on } \varphi\text{, i.e. } \varphi(p_I)(c) = 0\end{matrix}\right\}.$$

Sketch of proof: The fact follows from a more general result that the scheme $M(I)/\Gamma_1(I)$ represents the following contravariant factor $F(I)$ from the category of A-schemes in the category of sets:

$$Y_1(I)(S) \simeq \left\{\begin{matrix}\text{the set of isomorphism classes of pairs} \\ (\varphi,c),\ \text{where } \varphi \text{ is an elliptic } A\text{-module} \\ \text{over } S \text{ and } c \neq 0 \text{ is a point of} \\ \text{order } I \text{ on } \varphi\text{, i.e. } \varphi(p_I)(c) = 0\end{matrix}\right\}.$$

The last assertion follows from Theorem 4.2.17 together with the fact that pairs of the form (φ,c) have no non-trivial isomorphisms and that the set of orbits of $\Gamma_1(I)$-action on the set of structures of level I on an elliptic module can be canonically identified with the set of points $c \neq 0$ of order I on φ . ■

Exercise 4.2.31. Give a detailed proof of Proposition 4.2.30.

Now we can write out the equation of a plane model of $X_1(I)$. Let us introduce some notation. Let a_k be the coefficients of the polynomial p_I which we assume to be monic:

$$p_I(T) = \sum_{k=0}^{m} a_k \cdot T^k , \quad a_k \in \mathbb{F}_q , \quad a_m = 1 .$$

Let us define the polynomials $P_k \in \mathbb{F}_q[x,y]$ by induction, setting $P_0(x,y) = 1$, and

$$P_{k+1}(x,y) = x \cdot P_k(x,y)^{q^2} + y \cdot P_k(x,y)^q \quad \text{for} \quad k \geq 0 .$$

Exercise 4.3.32. Show (by induction over k) that if $\varphi : A \longrightarrow \overline{\mathbb{F}}_q\{\tau\}$ is an elliptic A-module over $\overline{\mathbb{F}}_q$ of characteristic (T) ,

$$\varphi(T) = a \cdot F^2 + b \cdot F ,$$

$a,b \in \overline{\mathbb{F}}_q$, then for the action of $\varphi(T^k)$ on the element $1 \in \overline{\mathbb{F}}_q$ we have:

$$\varphi(T^k)(1) = P_k(a,b) .$$

Let

$$F(x,y) = F_I(x,y) = \sum_{k=0}^{m} a_k \cdot P_k(x,y) .$$

The equation $F_I(x,y) = 0$ is an analogue of the modular

equation. To state the corresponding result more rigorously we use the following notation. If $I = (p_I)$ is a prime ideal in A, we say that the ideal I' is conjugate to I (over the prime field $\mathbb{F}_p$) if I is generated by a polynomial

$$p'_I = \sum_{k=0}^{m} a'_k \cdot T^k ,$$

where $a'_k = \sigma a_k$ for some element $\sigma \in Gal(\mathbb{F}_q/\mathbb{F}_p)$. In particular if $a_k \in \mathbb{F}_p$ for $k = 0,\ldots,m$ then $I = I'$.

Proposition 4.2.33. *Let* $\tilde{X}_1(I) \subset \mathbb{P}^2$ *be a plane curve with the affine equation* $F_I(x,y) = 0$. *Then*

a) *The curve* $\tilde{X}_1(I)$ *has exactly two singular points* $P_1 = (0:0:1)$ *and* $P_2 = (0:-1:1)$.

b) $\tilde{X}_1(I)$ *is birationally isomorphic over* $\mathbb{F}_q$ *to* $X_1(I')$, I' *being an ideal conjugate to* I *over* $\mathbb{F}_p$.

Proof: a) Let us check that $\tilde{X}_1(I)$ has no singularities in the finite part of the plane, i.e. that the system of equations

$$\begin{cases} F_I = 0 \\ \partial F_I/\partial x = 0 \\ \partial F_I/\partial y = 0 \end{cases}$$

has no solutions. Indeed, it is easy to check that

$$F - (x\cdot\partial F_I/\partial x + y\cdot\partial F_I/\partial y) = a_0 ;$$

since p_I is irreducible, and $a_0 \neq 0$, we are done. Let $(Z_0:Z_1:Z_2)$ be coordinates on $\mathbb{P}^2$ such that $x = Z_1/Z_0$, $y = Z_2/Z_0$. The homogeneous equation $G(Z_0:Z_1:Z_2) = 0$ of the curve $\tilde{X}_1(I)$ can be written as follows:

$$G(Z_0:Z_1:Z_2) = Z_0^{(q^{2m}-1)/(q^2-1)} \cdot F_I(Z_1/Z_0, Z_2/Z_0) =$$

$$= Z_1^{(q^{2m-2}-1)/(q^2-1)} \cdot (Z_1 + Z_2)^{q^{2m-2}} + Z_0 \cdot H(Z_0:Z_1:Z_2)$$

for a form $H(Z_0:Z_1:Z_2)$ of degree

$$deg\ H = (q^{2m}-1)/(q^2-1) - 1 .$$

Hence $\tilde{X}_1(I) \cap \{Z_0 = 0\}$ consists exactly of two points $\{P_1, P_2\}$, and it is easy to check that they are both singular.

b) Let us construct a map $\tilde{\nu} : X_1(I) \longrightarrow \tilde{X}_1(I)$. To do this, we use Proposition 4.2.30 and define this map on $Y_1(I)$. Any point from $Y_1(I)(\overline{\mathbb{F}}_q)$ is defined by a pair (φ, c), where φ is an elliptic A-module over $\mathbb{F}_q$, $\varphi(T) = a \cdot F^2 + b \cdot F$, and $c \neq 0$ is a point of order p_I on φ, $a, b, c \in \overline{\mathbb{F}}_q$. The point $Y_1(I)$ defined by (φ, c) we map to the point

$$(c^{q^2-1} \cdot a \ , \ c^{q-1} \cdot b) \in \mathbb{A}^2 .$$

Since $\varphi(p_I)(c) = 0$, one can easily deduce that

$$F_I(c^{q^2-1} \cdot a \ , \ c^{q-1} \cdot b) = 0,$$

so that this point lies on $\tilde{X}_1(I)$. It is easy to check that $\tilde{\nu}$ is well defined. Indeed note that a pair (φ, c) is isomorphic to a pair (φ', c'), where

$$\varphi(T) = a \cdot F^2 + b \cdot F ,$$

$$\varphi'(T) = a' \cdot F^2 + b' \cdot F ,$$

if and only if

$$a' = \lambda^{q^2-1}\cdot a \ , \ b' = \lambda^{q-1}\cdot b \ , \ c' = \lambda^{-1}\cdot c$$

for some $\lambda \in \overline{\mathbb{F}}_q$. This also shows that the map

$$\tilde{\nu} : Y_1(I)(\overline{\mathbb{F}}_q) \longrightarrow \tilde{X}_1(I) \cap (\mathbb{A}^2 - \{x = 0\})$$

is bijective. It is also clear that the map $\tilde{\nu}$ is defined over $\mathbb{F}_q$. Being bijective on $\mathbb{F}_q$-points, it is a composition of a birational isomorphism ν and some power of the Frobenius endomorphism, and the assertion of b) follows. ■

Thus, the curve $\tilde{X}_1(I)$ is biratonally isomorphic to the curve $X_1(I')$ for an ideal of the same type as I . Note that the birational isomorphism $\nu : X_1(I') \longrightarrow \tilde{X}_1(I)$ is the normalization map for the curve $\tilde{X}_1(I)$.

Exercise 4.2.34. Show that for $(m,a) = 1$, where $a = \log_p q$, there is a prime ideal $I \subset A$, generated by a polynomial p_I of degree m , whose coefficients lie in a prime field $\mathbb{F}_p$, so that $I = I'$.

Exercise 4.2.35. Check the following assertions which are used in the proof of Proposition 4.2.33:

a) We have $F - (x\cdot \partial F_I/\partial x + y\cdot \partial F_I/\partial y) = a_0$.

(*Hint*: $\partial P_k/\partial x = P_{k-1}^{q^2}$, $\partial P_k/\partial y = P_{k-1}^{q}$).

b) Points P_1 and P_2 are singular. (*Hint*: Write out local equations of $\tilde{X}_1(I)$ in neighbourhoods of these points).

c) We have

$$F_I(c^{q^2-1}\cdot a\ ,\ c^{q-1}\cdot b) = 0\ .$$

(*Hint*: Use Exercise 4.2.32).

d) Any regular map $\varphi : X \longrightarrow Y$ of the curves over $\mathbb{F}_q$, which maps $X(\overline{\mathbb{F}}_q)$ onto $Y(\overline{\mathbb{F}}_q)$ bijectively, is a composition of a birational isomorphism and some power of the Frobenius endomorphism.

The image of the set $Cusps_1(I)$ on $\tilde{X}_1(I)$ is described as follows.

Corollary 4.2.36. *The image of the set of* $\mathbb{F}_q$*-rational points from* $Cusps_1(I)$ *coincides with* $\{P_1,P_2\}$ *. The set of points from* $Cusps_1(I)$ *which are not* $\mathbb{F}_q$*-rational is mapped bijectively onto the intersection of* $\tilde{X}_1(I)$ *with the y-axis , i.e. onto* $\tilde{X}_1(I) \cap \{x = 0\}$ *.*

Proof: Since $Y_1(I)(\overline{\mathbb{F}}_q)$ is bijective to $\mathbb{F}_q$-points of $\tilde{X}_1(I)\cap(\mathbb{A}^2 - \{x = 0\})$, $\nu(Cusps_1(I))$ coincides with

$$(\tilde{X}_1(I)\cap\{x = 0\})\cup\{P_1,P_2\}\ .$$

The set of ordinates of points from $\tilde{X}_1(I)\cap\{x = 0\}$ is defined by the equation

$$F(y) = F_I(0,y) = \sum_{k=0}^{m} a_k\cdot y^{(q^k-1)/(q-1)} = 0\ .$$

This set contains $(q^m - 1)/(q - 1)$ elements no one of which lies in $\mathbb{F}_q$ since $p_I(T)$ is irreducible. Hence it coincides with the image of non-rational points of $Cusps_1(I)$, and $\{P_1,P_2\}$ with the image of rational ones. ∎

It is not difficult to give also an explicit description of the action of the group $H = (A/I)^*/\mathbb{F}_q^*$ on the curve $\tilde{X}_1(I)$:

Corollary 4.2.37. *Let*

$$r(T) = \sum_{k=0}^{m-1} r_k \cdot T^k \in A$$

be a polynomial of degree at most $m - 1$ *whose image in* $(A/I)^*/\mathbb{F}_q^*$ *is a generator* θ *of this group. Then the action of* θ *on the point* $(x,y) \in \tilde{X}_1(I)$ *is defined as*

$$\theta : (x,y) \longmapsto (x \cdot \lambda^{q^2-1} , y \cdot \lambda^{q-1})$$

where

$$\lambda = \sum_{k=0}^{m-1} r_k \cdot P_k(x,y) .$$

Proof: The assertion follows from the fact that the action of θ on triples of the form $(a,b,1)$ is given by

$$\theta : (a,b,1) \longmapsto (a,b,c) ,$$

where

$$c = \sum_{k=0}^{m-1} r_k \cdot P_k(a,b)$$

since $\varphi(T^k)(1) = P_k(a,b)$, $\varphi(T) = a \cdot F^2 + b \cdot F$. ■

$\mathbb{F}_{q^2}$-rational points. Similarly to the curves $X_0(N)/p$ and $X_p(N)/p$, the curves $X_0(I)$ and $C(I)$ have many $\mathbb{F}_{q^2}$-rational points.

Theorem 4.2.38. *Let $N(I)$ be the number of $\mathbb{F}_{q^2}$-rational points of the curve $C(I)$, $g(I)$ its genus, and $m(I)$ the degree of an irreducible polynomial p_I generating I. Then*

$$\lim_{m(I)\longrightarrow\infty} \frac{N(I)}{g(I)} = q - 1 .$$

Proof: It suffices to prove that

$$\lim N(I)/g(I) \geq q - 1 ,$$

since the inverse inequality follows from Theorem 2.3.22. According to Proposition 4.2.20,

$$g(I) = 1 + (q^{2m} - 1)\cdot(q^m - q - 1)/(q^2 - 1) ,$$

where $m = m(I)$. Let us consider the projection

$$C(I) \longrightarrow C(I)/PGL(2,A/I) = \mathbb{P}^1$$

of degree $q^m\cdot(q^{2m} - 1) = |PGL(2,A/I)|$. The points of $C(I)$, corresponding to the pairs of the form (φ_0,λ) , φ_0 being an elliptic A-module such that $\varphi_0(T) = F^2$ (it is natural to call this module *supersingular*) are ramification points of this projection with indices $(q + 1)$, which follows from the description of the groups $Aut(\varphi)$ for elliptic A-modules φ (cf. Exercise 4.2.5). Hence the number of such points is equal to

$$(q + 1)^{-1}\cdot|PGL(2,A/I)| = q^m\cdot(q^{2m} - 1)/(q + 1) .$$

It suffices to prove $\mathbb{F}_{q^2}$-rationality of these points, since it would yield

$$\frac{N(I)}{g(I)} \geq \frac{q^m \cdot (q^{2m} - 1)/(q + 1)}{1 + (q^{2m} - 1) \cdot (q^m - q - 1)/(q^2 - 1)} \geq q - 1 .$$

Let us consider a point of $M(I) \otimes \mathbb{F}_q$ defined by a pair (φ_0, λ), where λ is a structure of level I on φ. To show that (φ_0, λ) defines an $\mathbb{F}_{q^2}$-rational point $C(I)$, it is sufficient to show that under the action of the generator F^2 of the group $Gal(\overline{\mathbb{F}}_q/\mathbb{F}_{q^2})$ the structure λ is mapped to the structure λ' such that $\lambda = \lambda' \cdot h$ for some $h \in H(I)$. To do this, we remark that F^2 lies in the center of $End(\varphi_0)$, since, as it is easy to check,

$$End(\varphi_0) = \mathbb{F}_{q^2}\{F\} \quad \text{and} \quad Cent(End(\varphi_0)) = \mathbb{F}_q[F^2] .$$

Hence, the image of F^2 under the representation of $(End(\varphi_0) \otimes_A (A/I))^*$ in $GL(2, A/I)$, defined by the action on the set of structures of level I, belongs to the centre of $GL(2, A/I)$, i.e. to $H(I)$. ■

Remark 4.2.39. It is clear that the proof of Proposition 4.2.38 is completely analogous to that of Theorem 4.1.52 with the only simplification that there exists only one supersingular elliptic module φ_0.

Exercise 4.2.40. Show that the module φ_0 has no non-trivial points of period p. This justifies the term supersingular.

Remark 4.2.41. Similarly to Remark 4.1.54, for the number $N(I)$ for some $c > 0$ we have

$$N(I) \geq (q-1)\cdot g(I) + c\cdot g(I)^{2/3} + o(g(I)^{2/3}) .$$

The existence of a surjection $C(I) \longrightarrow X_0(I)$ yields

Corollary 4.2.42. *Let $N_0(I)$ be the number of $\mathbb{F}_{q^2}$-rational points of $X_0(I)$, and let $g_0(I)$ be the genus of the curve $X_0(I)$. Then*

$$\lim_{m(I)\longrightarrow\infty} \frac{N_0(I)}{g_0(I)} = q - 1 . \blacksquare$$

To get more precise information on $\mathbb{F}_{q^2}$-rational points of the curve $X_0(I)$, one has to refer to the modular equation $F_I(x,y)$. Since the only fact garanteed by Proposition 4.2.33 is that the equation $F_I(x,y) = 0$ gives the model of the curve $X_1(I')$ for *some* ideal I' conjugate to I, it is necessary to use the curve $X_0(I')$ instead of $X_0(I)$. Note however that if I runs over the complete system of ideals $I \subset A$ conjugate over $\mathbb{F}_p$, then I' also runs over the same system (may be in some other order), hence considering the curve $X_0(I')$ instead of $X_0(I)$ we do not loose generality.

Supersingular points of $X_0(I')$, i.e. images of points of the form (φ_0,λ) on $C(I')$, are pulled up to points with $y = 0$ on $\tilde{X}_1(I)$ whose abscissae satisfy the equation

$$F(x) = F_I(x,0) = \sum_{k=0}^{m} a_k\cdot x^{(q^{2k}-1)/(q^2-1)} = 0$$

Proposition 4.2.43. *The number of supersingular* $\mathbb{F}_{q^2}$*-points of* $X_0(I')$ *equals* $(q^m + 1)/(q + 1)$.

Proof: The polynomial $F(x)$ is separable, hence it has

$$deg\ F(x) = (q^{2m}-1)/(q^2-1)$$

roots. These roots correspond to the points of $X_1(I')$ lying over "supersingular" points of $X_0(I')$, hence the number of the latter equals

$$(q^{2m} - 1)\cdot(q^2 - 1)^{-1}\cdot|(A/I)^*/\mathbb{F}_q^*|^{-1} = (q^m + 1)/(q + 1)\ . \blacksquare$$

The action of $H = (A/I)^*/\mathbb{F}_q^*$ makes it possible to describe the decomposition of $F(x)$ into factors irreducible over $\mathbb{F}_{q^2}$. Let H_1 be the subgroup in H generated by the image of the element T and let $h_1 = |H_1|$ be its order and $h_2 = [H:H_1]$ its index in H , $h = h_1\cdot h_2$.

Proposition 4.2.44. *The polynomial* $F(x)$ *decomposes over* $\mathbb{F}_{q^2}$ *into* $n\cdot h_2$ *irreducible factors, each one of degree* h_1 .

Proof: Note that the generator θ_1 (the image of T in H) of the group H_1 acts on points of $\tilde{X}_1(I)$ in the following way:

$$\theta_1 : (x,y) \longmapsto (x\cdot(x + y)^{q^2-1}, y\cdot(x + y)^{q-1})\ ,$$

which follows from Corollary 4.2.37. Hence, on

supersingular points of $\tilde{X}_1(I)$ with $y = 0$, the action of θ_1 coincides with that of the (topological) generator F^2 of $Gal(\overline{\mathbb{F}}_q/\mathbb{F}_{q^2})$. Since the action of H_1 on the set of points of the form $(x,0)$ is free, we obtain the required assertion. ■

Corollary 4.2.45. *Let R be the set of roots of $F(x)$. Then:*

a) Each orbit of the action of H on R is of the form

$$H\cdot r = \bigcup_{i=1}^{h_2} M_i ,$$

where $r \in R$, and M_i is the set of roots of irreducible over $\mathbb{F}_{q^2}$ polynomial F_i of degree h_1 dividing $F(x)$.

b) The set R lies in the field $\mathbb{F}_{q^{2h_1}}$. ■

4.2.3. Codes

Now let us pass to the description of codes obtained on Drinfeld curves.

Parameters. Consider codes over $\mathbb{F}_{q^2}$ on curves $C(I)$ and $X_0(I)$. The information on the structure of the set of their $\mathbb{F}_{q^2}$-rational points, obtained in the previous section, makes it possible to estimate parameters of these codes. We still suppose the above assumptions about I valid.

Proposition 4.2.46. *Let $C = (C(I), \mathcal{P}, \mathcal{L})_H$ be an algebraic-geometric $[n,k,d]_q$-code constructed from the curve $C(I)$, the set $\mathcal{P}$ of its supersingular $\mathbb{F}_{q^2}$-rational points, and a line bundle $\mathcal{L}$ of degree a. Then for parameters of C we have:*

$$n = q^m \cdot (q^{2m} - 1)/(q + 1) ,$$

$$k \geq a - (q^{2m} - 1) \cdot (q^m - 1) \cdot (q^m - q - 1)/(q^2 - 1) ,$$

$$d \geq q^m \cdot (q^{2m} - 1)/(q + 1) - a . \blacksquare$$

Remark 4.2.47. The set of possible length values for codes obtained from curves $C(I)$ and the set of their "supersingular" points is quite rare in positive integers: passing from m to $(m + 2)$ (m should be odd), the length of such codes multiplies approximately by q^6. Such codes are of particular interest for $q \geq 7$ (i.e. $q^2 \geq 49$), and in this case n grows with fantastic speed. The first non-trivial code over $\mathbb{F}_{49}$, corresponding to $m = 5$ (for $m = 1$ the genus of $C(I)$ equals 0 and $m = 3$ is not co-prime with $(q - 1)$) is of length exceeding 6 billions. This simple estimate shows that codes constructed from the curves $C(I)$ are not for practical use.

Proposition 4.2.48. *Parameters of an $[n,k,d]_{q^2}$-code $C = (X_0(I), \mathcal{P}, \mathcal{L})_H$, constructed from the curve $X_0(I)$, the set $\mathcal{P}$ of its supersingular points, and a line bundle $\mathcal{L}$ of degree a, satisfy the conditions:*

$$n = (q^m + 1)/(q + 1) ,$$

$$k \geq a - q \cdot (q^{2\ell} - 1)/(q^2 - 1) + 1 ,$$

$$d \geq (q^m + 1)/(q + 1) - a ,$$

where $\ell = \lceil m/2 \rceil$.

Proof: It suffices to show that

$$g_0(I) \le q\cdot(q^{2\ell} - 1)/(q^2 - 1) \ ,$$

which follows from that fact that, by the Hurwitz formula, $g_0(I) \le \lceil g(I)/(q^{2m} - q^m)\rceil$ since the degree of the projection $C(I) \longrightarrow X_0(I)$ equals $q^m\cdot(q^m - 1)$. ■

Exercise 4.2.49. Using the formula for $g(I)$, check that

$$\lceil g(I)/(q^{2m} - q^m)\rceil = q\cdot(q^{2\ell} - 1)/(q^2 - 1) \ .$$

Remark 4.2.50. Remark 4.2.47 is valid, though to a lesser extent, for codes on curves $X_0(I)$. The length of such codes in passing from m to $(m + 2)$ is multiplied by q^2 . Codes whose length is not exceedingly high are obtained only for rather small m .

Example 4.2.51. The value $m = 3$ is admissible for $q \equiv 2(mod\ 3)$. In this case the q^2-ary codes on $X_0(I)$ have the parameters:

$$n = q^2 - q + 1 \ ,$$

$$k \ge a - q + 1 \ ,$$

$$d \ge q^2 - q + 1 - a \ .$$

One can see that the parameters of these codes are worse than those of Reed-Solomon codes.

Example 4.2.52. The value $m = 5$ is admissible for $q \not\equiv 1 (mod\ 5)$. For codes on $X_0(I)$ we have q^2-ary codes with

$$n = q^4 - q^3 + q^2 - q + 1 ,$$

$$k \geq a - q\cdot(q^2 + 1) + 1 ,$$

$$d \geq q^4 - q^3 + q^2 - q + 1 - a .$$

Parameters of these codes are already not so bad as for $m = 3$. In any case, for low values of q the code length n lies in reasonable limits. For example, for $q = 7$ we have: $n = 2101$, $k + d \geq 1852$ (for an $\mathbb{F}_{49}$-code); for $q = 8$ we have: $n = 3301$, $k + d \geq 2782$ (for an $\mathbb{F}_{64}$-code).

Group properties. Similarly to codes on reductions of $X(N)$, codes on $C(I)$ can be interpreted in terms of ideals in the group algebra. Moreover, because of uniqueness of the "supersingular" elliptic module φ_0, the situation for $C(I)$ is simpler than that for $X_0(N)$.

Let $\overline{Cusps(I)}$ be the image of $Cusps(I)$ in the curve $C(I)$; i.e. $\overline{Cusps(I)}$ is the set of cusps of the curve $C(I)$. The set $\overline{Cusps(I)}$ can be otherwise defined as the inverse image of $\infty \in \mathbb{P}^1$ under the natural projection $j : C(I) \longrightarrow \mathbb{P}^1$ defined by j-invariant of elliptic modules. Let us consider the divisor

$$F_\infty = \sum_{P\in\overline{Cusps(I)}} P .$$

The given interpretation of the set $\overline{Cusps(I)}$ shows that the divisor F_∞ is $\mathbb{F}_q$-rational.

Proposition 4.2.53. *The degree of* F_∞ *is equal to* $q^{2m} - 1$.

Proof: The degree of F_∞ is equal to the number of cusps on $C(I)$. This set can be identified with $\Gamma(I)\backslash\mathbb{P}^1(k)$. Since $\mathbb{P}^1(k)$ is identified with $PGL(2,A)/B$, where B is a subgroup of matrices of the form $\begin{pmatrix} 1 & a \\ 0 & 1 \end{pmatrix}$, $a \in A$, we have

$$\overline{Cusps(I)} = \Gamma(I)\backslash PGL(2,A)/B \; .$$

Noting that

$$\Gamma(I)\backslash PGL(2,A)/B = PGL(2,A/I)/\overline{B} \; ,$$

$\overline{B}$ being the image of B in $PGL(2,A/I)$, i.e. the subgroup of matrices of the form $\begin{pmatrix} 1 & \overline{a} \\ 0 & 1 \end{pmatrix}$, $\overline{a} \in A/I$, we see that

$$deg\; F_\infty = |\overline{Cusps(I)}| = |PGL(2,A/I)|\cdot|A/I|^{-1} = q^{2m} - 1 \; . \blacksquare$$

Since the projection $j : C(I) \longrightarrow \mathbb{P}^1$ is defined by factorizing over $PGL(2,A/I)$, the divisor F_∞ is $PGL(2,A/I)$-invariant. Let $D = a_1 \cdot F_\infty$, a_1 being a positive integer. Noticing that the group $PGL(2,A/I)$ acts transitively on the set $\mathcal{P}$ of "supersingular" $\mathbb{F}_{q^2}$-rational points of $C(I)$, we get

Proposition 4.2.54. *The code $C_{a_1} = (C(I),\mathcal{P},D)_L$ over $\mathbb{F}_{q^2}$ constructed from the curve $C(I)$, the set $\mathcal{P}$ and the divisor $D = a_1 \cdot F_\infty$ has a natural transitive action of the group $G(I) = PGL(2,A/I)$.* $\blacksquare$

Corollary 4.2.55. *The code C_{a_1} can be realized as an ideal in the group algebra $\mathbb{F}_{q^2}[G(I)]$ lying in the ideal $I_H = \mathbb{F}_{q^2}[G(I)/H]$, H being the image of $\mathbb{F}^*_{q^2}$ in $G(I)$.* $\blacksquare$

Problem 4.2.56. Give an explicit description of C_{a_1} as of an ideal in $\mathbb{F}_{q^2}[G(I)]$ and estimate its parameters without using the Riemann-Roch theorem.

Remark 4.2.57. Knowing no solution of Problem 4.2.56 we can nevertheless obtain (using the Riemann-Roch theorem) some information on the structure of the ideal in $\mathbb{F}_{q^2}[G(I)]$ obtained from C_{a_1}. Let

$$a_0 = \lceil (2\cdot g(I) - 2)/(q^{2m}-1) \rceil + 1 =$$

$$= 2q\cdot(q^{2\ell} - 1)\cdot(q^2 - 1)^{-1} + 1,$$

where $\ell = \lceil \frac{m}{2} \rceil$. For $b \geq a_0$ by the Riemann-Roch theorem we have

$$\ell(b\cdot F_\infty) = b\cdot(q^{2m} - 1) - g(I) + 1 .$$

Thus, $L((b + 1)\cdot F_\infty)/L(b\cdot F_\infty)$ for $b \geq a_0$ is a $G(I)$-module of dimension $q^{2m} - 1$. Hence, the code C_{a_1} as a $G(I)$-ideal has a filtration by $G(I)$-submodules of the form $C_{a_1} \supset C_{a_1-1} \supset \ldots \supset C_{a_0}$, where C_b for $a_1 \geq b \geq a_0$ is the image of $L(b\cdot F_\infty)$ in C_{a_1}, and all the factors C_{b+1}/C_b for $b = a_0,\ldots,a_1 - 1$ are of dimension $q^{2m} - 1$. Unfortunately the group algebra $\mathbb{F}_{q^2}[G(I)]$ is never semi-simple and thus the above filtration is here less useful than for classical modular curves.F

CHAPTER 4.3

POLYNOMIALITY

In this chapter we show that codes on modular curves and Drinfeld curves have polynomial construction complexity. To be precise, one has

Theorem 4.3.1. a) *There exists a polynomial algorithm constructing a generator matrix for the code* $C = (X,\mathcal{P},D)_L$ *over* $\mathbb{F}_r$, $r = p^2$, *where* $X = X_0(N)/p$ *is the reduction of* $X_0(N)$ *over* $p \nmid N$, $\mathcal{P}$ *being the set of supersingular* $\mathbb{F}_r$*-points of* X *and* D *being an* $\mathbb{F}_r$*-rational divisor on* X *such that* $\mathrm{Supp}\, D \cap \mathcal{P} = \emptyset$.

b) *There exists a polynomial algorithm constructing a generator matrix for the code* $C' = (X',\mathcal{P}',D')_L$ *over* $\mathbb{F}_r$, $r = q^2 = p^{2e}$, *where* $X' = X_0(I)$, *the ideal* $I \subset A = \mathbb{F}_q[T]$ *is generated by an irreducible polynomial of an odd degree* m *relatively prime to* $(q - 1)$, $\mathcal{P}'$ *is the set of "supersingular"* $\mathbb{F}_r$*-rational points of* X', *and* D *is an* $\mathbb{F}_r$*-rational divisor,* $\mathrm{Supp}\, D' \cap \mathcal{P} = \emptyset$.

Remark 4.3.2. We should remark that in the proof of the theorem the corresponding algorithms are written out explicitly, their time complexity being estimated by $T \le c_1 \cdot n^{c_2}$, where n is the code length. The exact values of c_1 and c_2 depend on the model of computation which is used (in contrast to the polynomiality result valid for any reasonable model of computation). For example, in the case of the Turing machine one can take $c_2 = 40$ and 60 for parts a) and b), respectively, with moderate values of c_1. For a model of computation close to a real computer one can take $c_2 = 20$ and 30, respectively. We should also point out two other facts: on the one hand more thorough estimates of time complexity of our algorithms can give lower values c_2 than above, on the other hand (which is more important) the algorithms constructed on the same principle and having a structure similar to that of ours, cannot be realized on a modern computer. Thus the polynomiality result is a purely theoretical one: it is not difficult to write a program constructing codes of Theorem 4.2.1 for a real computer, but its rate would surely be unsufficient to make this program work.

We believe that ways to obtain more effective algorithms lie in more thorough study of the structure of modular codes, especially of their group-theoretic properties (cf. Problems 4.1.65 and 4.2.56).

The rest of this chapter is devoted to the proof of Theorem 4.3.1. Here we restrict ourselves to the proof of some particular cases of parts a) and b) of the theorem in order to simplify reading. Exact additional assumptions in which we prove parts a) and b) of the theorem, are stated at the beginning of Sections 4.3.1 and 4.3.2, respectively.

In Section 4.3.1 we describe an algorithm of construction for codes on reductions of classical modular curves and in Section 4.3.2 an algorithm for Drinfeld curves. Section 4.3.3 contains a description of standard procedures, necessary for the work of the algorithm; we also explain there why these algorithms are polynomial (main conceptual ideas of these algorithms are summed up in Remark 4.3.33).

4.3.1. Classical case

Now we shall describe an algorithm of construction of a generator matrix for the code $C = (X,\mathcal{P},D)_L$ over $\mathbb{F}_{p^2}$ in the following particular case: $N = 11\cdot\ell$, where ℓ is a prime, $\ell \geq 133$, the divisor D being of the form

$$D = g\cdot F - h\cdot Q_2 - h\cdot Q_4 ,$$

where g and h are integers such that

$$1 \leq g \leq (p-1)/2 , \quad 0 \leq h \leq \ell - 2 ,$$

and

$$F = (\ell - 1)\cdot Q_2 + (\ell - 1)\cdot Q_4 ,$$

Q_2 and Q_4 being cusps of X (recall that on X there are four cusps Q_1, Q_2, Q_3, Q_4, see Proposition 4.3.3). The parameters of the code $C = C_{g,h}$ corresponding to the pair (g,h) are

$$n = (p-1)(\ell+1) ,$$

$$k \geq (2g-1)\cdot(\ell-1) - 2h ,$$

$$d \geq (p-1)\cdot(\ell+1) - 2h - \ell - (2g-1)\cdot(\ell-1)$$

Further on we also need the following notation: $\tilde{X}$ is a plane singular model of X (over $\mathbb{F}_p$), defined by the modular equation reduced modulo p, i.e. $F_N(x,j) = 0$, $F_N = \Phi_N (\mathrm{mod}\ p)$. By Z we denote the elliptic curve $X_0(11)/p$ over $\mathbb{F}_p$. The field of ratonal functions $\mathbb{F}_p(X)$ on X is of the form $\mathbb{F}_p(X) = \mathbb{F}_p(x,j)$, where j is j-invariant, and $x = j_N$ is the image of j under the canonical involution w_N.

Divisors and linear systems on X. We need some facts on divisors and linear systems on X. Let $I = (x)_\infty$ and $J = (j)_\infty$ be the divisors of poles of functions x and j.

Proposition 4.3.3.

$$I = 11\cdot\ell\cdot Q_1 + 11\cdot Q_2 + \ell\cdot Q_3 + Q_4 .$$

$$J = Q_1 + \ell\cdot Q_2 + 11\cdot Q_3 + 11\cdot\ell\cdot Q_4 .$$

Proof: j defines the map $j : X \longrightarrow \mathbb{P}^1$. The formula for J follows from the description of ramification of j (cf. Exercise 4.1.22 where this map was denoted by θ). Since the involution $w_{11\cdot\ell}$ permutes x and j and thus also I and J, it suffices to note that $w_{11\cdot\ell}$ acts on the 4-tuple (Q_1,Q_2,Q_3,Q_4) as follows:

$$w_{11\cdot\ell} : (Q_1,Q_2,Q_3,Q_4) \longmapsto (Q_4,Q_3,Q_2,Q_1) . \blacksquare$$

Exercise 4.3.4. Check the formula for the action of $w_{11\cdot\ell}$ on the 4-tuple (Q_1,Q_2,Q_3,Q_4). (*Hint*: Compute the action of $w_{11\cdot\ell}$ on $\Gamma_0(11\cdot\ell)\backslash\mathbb{P}^1(\mathbb{Q})$).

We also need an explicit description of the space $L(g\cdot F - h\cdot Q_2 - h\cdot Q_4)$ used in the definition of the code. Our immediate purpose is to show that this space contains a basis of some special form consisting of rational functions with bounded degrees of numerator and denominator. To be precise, set $s = 24\cdot\ell + 19$, and let

$$V_g = \left\{\begin{array}{l}\text{rational functions on } X\otimes\overline{\mathbb{F}}_p \text{ of the form } p(x,j),\\ \text{where } p(x,j) \in \overline{\mathbb{F}}_p[x,j]\ ,\ deg\ p(x,j) \le g\cdot s\end{array}\right\},$$

$$V_g^0 = V_g \cap \mathbb{F}_p(X) = \left\{\begin{array}{l}\text{rational functions on } X \text{ of the form}\\ p(x,j), \quad \text{where } p(x,j) \in \mathbb{F}_p[x,j]\ ,\\ deg\ p(x,j) \le g\cdot s\end{array}\right\}.$$

Proposition 4.3.5. *The space* $L(g\cdot F)$ *for* $g \ge 2$ *has a basis of the form* f_α/f_0 , where

$$f_\alpha \in V_g^0 \cap L(2g\cdot(6\ell + 5)\cdot(I + J) - g\cdot D)\ ,$$

$$f_0 \in V_g^0 \cap L(2g\cdot(6\ell + 5)\cdot(I + J) - g\cdot D - g\cdot F)\ ,$$

D being the conductor of $\tilde{X}$.

Proof: Since the divisors D, I, J, and F are $\mathbb{F}_{p^2}$-rational, it suffices to show that

$$L(g\cdot F) \otimes \overline{\mathbb{F}}_p = \{f/f_0 \in \overline{\mathbb{F}}_p(X) \mid f \in V_1\ ,\ f_0 \in V_2\}\ , \qquad (4.3.1)$$

where

$$V_1 = V_g \cap L(2g\cdot(6\ell + 5)\cdot(I + J) - g\cdot D)\ ,$$

$$V_2 = V_g \cap L(2g\cdot(6\ell + 5)\cdot(I + J) - g\cdot D - g\cdot F)\ .$$

Throughout the proof we denote we denote $X\otimes\overline{\mathbb{F}}_p$ simply by X . Since for any divisors D , and D' and functions $f \in L(D)$ and $f' \in L(D - D')$ the function f/f' lies in $L(D')$ (check!), the right hand set of (4.3.1) lies in the left hand set. Besides, by the Riemann-Roch theorem $\ell(g\cdot F) = (2g - 1)\cdot(\ell - 1)$. Hence, to verify (4.3.1) it suffices to check that

$$dim\ V_1 \geq (2g - 1)\cdot(\ell - 1) \qquad (4.3.2)$$

$$dim\ V_2 \geq 1 \qquad (4.3.3)$$

Then we shall get (4.3.1) (note also that (4.3.2) and (4.3.3) are in fact equalities rather than inequalities).

Let us consider the space $L(g\cdot K)$, where $K = (dj)$ is a divisor from the canonical class of X. We are going to show that $f \in L(g\cdot K)$ implies

$$f\cdot F'_X \in (V_g \cap L(2g\cdot(6\ell + 5)\cdot(I + J) - g\cdot D)$$

which yields (4.3.2), since $\ell(g\cdot K) = (2g - 1)\cdot(\ell - 1)$ by the Riemann-Roch theorem. Indeed let $f \in L(g\cdot K)$. Since X is non-hyperelliptic by Proposition 4.1.29 and Exercise 4.1.49, using Theorem 2.2.32 and Proposition 2.5.21, we prove that f is of the form $P/(F'_X)^g$, where $P \in V_g$, since the degree of a homogeneous equation of $\tilde{X}$ is less that $s + 4 = 24\cdot\ell + 23$. Indeed, by Theorem 2.2.32, f is a sum of functions of the form $f_1\cdot\ldots\cdot f_g$, where $f_i \in L(K)$, $i = 1,\ldots,g$. Hence, by Proposition 2.5.21 any $f \in L(K)$ is of the form P/F'_X, $P \in V_1$. Thus $f \in L(g\cdot K)$ implies $f \in V_g$. On the other hand, if $f = P/(F'_X)^g$, $f \in L(g\cdot K)$, then

$$(f) + g\cdot(dj) = (P) - g\cdot(F'_X) + g\cdot(dj) \geq 0\ .$$

Using Proposition 2.5.25 and the formula for the ramification divisor B of j (Exercise 4.1.22), we get

$$(P) - g\cdot D + 2g\cdot(6\ell + 5)\cdot(I + J) \geq 0,$$

which proves (4.3.2).

Now let us prove (4.3.3). Let $\omega \neq 0$ be a non-vanishing regular differential form on Z. Consider its pull-back ω' on X. From the description of the ramification of $X \longrightarrow Z$ (cf. Exercise 4.1.22) we get $(\omega') = F$ (check!). Proposition 2.5.21 gives $\omega' = f \cdot dj/F'_X$ for some

$$f \in V_1 \cap L(2\cdot(6\ell + 5)\cdot(I + J) - D) .$$

Hence

$$f_g \in V \cap L(2g\cdot(6\ell + 5)\cdot(I + J) - g\cdot D) ,$$

and thus the last space is non-trivial. ∎

The space W_g. Let

$$W_g = \{p(x,j) \in \mathbb{F}_p[x,j] \mid deg\, p(x,j) \leq g\cdot s\} ,$$

where as above $s = 24\cdot\ell + 19$;

$$dim_{\mathbb{F}_p} W_g = (g\cdot s + 1)\cdot(g\cdot s + 2)/2 .$$

To construct modular codes, it is necessary to consider some linear functionals on W_g. To define them one needs to use two types of operators which act in the ring of Laurent formal series. Let K be a commutative ring, let $K((t))$ be the ring of Laurent formal series over K, and let m be an integer. If

$$f = \sum_{i=s}^{\infty} a_i\cdot t^i \in K((t)) ,$$

$a_i \in K$, then by definition

$$R_n(f) = \sum_{i=s}^{n} a_i\cdot t^i \in K[t,t^{-1}] , \quad Q_n(f) = a_n \in K .$$

Thus, R_n and Q_n are morphisms of K-modules:

$$R_n : K((t)) \longrightarrow K[t,t^{-1}] \ , \quad Q_n : K((t)) \longrightarrow K \ .$$

Exercise 4.3.6. Prove that

a) If $R_{-m-1}(f_1) = R_{-n-1}(f_2) = 0$, then for any integer g

$$R_g(f_1 \cdot f_2) = R_g(R_{n+g}(f_1) \cdot R_{m+g}(f_2)) \ ,$$

$$Q_g(f_1 \cdot f_2) = Q_g(R_{n+g}(f_1) \cdot R_{m+g}(f_2)) \ ,$$

b) For any integers g and m

$$R_g(t^m \cdot f) = t^m \cdot R_{g-m}(f) \ , \quad Q_g(t^m \cdot f) = Q_{g-m}(f) \ .$$

In order to emphasize (when necessary) the dependence of R_n, Q_n on t (i.e. to indicate the variable of their action) sometimes we denote them by $R_{n,t}$, $Q_{n,t}$.

Let us now recall that for a singular point P of a plane curve X we have defined in Section 2.5.4 a labeled tree Γ_P of its infinitely close points (desingularization tree) such that the label of any of its nodes $s \in V_P$ is of the form

$$(P(x_s,y_s),(X_s,Y_s),\Lambda_s)$$

where $P(x_s,y_s) = 0$ is a local equation of the proper inverse image of X in a neighbourhood of an infinitely close point Q_s corresponding to s, (X_s,Y_s) are the polynomials expressing local coordinates in a neighbourhood of P in terms of local coordinates in a neighbourhood of Q_s, and Λ_s indicates whether Q_s is smooth, or not.

Let P be a singular point of $\tilde{X}$, i.e. either a point with local coordinates $(x - \alpha_P, j - \beta_P)$, or one of the points $P_1 = (0:0:1)$ and $P_2 = (0:1:0)$. We use $x_1 = 1/x$, $j_1 = j/x$ as local coordinates in a neighbourhood of P_1 and $x_2 = x/j$, $j_2 = 1/j$ in that of P_2 . Let Γ_P be the desingularization tree associated to P . The points of the normalization X of $\tilde{X}$, lying over the point P , correspond to free nodes of Γ_P (i.e. to the nodes s such that $\Lambda_s = 1$). For $Q = Q_t$ corresponding to a node t we can define functionals $A_{m,t}$, $m \in \mathbb{Z}$ on the set $W_g \otimes \mathbb{F}_{r_t}$, $r_t = p^{a_t}$, $\mathbb{F}_{r_t}$ being the field of definition of Q_t .

Let us fix a node t labelled by $(P_t, (X_t, J_t), 1)$ and an integer m , and let $H(x,j) \in W_g \otimes \mathbb{F}_{r_t}$. The polynomial $H(x,j)$ defines a rational function $\tilde{H}$ on X . Since the point Q_t is smooth, one of the functions x_t or j_t is a local parameter on X in a neighbourhood of Q_t . Let j_t , for example, be such a parameter. Then $\tilde{H}$ can be expanded into a Laurent formal series

$$\tilde{H} = \sum_{i=s}^{\infty} c_{i,t} \cdot j_t^i , \quad c_{i,t} \in \mathbb{F}_{r_t} .$$

By $A_{m,t}(H)$ we denote the m-th coefficient of this expansion: $A_{m,t}(H) = c_{m,t}$. In a more formal way one can define the functional $A_{m,t}$ as follows. Since j_t is a local parameter on X in a neighbourhood of Q_t , the value of $\partial P_t/\partial x_t$ at Q_t is not zero. Hence there exists the unique power series $f_t(j_t) \in \mathbb{F}_{r_t}[[j_t]]$ such that $P_t(f_t(x_t), j_t) = 0$, i.e. $f_t(j_t)$ is the expansion of the function x_t into a power series in j_t .

Exercise 4.3.7. Prove that

$$A_{m,t}(H) = Q_m(H(X_t(f_t(j_t), j_t), J_t(f_t(j_t), j_t))) .$$

Let us remark that for the points $P \in Sing\,X$ other than P_1 and P_2, i.e. for points in $Sing\,\tilde{X} \cap \mathbb{A}^2$, all the functionals $A_{m,t}$ with negative m vanish, since for $H \in W_g \otimes \overline{\mathbb{F}}_p$ the function of the form $\tilde{H}$ is regular on $\tilde{X} \cap \mathbb{A}^2$.

Exercise 4.3.8. Prove that the condition

$$A_{m,t}(H) = 0 \quad \text{for any} \quad m \le g$$

is equivalent to

$$ord_{Q_t}(\tilde{H}) \ge g + 1 .$$

It is the result of this exercise which motivates consideration of functionals $A_{m,t}$.

Proposition 4.3.9. *The set of functionals* $\{A_{m,t}\}$, $t \in \bigcup_{P \in Sing\,\tilde{X}} V'_P$, *is defined over* $\mathbb{F}_p$.

Proof: The action of the Galois group $Gal(\overline{\mathbb{F}}_p/\mathbb{F}_p)$ on $\tilde{X}$ generates its action on $\bigcup_{P \in Sing\,\tilde{X}} V_P$. It is clear that for any $t \in \bigcup V_p$ we have $\sigma(A_{m,t}) = A_{m,\sigma t}$, and $\sigma \in Gal(\overline{\mathbb{F}}_p/\mathbb{F}_p)$ which proves the proposition. ■

Construction of the codes. One can use functionals $\{A_{m,t}\}$ to describe codes on X. To begin with, we introduce some notation. For $Q \in X$ by n_Q we denote the multiplicity of Q in the divisor D of double points of $\tilde{X}$:

$$D = \sum_Q n_Q \cdot Q .$$

If $Q_i \in \{Q_1, Q_2, Q_3, Q_4\}$ we write n_i for n_{Q_i}. Since the equation $F_N(x,j) = 0$ of $\tilde{X}$ is invariant under the involution w_N, we get $n_1 = n_4$ and $n_2 = n_3$. Let

$$\mu = (11\ell + 1)\cdot g\cdot s ,$$

$$\nu = (\ell + 11)\cdot g\cdot s ,$$

$$\lambda = 2\cdot(11\ell + 1)\cdot(6\ell + 5)\cdot g ,$$

$$\kappa = 2\cdot(\ell + 11)\cdot(6\ell + 5)\cdot g ,$$

$$\kappa' = \kappa - g\cdot n_2 - g\cdot(\ell - 1) ,$$

$$\lambda' = \lambda - g\cdot n_2 - g\cdot(\ell - 1) ,$$

here, as above, $s = 24\cdot\ell + 19$, it is clear that $\mu > \lambda$, $\nu > \kappa$.

Let

$$V' = \bigcup_{P\in(Sing\ \tilde{X})\cap\mathbb{A}^2} V_{P'}$$

be the set of free nodes of the desingularization trees of singular points of $\tilde{X}$ in the finite part of the plane. We write t_1, t_2, t_3, and t_4 for free nodes of the trees V_{P_1} and V_{P_2}, where t_i corresponds to Q_i. We write $A_{m,i}$ for A_{m,t_i}, $i = 1,2,3,4$. Let us consider the following subspaces U_g and U'_g in W_g:

$$U_g = (\bigcap_{t\in V'}(\bigcap_{m=0}^{g\cdot n_t} Ker\ A_{m,t}))\cap(\bigcap_{m=\lambda-g\cdot n_1}^{\mu}((Ker\ A_{m,1})\cap(Ker\ A_{m,4}))\cap$$

$$\cap(\bigcap_{m=0}^{\nu}((Ker\ A_{m,2})\cap(Ker\ A_{m,2}))\cap W_g ,$$

$$U'_g = U_g \cap (\bigcap_{m=\kappa'}^{\nu} Ker\ A_{m,2}) \cap (\bigcap_{m=\lambda'}^{\mu} Ker\ A_{m,4}) ,$$

all the subspaces being considered as subsets in $W_g \otimes \overline{\mathbb{F}}_p$; by definition, $n_t = n_{Q_t}$.

Proposition 4.3.10. *The image of the space* U_g *in* V_g^0 *coincides with*

$$V_g^0 \cap L(2g\cdot(6\ell + 5)\cdot(I + J) - g\cdot D) ,$$

and the image of U'_g *with the one-dimensional subspace*

$$V_g^0 \cap L(2g\cdot(6\ell + 5)\cdot(I + J) - g\cdot D - g\cdot F) . \blacksquare$$

Exercise 4.3.11. Prove the proposition. (*Hint*: use the result of Exercise 4.3.8 and the description of divisors I , J , D and F).

Corollary 4.3.12. *Let* h *be an integer,* $0 \le h \le (\ell - 1)$. *Then the image of the subspace*

$$U_{g,h} = U_g \cap \Big(\bigcap_{m=\kappa-gn_2-h}^{\kappa-gn_2} Ker\ A_{m,2}\Big) \cap \Big(\bigcap_{m=\lambda-gn_1-h}^{\lambda-gn_1} Ker\ A_{m,4}\Big)$$

in V_g^0 *coincides with the subspace*

$$V_g^0 \cap L(2g\cdot(6\ell + 5)\cdot(I + J) - g\cdot D - h\cdot Q_2 - h\cdot Q_4) . \blacksquare$$

To construct the code we need the following notation. We write $ss(\tilde{X}) \subseteq \tilde{X}(\mathbb{F}_{p^2})$ for the set of points $P \in \tilde{X}(\mathbb{F}_{p^2})$ such that $\alpha_P = x(P)$ and $\beta_P = j(P)$ are supersingular values of j-invariant in characteristic p . If $P \in ss(\tilde{X})$ is a singular point, then the tree Γ_P and functionals $A_{m,t}$ are defined, where $t \in V'_P$ is the free node of Γ_P

corresponding to Q_t which lies over P. Now let $ss(\tilde{X}) \cap (\tilde{X} - Sing\ \tilde{X})$ be a smooth supersingular point. Let us extend the definition of Γ_P to this case. We define Γ_P as a tree consisting of one free node w_P labeled by the triple

$$(F_N(x - \alpha_P, j - \beta_P), (x - \alpha_P, j - \beta_P), 1) .$$

Thus, $V'_P = \{w_P\}$. Let us also define the functional

$$A_{1,w_P} : W_g \otimes \mathbb{F}_{p^2} \longrightarrow \mathbb{F}_{p^2} ; \quad A_{1,w_P}(H) = H(\alpha_P, \beta_P) ,$$

for $H \in W_g \otimes \mathbb{F}_{p^2} \subset \mathbb{F}_{p^2}[x, j]$.

Noting that in such a definition for all $P \in ss(\tilde{X})$ the set V'_P of free nodes of the tree Γ_P is in a bijection with the set of points $Q \in X$ lying over P, we can rewrite the formula for the number of supersingular points of X as follows:

$$\sum_{P \in ss(\tilde{X})} v'_P = (p - 1) \cdot (\ell + 1) ,$$

where $v'_P = |V'_P|$ is the cardinality of V'_P.

Collecting the results proved in this section together, we come to the following

Theorem 4.3.13. *Let us fix a non-vanishing element* $p_0 \in U'_g$ *and consider the following* $\mathbb{F}_{p^2}$*-linear map:*

$$\tilde{\ell} : U_{g,h} \otimes \mathbb{F}_{p^2} \longrightarrow \mathbb{F}^n_{p^2} ,$$

$$\tilde{\ell} : H \longmapsto (A_{g \cdot n_t + 1, t}(H) / A_{g \cdot n_t + 1, t}(H_0)) ,$$

t *running over the set* $\bigcup_{P \in ss(\tilde{X})} V'_P$ *of cardinality*

$n = (p-1)\cdot(\ell+1)$. *Then the image of* $\tilde{\ell}$ *coincides with* $C_{g,h}$.

Proof: Let us begin with $h = 0$, so that $U_{g,h} = U_g$. Then the theorem follows from Propositions 4.3.5 and 4.3.10, from the definition of the code, and from the coincidence of $A_{g\cdot n_t+1,t}(H)$ with the value of $\tilde{H}$ in Q_t . For $h \neq 0$ one has to apply also Corollary 4.3.12. ∎

Formulae. Let us give some formulae which are necessary to construct a generator matrix of $C_{g,h}$.

First of all, we should work with the curve $\tilde{X}$ defined by the modular equation $F_N(x,j) = 0$, considered in characteristic p , so that

$$F_N(x,j) = x^{12\cdot(\ell+1)} + j^{12\cdot(\ell+1)} + \sum_{m,n=0}^{12\cdot\ell+11} c_{mn}\cdot x^m\cdot y^n ,$$

where $c_{ba} = c_{ab}$ are elements of $\mathbb{F}_p$.

Coefficients c_{ab} can be expressed in terms of the coefficients of q-expansion of j-invariant.

Proposition 4.3.14. *The* q*-expansion* $j(q)$ *of the function* j *equals*

$$j(q) = \frac{\left(1 + 504\cdot\sum_{n=1}^{\infty}\sigma_3(n)\cdot q^n\right)^3}{q\cdot\prod_{n=1}^{\infty}(1-q^n)^{24}}$$

where $\sigma_3(n) = \sum_{d|n} d^3$. ∎

Thus we see that

$$j(q) = q^{-1}\cdot(1 + \sum_{m=1}^{\infty} c_m\cdot q^m) ,$$

where c_m are integers. Now let us introduce the following notation. We write $\mathbb{F}_{p^r}$ for the field $\mathbb{F}_p[\zeta]$, where $\zeta = \zeta_{11\cdot\ell}$ is a primitive root of unity of degree $11\cdot\ell$ in $\overline{\mathbb{F}}_p$. Set

$$A(x,t) = (t^{121\cdot\ell^2}\cdot x - 1 - c_1\cdot t^{121\cdot\ell^2}) ,$$

$$B(x,t) = \prod_{k=0}^{10} t^{\ell^2}\cdot x - \zeta^{-\ell\cdot k}\cdot(1 + \sum_{m=1}^{132} c_m\cdot\zeta^{\ell^2\cdot k\cdot m}\cdot t^{\ell^2\cdot m})) ,$$

$$C(x,t) = \prod_{k=0}^{\ell-1}(t^{121}\cdot x - \zeta^{-11\cdot k}(1 + \sum_{m=1}^{u} c_m\cdot\zeta^{121k\cdot m}\cdot t^{121\cdot m})) ,$$

$$D(x,t) = \prod_{k=0}^{11\cdot\ell-1} (t\cdot x - \zeta^{-k}(1 + \sum_{m=1}^{v} c_m\cdot\zeta^{k\cdot m}\cdot t^m)) ,$$

where $c_m \in \mathbb{F}_p$ are coefficients of the q-expansion of j-invariant, reduced modulo p , and the positive integers u and v are defined as

$$u = \lceil 12\ell\cdot(\ell + 1)/11\rceil , \quad v = 132\cdot\ell\cdot(\ell + 1) .$$

Thus $A(x,t) \in \mathbb{F}_p[x,t]$; $B(x,t), C(x,t), D(x,t) \in \mathbb{F}_s[x,t]$, $s = p^r$. Consider an embedding of $F(x,j)$ into the ring $\mathbb{F}_p[x]((q))$ defined by the q-expansion of j . If we set $q = t^{11\cdot\ell}$, then the ring $\mathbb{F}_p[x]((q))$ is a subring in $\mathbb{F}_s[x]((t))$. On the other hand it is clear that $\mathbb{F}_s[x,t] \subset \mathbb{F}_s[x]((t))$. In the following exercise by R_m we denote the operator $R_{m,t}$ acting in $\mathbb{F}_s[x]((t))$.

Exercise 4.3.15. Let $\ell \geq 133$. Show that

$$R_0(F(x,j)) = t^{-V} \cdot R_V(A \cdot B \cdot C \cdot D) , \qquad (4.3.6)$$

where $A = A(x,t)$, $B = B(x,t)$, $C = C(x,t)$, $D = D(x,t)$ are polynomials introduced above. (*Hint*: Prove this equality in the ring $\mathbb{Z}[\zeta_{11\cdot\ell}]((t))$, where $\zeta_{11\cdot\ell} = e^{2\pi i/11\cdot\ell}$, and then consider the reduction modulo p . According to Theorem 4.1.33,

$$F_N(x,j) = \prod_{\alpha\in A}(x - j(\alpha(z)))(mod\ p) ,$$

α running over the set A of all matrices of the form $\begin{pmatrix} a & b \\ c & d \end{pmatrix}$, $a,b,d \in \mathbb{Z}$, $a > 0$, $d > 0$, $a \cdot d = 11 \cdot \ell$, $0 \leq b < d$. Since for $\alpha = \begin{pmatrix} a & b \\ c & d \end{pmatrix} \in A$ the function $j(\alpha(z))$ has the following expansion in t :

$$j(\alpha(z)) = \zeta^{-b\cdot a} \cdot t^{-a^2}(1 + \sum_{m=1}^{\infty} \zeta^{m\cdot a\cdot b} \cdot c_m \cdot t^{m\cdot a^2})$$

taking into account the properties of R_m described in Exercise 4.3.6, one obtains (4.3.6)).

Let

$$R_0(F(x,j)) = x^{12\cdot(\ell+1)} + \sum_{a=0}^{12\cdot\ell+11} Q_a(t) \cdot x^a ,$$

where $Q_a(t) \in \mathbb{F}_p[t,t^{-1}]$.

Since $F(x,j)$ lies in $\mathbb{F}_p[x]((q))$ and $q = t^{11\cdot\ell}$, the Laurent polynomials $Q_a(t)$ lie in fact in $\mathbb{F}_p[q,q^{-1}]$, and we can therefore write

$$Q_a(t) = \sum_{m=0}^{12\cdot\ell+11} r_{am} \cdot t^{-11\cdot\ell\cdot m} = \sum_{m=0}^{12\cdot\ell+11} r_{am} \cdot q^{-m} ,$$

for $a = 1,\dots,12\cdot\ell + 11$, $r_{am} \in \mathbb{F}_p$;

$$Q_0(t) = \sum_{m=0}^{12\cdot(\ell+1)} r_{om}\cdot q^{-m} .$$

Thus, the coefficients r_{am} can be written out as follows:

$$r_{am} = Q_{m,q}(Q_{a,x}(R_0(F(x,j)))) ,$$

the operator $Q_{m,q}$ acting in $\mathbb{F}_p((q))$ and $Q_{a,x}$ in $\mathbb{F}_p[q,q^{-1}]((x))$. For $1 \le i \le 12\cdot\ell + 11$, $1 \le b \le 12\cdot\ell + 11$, let

$$v_{i,b} = Q_{i,q}(q^{-b}\cdot(1 + \sum_{m=1}^{\infty} c_m\cdot q^m)^b) \in \mathbb{F}_p .$$

Proposition 4.3.16. *For the coefficients* $c_{ab} \in \mathbb{F}_p$ *of the modular equation*

$$F_N(x,j) = x^{12\cdot(\ell+1)} + j^{12\cdot(\ell+1)} + \sum_{a,b=0}^{12\cdot\ell+1} c_{ab}\cdot x^a\cdot j^b \in \mathbb{F}_p[x,j]$$

we have:

$$c_{a,12\cdot\ell+11} = r_{a,12\cdot\ell+11} , \quad a = 1,\dots,12\cdot\ell + 1 ;$$

$$c_{a,12\cdot\ell+11-i} = r_{a,12\cdot\ell+11-i} - \sum_{m=12\cdot(\ell+1)-i}^{12\cdot\ell+11} c_{am}\cdot v_{m,12\cdot\ell+11-i}$$

for $1 \le a \le 12\cdot\ell + 11$, *the last expression implying* $c_{0,12\cdot(\ell+1)} = 1$.

Proof: By definition, we have

$$Q_a(t) = \sum_{b=1}^{12\cdot\ell+1} c_{ab}\cdot R_0(j^b) \quad \text{for} \quad 1 \le a \le 12\cdot\ell + 11 ;$$

$$Q_0(t) = \sum_{b=1}^{12\cdot\ell+1} c_{ab}\cdot R_0(j^b) + R_0(j^{12\cdot\ell+12}) ,$$

thus for any $1 \le i \le 12\cdot\ell + 11$ and $1 \le a \le 12\cdot\ell + 11$ we have:

$$r_{ai} = \sum_{m=i}^{12\cdot\ell+11} c_{am}\cdot v_{mi} ,$$

and for any $i = 1,\dots,12\cdot\ell + 1$:

$$r_{oi} = \sum_{m=i}^{12\cdot\ell+1} c_{om}\cdot v_{mi} .$$

Using these relations for c_{am}, we obtain the proposition. ■

Now we shall describe singular points of $\tilde{X}$. To do this, it is necessary to know the coordinates (α_P,β_P) of all singular points $P \in \tilde{X}$ lying in $\mathbb{A}^2$ (the points $P_1,P_2 \in \tilde{X} \cap (\mathbb{P}^2 - \mathbb{A}^2)$ are both singular) and to have a description of the divisor D of double points of $\tilde{X}$.

To determine coodinates of singular points of $\tilde{X}$ in the finite part of the plane we can use resultants. Let us define the polynomials $R_1(x)$ and $R_2(x) \in \mathbb{F}_p[x]$ as follows:

$$R_1(x) = Res_j(F_N,\partial F_N/\partial x) \in \mathbb{F}_p[x] ,$$

$$R_2(x) = Res_j(F,\partial F_N/\partial j) \in \mathbb{F}_p[x] ,$$

Res_j being the resultant of polynomials from $\mathbb{F}_p[x,j]$

considered as polynomials in j with coefficients from $\mathbb{F}_p[x]$ (we beg your pardon: they have nothing to do with residues denoted in the same way). Let $R(x)$ be the greatest common divisor of polynomials $R_1(x)$ and $R_2(x)$.

Proposition 4.3.17. *The set of roots of $R(x)$ coincides with the set of first coordinates α_P of singular points (α_P, β_P) of $\tilde{X}$ in the finite part of the plane.*

Proof: The assertion follows from the principal property of the resultant together with the fact that singular points $P = (\alpha_P, \beta_P) \in Sing\ \tilde{X} \cap \mathbb{A}^2$ are defined as solutions of the following system of equations:

$$\begin{cases} F_N = 0 \\ \partial F_N/\partial x = 0 \\ \partial F_N/\partial j = 0 \ . \blacksquare \end{cases}$$

Corrolary 4.3.18. *Let a_0 be a root of the equation $R(x) = 0$. Then the second coordinates β_P of the singular points $P \in Sing\ \tilde{X} \cap \mathbb{A}^2$ with $\alpha_p = \alpha_0$ are defined from the system*

$$F_N(\alpha_0, \beta) = \partial F_N/\partial x\ (\alpha_0, \beta) = \partial F_N/\partial y\ (\alpha_0, \beta) = 0 \ . \blacksquare$$

To describe the conductor D of $\tilde{X}$, we must find an expression for numbers n_t , where

$$D = \sum n_t \cdot Q_t \ .$$

Since Q_t is a smooth point of X , either $\partial P_t/\partial x_t$ or $\partial P_t/\partial j_t$ does not vanish at the origin. Let us suppose that $\partial P_t/\partial x_t|_{x_t=j_t=0} \neq 0$ (if it is not the case, one just has to interchange everywhere x_t and j_t). Thus j_t is a local parameter on X at Q_t . Let $f_t(j_t)$ be an expansion of x_t into a power series in j_t ,

$$x_t = f_t(j_t),\ f_t(j_t) \in \mathbb{F}_{r_t}[[j_t]]\ ,\ r_t = p^{a_t}\ .$$

Let $d = (24\cdot\ell + 24)^2$. Let us define $R_t \in \mathbb{F}_{r_t}[j_t]$ by

$$R_t(j_t) = R_d(f_t(j_t))\ ;$$

it is clear that $deg\ R_t(j_t) \leq d$. In this notation we have

Proposition 4.3.19. *For* $m = 1,\ldots,d$ *let*

$$b_{m,t} = Q_{m,j_t}(\partial F_N/\partial x\ (X_t(R_t(j_t),j_t),J_t(R_t(j_t),j_t)))$$

and

$$\nu_t = max\ \{m \mid b_m \neq 0\}\ .$$

Then

$$n_t = \begin{cases} \nu_t & \text{for} \quad j(Q_t) \neq 0,1728,\infty \\ \nu_t - 1 & \text{for} \quad j(Q_t) = 0 \\ \nu_t - 2 & \text{for} \quad j(Q_t) = 1728 \\ \nu_t - e_i + f_i & \text{for} \quad Q_t = Q_i \in \{Q_1,\ldots,Q_4\}, \end{cases}$$

where

$$e_1 = 0,\ e_2 = \ell - 1\ ,\ e_3 = 10,\ e_4 = 11\cdot\ell - 1,$$

$$f_1 = 132\cdot\ell^2 + 122\cdot\ell + 12\ ,\ f_2 = 12\cdot\ell^2 + 144\cdot\ell + 10\ ,$$

$$f_3 = 12\cdot\ell^2 + 142\cdot\ell + 132\ ,\ f_4 = 132\ell^2 + 144\cdot\ell + 10\ .$$

Proof: By Proposition 2.5.25,

$$(F'_x) = B_\theta + D - 2\cdot(6\ell + 5)\cdot I - 12\cdot(\ell + 1)\cdot J \ .$$

Using the formulae for B_θ (cf. Exercise 4.1.22), for I, and for J (cf. Proposition 4.3.3), we can conclude that the proposition is valid if we substitute

$$\mu_t = ord_{Q_t}(F'_x) = max\ \{m \mid b'_m \neq 0\}\ ,$$

for ν_t, where

$$b'_m = Q_{m,j_t}(\partial F_N/\partial x\ (X_t(f_t(j_t),j_t),J_t(f_t(j_t),j_t)))\ .$$

From the properties of R_d it follows immediately that $b_m = b'_m$ for $m \leq d$. Now it suffices to note that

$$\mu_t \leq (deg\ hom\ \partial F_N/\partial x)\cdot(deg\ hom\ F) \leq (24\cdot\ell + 24)^2 = d\ .$$

here by *deg hom P* for $P \in \mathbb{F}_p[x,j]$ we denote the degree of the corresponding form on $\mathbb{P}^2$. ∎

Exercise 4.3.20. Check that

$$\mu_t \leq (deg\ hom\ \partial F_N/\partial x)\cdot(deg\ hom\ F)\ .$$

(*Hint*: Show that the degree of the divisor $(\partial F_N/\partial x)$ is equal to $(deg\ hom\ \partial F_N/\partial x)\cdot(deg\ hom\ F)$).

The information obtained also yields formulae for coefficients of functionals $A_{m,t}$. For a basis in the space W_g we choose a natural basis of the form $\{x^a\cdot j^b\}$, $a \geq 0$, $b \geq 0$, $a + b \leq g\cdot s$. Coefficients of $A_{m,t}$ in this basis are denoted by a_{mtab} :

$$a_{mtab} = A_{m,t}(x^a \cdot j^b) .$$

The above argument immediately yields

Proposition 4.3.21. *Let* $t \in V'_p$. *Set* $e_t = g \cdot n_t$,

$$S_t(j_t) = R_{e_t}(f_t(j_t)) \in \mathbb{F}_{r_t}[j_t] .$$

Then

$$a_{mtab} = Q_{m,j_t}(X_t(S_t(j_t)^a \cdot J_t(S_t(j_t), j_t)^b) .$$ ∎

Exercise 4.3.22. Prove Proposition 4.3.21.

Algorithm. Now we can describe a polynomial algorithm of construction of a generator matrix for the code $C_{g,h}$. We are about to describe its successive steps, pointing out the input information needed for the step, information obtained as a result of its work, and the computation method.

Step 1. Calculation of the coefficients of the modular equation.

Input: positive integers ℓ , p , g ; ℓ and p being primes, $\ell \geq 133$, $p \neq 11$, $p \neq \ell$, and $g \leq (p-1)/2$.

Output: The set $\{c_{ab} \in \mathbb{F}_p\}$ of coefficients of the modular equation (*mod p*):

$$F_N(x,j) = x^{12\cdot(\ell+1)} + j^{12\cdot(\ell+1)} + \sum_{a,b=0}^{12\cdot\ell+11} c_{ab} \cdot x^a \cdot j^b .$$

Method: First of all compute the polynomial

$G(t) \in \mathbb{F}_p[t]$ of degree $\mu = deg\, G(t) = (11\cdot\ell + 1)\cdot g\cdot s$, $s = 24\cdot\ell + 19$ such that

$$G(t) = t\cdot R_{\mu-1}(j(t)) ,$$

where $j(t) = t^{-1}\cdot(1 + \sum_{m=0}^{\infty} c_m\cdot t^m)$ is the Fourier expansion of j . Here the formula (4.3.5) for $j(t)$ is used. Then we can use the formulae of Proposition 4.3.16 to calculate $\{r_{a_i}\}$ and $\{v_{i,b}\}$ for $a, i, b = 0,1,\ldots,12\cdot\ell + 11$ and coefficients $\{c_{ab}\}$.

Step 2. Computing the coordinates of singular points of $\tilde{X}$.

Input: The matrix $\{c_{ab}\}$ of coefficients of the polynomial $F(x,j)$.

Output: The set $\{(\alpha_P,\beta_P)\}$ of coordinates of singular points of $\tilde{X}$ in the finite part of the plane.

Method: Computation of resultants, as in Proposition 4.3.17.

Step 3. Construction of the set of labeled trees $\{\Gamma_P\}$.

Input: The matrix $\{c_{ab}\}$, and the set $\{(\alpha_P,\beta_P)\}$ of coordinates of points $P \in Sing\, \tilde{X} \cap \mathbb{A}^2$.

Output: The set $\{\Gamma_P\}$, $P \in Sing\, \tilde{X}$ of labeled desingularization trees for singular points of $\tilde{X}$.

Method: Each tree Γ_P is constructed by induction. Namely, a subtree $\Gamma \subset \Gamma_P$ is constructed, at the beginning of the process coinciding with the root w_P of the tree Γ_P . Then the condition $\Lambda_t = 1$ is verified for all the free nodes $t \in V'_\Gamma$ of the tree Γ . If this condition is satisfied, then $\Gamma = \Gamma_P$, and the tree is already constructed. If it is not the case, the construction of Γ is continued by blowing up Q_t such that $\Lambda_t = 0$, and of computing the labels by the formulae (2.5.7) to (2.5.14).

Step 4. Computation of the divisor D.

Input: The set $\{\Gamma_P\}$ of desingularization trees of singular points of $\tilde{X}$.

Output: The set of numbers $\{n_t\}$, $t \in \bigcup_P V'_P$, where

$$D = \sum_t n_t \cdot Q_t .$$

Method: Follows immediately from Proposition 4.3.19.

Step 5. Computation of the coefficients of linear forms $\{A_{m,t}\}$.

Input: The set $\{\Gamma_P\}$, $P \in Sing\, \tilde{X}$ of desingularization trees of singular points of $\tilde{X}$, and the set $\{n_t\}$;

Output: The set of coefficients $\{a_{mtab}\}$ of linear forms $\{A_{m,t}\}$ in the basis $\{x^a \cdot j^b\}$ of the space $W_g \otimes \mathbb{F}_{r_t}$, for $a,b \geq 0$, $a + b \leq g \cdot S$, $m = 0,1,\ldots,g \cdot n_t$, $t \in \bigcup_P V'_P$, .

Method: Follows from Proposition 4.3.21.

Step 6. Computation of the set $ss(\tilde{X})$.

Input: The matrix $\{c_{ab}\}$ of coefficients of $F_N(x,j)$.

Output: The set of supersingular points $\{(\alpha_P, \beta_P)\}$ on the curve $\tilde{X}$.

Method: Coordinates $\beta_P = j(P)$ are obtained

solving the equation (2.4.8) and finding the corresponding values of $j = j(\lambda)$. If the coordinate $\beta_P = \beta_0$ is fixed, the corresponding coordinates $\alpha_P = x(P)$ are found from the equation $F(\alpha, \beta_0) = 0$.

Step 7. Computing a generator matrix of the code.

Input: The set $ss(\tilde{X})$, the set of coefficients $\{a_{mtab}\}$ of linear forms $\{A_{m,t}\}$.

Output: The generator matrix $M = M_{g,h}$ of the modular code $C_{g,h}$:

$$M \in Mat((2\cdot(\ell - 1)\cdot g - 2\cdot h)\times((p - 1)(\ell + 1)), \mathbb{F}_{p^2}) .$$

Method: Solving the systems of linear equations one finds the bases of the spaces $U_{g,n}$ and U'_g . Then use Theorem 4.3.13.

Remark 4.3.22. Note that codes obtained by reductions of $X_0(11\cdot\ell)$, cannot be $\mathbb{F}_{121}$-codes. To obtain $\mathbb{F}_{121}$-codes, one has to consider curves of the form $X_0(24\cdot\ell)/11$, $\ell \neq 11$ being a prime.

Exercise 4.3.23. Give an algorithm for the code on the curve $X_0(24\cdot\ell)/11$ similar to the described one for $X_0(11\cdot\ell)/p$. (*Hint*: The curve $X_0(24)$ is elliptic).

Remark 4.3.24. To obtain the proof of part a) of Theorem 4.3.1 without any restrictions on N and on the divisor D used in this section, instead of W_m one has to consider the space of the form

$$W'_m = \frac{1}{P_m(x,y)}\cdot W_m ,$$

i.e. the space of fractions of the form $f/P_m(x,y)$, $f \in W_m$, $P_m(x,y)$ being a polynomial whose degree and exact form depend in a controlable way on N and D . Substituting W'_m for W_m , we can construct a polynomial algorithm for calculating the generator matrix of the corresponding code which is entirely parallel to the above one.

4.3.2. The case of Drinfeld curves

Now let us give an algorithm to construct a generator matrix of the code

$$C' = (X', \mathcal{P}', D')_L$$

over $\mathbb{F}_{q^2}$ for a Drinfeld curve $X' = X_0(I)$, $\mathcal{P}$ being the set of supersingular $\mathbb{F}_{q^2}$-rational points of X' , and the divisor D' being of the form $b \cdot Q_1$, where Q_1 is a cusp point of X' such that all its inverse images on $X_1(I)$ are $\mathbb{F}_{q^2}$-rational points.

In this section we use the following notation: $p_I(T) \in A = \mathbb{F}_q[T]$ is a monic irreducible polynomial of an odd degree m relatively prime to $(q - 1)$, generating the ideal $I = p_I \cdot A$; $r(T)$ is a polynomial of degree at most $m - 1$, whose image in $H = (A/I)^* / \mathbb{F}_q^*$ is a fixed generator θ of this group; X denotes the curve $X_1(I)$, π is the natural projection $X \longrightarrow X'$; $\tilde{X}$ is the singular plane model of X defined by the modular equation $F_I(x,y) = 0$; its singular points are P_1 and P_2 ; ν is the normalization map $\nu : X \longrightarrow \tilde{X}$; $Cusps_1(I) = \nu^{-1}(\{P_1, P_2\})$ is the set of cusp points on X ; $Cusps' = Cusps_1(I)(\mathbb{F}_q)$ is the set of cusps on X defined over $\mathbb{F}_q$, the latter set coincides with $\pi^{-1}(Q_1)$, where $\{Q_1, Q_2\}$ is the set of cusp points of X'. For an integer b by D' we denote the divisor $b \cdot Q_1$ on X'. By $C' = C'_{m,b}$ we denote a q^2-ary code on X' defined by the divisor D' and the set of supersingular $\mathbb{F}_{q^2}$-points on X'. The code $C'_{m,b}$ has the following parameters:

$$n = n(C) = (q^m + 1)/(q + 1) ,$$

$$k = k(C) \geq b - g' + 1 ,$$

$$d = d(C) \geq n - b ,$$

where $g' = g(X')$ is the genus of X_0, $g' \leq n/(g - 1)$. Desingularization trees of the points P_1 and P_2 are denoted, respectively, by Γ_1, Γ_2, we write $\Gamma = \Gamma_1 \cup \Gamma_2$. We use the notation

$$V_1 = V_{P_1} , \; V_2 = V_{P_2} , \; V'_1 = V'_{P_1} , \; V'_2 = V'_{P_2} , \text{ etc.}$$

Linear system $L(D')$. To construct the code, one has to construct the system $L(D')$ in terms of polynomials in x and y. To do this, one uses the following fact:

Proposition 4.3.25. *Let $f \in \mathbb{F}_q(x)$. Then the condition $f \in \pi^*(L(D'))$ is equivalent to the following one:*

$$f = \sum_i \theta^i g$$

for some $g \in L(\pi^* D')$, where $h = |H|$.

Proof: It is easy to show (Exercise 2.2.16) that

$$\pi^*(L(F)) = L(\pi^* F)^H ,$$

where by V^H for $H \subset GL(V)$ we denote a subspace of elements fixed by H. Since it is clear that any element of the form $\sum_i \theta^i g$ is invariant under H, the only fact we need is that the element $f \in L(\pi^* F)$ such that $\gamma \cdot f = f$ for $\gamma \in H$ can be represented in such a form. To do this, it suffices to take $g = (\bar{h})^{-1} \cdot f$, where $\bar{h} \in \mathbb{F}_p$, $\bar{h} \equiv h (mod\ p)$. ■

On the other hand, the set $L(\pi^* D')$ can be given by polynomials of bounded degree:

Proposition 4.3.26. *Let*

$$W_m = \{p(x,y) \in \mathbb{F}_q[x,y] \mid deg\ p(x,y) \le q^{4m}\}$$

and let V_m *be the image of* W_m *in* $\mathbb{F}_q(X)$. *Then*

$$L(\pi^* D') \subseteq V_m .$$

Proof: Let $f \in L(\pi^* D')$. Since the curve X_1 is smooth in the affine part of the plane, there exists an element $P \in \mathbb{F}_q[x,y]$ such that $f = \tilde{P}$ (check!). Let $s = deg\ P$. Let us show that if $s > q^{4m}$, there exists a polynomial P' of degree $s' < s$ such that $\tilde{P}' = \tilde{P} = f$. In homogeneous coordinates $(Z_0:Z_1:Z_2)$ (such that $x = Z_1/Z_0$ and $y = Z_2/Z_0$) we have:

$$P(x,y) = Z_0^{-s} \cdot H(Z_0:Z_1:Z_2) ,$$

where $H(Z_0:Z_1:Z_2)$ is a form of degree s. The form H defines the divisor (H) on X. Let $Q \in Cusps'(I)$. Since $f \in L(\pi^* F)$, $ord_Q(f) \ge -b$, whence

$$ord_Q((H)) = s \cdot ord_Q(Z_0) + ord_Q(f) \ge$$

$$\ge s \cdot ord_Q((Z_0)) - b \ge (s - 1) + ord_Q((Z_0)) - b .$$

Therefore we have:

$$(H) \geq (s - b - 1)\cdot \sum_{Q\in Cusps_1(I)} Q + (Z_0) \ .$$

On the other hand, for the conductor $D \in Div(X)$ of $\tilde{X}$ we have

$$D = \sum_{Q\in Cusps'} n_Q\cdot Q \leq \sum_{Q\in Cusps'} (deg\ D)\cdot Q \ ,$$

since $Supp\ D = Cusps'$.

By Proposition 2.5.26,

$$deg\ D < (q^{2m} - 1)\cdot q^{2m-2}/(q^2 - 1) < q^{4m} \ ,$$

we have

$$(H) \geq D + (Z_0) \ ,$$

whence by the Noether theorem (Theorem 2.5.24) there exist forms A and B , $deg\ A = s - 1$, $deg\ B = s - deg\ H$ such that

$$H = Z_0\cdot A + F\cdot B \ .$$

Now it is clear that we can put

$$P'(x,y) = A(Z_0:Z_1:Z_2)/Z^{s-1} \ .$$

■

To construct $L(\pi^* D')$, we have to consider functionals $A_{r,t}$ defined in 4.3.1, t being a free node of one of the trees Γ_1 or Γ_2 . Note that since all the points from $Cusps'$ are defined over $\mathbb{F}_q$, the functionals $A_{r,t}$ are defined on the space W_m . Note also that the functionals $A_{r,t}$ are trivial for $r < -q^{4m}\cdot n_t$, where $n_t = ord_{Q_t}(Z_0)$, since any $P \in W_m$ can be written in the form $P = G(Z_0,Z_1,Z_2)/Z_0^s$ for $s \leq q^{4m}$.

Proposition 4.3.27. *Let*

$$W_b = \bigcap_{t\in V'} \bigcap_{r=-n\cdot q^{4m}}^{-b} Ker(A_{r,t}) \subset W_m .$$

Then the image of the set W_b *in* V_m *coincides with* $L(\pi^* D')$.

Proof: Since $L(\pi^* D') \subseteq V_m$, our only purpose is to sort out from W_m the elements $p \in W_m$ such that $ord_{Q_t}(\tilde{p}) \geq -b$ for any $t \in V'$. These are singled out by the conditions $A_{r,t}(p) = 0$ for $t \in V'$ and $r = -n\cdot q^{4m},\dots,-b$. ■

Realization of C' . The algebraic-geometric modular code $C' = C'_{m,b}$ over $\mathbb{F}_{q^2}$ is given by its generator matrix $M = (m_{\alpha,\beta})$, α running over the set of indices of $\mathbb{F}_{q^2}$-rational supersingular points $\{P_\alpha\} \in X_0(I)(\mathbb{F}_{q^2})$ and β over that of the basis $\{f_\beta\}$ of $L(D')$:

$$m_{\alpha,\beta} = f_\beta(P_\alpha) \in \mathbb{F}_{q^2} .$$

Let S be the set of orbit representatives for the action of H on R , where R is the set of roots of the polynomial $F(x) = F_I(x,0)$. The cardinality of R equals $n\cdot h$, that of S equals n ; both sets lie in the field $L = \mathbb{F}_s$, $s = p^{2h_1}$. By R_σ for $\sigma \in S$ we denote the orbit of σ , $|R_\sigma| = h$. Consider the following $\mathbb{F}_{q^2}$-linear map:

$$\ell : W_b \otimes \mathbb{F}_{q^2} \longrightarrow (\mathbb{F}_{q^2})^m ,$$

$$\ell : p(x,y) \longmapsto h^{-1}\cdot \sum_{\alpha\in R_\sigma} p(\alpha,0) .$$

Theorem 4.3.28. *The image of* ℓ *in* $(\mathbb{F}_{q^2})^n$ *coincides with the modular code* $C'_{m,b}$.

Proof: Let us consider the following map:

$$\tilde{\ell} : V_m \otimes \mathbb{F}_{q^2} \longrightarrow (\mathbb{F}_{q^2})^n ,$$

$$\tilde{\ell} : f \longmapsto h^{-1} \cdot \sum_{\alpha \in R_\sigma} F(P_\alpha) ,$$

P_α being a point $(\alpha, 0)$ on X_1 . It is clear that the diagram

$$\begin{array}{ccc} W_b \otimes \mathbb{F}_{q^2} & \xrightarrow{\ \ell\ } & \mathbb{F}_{q^2}^n , \\ & \searrow \qquad \swarrow \tilde{\ell} & \\ & V_m \otimes \mathbb{F}_{q^2} & \end{array}$$

commutes. By Proposition 4.3.27 the image of $W_\ell \otimes \mathbb{F}_{q^2}$ in $V_m \otimes \mathbb{F}_{q^2}$ coincides with $L(\pi^* D')$. Since $\tilde{\ell}$ is invariant under H ,

$$\ell(W_b \otimes \mathbb{F}_{q^2}) = \tilde{\ell}(L(\pi^* D')) = \tilde{\ell}(L(\pi^* D')^H) .$$

If $f \in L(\pi^* F)^H$, it is clear that

$$h^{-1} \cdot \sum_{\alpha \in R_\sigma} f(P_\alpha) = F(\pi(P_\alpha)) ,$$

so that $\tilde{\ell}(L(\pi^* D')^H)$ coincides with the image of $L(D')$ under the map defined by values at supersingular points on X' . ∎

Algorithm. Let us give an algorithm to construct a generator matrix of $C' = C'_{m,b}$. It is similar to that for the code $C_{g,h}$ described in Section 4.3.1.

Step 1. Choice of $p_I(T)$ and construction of a generator in H .

Input: A positive odd integer m relatively prime to $(q - 1)$.

Output: An irreducible polynomial $p_I(T)$, $\deg p_I = m$, and a polynomial $r(T)$, $\deg r(T) < m$, such that its image θ is a generator in $H = (A/(p_I))^*/\mathbb{F}_q^*$.

Method: To find $p_I(T)$ it suffices to sort all monic polynomials of degree m , checking if they are irreducible, and to find $r(T)$ one has to sort all the polynomials of degree at most $(m - 1)$ checking the order of image of every polynomial in the group H .

Step 2. Construction of Γ .

Input: $p_I(T)$.

Output: The graph $\Gamma = \Gamma_1 \cup \Gamma_2$, where Γ_1 and Γ_2 are desingularization trees of P_1 and P_2 , respectively.

Method: The same as in *Step* 3 of the algorithm from Section 4.3.1.

The next two steps are completely parallel to Steps 4 and 5 of the algorithm from 4.3.1, for this reason we mention only their titles.

Step 3. Computing the conductor D .

Step 4. Computing coefficients of linear forms $\{A_{r,t}\}$.

The next step substitutes for Step 6 of the algorithm of Section 4.3.1.

Step 5. Construction of orbits of H on R .

Input: Polynomials $p_I(T)$ and $r(T)$.

Output: The set S of orbit representatives for the action of H on R , and for every $\sigma \in S$ a list of elements of R lying in σ .

Method: Decompose $F(x)$ into $n \cdot h_2$ factors irreducible over $\mathbb{F}_{q^2}$, find the roots of these factors and use the explicit action of θ on points of $\tilde{X}$.

Step 6. Construction of a generator matrix for $C'_{m,b}$.

Input: Coefficients of linear forms $\{A_{r,t}\}$, the set S , and the set $\{R_\sigma\}$.

Output: A generator matrix $M' = M'_{m,b}$ of $C'_{m,b}$.

Method: Construct a basis in V_b solving the linear system and apply Theorem 4.3.28.

4.3.3. Complexity

Now let us describe standard procedures necessary to realize our algorithms and make us sure that they indeed require polynomial resourses.

We should notice that we did not aim to minimize time complexity when choosing standard procedures. A more subtle computing techniques (preserving the general algorithm structure) does not allow any essential economy of computing resources.

Standard procedures. To realize the algorithm one has to work with polynomials over finite fields which is reduced to operations with their coefficients. To do this it is convenient to describe $\mathbb{F}_{p^r}$ by its set of structure coefficients $\{c_{ijk}\}$, $c_{ijk} \in \mathbb{F}_p$, $i,j,k = 1,\dots,r$ in some basis $\{\alpha_1,\dots,\alpha_2\}$ of $\mathbb{F}_{p^r}$ over $\mathbb{F}_p$:

$$\alpha_i \cdot \alpha_j = \sum_{k=1}^{r} c_{ijk} \cdot \alpha_k \ ;$$

using this representation one can easily work with finite fields.

Besides the procedures for arithmetical operations and for the Euclid algorithm for polynomials, in our algorithms we have to solve equations of the form $P(Z) = 0$, where $P \in \mathbb{F}_{p^r}[Z]$ is an irreducible polynomial of degree $s = deg\ P$. To find its roots lying in $\mathbb{F}_{p^{r\cdot s}}$ it suffices to construct a set of structure constants of $\mathbb{F}_{p^{r\cdot s}} = \mathbb{F}_{p^r}[Z]/(P(Z))$ in the power basis $1, \beta, \dots, \beta^{s-1}$, where $\beta = Z(mod\ P(Z))$, which is easy to do using the relation

$$\beta^s = -(a_{s-1}\beta^{s-1} + \dots + a_0) \ ,$$

where $P(Z) = Z^s + a_{s-1} \cdot Z^{s-1} + \dots + a_0$. The roots of the polynomial $P(Z)$ are $\{\beta, \beta^p, \dots, \beta^{p^{s-1}}\}$.

Exercise 4.3.29. Check that the described procedure of solving the equation $P(Z) = 0$ is of polynomial complexity.

When realizing algorithms, one has to decompose a polynomial $P \in \mathbb{F}_{p^b}[Z]$ without multiple roots of degree $deg\, P = s$ into factors irreducible over $\mathbb{F}_{p^b}$. Let us describe a method to realize such a decomposition with complexity polynomial in b and s. Consider a semi-simple algebra $R = \mathbb{F}_{p^b}[Z]/(P)$ and let $\{c_{ijk}\}$, $i,j,k = 1,\dots,b\cdot s$ be its structure constants. Let τ be a Frobenius operator on R over the field $\mathbb{F}_p$:

$$\tau : \sum_{i=0}^{s-1} a_i \cdot Z^i \,(mod\ P) \longmapsto \sum_{i=0}^{s-1} a_i^p \cdot Z^{i\cdot p} \,(mod\ P) .$$

Let M_τ be the matrix of τ in the basis $\{\alpha_i \cdot X^j (mod\ P)\}$; $i = 1,\dots,b$, $j = 0,1,\dots,s-1$ of R over $\mathbb{F}_p$, where $\{\alpha_i\}$, $i = 1,\dots,b$ is the basis of $\mathbb{F}_{p^b}$ over $\mathbb{F}_p$. Solving the linear system $M_\tau \cdot x = x$, we find an $\mathbb{F}_p$-basis in the τ-invariant space R^τ. Its dimension equals to the number of irreducible divisors of P. If $r_1, r_2 \in R^\tau$ are linearly independent over $\mathbb{F}_p$, then $r_1 + a\cdot r_2$ is a divisor of zero for some $a \in \mathbb{F}_p^*$. To find whether this is the case, or not, it suffices to calculate the determinant of the multiplication matrix by $r_1 + a\cdot r_2$ in R^τ; if it vanishes, then $r_1 + a\cdot r_2$ is a divisor of zero. If $H(Z) \in \mathbb{F}_{p^b}[Z]$ is a polynomial such that

$$deg\, H(Z) < deg\, P(Z) \quad \text{and} \quad H(Z) \equiv r_1 + a\cdot r_2 (mod\ P) ,$$

$H(Z)$ and $P(Z)$ have a non-trivial common divisor which makes it possible to decompose $P(Z)$ using induction.

When calculating the coefficients of the forms $\{A_{m,t}\}$ it is also necessary to expand an implicit function Y defined by $F(X,Y) = 0$ into power series in X. To be precise, let $F(X,Y) \in \mathbb{F}_{p^b}[X,Y]$ be an absolutely irreducible polynomial of degree $deg\, F = s$ such that $F(0,0) = 0$, $F'_Y(0,0) \neq 0$.

Suppose that x and y satisfy $F(x,y) = 0$. Since $F'_Y(0,0) \neq 0$, y allows a power expansion in x. Formulae given in Section 2.1.2 for power expansion of functions in local parameters make it possible to construct segments of corresponding power series bounded by a polynomial in d, s, and b with polynomial complexity.

Exercise 4.3.30. Write out a detailed algorithm for the calculation of segments of power expansion in x and y and estimate its complexity.

Polynomiality. To show that above algorithms are polynomial, it suffices to show that the information used by each step of the algorithm is bounded by a polynomial in n. Indeed the procedures used in the algorithm are polynomial, i.e. use a polynomial time and memory when treating an information unit.

Let us give the required estimates for $C_{g,h}$ and leave those for $C'_{m,b}$ as an exercise for the reader.

For the majority of objects used by the algorithm it is clear that they are polynomial. Let us check the polynomiality of the set of trees $\{\Gamma_P\}$, of the divisor D, and of the set $\{A_{m,t}\}$. It is clear that it suffices to estimate the whole number of nodes in all trees $\{\Gamma_P\}$ which is equal to $\sum_{P \in Sing\, \tilde{X}} |V_P|$, as well as the degrees a_t of fields of definition of Q_t.

Proposition 4.3.31. *We have*

a) $$\sum_{P \in Sing\, \tilde{X}} v_P \leq 18 \cdot (6\ell + 5) \cdot (8\ell + 7) \ .$$

b) $$a_t \leq \sum_{P \in Sing\, \tilde{X}} v'_P \leq 12 \cdot (6\ell + 5) \cdot (8\ell + 7) \ .$$

Proof: a) Let $m = deg\ hom\ F_N(x, j)$ be the degree of homogeneous equation defining the curve $\tilde{X}$ in $\mathbb{P}^2$; as specified above, $m \leq 2 \cdot (12 \cdot \ell + 11)$. Hence

$$p_a(\tilde{X}) = \frac{(m - 1) \cdot (m - 2)}{2} \leq 6 \cdot (6\ell + 5) \cdot (8\ell + 7)$$

By (2.5.3), we have

$$0 \leq \ell = g(X) = p_a(\tilde{X}) - \sum_{P \in Sing\, \tilde{X}} \delta_P \ .$$

Using (2.5.5) for δ_P we obtain

$$\sum_{P \in Sing\, \tilde{X}} \ \sum_{Q \longmapsto P} r_Q \cdot (r_Q - 1)/2 =$$

$$= \sum_{P \in Sing\, \tilde{X}} \delta_P \leq 6 \cdot (6\ell + 5) \cdot (8\ell + 7) \ .$$

Note that for non-free nodes $v \in V_P - V'_P$ we have $r_{Q_v} \geq 2$, so that

$$\sum_{P \in Sing\, \tilde{X}} (v_P - v_P') = \sum_P \ \sum_{v \in V_P - V'_P} 1 \leq$$

$$\leq \sum_P \ \sum_{Q \longmapsto P} r_Q \cdot (r_Q - 1) \leq 6 \cdot (6\ell + 5) \cdot (8\ell + 7) \ ,$$

hence

$$\sum_{P \in Sing\, \tilde{X}} v_P \leq 6 \cdot (6\ell + 5) \cdot (8\ell + 7) + \sum_{P \in Sing\, \tilde{X}} v'_P \ .$$

Moreover, we see that

$$v'_P \le \sum_{\substack{Q \mapsto P \\ r_Q \ge 2}} r_Q \le \sum_{Q \mapsto P} r_Q \cdot (r_Q - 1) \ ;$$

therefore

$$\sum_{P \in Sing\, \tilde{X}} v'_P \le \sum_{P \in Sing\, \tilde{X}} v_P \le$$

$$\le \sum_{P \in Sing\, \tilde{X}} \ \sum_{Q \longrightarrow P} r_Q \cdot (r_Q - 1) \le 12 \cdot (6\ell + 5) \cdot (8\ell + 7),$$

which proves the first part of the proposition.

b) Since X is defined over $\mathbb{F}_p$, the Galois group $Gal(\overline{\mathbb{F}}_p/\mathbb{F}_p)$ acts on the set $\bigcup_P V'_P$ of free nodes of the trees $\{\Gamma_P\}$. The orbit length of Q_t equals a_t. Thus $a_t \le \sum_{P \in Sing\, \tilde{X}} v'_P \le 12 \cdot (6\ell + 5) \cdot (8\ell + 7)$ by the proof of part a). ■

For D it suffices to estimate its degree.

Proposition 4.3.32.

$$deg\ D \le (12 \cdot \ell + 11)^2 \ .$$

Proof: The assertion follows immediately from Proposition 2.5.26. ■

Remark 4.3.33. The method to prove polynomiality for families of modular codes exposed in this section is valid also for a wide class of AG-codes. To be more precise, let $C_i = (X_i, \mathcal{P}_i, D_i)_L$ be a family of AG-codes over $\mathbb{F}_q$, where X_i is an absolutely irreducible smooth curve of genus g_i over $\mathbb{F}_q$, $\mathcal{P}_i \subseteq X_i(\mathbb{F}_q)$, D_i is a divisor on X_i defined over $\mathbb{F}_q$ so that X_i has a "polynomial" plane model $\tilde{X}_i$, and the set $\mathcal{P}_i$ is "polynomial" as well as D_i. Polynomiality conditions mean the following: there exists an algorithm polynomial in g_i for computing the polynomial $F_i(x,y) \in \mathbb{F}_q[x,y]$ defining a model $\tilde{X}_i$ of X_i over $\mathbb{F}_q$ (i.e. $\tilde{X}_i$ is birationally isomorphic to X_i over $\mathbb{F}_q$), the degree of F_i being bounded by a polynomial in g_i; there exists an algorithm polynomial in g_i finding out whether the point $P \in X_i(\mathbb{F}_q)$ belongs to $\mathcal{P}_i$ or not, and finally there exists an algorithm polynomial in g_i for calculating D_i (i.e. the set $Supp\, D_i$ and the corresponding multiplicities of $P \in Supp\, D_i$ in D_i).

These conditions being satisfied, the method used to construct modular codes turns into a polynomial in g_i algorithm of construction of generator matrices M_i for the codes C_i; if $n_i \geq c \cdot g_i$, c being an absolute constant (in fact only codes with $n_i \geq g_i$ are reasonable to consider), this algorithm is polynomial in n_i, so that the family of codes C_i is polynomial.

The algorithm to construct C_i consists of:

a) calculation of the coordinates of singularities on $\tilde{X}_i$;

b) construction of the set of labeled desingularization trees for the singularities;

c) calculation of the conductor;

d) computation of the coefficients of suitable linear forms on some linear space (the forms depend on the equation of $\tilde{X}_i$ and on the divisor D_i);

e) calculation of the set $\mathcal{P}_i$;

f) computation of the generator matrix M_i.

Steps a) to e) correspond to Steps 2 to 7 of the algorithm for codes on reductions of classical modular curves. Step 1 of the latter algorithm garantees the polynomiality of the plane model of modular curve.

The algorithm of construction of codes on Drinfeld curves makes it possible to generalize the previous algorithm to codes of the form $C'_i = (X'_i, \mathcal{P}'_i, D'_i)_L$, where X_i is an absolutely irreducible smooth curve over $\mathbb{F}_q$, $\mathcal{P}'_i \subseteq X'_i(\mathbb{F}_q)$, D'_i is a divisor on X'_i, and the curve X'_i is of the form X_i/G_i, X_i being a curve with a polynomial plane model, and G_i being a subgroup of the automorphism group of X_i of order relatively prime to q with a polynomial action, i.e. such that for any $P \in X_i$ and $g \in G_i$ the point $g \cdot P$ can be obtained as a result of polynomial calculations such that the genus g_i of X_i is also bounded by a polynomial in g'_i. Here we suppose the set $\mathcal{P}'_i$ and the divisor D'_i to be polynomial in the above sense.

The construction algorithm for the code sequence C'_i is similar to that for codes C_i; the only exception is that the calculations are performed for $\tilde{X}_i$ which is a polynomial plane model of X_i, hence in Step d one has to work with the set of orbits for the action of G_i on $\mathcal{P}_i \subseteq X_i$, which is the inverse image of $\mathcal{P}'_i$.

Note as well that the algorithm of construction of codes on Drinfeld curves has a simpler structure than that for classical modular curves, since the plane model $\tilde{X}_i$ of X_i has in the former case a more explicit definition; in particular, singularities of the curve $\tilde{X}_1(I)$ are known *a priori*.

Historical and bibliographic notes to Part 4

Classical modular curves are one of those subjects which from the viewpoint of their origins are just a problem for a historian of mathematics. A reader interested in the history of this field can refer to Chapter VII of the book [Ab]; note however that even there he will not find a complete description. Some information on the history of modular curves and functions, and first of all on their connections with arithmetics, is contained in the book [Vl 3], the reader will also find there a more detailed exposition of the theory of modular curves. Now we shall restrict ourselves to the principle landmarks of the history of modular curves and functions.

Modular curves arose while studying elliptic integrals. One of the main problems of the theory of elliptic integrals was that of transformation. In modern language this problem corresponds to that of search for an explicit expression for isogenies of elliptic curves. The problem of transformation was solved independently by N.H.Abel and K.G.Jacobi in the end of twenties of the nineteenth century. An equation for moduli of elliptic integrals related by transformation of order n, is called the modular equation of order n. Examples of such equations for small n were written out by A.M.Legendre, N.H.Abel, and K.G.Jacobi. Modular functions were studied from different points of view by G.F.B.Riemann, L.Kronecker, C.Hermit and by many others. Later these functions proved to be rather a powerful instrument in some questions of analysis and number theory. In particular, E.Galois calculated the Galois group of the modular equation. Up to the seventies of the last century, the principle object to study were functions obtained from the Legendre modulus, the latter being a modular function for $\Gamma(2)$ and not for the complete modular group, which lead to a number of inconveniences in manipulating them. Due to R.Dedekind from the beginning of seventies of the last century the absolute invariant j was studied and widely used. A valuable contribution to the theory of modular functions was made by F.Klein; he found out that they give a uniformization of algebraic curves which were later called modular curves. In papers of L.Kronecker, H.Weber, R.Fueter, H.Hasse, and M.Deuring modular functions were applied to construct class fields for complex quadratic fields (1880-1930). E.Hecke greatly contributed to the theory introducing operators now named after him. After Hecke the theory did not attract much attention up to 1960-th when deep relations were found connecting modular functions to various fields of mathematics, first of all to infinite-dimensional representations of Lie groups (the Langlands program). The latter gave an explosion of activity in modular functions, which was reflected in its principle moments in the papers of conferences on modular functions [MF 1-6].

The best reading on modular functions are beautiful books by G.Shimura [Shi] and S.Lang [Lan 3]. Here the reader can find a detailed exposition of the material exposed in Section 4.1.1 except for the section on Weierstrass points where we borrowed from the paper by A.Atkin [At] and for Proposition 4.1.29 due to R.Ogg [Ogg]. The theory of moduli schemes of elliptic curves was developed by P.Deligne and M.Rapoport; it was essentially improved later by N.Katz and B.Mazur [Kat/Maz]. In Section 4.1.2 we follow [De/Ra]. Calculation of the number of $\mathbb{F}_q$-points on reductions of modular curves which is a simple exercise on application of Deligne-Rapoport results, is taken from [Ts/Vl/Z]. The material of Section 4.1.3 is mostly original.

In contrast to classical elliptic and modular curves, elliptic modules and their moduli varieties have a unique author (V.G.Drinfeld), and all results described in Sections 4.2.1 and 4.2.2 are due to him (consideration of elliptic modules is indispensable if we want to prove the Langlands conjecture for $GL(2)$ over a functional field). A great deal of these results is published in [Drf 1] and [Drf 2] while the rest were learned by the authors from their private conversations with Drinfeld. The only exceptions are Propositions and Corollaries 4.2.23 to 4.2.45 which are just a simple application of Drinfeld's results. These results are due to Vladut, see [Ma/Vl]. Section 4.2.3 is original.

The results of Chapter 4.3 are due to Vladut (cf.[Ma/Vl] and [Vl 2]).

PART 5

SPHERE PACKINGS

This part is mostly devoted to another kind of problems which resemble greatly the questions about codes - namely, to quite a classical problem of dense packing of equal non-overlapping spheres in $\mathbb{R}^N$. It comes out that both direct application of algebraic-geometric codes and the use of intuition developped while studying them are quite useful. Moreover, in this part - especially in the last chapter - one can see even better then before the marvelous integrity of mathematics, two more parts of which - number theory and that of packings being added to coding theory and algebraic geometry.

In Chapter 5.1 we give basic definitions, put the packing problem (both for lattice and non-lattice packings), look at examples and discuss asymptotical results. Chapter 5.2 is devoted to some constructions of sphere packings using codes; algebraic-geometric codes happen to be rather useful, especially (as one would expect) for asymptotic problems. Chapter 5.3 is a very short analogue of Part 2, we give a brief textbook on number theory, mostly without proofs, concentrating on the needs of what follows. Chapter 5.4 is the core of this part: we consider analogues of algebraic-geometric codes from many different points of view. We come to explicit algebraic-geometric and number-theoretic constructions of asymptotically good lattice packings. Algebraic-geometric approach should be also quite useful to construct examples of dense packings "on a finite level", but we do not say much about that here.

CHAPTER 5.1

Definitions and Examples

How can one pack equal non-overlapping spheres in $\mathbb{R}^N$? What is the density of such packing and how the density behaves for $N \longrightarrow \infty$?

In Section 5.1.1 we give necessary definitions and state the problem rigorously. Section 5.1.2 is devoted to some important examples, including the Leech lattice. In Section 5.1.3 we put asymptotic problems. In Section 5.1.4 we discuss some analogies between codes and packings.

5.1.1. Parameters

Packings. Let us consider the classical problem of *packing* equal non-overlapping spheres in $\mathbb{R}^N$. Let P be the set of centers and let

$$d = d(P) = \inf_{v,u \in L,\ v \neq u} |u - v| ,$$

d is the *minimum distance* of the packing, which equals the maximum possible diameter of non-overlapping spheres centered in P.

The *density* of P is the part of $\mathbb{R}^N$ covered by spheres; to be precise, it can be defined as

$$\Delta(P) = \limsup_{u \longrightarrow \infty} v(S \cap B_u)/v(B_u) \quad ,$$

where

$$S = \{x \in \mathbb{R}^N \mid \exists\, y \in P \, , \, |x - y| < \frac{d}{2}\}$$

$$B_u = \{x \in \mathbb{R}^N \mid |x| \le u\}$$

and $v(\cdot)$ is the standard volume in $\mathbb{R}^N$.

Lattices. If P is an additive subgroup of $\mathbb{R}^N$, we call the packing P a *lattice packing* (or just a *lattice*; in this case we use the letter L rather than P). Further on we suppose that the rank of L equals N since otherwise $\Delta(L) = 0$. If L is a lattice then any choice of a basis $e_1,\ldots,e_N$ in L defines a map $\mathbb{Z}^N \longrightarrow \mathbb{R}^N$; its matrix is called a *generator matrix* of the lattice.

Exercise 5.1.1. Check that for lattices the definition of $\Delta(L)$ does not depend on the choice of origin and does not change if we replace the ball B_u by a cube (or by any homotetically increasing solid containing a neighbourhood of the origin).

Exercise 5.1.2. Check that the volume of the fundamental domain

$$F = \{\sum_{i=1}^{N} x_i \cdot e_i \mid 0 \le x_i < 1\} \subset \mathbb{R}^N$$

equals the absolute value of the determinant of the generator matrix. This volume is called the *determinant* of the lattice and is denoted by $det\ L$; we define the *discriminant* $discr\ L$ of L as the determinant of the matrix of inner products $\|(e_i, e_j)\|$, $i,j = 1,\ldots,N$.

Exercise 5.1.3. Check that

$$discr\ L = (det\ L)^2 .$$

Let $V_N = \pi^{N/2}/\Gamma(N/2 + 1)$ be the volume of unit ball in $\mathbb{R}^N$.

Exercise 5.1.4. Check that

$$\Delta(L) = \frac{d(L)^N \cdot V_N}{2^N \cdot det\ L} .$$

Note that by the Stirling formula we have

$$log_2 V_N = \frac{N}{2} \cdot log_2\left(\frac{2\pi e}{N}\right) - log_2\sqrt{\pi \cdot N} + o(1) ;$$

we write this as

$$\frac{1}{N} \cdot log_2 V_N \sim \frac{1}{2} \cdot log_2\left(\frac{2\pi e}{N}\right) .$$

Thus for $N \longrightarrow \infty$ we get

$$- \frac{1}{N} \cdot log_2 \Delta(L) \sim - log_2\sqrt{\frac{\pi \cdot e}{2}} + log_2\sqrt{N} -$$

$$- log_2 d(L) + \frac{1}{N} \cdot log_2(det\ L) .$$

Other parameters. Let us define some other parameters of packings (which are often more convenient than Δ) setting

$$\delta(P) = \Delta(P)/V_N ,$$

$$\lambda(P) = -(\log_2 \Delta(P))/N ,$$

$$\nu(P) = \log_2 \delta(P) ;$$

we call $\delta(P)$ the *centre density*, and $\lambda(P)$ the *density exponent*. Clearly,

$$\Delta(P) = 2^{-\lambda(P)\cdot N} .$$

For root lattices it is convenient to use $\delta(P)$; $\nu(P)$ is useful to compute the density of lattices obtained by constructions of Section 5.2.1. The density exponent $\lambda(P)$ is especially important for asymptotic problems.

Densest packings. Set

$$\lambda(N) = \inf_{P\subset\mathbb{R}^N} \lambda(P) , \quad \Delta(N) = \sup_{P\subset\mathbb{R}^N} \Delta(P), \ldots$$

A natural problem of finding the densest possible packing in a given dimension can be decomposed into two problems:

A. Find the precise value of $\lambda(N)$ (or, what is the same, of $\Delta(N)$ or of $\delta(N)$, ...).

B. Find a packing P with $\lambda(P) = \lambda(N)$.

These problems are completely solved only for $N = 1$ and $N = 2$.

Since $\Delta(P) \le 1$, we get

$$\lambda(P) \ge 0$$

for any packing.

For $N = 1$ the answer is obvious: equal segments cover the whole line, and hence for this packing L_1 one has $\Delta(L_1) = 1$, i.e.

$$\Delta(1) = 1 \ , \ \lambda(1) = 0 \ .$$

For $N = 2$ the problem is not so simple but one can prove that

$$\Delta(2) = \pi/2\sqrt{3} \ .$$

Exercise 5.1.5. Check that for the lattice $L_2 \subset \mathbb{R}^2 = \mathbb{C}$ generated by 1 and $\frac{1+\sqrt{-3}}{2}$ we have $\Delta(L_2) = \Delta(2)$; L_2 is called the *hexagonal* lattice.

Strangely enough, $\lambda(N)$ is unknown for any $N \ge 3$.

Densest lattices. For lattice packings we know slightly more. Let

$$\lambda_\ell(N) = \inf_{L\subset\mathbb{R}^N} \lambda(L) \quad , \quad \Delta_\ell(N) = \sup_{L\subset\mathbb{R}^N} \Delta(L) \ , \ \ldots$$

Clearly,

$$\lambda_\ell(N) \ge \lambda(N) \quad , \quad \Delta_\ell(N) \le \Delta(N) \quad , \ \ldots$$

Recall that a lattice L is called *unimodular* iff $\det L = 1$. The *dual lattice*

$$L^\perp = \{x \in \mathbb{R}^N \mid (x,\ell) \in \mathbb{Z} \ \text{ for any } \ \ell \in L\}$$

is in this case also unimodular.

Recall also that the inner product in $\mathbb{R}^N$ induces on L a positively definite bilinear form.

Exercise 5.1.6. Show that any integral positively definite bilinear form can be obtained from a lattice. (*Hint*: Let A be the matrix of the form; it is sufficient to find a matrix M such that ${}^tM \cdot M = A$ and to generate a lattice by its rows).

Let now $\varphi(x,y)$ be a positively definite bilinear form in N integral variables, and let $f(x) = \varphi(x,x)$ be the corresponding quadratic form. Suppose that φ is unimodular, i.e. $discr\ \varphi = 1$. Such forms are in bijection with unimodular lattices L in $\mathbb{R}^N$. Set

$$\gamma(L) = \gamma(\varphi) = \min_{x \in \mathbb{Z}^N - \{0\}} f(x) \quad .$$

In lattice terms it is the squared length of the shortest non-zero vector.

Exercise 5.1.7. Check that

$$\Delta(L) = V_N \cdot \left(\frac{\gamma(L)}{4}\right)^{N/2} \quad ;$$

$$\gamma(L) = 4 \cdot \left(\frac{\Delta(L)}{V_N}\right)^{2/N} = 4 \cdot (\delta(L))^{2/N} \quad .$$

One can naturally extend the definition of $\gamma(\varphi)$ to non-unimodular case:

$$\gamma(L) = \gamma(\varphi) = \min_{x \in \mathbb{Z}^N - \{0\}} (f(x) / \sqrt[N]{discr\ \varphi}) \quad .$$

Now let us put

$$\gamma(N) = \max_{\varphi} \gamma(\varphi) \quad ,$$

where maximum is taken over all positively definite bilinear forms in N variables; $\gamma(N)$ and $\Delta_\ell(N)$ are related by formulae similar to those of Exercise 5.1.7.

Note that for $N \longrightarrow \infty$ we obtain

$$\lambda(N) \sim \log_2 \sqrt{\frac{2N}{\pi \cdot e \cdot \gamma(N)}} \quad ,$$

$$\log_2(\gamma(N)) \sim \log_2\left(\frac{2N}{\pi e} \cdot \Delta^{2/N}\right) = -2 \cdot \lambda(N) + \log_2\left(\frac{2N}{\pi e}\right) \quad .$$

Precise values of $\Delta_\ell(N)$ are known for $1 \le N \le 8$; see Table A.4.1 where we have collected the values of all the above parameters for these N .

Note that within the table $\Delta(N)$ increases and $\gamma(N)$ decreases. It is interesting to know whether it is the case for any N .

In Table A.4.1 the integrality of $\delta(N)^{-2}$ attracts attention. We do not know whether $\delta(N)^{-2}$ is integral for any N ; note however that the densest lattice of a given rank can be generated by a matrix with rational entries and the rationality of $\delta(N)^{-2}$ follows.

5.1.2. Examples

Now we describe the densest lattices for $N \le 8$ and introduce some interesting lattice families.

We construct families $L \subset \mathbb{R}^N$ and give the values of

$d(L)$ and $det(L)$. Other density parameters for these families are given in Table A.4.2 of Appendix A.

The simplest family is

$$\mathbb{Z}^N \subset \mathbb{R}^N \quad ;$$

for these lattices

$$d(\mathbb{Z}^N) = 1 \ , \quad det(\mathbb{Z}^N) = 1 \ .$$

Root lattices. Let us consider in $\mathbb{R}^{N+1}$ the following lattice A_N of rank N :

$$A_N = \left\{ \sum_{i=1}^{N+1} a_i \cdot e_i \mid a_i \in \mathbb{Z} \ , \ \sum a_i = 0 \right\} \ ,$$

$\{e_i\}$ being the standard basis in $\mathbb{R}^{N+1}$.

Exercise 5.1.8. Show that the lattice A_N is generated by vectors

$$\alpha_1 = e_1 - e_2 \ , \ \alpha_2 = e_2 - e_3 , \ \ldots \ , \ \alpha_N = e_N - e_{N+1} \quad ;$$

and that

$$d(A_N) = \sqrt{2} \ , \ det \ A_N = \sqrt{N+1} \ .$$

The family $D_N \subset \mathbb{R}^N$ is defined by

$$D_N = \left\{ \sum_{i=1}^{N} a_i \cdot e_i \mid a_i \in \mathbb{Z} \ , \ \sum a_i \equiv 0 \ (mod\ 2) \right\} \ .$$

Exercise 5.1.9. Show that D_N is generated by $\alpha_1 = e_1 - e_2$, $\alpha_2 = e_2 - e_3$, ... , $\alpha_{N-1} = e_{N-1} - e_N$, and $\alpha_N = e_{N-1} + e_N$; and that

$$d(D_N) = \sqrt{2} \quad , \quad det \ D_N = 2 \ .$$

The following important family does exist only for $N = 4, 5, 6, 7, 8$. For such N define the lattice E_N in $\mathbb{R}^8$ by its basis:

$$\alpha_1 = \frac{1}{2}\cdot(e_1 + e_8) - \frac{1}{2}\cdot(e_2 + \ldots + e_7) ,$$

$$\alpha_2 = e_1 + e_2 ,$$

$$\alpha_i = e_i - e_{i-1} \quad \text{for} \quad i = 3,\ldots,N .$$

Exercise 5.1.10. Show that E_8 can be given by

$$E_8 = \left\{\sum_{i=1}^{8} a_i\cdot e_i \mid 2\cdot a_i \in \mathbb{Z} , a_i - a_j \in \mathbb{Z} , \sum_{i=1}^{8} a_i \in 2\cdot\mathbb{Z}\right\} ;$$

and that the rest E_N are intersections of E_8 with planes of codimension $(8 - N)$. In particular show that

$$E_7 = \left\{x = \sum_{i=1}^{8} a_i\cdot e_i \mid x \in E_8 , a_7 = -a_8\right\} ,$$

$$E_6 = \left\{x = \sum_{i=1}^{8} a_i\cdot e_i \mid x \in E_7 , a_6 = a_7\right\} .$$

Show that $d(E_N) = \sqrt{2}$ and $det(E_N) = 9 - N$.

Note that $A_1 = \mathbb{Z}$, $D_3 = A_3$, $E_4 = A_4$, $E_5 = D_5$. The lattice families A , D , and E are root lattices which arise in many questions: in the theory of Lie groups and algebras, in the singularity theory, in the theory of rational surfaces, etc.

The lattices Γ. Let now $N \geq 8$, $N \equiv 0 (mod\ 4)$. Set

$$\Gamma_N = \left\{ \sum_{i=1}^{N} a_i\cdot e_i \mid 2\cdot a_i \in \mathbb{Z} , a_i - a_j \in \mathbb{Z} , \sum_{i=1}^{N} a_i \in 2\mathbb{Z}\right\} .$$

Exercise 5.1.11. Show that Γ_N is generated by vectors $e_i + e_j$ and the vector $\frac{1}{2} \cdot \sum_{i=1}^{N} e_i$; and that

$$d(\Gamma_N) \geq \sqrt{2} \quad , \quad det\ \Gamma_N = 1 \quad .$$

Note that $\Gamma_8 = E_8$.

Exercise 5.1.12. Calculate the densities of the described lattices.

The lattices $A_1 = \mathbb{Z}$, A_2 , $A_3 = D_3$, D_4 , D_5 , E_6 , E_7 , $E_8 = \Gamma_8$ have the density coinciding with that from Table A.4.1; they are the densest lattices in their dimensions. A proof of this fact can be obtained by the reduction theory of quadratic forms; we do not give it here.

Note that for all the described families

$$\lambda(L_N) \sim log_2\sqrt{N} \longrightarrow \infty \quad \text{for} \quad N \longrightarrow \infty \ .$$

We shall see that there are lattices which asymptotically behave significantly better.

For $N \geq 9$ we do not know the precise value of $\lambda_\ell(N)$ and only some bounds are known. As in the case of codes it is natural to call upper bounds for $\Delta(N)$ and $\Delta_\ell(N)$ *possibility bounds* and lower ones *existence bounds* (note however that for $\lambda(N)$ possibility bounds are lower ones, and existence bounds are upper ones).

We do not describe here various methods of constructing dense packings in dimensions from 9 up to 100000. We need here only the Leech lattice which is a very beautiful object arising in many questions.

The Leech lattice. There exists a unique integral even unimodular lattice of dimension 24 which has no vector of length $\sqrt{2}$ (recall that a lattice is called *even* iff the scalar square of any its vector is even). This lattice is called the *Leech lattice* and is denoted by Λ_{24} ; it is closely connected with Golay $[24,12,8]_2$-code C_{24} . It can be constructed in many ways. Here is one of the simplest.

The lattice Λ_{24} is generated by vectors

$$v_{i,c} = \frac{1}{\sqrt{8}}\cdot u_{i,c} \,, \quad 1 \le i \le 24 \,, \; c \in C_{24} \,,$$

where $u_{i,c}$ has ∓ 3 in i-th position and ± 1 in all other positions, and the upper sign is chosen for a position where the codeword c has 1 .

The vectors $v_{i,c}$ are shortest vectors of Λ_{24} , their length being equal to 2 (their number equals $2^{12}\cdot 24$). There are two more types of shortest vectors, namely $t_{c,\varepsilon}$ and $w_{i,j,\delta}$, where

$$t_{c,\varepsilon} = \frac{1}{\sqrt{8}}\cdot s_{c,\varepsilon} \quad \text{for} \quad c \in C_{24} \,, \; \|c\| = 8 \,, \; \varepsilon \in (\mathbb{Z}/2)^7 \,,$$

vector $s_{c,\varepsilon}$ has ± 2 in positions where c has 1 , the number of -2 is even, and the distribution of signs is defined by ε ; in all other positions $s_{c,\varepsilon}$ has zero entries (there are $2^7\cdot 759$ vectors of this type). Vectors $w_{i,j,\delta}$ are defined by

$$w_{i,j,\delta} = \frac{1}{\sqrt{8}}\cdot x_{i,j,\delta} \,, \quad 1 \le i,j \le 24 \,, \; i \ne j \,, \; \delta \in (\mathbb{Z}/2)^2 \,,$$

where $x_{i,j,\delta}$ has ± 4 on positions i and j with arbitrary sign distribution which is defined by δ , and zero entries in all other positions (there are 1104 vectors of this type). These vector types are denoted $(3^1 1^{23})$,

$(2^8 0^{16})$ and $(4^2 0^{22})$, respectively. There are 196560 shortest lattice vectors, i.e. any sphere in corresponding packing kisses exactly 196560 other spheres.

Exercise 5.1.13. Prove that

$$det(\Lambda_{24}) = 1 \ , \ d(\Lambda_{24}) = 2 \ ;$$

and hence

$$\delta(\Lambda_{24}) = 1 \ , \ \nu(\Lambda_{24}) = 0 \ , \ \gamma(\Lambda_{24}) = 4 \ ,$$

$$\Delta(\Lambda_{24}) \approx 0.00193 \quad \text{and} \quad \lambda(\Lambda_{24}) \approx 0.376 \ .$$

The covering radius of the Leech lattice equals $2\sqrt{2}$, i.e. balls of radius $2\sqrt{2}$ centered at lattice points cover the whole space $\mathbb{R}^{24}$. One can describe "deep holes" of the Leech lattice, i.e. points with distance $2\sqrt{2}$ from the nearest lattice point.

The automorphism group Co_0 of the Leech lattice is enormous:

$$|Co_0| = 8315553613086720000.$$

The maximal sporadic simple group , the Fischer-Gries group (the Monster), can be realised as the automorphism group of an algebra closely connected to the Leech lattice.

5.1.3. Asymptotic problems

For asymptotic problems it is convenient to consider

$$\tilde{\lambda} = \liminf_{N \longrightarrow \infty} \lambda(N) \quad .$$

To compute $\tilde{\lambda}$, i.e. to understand which is the maximum asymptotic density $\tilde{\Delta} = 2^{-\tilde{\lambda}N}$ of a high-dimensional packing, is most likely a very hard problem. We are interested in bounds for this value. The situation here is similar to that in coding theory, and $\tilde{\lambda}$ is an analogue of $\alpha_q(\delta)$.

A *family of packings* is a set $\{P_N\}$ of packings, $P^N \subset \mathbb{R}^N$ where N runs over an infinite subset of $\mathbb{N}$.

Let

$$\lambda(\{P_N\}) = \liminf \lambda(P_N) \quad .$$

We call families with $\lambda(\{P_N\}) < \infty$ *good families* (they are analogues of good families of codes).

Exercise 5.1.14. Prove that $\inf_{\{P_N\}} \lambda(\{P_N\}) = \tilde{\lambda}$.

Similarly for lattices we set

$$\tilde{\lambda}_\ell = \liminf_{N \longrightarrow \infty} \lambda_\ell(N) = \inf_{\{L_N\}} \lambda(\{L_N\}) \quad .$$

Bounds. Here are the best known estimates of $\tilde{\lambda}$:

Theorem 5.1.15. $1 \geq \tilde{\lambda}_\ell \geq \lambda \geq 0.599$. ■

The upper bound which is the existence bound is called the *Minkowski bound*, the lower one (the possibility bound) the *Kabatyansky-Levenstein bound*.

The Kabatyansky-Levenstein bound can be obtained by technique similar to that of the McEliece-Rodemich-Ramsey-Welch (cf. Theorems 1.3.11 and 1.3.12). The proof of the former consists of two parts: the first is the linear programming bound for packing of spheres on $S^N \subset \mathbb{R}^{N+1}$ and the second provides a way to pass from S^N to $\mathbb{R}^N$, which is based on the following simple construction. Let Λ_N be a packing in $\mathbb{R}^N$ and let us embed $\mathbb{R}^N$ into $\mathbb{R}^{N+1}$ in the natural way (i.e. assuming that vectors from $\mathbb{R}^N$ have zero for the last coordinate). Thus $\mathbb{R}^N \cap S^N = S^{N-1}$; let us consider those balls from Λ_N which are contained in the unit $(N-1)$-ball. Lifting their centers to S^N we obtain a packing of S^N and its parameters can be estimated through the parameters of Λ_N.

Here is another bound:

Proposition 5.1.16.

$$\Delta(N) \le \sigma_N ,$$

where σ_N *is the ratio of the volume of the intersection* $(\bigcup_{i=1}^{N+1} B_i \cap \Sigma_N)$ *to the volume of* Σ_N, *where* Σ_N *is the perfect simplex of edge length* 2 *and* $B_1, \ldots, B_{N+1}$ *are unit balls centered at the vertices of* Σ_N. ∎

This bound gives $\tilde{\lambda} \ge 0.5$ but it is quite useful for moderate values of N.

The Minkowski bound (which is an analogue of the Gilbert-Varshamov bound) can be obtained by a techinque similar to the code-theoretic one. A family of (non-lattice) packings with $\lambda(\{P_N\}) = 1$ is constructed similarly to Exercise 1.1.54 (cf. Theorem 5.1.17); the inequality $\tilde{\lambda}_\ell \le 1$ can be proved by computation of the mean value of $\lambda(L_N)$ on a sufficiently large set of lattices. As in the case of codes (see Remark 1.3.17), almost all lattice families have $\lambda(\{L_N\}) = 1$.

Thus it is known that there exist lattices of density $\Delta \sim 2^{-N}$ but we do not know how to construct them explicitly. The problem of explicit construction of dense packings naturally arises. Formalization of the explicity requirement leads to the following problem.

We call a packing family $\{P_N\}$ *polynomial* if there exists an algorithm which requires a polynomial in N number of operations over real numbers and for an output has an algorithm of consecutive construction of corresponding sphere centers (for lattices its output can be just the lattice basis).

Let

$$\tilde{\lambda}^{pol} = \liminf_{\{P_N\}\ polynomial} \lambda(\{P_N\}) \ ;$$

similarly one defines $\tilde{\lambda}_{\ell}^{pol}$ for lattices, and also $\tilde{\lambda}^{exp}$ and $\tilde{\lambda}_{\ell}^{exp}$ (for *exponential* families).

We shall look for upper bounds for these parameters. Note that *a priori* it is not even clear that they are finite.

Theorem 5.1.17.

$$\tilde{\lambda}^{exp} \le 1 \ .$$

Sketch of proof: For a given N let us consider a cube K with edge length $N^{1/4}$ centered at the origin and the cube K_1 of edge length $N^{1/4} + 2$ homotetic to K. Let us pack the cube K by unit spheres centered in $N^{-3/4}\mathbb{Z}$. To do this, arrange all the obtained centers in some order; note that their number is approximately N^N. Using this order, we can apply expurgation to construct the desired packing. Then we cover $\mathbb{R}^N$ by translations of K_1. ■

Exercise 5.1.18. Check that the described procedure

yields a packing P_N with the density

$$\lambda(P_N) \le \log_2(2 + N^{-1/4}) + \log_2(1 + 2N^{-1/4}) ,$$

note that $\lambda(P_N) \longrightarrow 1$ for $N \longrightarrow \infty$. Check that our algorithm is exponential in N. ∎

The situation with $\tilde{\lambda}_\ell^{exp}$ is more complicated. One can prove

Theorem 5.1.19.

$$\tilde{\lambda}_\ell^{exp} \le 1 + 8\cdot 10^{-9} .$$

We do not prove this result here, we only give a brief description of the method.

Method of proof: Let us identify the field $\mathbb{F}_p$ for $p \ne 2$ with the set

$$I_p = \left\{-\frac{p-1}{2}, \ldots, 0, \ldots, \frac{p-1}{2}\right\} \subset \mathbb{Z} .$$

For a positive integer σ and for $x \in \mathbb{F}_p^n$ set $\|x\|_\sigma = \sum|x_i|^\sigma$, where $|a|$ for $a \in \mathbb{F}_p$ denotes the absolute value of x as of an integer. Let C be a linear subspace and let $k = \dim C$, $w = \min \|c\|_\sigma$ where minimum is taken over all $c \in C - \{0\}$; C is called an $[n,k,w]_p$-"σ-code". The usual linear codes correspond to the degenerate case $\sigma = 0$. A proof of Theorem 5.1.19 can be obtained by solving the following three exercises:

Exercise 5.1.20. Establish for "σ-codes" an analogue of the Gilbert-Varshamov bound.

Exercise 5.1.21. Apply the "simplest construction" of Section 5.2.1 to "2-codes". Show that

$$det\ L = p^{n-k} ,$$

$$d(L) = min\ \{p,\sqrt{w}\} .$$

Exercise 5.1.22. Prove Theorem 5.1.19 applying the result of Exercise 5.1.21 to "2-codes" lying on the analogue of the Gilbert-Varshamov bound (for $n = w = p^2 \longrightarrow \infty$). ■

One can ameliorate Theorem 5.1.19 using nearly the same method in a more subtle way and prove

Theorem 5.1.23.

$$\tilde{\lambda}_{\ell}^{exp} \le 1 .$$ ■

Remark 5.1.24. Moreover the same result is also valid for packing any bounded convex body whose automorphisms include the reflections through the coordinate hyperplanes.

5.1.4. Codes and packings

Between codes and packings there exists a system of beautiful analogies. Indeed, one can consider an $[n,k,d]_q$-code $C \subseteq \mathbb{F}_q^n$ as the set of centers of a sphere packing (of radius $t = \left\lceil \frac{d-1}{2} \right\rceil$) in the Hamming metric. Minimum distance of a code corresponds to the diameter $d(L)$ of a sphere packing.

Linear codes correspond to lattice packings. Indeed, a linear code is a subset in $\mathbb{F}_q^n$ which is closed under addition and under multiplication by elements of $\mathbb{F}_q$, and lattice is closed under addition and multiplication by integers. Strictly speaking, we can consider "quasi-linear" codes, i.e. subsets which are closed under addition and under multiplication by elements of $\mathbb{F}_p$ (rather than $\mathbb{F}_q$) as an analogue of lattices, but we do not pursue this idea here.

Let C be a linear $[n,k,d]_q$-code. Then the volume (the cardinality) of the factor-space $\mathbb{F}_q^n/C$ equals q^{n-k}. For a lattice $L \subset \mathbb{R}^N$ the volume of the factor-space $\mathbb{R}^N/L$ equals $det\ L$, i.e. $log(det\ L)$ is an analogue of the code codimension $(n-k)$. To be definite we shall assume that in the expression $log(det\ L)$ the log symbol corresponds to the binary logarithm.

There are two possible analogues of the dimension N of a lattice (which equals its rank): the length n and the dimension k of a code. We use the first one; nevertheless we think that the second can be also of some use.

The density of a packing corresponds to the density of a packing in the Hamming metric. Note that the density of a lattice packing equals the volume of the ball of radius d divided by $det\ L$. For the density of a packing in the Hamming metric the analoguous statement is also true if we assume the ball volume to be normalized:

$$\text{the ball volume} = (\text{number of points in the ball})/q^n .$$

An analogy between code and lattice parameters is not complete. Indeed, the density of packing does not change under a homotety $L \longmapsto a \cdot L$. Hence one can assume that $d(L) = 1$ (or $det\ L = 1$) and thus a packing has two essential parameters N and Δ, whence a code has three essential parameters n, k, and d. Thus the unique asymptotic parameter λ is an analogue of the pair of code asymptotic parameters (δ, R).

An asymptotically good packing family (i.e. with $\lambda < \infty$ for $N \longrightarrow \infty$) is an analogue of an asymptotically good code family (i.e. with $R \cdot \delta > 0$ for $n \longrightarrow \infty$).

The Gilbert-Varshamov bound corresponds to the Minkowski bound; and the Hamming bound to the condition $\lambda \geq 0$. It is not clear which is a reasonable analogue of the Plotkin bound (this is an interesting question). The Kabatyansky-Levenstein bound corresponds to the McElice-Rodemich-Ramsey-Welch bound.

Packings on a sphere correspond to constant-weight codes.

An interesting question about analogies between concrete code families and lattice families is mostly open. For instance, parity check codes correspond either to lattices A_N or to D_N.

The θ-function of a lattice corresponds to the code enumerator; this analogy is quite useful.

Unimodular lattices correspond to self-dual codes.

We are interested in analogues of AG-codes. We describe some of them below in Chapter 5.4. These analogies are closely connected to a very deep analogy between algebraic curves over finite fields and algebraic number fields which we discuss in Section 5.3.3.

CHAPTER 5.2

ASYMPTOTICALLY DENSE PACKINGS

In the previous chapter we have seen that there are numerous analogies between linear codes and lattices; we have also constructed the Leech lattice using the Golay code. Here we continue to use codes to obtain packings. In Section 5.2.1 we describe some beautiful constructions which give various dense packings using various code families. Section 5.2.2 is devoted to asymptotically good packings obtained by this construction.

5.2.1. Constructions

The simplest construction. Let p be a prime, and let us embed $\mathbb{F}_p$ into $\mathbb{Z}$ as the subset $\{0, 1, \ldots, p-1\}$. Let C be a linear $[N,k,d]_p$-code; denote its image under the composition map $\mathbb{F}_p^N \hookrightarrow \mathbb{Z}^N \hookrightarrow \mathbb{R}^N$ also by C. The lattice $C\cdot\mathbb{Z} = \{a\cdot c \mid a \in \mathbb{Z},\ c \in C\}$ is of rank k. Set

$$L = C\cdot\mathbb{Z} + p\cdot\mathbb{Z}^N ,$$

i.e.

$$L = \{x \in \mathbb{Z}^N \mid x(mod\ p) \in C\} .$$

The length of $x \in L$ is either at least p (for $x \in p\cdot\mathbb{Z}^N$), or at least $\sqrt{d}$ (for $x \notin p\cdot\mathbb{Z}^N$, since in this case x has at least d non-zero entries). Thus

$$d(L) \geq min\ \{p, \sqrt{d}\} ,$$

and for $p = 2$ we have the equality.

Let as compute $det\ L$; since $det(p\cdot\mathbb{Z}^N) = p^N$ and $L/p\cdot\mathbb{Z}^N \simeq C \subseteq \mathbb{F}_p^N \simeq \mathbb{Z}^N/p\cdot\mathbb{Z}^N$, we have $|L/p\cdot\mathbb{Z}^N| = p^k$ and

$$det\ L = p^{N-k} .$$

Exercise 5.2.1. Let C be the even weight $[N,N-1,2]_2$-code. Show that the described construction gives the lattice D_N.

Exercise 5.2.2. Let C be the $[8,4,4]_2$-code obtained from the Hamming code by adding the parity check. Show that $L \simeq E_8$.

Exercise 5.2.3. Compute the density of packings obtained from

a) Golay $[24,12,8]_2$-code (compare it with that of the Leech lattice);

b) Golay $[12, 6, 6]_3$-code;

c) Genus zero codes (for various values of p and d).

Exercise 5.2.4. Show that the described simplest construction cannot give an asymptotically good family of packings.

T-lattices. Let $L \subset \mathbb{R}^m$ be a lattice and let T be the composition of an orthogonal transformation of $\mathbb{R}^m$ and of a homotety with coefficient $t = \|T\|$. We call L a *(p,b)-lattice* iff the following conditions hold:

a) T maps L into $\frac{1}{p}\cdot L$;

b) T^{-1} maps L into L ;

c) $p\cdot T = F(T^{-1})$ for some $F \in \mathbb{Z}[x]$;

d) $b = -m\cdot log_p t$ is an integer.

Remark 5.2.5. Any $L \subset \mathbb{R}^m$ is a (p,m)-lattice. Indeed, we can set $T = \frac{1}{p}$.

Example 5.2.6. For $\mathbb{Z}^2$ we can set $T = \frac{1}{2}\cdot\begin{pmatrix}1 & 1\\ 1 & -1\end{pmatrix}$. Hence $2\cdot T = T^{-1} = \begin{pmatrix}1 & 1\\ 1 & -1\end{pmatrix}$, $p = 2$, $m = 2$ and $b = 1$.

The following definition generalizes the last example.

We call a $(2,m/2)$-lattice of an even rank m a *T-lattice*. We shall see below that lattices D_4, E_8 , and Λ_{24} are T-lattices.

Remark 5.2.7. For T-lattices one can define the values $\tilde{\lambda}_T$, $\tilde{\lambda}_T^{exp}$, and $\tilde{\lambda}_T^{pol}$ in a way similar to that of Section 5.1.3. One has obvious inequalities illustrated by the following diagram (where $\longrightarrow$ stands for $\geq$).

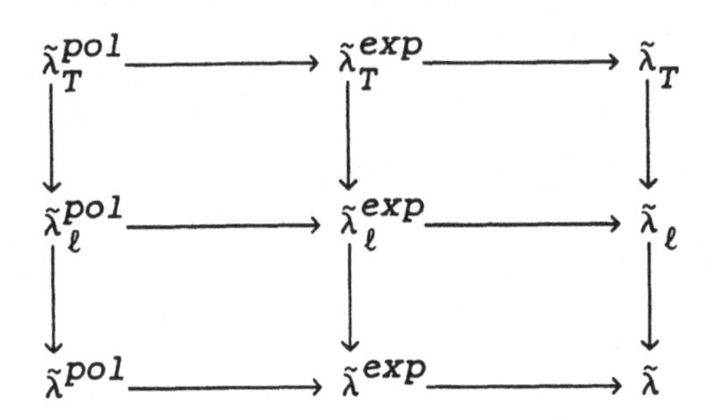

Basic construction. Let us consider a (p,b)-lattice $L \subset \mathbb{R}^m$. It is clear that L has a sublattice K of rank b such that

$$q = |K/(T^{-1}L \cap K)| = p^b .$$

Define $\psi_0 : \mathbb{F}_q \longrightarrow \mathbb{R}^m$ as the composition

$$\mathbb{F}_q \xrightarrow{\sim} \mathbb{F}_p^b \xrightarrow{\lambda^b} \mathbb{Z}^b \xrightarrow{\beta} K \hookrightarrow \mathbb{R}^m$$

where

$$\lambda : \mathbb{F}_p \xrightarrow{\sim} \{0, 1, \ldots , p-1\} \subset \mathbb{Z} ,$$

and β corresponds to a fixed basis of K. Set $\psi_i = T^i\psi_0$. We can assume that $\rho_0 = d(L)^2 \geq 1$ (rescaling if necessary).

For $q = p^b$ consider a family of nested $[n_i,k_i,d_i]_q$-codes $\mathbb{F}_q^n \supset C_1 \supset \ldots \supset C_a$ such that $d_i \cdot t^{2i} \cdot \rho_0 \geq 1$ and $C_i = \mathbb{F}_p\langle c_1,\ldots,c_{bk_i}\rangle$. Set $L_0 = L^n$, and

$$L_i = \mathbb{Z}\langle L_{i-1}, \psi_i(c_1), \ldots, \psi_i(c_{bk_i})\rangle .$$

Let us compute the parameters of these lattices. Computing the density it is convenient to use here the parameter $\nu(L) = log_2\delta(L)$.

Theorem 5.2.8. *Parameters of the lattices* L_i *obtained from a lattice* L *and a family of* $[n,k_j,d_j]_q$-*codes for* $j = 1,\ldots,a$ *and* $q = p^b$ *are :*

a) $$\min_{x\in L_i-\{0\}} |x|^2 \geq 1 ,$$

b) $$\det L_i = (\det L)^n \cdot p^{-b\cdot\sum_{j=1}^{i} k_j} ,$$

c) $$\nu(L_i) \geq n\cdot\nu(L) - b\cdot\sum_{j=1}^{i} k_j .$$

Sketch of proof: Let us use the induction over i (for $i = 0$ the theorem is obvious). On the way we check that $T^{-1} : L_i \longrightarrow L_{i-1}$ and $pT : L_i \longrightarrow L_i$. Let $x \in L_i$ and let

$$x = y + \sum_{j=1}^{b\cdot k_i} a_j\cdot T^i\psi_0(c_j)$$

for $a_j \in \mathbb{Z}$, $c_j \in C_i$ and $y \in L_{i-1}$. Since $p\cdot T$ sends L_{i-1} into itself we can assume that $0 \leq a_j \leq p-1$ for any j , i.e. that $x = y + T^i\psi_0(c)$ with $c \in C_i$. If $z \in \mathbb{F}_q^*$ then $\psi_0(z) \in K \leq L$ and since $K/(T^{-1}L \cap K) \simeq \mathbb{F}_q$ we get $\psi_0(z) \notin T^{-1}L$. Hence $x = T^i(T^{-i}y + \psi_0(c))$ where $T^{-1}y \in T^{-1}(L_0) = (T^{-1}L)^n$; let $c = (z_1,\ldots,z_m) \neq 0$ then $\psi_0(x_j)$ has at least d_i entries which do not lie in $T^{-1}L$. Therefore

$$\|x\| \geq t^{2i}\cdot d_i\cdot\rho_0 \geq 1$$

which gives a). The statement b) follows from the isomorphism $L_i/L_{i-1} \simeq C_i$; the proof of c) is straightforward. ∎

Exercise 5.2.9. Give a detailed proof of the theorem.

Now let $P \subset \mathbb{R}^m$ be a packing (may be a non-lattice one) with $\rho_0 = d(P)^2 \geq 1$. Set $\nu_0 = \nu(P)$, and let P be the union of disjoint packings P_x for $x \in \mathbb{F}_q^a$, $q = 2^b$. For $v = (v_1,\ldots,v_i) \in \mathbb{F}_q^i$, $i \leq a$ set

$$X_v = \{x = (x_1,\ldots,x_a) \in \mathbb{F}_q^a \mid (x_1,\ldots,x_i) = v\} \quad ,$$

$$P_v^{(i)} = \bigcup_{x \in X_v} P_x \quad , \quad \rho_i = d(P_v^{(i)})^2 \quad .$$

Let $C_1,\ldots,C_a \subseteq \mathbb{F}_q^n$ be a family of $[n,k_i,d_i]$-codes with $d_i \cdot \rho_i \geq 1$.

Let us consider a new packing

$$Q = \bigcup_{c_1 \in C_1} \bigcup_{c_2 \in C_2} \cdots \bigcup_{c_a \in C_a} \prod_{j=1}^{n} P_{(c_1^{(j)},\ldots,c_a^{(j)})} \subseteq \mathbb{R}^{mn} \quad ,$$

where $c_i = (c_i^{(1)},\ldots,c_i^{(n)})$.

Exercise 5.2.10. Prove that $\nu(Q) \geq n \cdot \nu_0 - b \cdot \sum_{i=1}^{a} k_i$.

If $P = L \subset \mathbb{R}^m$ is a (p,b)-lattice, we can set

$$P_x = L_x = L + \sum_{i=1}^{a} \psi_i(x_i) \quad , \quad x = (x_1,\ldots,x_a) \in \mathbb{F}_q^a \quad .$$

Exercise 5.2.11. Check that if the codes C_i are nested then we obtain the lattice of Theorem 5.2.8.

Remark 5.2.12. If the codes C_i are not nested then Q is not (in general) a lattice packing.

Experiments show that the densest lattices are obtained

from $(2,m)$-lattices; moreover T-lattices are most useful. Note that for a T-lattice the transformation T lifts as $(T,\ldots,T)$ to the obtained lattices and satisfies there all the requirement of the definition of T-lattices; i.e. all obtained lattices are also T-lattices.

Three examples. Many interesting lattices are in fact T-lattices.

Exercise 5.2.13. Apply the construction to T-lattice $\mathbb{Z}^2$ and to $[2,1,2]_2$-code $C_1 \subset \mathbb{F}_2^2$. Check that we obtain D_4. Thus D_4 is a T-lattice. Show that applying the construction to D_4 and C_1 we get E_8.

Exercise 5.2.14. Check that the following matrices define the T-lattice structure on D_4, E_8, and Λ_{24}, respectively:

$$\frac{1}{2}\cdot\begin{pmatrix} 1 & 1 & 0 & 0 \\ 1 & -1 & 0 & 0 \\ 0 & 0 & 1 & 1 \\ 0 & 0 & 1 & -1 \end{pmatrix}$$

$$\frac{1}{4}\cdot\begin{pmatrix} 1 & 1 & 1 & 1 & 1 & 1 & 1 & 1 \\ 1 & -1 & 1 & -1 & 1 & -1 & 1 & -1 \\ 1 & 1 & -1 & -1 & 1 & 1 & -1 & -1 \\ 1 & -1 & -1 & 1 & 1 & -1 & -1 & 1 \\ 1 & 1 & 1 & 1 & -1 & -1 & -1 & -1 \\ 1 & -1 & 1 & -1 & -1 & 1 & -1 & 1 \\ 1 & 1 & -1 & -1 & -1 & -1 & 1 & 1 \\ 1 & -1 & -1 & 1 & -1 & 1 & 1 & -1 \end{pmatrix}$$

$$\frac{1}{4}\cdot\left(\begin{array}{r|rrrrcr} 3 & -1 & -1 & -1 & -1 & \ldots & -1 \\ \hline -1 & -3 & 1 & 1 & 1 & \ldots & -1 \\ -1 & 1 & 1 & 1 & 1 & \ldots & -3 \\ -1 & 1 & 1 & 1 & -1 & \ldots & 1 \\ -1 & 1 & 1 & -1 & 1 & \ldots & 1 \\ \vdots & \vdots & \vdots & \vdots & \vdots & \ddots & \vdots \\ -1 & -1 & -3 & 1 & 1 & \ldots & -1 \end{array}\right),$$

where the last (24×24)-matrix is defined as follows: all the rows within the (23×23)-square are cyclic permutations of the first one which has -3 at the first position and quadratic residue symbol $\left(\frac{i-1}{23}\right)$ at i-th position.

5.2.2. Results

Let us look at asymptotic properties of lattices obtained by the basic construction.

Let there be given an infinite family of lattices $\{\Lambda_m\}$ and for any $\Lambda_m \subset \mathbb{R}^m$ a system of nested $[4^\ell, k_i, 4^{\ell-i+1}]_{2^m}$-codes C_i for $i = 1,\dots,\ell$. Then, considering Λ_m as $(2,m)$-lattices, one can construct a lattice family $\{L_N\}$, $L_N \subset \mathbb{R}^N$, $N = m \cdot 4^\ell$.

Corollary 5.2.15. *We have*

$$\lambda(\{L_N\}) \le \lambda(\{\Lambda_m\}) + \sum_{i=1}^{\infty} (1 - R_i) \quad . \blacksquare$$

Exercise 5.2.16. Check this fact.

Similarly for a family $\{\Lambda_m\}$ of T-lattices and for nested systems of $[2^\ell, k_i, 2^{\ell-i+1}]_{2^b}$-codes we obtain a family $\{L_N\}$ of T-lattices for $N = m \cdot 2^b$.

Corollary 5.2.17. *For these T-lattices*

$$\lambda(\{L_N\}) \le \lambda(\{\Lambda_m\}) + \frac{1}{2} \cdot \sum_{i=1}^{\infty} (1 - R_i) \quad . \blacksquare$$

This corollary is completely analogous to the previous one; one notes that the transformation $(T,\ldots,T)$ in $\mathbb{R}^N$ has the same properties as T .

Remark 5.2.18. If the codes are not nested then a similar fact is valid for packings.

Complexity reduction. One can use the above results to reduce the construction complexity.

Theorem 5.2.19.

a) $$\tilde{\lambda}_{\ell}^{pol} \le \tilde{\lambda}_{\ell}^{exp} + 4/3 \quad ,$$

b) $$\tilde{\lambda}_{T}^{pol} \le \tilde{\lambda}_{T}^{exp} + 1 \quad .$$

Proof: For any $\varepsilon > 0$ there exists a lattice family $\{\Lambda_m\}$ of exponential complexity with $\lambda(\{\Lambda_m\}) \le \tilde{\lambda}_{\ell}^{exp} + \varepsilon$. Applying Corollary 5.2.15 to $\{\Lambda_m\}$ and to the nested family of Reed-Solomon codes with parameters

$$[4^{m/2},\ 4^{m/2} - 4^{m/2-i+1} + 1,\ 4^{m/2-i+1}]_{2^m}$$

and taking into account that $\sum_{i=1}^{\infty} (1 - R_i) = 4/3$ we see that

$$\lambda(\{L_N\}) \le \tilde{\lambda}_{\ell}^{exp} + \varepsilon + 4/3 \quad .$$

Since $\{\Lambda_m\}$ is an exponential family and hence it is polynomial in $N = m\cdot 2^m$, the Reed-Solomon codes are polynomial, and our construction is also polynomial (i.e. the construction complexity of L_N is bounded by a polynomial in the construction complexities of Λ_m and of codes), the family we obtain is polynomial. Tending ε to 0 , we obtain a). Similarly for T-lattices Λ_m we can take nested Reed -Solomon codes with parameters

$$[2^{m/2},\ 2^{m/2} - 2^{m/2-i+1} + 1,\ 2^{m/2-i+1}]_{2^{m/2}} \ ,$$

and since $\frac{1}{2}\cdot \sum_{i=1}^{\infty} (1 - R_i) = 1$, we get b). ■

Asymptotic parameters. Let us look at the asymptotics.

Exercise 5.2.20. Using Theorem 5.1.17 check that $\tilde{\lambda}^{pol} \le 2.34$; using Theorem 5.1.23 check that $\tilde{\lambda}_\ell^{pol} \le 2.34$.

Exercise 5.2.21. Using the lattice $\mathbb{Z} \subset \mathbb{R}^1$ and nested code families lying on the Varshamov-Gilbert bound prove the following result:

Theorem 5.2.22.

$$\tilde{\lambda}_\ell^{exp} \le 1.27 \ .$$ ■

Further on we use AG-codes.

Theorem 5.2.23.

$$\tilde{\lambda}_T^{exp} \le 1.30 \ .$$

Sketch of proof: We only give a construction here. Let us apply the construction to the Leech lattice Λ_{24} (recall that it is a T-lattice) and to the following nested family of 2^{12}-ary codes. For a fixed length n let d_0 be the maximal integer such that for $d_0 \le d$ the dimension of the corresponding AG-code is more than that given by the Gilbert-Varshamov bound. We start with embedded AG-codes $C_1 \subset C_2 \subset \ldots \subset C_a$ such that $d \geq d_0$ and continue this family by the Varshamov procedure (Exercise 1.1.53) getting $C_a \subset C_{a+1} \subset \ldots \subset C_\ell$ with $d \leq d_0$. Direct calculation yields the statement. ■

Exercise 5.2.24. Give a detailed proof, in particular, compute the exact constant.

Corollary 5.2.25.

$$\tilde{\lambda}_T^{pol} \leq 2.30 \ .$$ ■

Exercise 5.2.26. Check that the construction of (non-lattice) packings of Exercise 5.2.10 being applied to the Leech lattice and to the concatenated codes with outer AG-codes and inner trivial $[n,n,1]$-codes yields

Theorem 5.2.27.

$$\tilde{\lambda}^{pol} \leq 1.31 \ .$$ ■

Asymptotical results known to us now are displayed on Diagram A.4.3.

CHAPTER 5.3

NUMBER FIELDS

In this chapter we briefly describe some properties of algebraic number fields. In Section 5.3.1 the basic definitions are given. Section 5.3.2 deals with extensions of algebraic number fields. In Section 5.3.3 we describe some analogies between algebraic number fields and fields of rational functions on curves over finite fields. These two types of fields are called *global fields;* they can be investigated by similar methods.

5.3.1. Algebraic integers

A finite extension k of $\mathbb{Q}$ is called an *algebraic number field*. Its *degree* $n = [k:\mathbb{Q}]$ equals the dimension of k as of a $\mathbb{Q}$-vector space.

If $x \in k$ satisfies the relation

$$x^m + a_{m-1} \cdot x^{m-1} + \ldots + a_1 \cdot x + a_0 = 0 \ , \quad a_i \in \mathbb{Z}$$

then x is called an *algebraic integer* or an *integral element* of k .

Proposition 5.3.1. *The sum and the product of algebraic integers are also algebraic integers.*

Proof: To begin with let us show that x is integral iff there exists a non-trivial finitely generated $\mathbb{Z}$-module $M \subset k$ such that $x \cdot M \subseteq M$. Indeed if x is integral then we can set $M = \mathbb{Z} + x \cdot \mathbb{Z} + \ldots + x^{n-1} \cdot \mathbb{Z}$; conversely if $x \cdot M \subseteq M$ and $v_1, \ldots, v_m$ is a basis in M then multiplication by x is given in this basis by a matrix A_x , and hence by the Caley-Hamilton theorem x is a root of the equation $det(\lambda - A_x) = 0$ of degree m with the leading coefficient 1 .

Let now x and y be integers, and let M and N be finitely generated $\mathbb{Z}$-submodules of k such that $x \cdot M \subseteq M$ and $y \cdot N \subseteq N$. Set

$$L = M \cdot N = \{a \cdot b \mid a \in M \ , \ b \in N\} \ .$$

It is clear that L is also a finitely generated $\mathbb{Z}$-submodule of k . Then $x \cdot y \cdot L \subseteq L$, $(x \pm y) \cdot L \subseteq L$. ∎

Corollary 5.3.2. *The subset of integral elements of* k *is a ring.* ∎

This ring O_k is called *the ring of integers* of k or its *maximal order*. Any subring $O \subseteq O_k$ of finite index $[O_k : O]$ is called an *order*.

Proposition 5.3.3. *For any $z \in k$ there exists $c \in \mathbb{Z}$ such that $c \cdot z$ is an algebraic integer.*

Proof: Since z is an algebraic number,

$$a_m z^m + a_{m-1} z^{m-1} + \ldots + a_0 = 0$$

for some $a_i \in \mathbb{Z}$, $0 \le i \le m$. Now we can take $c = a_m$; indeed one has

$$(a_m \cdot z)^m + a_m \cdot a_{m-1} \cdot (a_m \cdot z)^{m-1} + \ldots + a_m^{m-1} \cdot a_0 = 0 \quad .$$ ■

Proposition 5.3.4. *The ring O_k is finitely generated as a $\mathbb{Z}$-module.*

Sketch of proof: Let $\{w_1, \ldots, w_n\}$ be a basis of k over $\mathbb{Q}$; we can assume that $w_i \in O_k$ for $i = 1, \ldots, n$. Let us consider the linear form (the *$k/\mathbb{Q}$-trace*)

$$Tr = Tr_{k/\mathbb{Q}} : k \longrightarrow \mathbb{Q} ;$$

one can show that it is non-degenerate (see Exercise 5.3.6 below). Let $\{w'_1, \ldots, w'_n\}$ be the dual basis:

$$Tr(w'_i \cdot w_j) = \delta_{ij} = \begin{cases} 1 & \text{for } i = j \\ 0 & \text{for } i \neq j \end{cases} .$$

Let $c \in \mathbb{Z}$ be such that $c \cdot w'_i \in O_k$ for any $i = 1, \ldots, n$. Then for any $x \in O_k$ one has $c \cdot w'_i \cdot x \in O_k$ and $Tr(c \cdot w'_i \cdot x) \in \mathbb{Z}$. Hence $O_k \subseteq c^{-1} \cdot (\mathbb{Z}w_1 + \ldots + \mathbb{Z}w_n)$ and thus O_k is finitely generated. ■

Corollary 5.3.5. *Any order O is a free abelian group of rank n.*

Proof: The rank of O equals n since k/O is a torsion group. ■

Therefore we have

$$O_k = \mathbb{Z}w_1 + \ldots + \mathbb{Z}w_n$$

for some basis $\{w_1,\ldots,w_n\}$ of k over $\mathbb{Q}$: such a basis is called a *fundamental basis* of k .

Trace and norm. Let now $\Sigma = \{\sigma_1,\ldots,\sigma_n\}$ be the set of distinct embeddings of k into $\mathbb{C}$. Since for any embedding σ_i such that $\sigma_i(k)$ does not lie in $\mathbb{R}$ the embedding $\bar{\sigma}_i$ does not coincide with σ_i , these embeddings are present in the set Σ in pairs $(\sigma_i,\bar{\sigma}_i)$. Thus if s is the number of embedding $\sigma_i : k \hookrightarrow \mathbb{C}$ with $\sigma_i(k) \subset \mathbb{R}$ (such embeddings are called *real*) and t is the number of pairs $(\sigma_i,\bar{\sigma}_i)$ where $\sigma_i \neq \bar{\sigma}_i$ (such embeddings are called *complex*) then $s + 2t = n$.

Let us set

$$Tr(x) = Tr_{k/\mathbb{Q}}(x) = \sum_{i=1}^{n} \sigma_i(x) \quad ,$$

$$N(x) = N_{k/\mathbb{Q}}(x) = \prod_{i=1}^{n} \sigma_i(x) \quad .$$

$Tr(x)$ is called the $(k/\mathbb{Q})$*-trace* of x , and $N(x)$ the $(k/\mathbb{Q})$*-norm* of x .

Exercise 5.3.6. Check that if $a_m\cdot x^m + \ldots + a_0 = 0$ is the minimal equation of x over $\mathbb{Q}$ then $m|n$,

$$Tr(x) = -\,n\cdot a_{m-1}/(m\cdot a_m) \quad \text{and} \quad N(x) = (-1)^n(a_0/a_m)^{n/m} .$$

Prove that the bilinear form $Tr(x\cdot y)$ is non-degenerate. Check that $N(x) \in \mathbb{Z}$ iff $x \in O_k$.

Discriminant. Let k be an algebraic number field of degree n and let $\{w_1,\ldots,w_n\}$ be its fundamental basis. The integer $D_k = det(Tr(w_i \cdot w_j))$ is called the (absolute) *discriminant* of k.

Exercise 5.3.7. Check that this definition does not depend on the choice of $\{w_1,\ldots,w_n\}$.

Theorem 5.3.8. *If* $n > 1$, *i.e.* $k \neq \mathbb{Q}$, *then* $|D_k| > 1$. ■

One can give another definition of D_k which follows. Let s be the number of real embeddings σ_i and t be the number of conjugate pairs $(\sigma_j,\bar{\sigma}_j)$ of complex embeddings of k. Let $A = \mathbb{R}^s \times \mathbb{C}^t$ be a commutative $\mathbb{R}$-algebra of rank $n = s + 2t$, and let σ be the following ring embedding

$$\sigma : k \longrightarrow \mathbb{R}^s \times \mathbb{C}^t ,$$

$$\sigma : a \longmapsto (\sigma_1(a),\ldots,\sigma_s(a);\sigma_{s+1}(a),\ldots,\sigma_{s+t}(a)).$$

Exercise 5.3.9. Check that $\sigma(k)$ generates A (over $\mathbb{R}$); check also that $\sigma(O_k)$ is a lattice in $A \simeq \mathbb{R}^n$.

Proposition 5.3.10.

$$|det\ \sigma(O_k)| = 2^{-t}\cdot\sqrt{|D_k|} .$$

Proof: Let $\{w_1,\ldots,w_n\}$ be a fundamental basis of k, let $\sigma_j(w_i) = x_{ji} \in \mathbb{R}$ for any i, $j = 1,\ldots,s$, and let $\sigma_{s+j}(w_i) = y_{ji} + \sqrt{-1}\cdot z_{ji}$ for any $i,j = 1,\ldots,t$, where y_{ji} and $z_{ji} \in \mathbb{R}$. Then

$$d = det\ \sigma(O_k) = det\begin{pmatrix} x_{11},\ldots,x_{s1},y_{11},z_{11},\ldots,y_{t1},z_{t1} \\ \cdots\cdots\cdots\cdots \\ x_{1n},\ldots,x_{sn},y_{1n},z_{1n},\ldots,y_{tn},z_{tn} \end{pmatrix}.$$

It is clear that $d = d^*/(-2\sqrt{-1})^t$ where

$$d^* = det\begin{pmatrix} x_{11},\dots,x_{s1},y_{11} + \sqrt{-1}\cdot z_{11},y_{11} - \sqrt{-1}\cdot z_{11},\dots \\ \dots\dots\dots\dots\dots\dots\dots\dots\dots \\ x_{1n},\dots,x_{sn},y_{1n} + \sqrt{-1}\cdot z_{1n},y_{1n} - \sqrt{-1}\cdot z_{1n},\dots \end{pmatrix},$$

i.e. $d^* = det(\sigma_i(w_j))$, where $\{\sigma_1,\dots,\sigma_n\}$ is the full set of embeddings of k into $\mathbb{C}$. Since by the definition of trace

$$Tr(w_i\cdot w_j) = \sum_{\ell=1}^{n} \sigma_\ell(w_i)\cdot\sigma_\ell(w_j)$$

for $1 \le i,j \le n$, one has a matrix equality

$$(Tr(w_i\cdot w_j)) = {}^{trans}(\sigma_\ell(w_i))\cdot(\sigma_\ell(w_j))$$

where *trans* denotes transposition, whence

$$D_k = det(Tr(w_i\cdot w_j)) = (d^*)^2$$

and we are done. ∎

Units. An element $a \in O_k$ is called a *unit* iff $a^{-1} \in O_k$. Clearly all the units form a group which is denoted O_k^* . Torsion elements of O_k^* are roots of unity.

Exercise 5.3.11. Check that $a \in O_k^*$ iff $N_{k/\mathbb{Q}}(a) = \pm 1$.

The structure of the group O_k^* is rather simple:

Theorem 5.3.12. *O_k^* is the product of its finite torsion subgroup by a free abelian group of rank $r = s + t - 1$.*

Sketch of proof: Let us consider the map

$$log : O_k^* \longrightarrow \mathbb{R}^{s+t} ,$$

$$a \longmapsto (log|\sigma_1(a)|,\ldots,log|\sigma_s(a)|;log|\sigma_{s+1}(a)|^2,\ldots)$$

Its kernel is the torsion subgroup of O_k^*, and its image $log(O_k^*)$ is contained in the hyperplane $H \subset \mathbb{R}^{s+t}$ defined by $x_1 + \ldots + x_{s+t} = 0$. Indeed, Exercise 5.3.11 implies that

$$|\sigma_1(a)|\cdot\ldots\cdot|\sigma_s(a)|\cdot|\sigma_{s+1}(a)|^2\cdot\ldots\cdot|\sigma_{s+t}(a)|^2 = |N(a)| = 1 .$$

One can show that $log\ (O_k^*)$ is a lattice in H (of full rank) which gives the theorem. ∎

The determinant of this lattice

$$R = R_k = det\begin{pmatrix} log|\sigma_1(u_1)| & ,\ldots,log|\sigma_1(u_{s+t-1})| \\ \cdot\ \cdot\ \cdot\ \cdot & \cdot\ \cdot\ \cdot\ \cdot \\ log|\sigma_{s+t-1}(u_1)|, & \ldots,log|\sigma_{s+t-1}(u_{s+t-1})| \end{pmatrix}$$

where $\{u_1,\ldots,u_{s+t-1}\}$ is a basis of O_k^* modulo torsion, is called the *regulator* of k.

Places. A map $\|\cdot\| : k \longrightarrow \mathbb{R}$ is called an *absolute value* iff the following conditions hold:

$\|0\| = 0$, $\|x\| > 0$ if $x \neq 0$;

there exist $x, y \in k^*$ such that $\|x\| \neq \|y\|$;

$\|x\cdot y\| = \|x\|\cdot\|y\|$;

there exists a positive real λ such that

$$\|x + y\| \le \lambda \cdot (\|x\| + \|y\|) .$$

Two absolute values $\|\cdot\|_1$ and $\|\cdot\|_2$ are *equivalent* iff there exists a positive real θ such that $\|\cdot\|_1 = \|\cdot\|_2^{\theta}$. An equivalence class of absolute values is called a *place* of k .

There is a beautiful description of all places of a number field.

Let $\sigma : k \hookrightarrow \mathbb{C}$ be an embedding of fields. Let us put $\|x\|_{\sigma} = |\sigma(x)|$ iff σ is a real embedding (i.e. $Im\, \sigma \subset \mathbb{R}$), and $\|x\|_{\sigma} = |\sigma(x)|^2$ iff σ is a complex embedding). These are absolute values. One can check that two such absolute values $\|\cdot\|_{\sigma}$ and $\|\cdot\|_{\sigma'}$ are equivalent iff either $\sigma' = \sigma$, or $\sigma' = \overline{\sigma}$. Thus we obtain s *real* and t *complex places* of k . These places are called *infinite* or *archimedean*, the set of infinite places is denoted by S_{∞} .

Let then $\mathfrak{p}$ be a maximal ideal of O_k . For $x \in k^*$ let

$$ord_{\mathfrak{p}}(x) = max\ \{n \mid x \in \mathfrak{p}^n\} .$$

One easily checks that

$$ord_{\mathfrak{p}}(x \cdot y) = ord_{\mathfrak{p}}(x) + ord_{\mathfrak{p}}(y)$$

for any $x, y \in k^*$, and if also $x + y \in k^*$ then

$$ord_{\mathfrak{p}}(x + y) \ge min\ \{ord_{\mathfrak{p}}(x), ord_{\mathfrak{p}}(y)\} .$$

Let us define the corresponding absolute value: for $x \in k^*$ let

$$\|x\|_{\mathfrak{p}} = N(\mathfrak{p})^{-ord_{\mathfrak{p}}(x)} ,$$

where $N(\mathfrak{p}) = |O_k/\mathfrak{p}|$. For such an absolute value (and for any one equivalent to it) a stronger condition holds:

$$\|x + y\| \le \lambda \cdot max \{\|x\|, \|y\|\}$$

(which is wrong for archimedean absolute values). Such absolute values are caled *non-archimedean*. If $\mathfrak{p} \ne \mathfrak{p}'$ then the corresponding absolute values are not equivalent, i.e. each maximal ideal (each closed point of $Spec\ O_k$) corresponds to a place of k . Such places are called *finite* or *non-archimedean*.

It comes out that each place of a number field is either infinite or finite. If v is a place of k then the absolute values defined above are called *normalized* and denoted $\|\cdot\|_v$.

Let us recall that if there are no complex places, the number field is called *totally real*, if there are no real places, it is called *totally complex*.

Class group. Let $\mathfrak{a}$ be an ideal of O_k , and let $a \in k^*$. The set $\mathfrak{c} = a^{-1}\cdot\mathfrak{a}$ is called a *fractional ideal*. The set of non-zero fractional ideals is a group with the composition defined by

$$\mathfrak{a}\cdot\mathfrak{b} = \{x\cdot y \mid x \in \mathfrak{a} , y \in \mathfrak{b}\} .$$

Note that the inverse element is given by

$$\mathfrak{a}^{-1} = \{x \mid x^{-1} \in \mathfrak{a} - \{0\}\} \cup \{0\} .$$

We call fractional ideals $\mathfrak{c}$ and $\mathfrak{c}_1$ *equivalent* iff $\mathfrak{c}_1 = a\cdot\mathfrak{c}$ for some $a \in k^*$. Equivalence classes of non-zero fractional ideals form a group Cl_k which is called the (ideal) *class group* of k .

Theorem 5.3.13. *The group Cl_k is finite for any algebraic number field k .* ∎

The order of the class group $h = h_k = |Cl_k|$ is called the *class number* of k.

5.3.2. Extensions

Sometimes it is necessary to consider field extensions K/k, K and k being algebraic number fields. Let $[K:k] = dim_k K = n$ and let O_K and O_k be the rings of integers in K and k, respectively. Any $x \in K$ is a root of an irreducible over k equation of the form

$$a_m \cdot x^m + \ldots + a_0 = 0$$

where $a_i \in O$ for $i = 0,\ldots,m$.

Let us define the (*relative*) *trace* and *norm* as

$$T_{K/k}(x) = -a_{m-1}/a_m ,$$

$$N_{K/k}(x) = (-1)^m \, a_0/a_m .$$

Different. Let us consider the following subset in K:

$$\mathfrak{B}_{K/k} = \{x \in K \mid Tr_{K/k}(x \cdot y) \in O_k \text{ for any } y \in O_K\}.$$

One can easily check that $\mathfrak{B}_{K/k}$ is a O_K-submodule in K which contains O_K. Hence there exist a unique ideal $\mathfrak{D}_{K/k}$ in O_K such that $\mathfrak{D}_{K/k} \cdot \mathfrak{B}_{K/k} = O_K$. The ideal $\mathfrak{D}_{K/k}$ is called the *different* of the extensions K/k. The ideal

$$D_{K/k} = \{N_{K/k}(x) \mid x \in \mathfrak{D}_{K/k}\}$$

in O_k is called the (relative) *discriminant* of the extension K/k.

Exercise 5.3.14. Show that the relative discriminant $D_{K/\mathbb{Q}}$ equals the ideal in $\mathbb{Z}$ generated by the absolute discriminant D_k .

Thus D_k is defined by $D_{k/\mathbb{Q}}$ up to a sign.

Exercise 5.3.15. Let $L \supset K \supset k$ be algebraic number fields. Check that $\mathfrak{D}_{L/k} = \mathfrak{D}_{L/k} \cdot \mathfrak{D}_{K/k}$. Derive from this the following fact:

Proposition 5.3.16. *Let the degree of the extension* L/K *be equal to* m . *Then*

$$D_{L/k} = D_{K/k}^{m} \cdot N_{K/k}(D_{L/K}) \ . \blacksquare$$

Unramified extensions. An algebraic field extension is called *unramified* iff $D_{K/k} = (1)$. Theorem 5.3.8 (together with Exercise 5.3.14) implies that $\mathbb{Q}$ has no unramified extensions.

The rule $\mathfrak{a} \longmapsto \mathfrak{a} \cdot O_K$ defines a group homomorphism $Cl_k \longrightarrow Cl_K$; the norm map defines a homorphism $Cl_K \longrightarrow Cl_k$. If an extension K/k is unramified and abelian (i.e. normal with an abelian Galois group $Gal(K/k)$) then the (global) class field theory gives

Theorem 5.3.17. *The group* $Gal(K/k)$ *is isomorphic to the factor-group* $Cl_k/N_{K/k}(Cl_K)$. $\blacksquare$

Moreover there exists the maximal unramified abelian extension K_1 which is called the *Hilbert* or *absolute class-field* of k ; $Gal(K_1/k)$ is isomorphic to Cl_k .

Theorem 5.3.18. *Let* K_1 *be the absolute class field of an algebraic number field* k *. Then the canonical homomorphism* $Cl_k \longrightarrow Cl_{K_1}$ *is trivial, i.e. all the ideals of* O_k *become principal in* O_{K_1} *.* ■

Class field towers. As we have seen above $\mathbb{Q}$ has no unramified extensions. There exists many algebraic number fields k with $h_k > 1$; for these fields the absolute class field K_1 is an unramified extension of degree h_k . If $h_{K_1} > 1$ we get the field $K_2 = (K_1)_1$ which is by Proposition 5.3.16 an unramified extension K_2/k (note that the extension K_2/k cannot be abelian). Iterating this construction we get either

a) $h_{K_n} = 1$ for some n ; hence we cannot obtain a larger unramified extension of k by our construction; or

b) $h_{K_n} > 1$ for any n and hence we obtain an infinite unramified tower $k \subset K_1 \subset K_2 \subset \ldots \subset K_n \subset \ldots$.

An algebraic number field which satisfied the last condition is called a field with an *infinite class field tower*.

Theorem 5.3.19. *There exists a function* $f : \mathbb{N} \longrightarrow \mathbb{N}$ *such that if* k *is an algebraic number field of degree* n *and* D_k *has at least* $f(n)$ *distinct prime divisors then* k *has an infinite class field tower.* ■

One can give a precise formula for $f(n)$ but we do not need it here.

The discriminant D_k of a field satisfying the conditions of this theorem cannot be small. One can ask how to construct fields with infinite class field towers and small discriminants. To compare fields of various degrees one should use the parameter $|D_k|^{1/n}$ (note that it is constant in unramified towers). Here are the best:

Theorem 5.3.20. *The field*

$$k = \mathbb{Q}(\cos\frac{2\pi}{11}, \sqrt{-46})$$

of degree 10 *over* $\mathbb{Q}$ *has an infinite class field tower;*

$$|D_k| = 2^{15}\cdot 11^8\cdot 23^5 \quad \text{and} \quad |D_k|^{1/n} \approx 92.37 \ .$$

The field

$$k = \mathbb{Q}(\sqrt{2}, \sqrt{3\cdot 5\cdot 7\cdot 23\cdot 29})$$

of degree 4 *has an infinite class field tower of totally real fields;*

$$|D_k| = 2^8\cdot 3^2\cdot 5^2\cdot 7^2\cdot 23^2\cdot 29^2 \quad \text{and} \quad |D_k|^{1/n} \approx 1058.57. \blacksquare$$

On the other hand using "explicit formulae" (cf. Remark 2.3.23) one can obtain a lower bound for $|D_k|^{1/n}$:

Theorem 5.3.21. *Let* k_i *be algebraic number field and let* $n_i = [k_i:\mathbb{Q}] \longrightarrow \infty$. *Let* s_i *be the number of real embeddings and* t_i *the number of pairs of complex embeddings of* k_i . *Suppose that the limits* $\sigma = \lim s_i/n_i$ *and* $\tau = \lim t_i/n_i$ *do exist. Then*

$$\liminf |D_{k_i}|^{1/n_i} \geq (4\pi e^{\gamma+1})^{\sigma}\cdot(4\pi e^{\gamma})^{2\tau}$$

γ *being the Euler constant. If the generalized Riemann hypothesis is valid then*

$$\liminf |D_{k_i}|^{1/n_i} \geq (8\pi e^{\gamma+\pi/2})^{\sigma}\cdot(8\pi e^{\gamma})^{2\tau} \ . \blacksquare$$

5.3.3. Curves and number fields

Algebraic number fields and fields of rational functions on curves over finite fields are called *global fields*. They have many features in common. Here we briefly describe some of them.

Let k be an algebraic number field and let O_k be its ring of integers. Let X be a curve over $\mathbb{F}_q$, $K = \mathbb{F}_q(X)$, let F be a finite set of closed points of X, $U = X - F$, and let $O_F = \mathbb{F}_q[U]$ be the ring of rational functions which are regular on U.

For both rings O_k and O_F any factor over a maximal ideal is a finite field.

For O_F all these fields contain $\mathbb{F}_q$ (the so-called "case of equal characteristics"), in the number field case among these fields there is an extension of $\mathbb{F}_p$ for any prime p (the "case of different characteristics").

The notion of a place is in fact good for any global field. One can show that any place of $K = \mathbb{F}_q(X)$ is finite and corresponds to a closed point of X.

We can choose various finite sets F and get various rings O_F. In the number case we can choose a finite set S of maximal ideals of O_k and consider the ring O_S which is obtained from O_k by inverting non-zero elements of ideals from S; note that $Spec\, O_S = Spec\, O_k - S$ and $O_\emptyset = O_k$. Rings of the form O_S or O_F can be characterized as those having one-dimensional irreducible regular spectra of finite type over $\mathbb{Z}$.

The number field case is mostly more difficult than the

function field case. Indeed, $Spec\ O_F$ can be embedded into a proper scheme X and $Spec\ O_k$ has no "good" embedding into a proper scheme. The last fact makes it indispensable to study infinite places of $Spec\ O_k$.

The field $\mathbb{F}_q(T)$, T being a variable, is an analogue of $\mathbb{Q}$ since $k = \mathbb{F}_q(X)$ (where X is a curve over $\mathbb{F}_q$) is a finite extension of $\mathbb{F}_q(T)$; the ring $\mathbb{F}_q[T]$ is an analogue of $\mathbb{Z}$. Note, nevertheless, that there is no canonical embedding of $\mathbb{F}_q(T)$ into $\mathbb{F}_q(X)$ and hence we cannot say that $[\mathbb{F}_q(X):\mathbb{F}_q(T)]$ is an analogue of the degree of an algebraic number field.

One can suggest another analogue of the degree, namely, the number of $\mathbb{F}_q$-points of X . Indeed, if $|X(\mathbb{F}_q)| = N$, then the degree of a map $f : X \longrightarrow \mathbb{P}^1$ (i.e. the degree of an extension $[\mathbb{F}(X):\mathbb{F}_q(T)]$) can not be too small: $deg\ f \geq N/(q + 1)$, since any $\mathbb{F}_q$-point of X is mapped to an $\mathbb{F}_q$-point of $\mathbb{P}^1$ and any fiber of f contains at most $deg\ f$ $\mathbb{F}_q$-rational points.

Let a map $f : X \longrightarrow \mathbb{P}^1$ be fixed, and let us fix an $\mathbb{F}_q$-point ∞ on $\mathbb{P}^1$. Then we have $\mathbb{P}^1 - \{\infty\} = \mathbb{A}^1$, $\mathbb{F}_q[\mathbb{A}^1] = \mathbb{F}_q[T]$ and we can regard the integral closure of $\mathbb{F}_q[T]$ in $\mathbb{F}_q(X)$ as an analogue of O_k . Note that this closure coincides with $O_{F_\infty} = \mathbb{F}_q[X - F_\infty]$ where $\mathbb{F}_\infty = f^{-1}(\infty)$.

The ramification divisor B_f of the map f is an analogue of the different. The discriminant corresponds to the divisor $D = \sum e_P \cdot P$ where e_P is the ramification index of $P \in \mathbb{P}^1$. The Hurwitz formula (Theorem 2.2.36) corresponds to Proposition 5.3.16. The value $log \sqrt{|D_k|}$ is an analogue of the genus of a curve.

A fractional ideal $\mathfrak{a} = \prod_{\mathfrak{p}} \mathfrak{p}^{n_\mathfrak{p}}$, where $\mathfrak{p}$ runs over prime ideals of O_k , corresponds to a divisor on X , $\mathfrak{a}^{-1}$

corresponds to $L(D)$ and the group $Pic\ X$ is an analogue of Cl_k .

Note also that $\mathbb{F}_q(T)$ has no unramified extensions (just as $\mathbb{Q}$).

The value $\lim\inf\,(g/N)$ is an analogue of $\lim\inf \frac{\log|D_k|}{n}$. The "explicit formulae" technique gives estimates for both these values (Theorem 2.3.22, Remark 2.3.23, and Theorem 5.3.21).

In case of function fields there exists an analogue of Theorem 5.3.19, i.e. there exist infinite unramified (in some strict sense) towers (see Theorem 2.3.24.).

The question about an adequate analogue of the number of rational points on the Jacobian is rather delicate. One can suggest that its "genuine" analogue is the product $h_k \cdot R_k$ rather than h_k .

Units O_S^* of the ring O_S correspond to units O_F^* of O_F ; the group μ_k is an analogue of $\mathbb{F}_q^*$. Moreover, just as in the number field case, $O_F^*/\mathbb{F}_q^*$ is a free abelian group of rank $|F| - 1$ (note that an easy generalization of Theorem 5.3.12 says that $O_S^*/\mu_k \simeq \mathbb{Z}^{s+t+|S|-1}$).

There are some other analogies which are less clear but also useful, and we use them in the next chapter.

CHAPTER 5.4

ANALOGUES OF AG-CODES

AG-codes are constructed using curves over finite fields. As we have seen in Section 5.3.3 there is a system of analogies between curves over finite fields and algebraic number fields. These analogies lead to a number of constructions of codes and lattices. We discuss here seven constructions of this kind (the construction of AG-codes being the eighth). Each of them can be characterized by the following data: a) we use either number (N), or function (F) fields; b) we use either additive (A), or multiplicative (M) structure; c) we obtain either lattices (L), or codes (C); d) the construction either depends on a divisor (D), or not. These are the meanings of abbreviations we use below. Note that the construction of AG-codes can be abbreviated as FACD. For each construction we estimate parameters and try to produce asymptotically good families.

Section 5.4.1 is devoted to additive constructions (NAL and NAC) arising in the number field case, Section

5.4.2 contains multiplicative constructions of lattices (NML and FML). In Section 5.4.3 we discuss congruence constructions, i.e. those depending on a divisor (FMLD, NMLD, and FMCD). Section 5.4.4 contains some remarks and open problems.

In this chapter we write *log* for log_2 and *ln* for log_e .

5.4.1. Additive constructions

Additive lattices (NAL). Let k an be algebraic number field, of degree $N = s + 2t$, let O_k be its ring of integers, and let

$$\sigma : k \hookrightarrow \mathbb{R}^s \times \mathbb{C}^t$$

be the standard embedding. By Exercise 5.3.9 $L = \sigma(O_k)$ is a lattice of rank N .

Parameters. Let us compute the density of L . We have seen above (Proposition 5.3.10) that

$$det\ L = 2^{-t} \cdot \sqrt{|D_k|} .$$

Proposition 5.4.1.

$$\sqrt{s + t} \geq d(L) \geq \sqrt{s/2 + t}$$

and if $t = 0$ *then*

$$d(L) = \sqrt{N} .$$

Proof: Let

$$x = \sigma(f) = (x_1,\dots,x_s;y_1 + \sqrt{-1}\cdot z_1,\dots,y_t + \sqrt{-1}\cdot z_t).$$

We have

$$|\sigma(f)| = \sqrt{\sum_{j=1}^{s} x_j^2 + \sum_{j=1}^{t} (y_j^2 + z_j^2)} \ .$$

For $f = 1$, $|\sigma(1)| = \sqrt{s + t}$.

The arithmetic mean geometric mean inequality yields

$$\sqrt{\sum_{j=1}^{s} x_j^2 + \sum_{j=1}^{t} (y_j^2 + z_j^2)} \geq \frac{1}{\sqrt{2}}\cdot\sqrt{\sum_{j=1}^{s} x_j^2 + 2\cdot\sum_{j=1}^{t} (y_j^2 + z_j^2)} \geq$$

$$\geq \sqrt{\frac{s + 2t}{2}}\cdot\left(\prod_{j=1}^{s} x_j^2 \cdot \prod_{j=1}^{t} (y_j^2 + z_j^2)^2\right)^{1/2N} =$$

$$= \sqrt{\frac{s}{2} + t}\cdot|N_{K/\mathbb{Q}}(f)|^{1/N} \geq \sqrt{\frac{s}{2} + t} \ ,$$

since $N_{K/\mathbb{Q}}(f) \in \mathbb{Z}$. In the totally real case

$$\sqrt{\sum_{j=1}^{N} x_j^2} \geq \sqrt{N}\cdot\left(\prod_{j=1}^{N} x_j^2\right)^{1/2N} = \sqrt{N}\cdot|N_{K/\mathbb{Q}}(f)|^{1/N} \geq \sqrt{N} \ . \blacksquare$$

Unramified towers. Now let the field K vary so that $N \longrightarrow \infty$, and K is either totally real, or totally complex. Then

$$\lambda(L) \sim -log\sqrt{\frac{\pi e}{2}} + \frac{1}{N}\cdot log\sqrt{|D_K|} \ .$$

If we want to construct good lattices the last term should be bounded. It is definitely so if K runs over an unramified tower of fields over some K_0, in which case it is just constant. We get

Theorem 5.4.2. *If a number field K_0 of degree N_0 has an infinite unramified tower of fields $K \supset K_0$ which are either totally real, or totally complex, then it yields an asymptotically good family of lattices $\{L_N \subset \mathbb{R}^N\}$ with*

$$\lambda(\{L_N\}) \sim -\log\sqrt{\frac{\pi e}{2}} + \frac{1}{N_0}\cdot\log\sqrt{|D_{K_0}|}\ . \blacksquare$$

For $K_0 = \mathbb{Q}(\cos\frac{2\pi}{11}, \sqrt{-46})$ we get $\lambda \sim 2.2218$ (by Theorem 5.3.20 K_0 has an infinite class field tower).

On the other hand, Theorem 5.3.21 shows that for any family of fields K we cannot get asymptotically less than 1.193... (and 1.694... assuming the generalized Riemann hypothesis).

Exercise 5.4.3. Give an algorithm of construction of the described lattice family and estimate its complexity.

Remark 5.4.4. In contrast to the polynomial and exponential algorithms from Section 5.2.2 for lattices from Theorem 5.4.2 we know but a three times exponential algorithm. Nevertheless these lattices are interesting (their construction being quite natural).

Number field codes (NAC). The above construction can be generalized using non-archimedean places. Let $S = S_f \cup S_\infty$ be a fixed finite set of places of a number field K. For $v \in S_f$ let $k(v)$ be the residue field, for $v \in S_\infty$ let

$k(v) = \mathbb{R}$ for real places and $k(v) = \mathbb{C}$ for complex ones. Let $\mathfrak{a}$ be a fractional ideal such that $S_f \cap Supp(\mathfrak{a}) = \varnothing$, where $Supp(\mathfrak{a}) = \{\mathfrak{p} \in Spec\, O_k \mid \mathfrak{p} \supseteq \mathfrak{a}\}$. For any $v \in S$ there is a natural map

$$\sigma_v : \mathfrak{a}^{-1} = L(\mathfrak{a}) \longrightarrow k(v) .$$

Together they form a map

$$\sigma_S : L(\mathfrak{a}) \longrightarrow \bigoplus_{v \in S} k(v) .$$

Everything is quite natural, but what we get is neither a code, nor a lattice (except for the case $S = S_\infty$ discussed above). Here is a way out.

Construction. Let $[K:\mathbb{Q}] = N = s + 2t$. Fix two integers $q \geq r > 1$. Let S_f be the set of finite places v of K such that for some $c(v) \in \mathbb{Z}$

$$r \leq N(v)^{c(v)} \leq q ,$$

where $N(v) = |k(v)|$. Let $S = S_f \cup S_\infty$, S_∞ being the set of all archimedean places, let

$$n = |S| = |S_f| + s + t .$$

To make our considerations simpler we suppose that the field is totally real, i.e. $t = 0$.

Let

$$U = \{x \in \mathbb{R}^N \mid 0 < x_i < r^{a/N}\} .$$

There exists a shift U' of U such that

$$|U' \cap \sigma(O_K)| \geq r^a/\sqrt{|D_K|} ,$$

$\sigma(O_K)$ being the lattice studied above. Divide each side of the cube U' into q equal parts, and identify the set of q^N small cubes with $\mathbb{F}_q^N$. Define

$$\varphi_\infty : U' \cap \sigma(O_K) \longrightarrow \mathbb{F}_q^N$$

mapping each point to the small cube it lies in; let φ_v be the component of φ_∞ corresponding to $v \in S_\infty$. For $v \in S_f$ let φ_v be the map

$$U' \cap \sigma(O_K) \hookrightarrow O_K \longrightarrow O_K/v^{c(v)} \hookrightarrow \mathbb{F}_q \quad ,$$

where we identify the place $v \in S_f$ with the corresponding prime ideal, and $O_K/v^{c(v)} \hookrightarrow \mathbb{F}_q$ is some fixed embedding of sets.

Let

$$\varphi_S : U' \cap \sigma(O_K) \longrightarrow \mathbb{F}_q^n$$

be defined as

$$\varphi_S = (\varphi_{v_1}, \varphi_{v_2}, \ldots, \varphi_{v_n})$$

for all $v \in S$. The image

$$C = \varphi_S(U' \cap \sigma(O_K)) \subseteq \mathbb{F}_q^n$$

is a (non-linear) code of length n.

Parameters. The parameters are estimated by the following result reminding one of Theorem 3.1.1.

Proposition 5.4.5. *Let* $a \le n$. *Then*

a) $d \ge n + 1 - a$,

b) $k \ge a \cdot \log_q r - \log_q \sqrt{|D_K|}$.

Proof: a) Let $f_1, f_2 \in U' \cap \sigma(O_K)$. Set

$$A = \{v \in S_\infty \mid \varphi_v(f_1) = \varphi_v(f_2)\} ,$$

$$B = \{v \in S_f \mid \varphi_v(f_1) = \varphi_v(f_2)\} .$$

On one hand

$$|f_1 - f_2|_v \le r^{a/N}$$

for any $v \in S_\infty$, and

$$|f_1 - f_2|_v \le \frac{r^{a/N}}{q}$$

for $v \in A$. On the other hand

$$f_1 - f_2 \in v^{c(v)}$$

for any $v \in B$, and $N(v^{c(v)}) \ge r$. Let $\alpha = |A|$, $\beta = |B|$. We have

$$r^\beta \le N_{K/\mathbb{Q}}(f_1 - f_2) < \frac{r^a}{q^\alpha} \le r^{a-\alpha} .$$

Therefore $\alpha + \beta < a$, i.e. $d > n - a$.

b) We see that if $a \le n$ then φ_S is an embedding, and

$$|U' \cap \sigma(O_K)| \ge r^a/\sqrt{|D_K|} . \blacksquare$$

Proposition 5.4.5 is also valid for $t \ne 0$, but the proof is slightly more difficult.

Asymptotic behaviour. Fix q and r and consider a family of fields K of growing degree. Let

$$\gamma = \liminf_{K} \frac{\log_q \sqrt{|D_K|}}{n} ,$$

where $n = s + t + |S_f|$. We get a family of non-linear codes with $n \longrightarrow \infty$ and

$$R \gtrsim (1 - \delta) \cdot \log_q r - \gamma ,$$

if $\gamma < \log_q r$ then among them there exist asymptotically good codes.

It is possible to prove modulo the generalized Riemann hypothesis that $\gamma > (\sqrt{q} - 1)^{-1}$ (using the "explicit formulae" again), which shows that the parameters of these codes are worse than those of modular AG-codes.

On the other hand (modulo the generalized Riemann hypothesis again) for $r = \frac{q+1}{2}$ there exist fields with

$$\gamma \le const \cdot \frac{\log q}{q^{1/4}} ,$$

where *const* does not depend on q. Summing up we get

Theorem 5.4.6. *A family of number fields K of growing degree with*

$$\lim \frac{\log_q \sqrt{|D_K|}}{n} = \gamma < \log_q r ,$$

where $n = s + t + |S_f|$, S_f being the set of non-archimedean places v such that for some $c(v) \in \mathbb{Z}$ we have $r \le N_{K/\mathbb{Q}}(v)^{c(v)} \le q$ (r and q being fixed), yields *a family of asymptotically good non-linear q-ary codes with any given* δ *and*

$$R \gtrsim (1 - \delta)\cdot \log_q r - \gamma \quad .$$

If the generalized Riemann hypothesis is valid then there exist fields with

$$\gamma = const \cdot \frac{\log q}{q^{1/4}} \quad . \blacksquare$$

It is not difficult to see that for large q on some segment of δ-axis these codes are better than the Gilbert-Varshamov bound.

5.4.2. Multiplicative constructions

Multiplicative number field lattices (NML). Up to this moment we have used the additive groups of global fields. Now we are going to exploit their multiplicative structure.

Construction. We start with a number field K of degree $N = s + 2t$ and a finite number of its places $S = S_\infty \cup S_f$ which includes all archimedean ones, let $n = |S|$. Let O_S^* be the set of S-units, i.e. $a \in O_S^*$ iff all the prime divisors of its numerator and denominator belong to S_f .

There is a natural map

$$\varphi_S : O_S^* \longrightarrow \mathbb{R}^n \quad ,$$

$$\varphi_S : f \longmapsto \{\ln \|f\|_v\} \quad ,$$

where $v \in S$, and $\|\cdot\|_v$ is the normalized absolute value, i.e. $\|f\|_v = |\sigma_v(f)|$ for real places, $\|f\|_v = |\sigma_v(f)|^2$ for complex ones, and $\|f\|_v = N(v)^{-ord_v(f)}$ for $v \in S_f$. It is clear that

$$Ker\ \varphi_S = \mu_K$$

is the group of roots of 1 in K, and that

$$Im\ \varphi_S \subset H = \{x \in \mathbb{R}^n \mid \sum x_i = 0\}$$

because of the product formula.

Parameters. Let R be the regulator of K and let $h = h_K$ be its class number. Set $h(f) = \sum_v |ln\ \|f\|_v|$ for $f \in K^*$, this is the *height function* (sorry that it is denoted by the same letter as the class number); $h(f) = 0$ iff $f \in \mu_K$. We set

$$h(K) = \min_{f \in K^* - \mu_K} h(a)$$

and call it the *height of the field* K.

Proposition 5.4.7. *Let* $L_S = \varphi_S(O_S^*)$. *Then*

a) $d(L_S) \geq \frac{1}{\sqrt{n}} \cdot h(K)$,

b) $rk\ L_S = n - 1$ *and*

$$det\ L_S \leq \sqrt{n} \cdot R \cdot h \cdot \prod_{v \in S_f} ln\ N(v) .$$

Proof: a) is obvious since $\sqrt{\sum_{i=1}^{n} x_i^2} \geq \frac{1}{\sqrt{n}} \cdot \sum_{i=1}^{n} |x_i|$.

b) Let the first coordinates in $\mathbb{R}^n$ correspond to $v \in S_\infty$. Consider the orthogonal projection

$$T : H \longrightarrow H_0 = \{x \in \mathbb{R}^n \mid \sum_{i=1}^{s+t} x_i = 0\} = H_1 \oplus \mathbb{R}^{n-s-t} ,$$

where

$$H_1 = \{x \in \mathbb{R}^{s+t} \mid \sum x_i = 0\} ,$$

T multiplies volumes by $\sqrt{s+t}/\sqrt{n}$. Since $H_1 \cap T(L_S)$ is the lattice of units,

$$det\ (T(L_S) \cap H_1) = \sqrt{s+t} \cdot R$$

and

$$det\ T(L_S) = \sqrt{s+t} \cdot R \cdot det\ (pr_2 T(L_S)) .$$

Using the obvious inequality

$$[\sum_{v \in S_f} \mathbb{Z} \cdot ln\ N(v) : pr_2 T(L_S)] \le h ,$$

we get the answer. ■

Asymptotic behaviour. To obtain asymptotically good families of lattices we are going to consider unramified towers of fields. In such towers $\frac{1}{N} \cdot log \sqrt{|D_K|}$ is constant. Let us for simplicity assume that all the fields in the tower are totally real.

Theorem 5.4.8. *If a number field* K_0 *of degree* n_0 *has an infinite unramified tower of totally real fields then the above construction with* $S = S_\infty$ *yields a family of* *asymptotically good multiplicative lattices* $\{L_N = L_S \subset \mathbb{R}^N\}$ *with* $N \longrightarrow \infty$ *and*

$$\lambda(\{L_N\}) \le -\log\sqrt{\pi^3 \cdot e/2} - \log \ln\left(\frac{1+\sqrt{5}}{2}\right) + \frac{1}{n_0}\cdot \log |D_{K_0}| \ .$$

Proof: Let us first estimate $\det L_S$. Since in our case $S_f = \emptyset$, $|\mu_K| = 2$, to bound $R \cdot h$ in the tower we can use standard estimates for the residue of ζ-function of K (see Section 5.4.3 below), we get

$$R \cdot h \le 2 \cdot |D_K|/\pi^m \ ,$$

where $m = [K:\mathbb{Q}]$.

To estimate $d(L_S)$ it is sufficient to estimate the product $\prod_{v \in S} \max\{1, \|f\|_v\}$ for $f \in K^* - \{\pm 1\}$. It is known that for any totally real field K of degree m

$$\prod_{v \in S} \max\{1, \|f\|_v\} \ge \left(\frac{1+\sqrt{5}}{2}\right)^m . \tag{5.4.1}$$

Substituting these estimates into the formula for $\lambda(\{L_N\})$ we obtain the result. ■

For $K_0 = \mathbb{Q}(\sqrt{2}, \sqrt{3\cdot 5\cdot 7\cdot 23\cdot 29})$ we get $\lambda \lesssim 8.41$.

Function field lattices (FML). Here is a direct function field analogue of the previous construction.

Construction. In the notation of Part 3, let

$$O_{\mathcal{P}}^* = \{f \in K^* \mid Supp(f) \subseteq \mathcal{P}\} \ .$$

Recall that $\mathcal{P} \subseteq X(\mathbb{F}_q)$ for a curve X over $\mathbb{F}_q$ and $K = \mathbb{F}_q(X)$. Let $Div_{\mathcal{P}}(X)$ denote the group of divisors supported in $\mathcal{P}$, $Div^0_{\mathcal{P}}(X)$ of those of degree 0, $P_{\mathcal{P}}(X)$ the subgroup of principal divisors. Let $J_X = Div^0(X)/P(X)$ be the Jacobian of X.

There is a natural map

$$\varphi_{\mathcal{P}} : O^*_{\mathcal{P}} \longrightarrow Div_{\mathcal{P}}(X) \simeq \mathbb{Z}^n \quad ,$$

$$\varphi_{\mathcal{P}} : f \longmapsto (f) \quad .$$

It is clear that $Ker\ \varphi_{\mathcal{P}} = \mathbb{F}^*_q$ is again the group of roots of 1 in K, and that

$$Im\ \varphi_{\mathcal{P}} \subseteq Div^0_{\mathcal{P}}(X) \simeq A_{n-1} = \{x \in \mathbb{Z}^n \mid \sum x_i = 0\} \ .$$

We set

$$L_{\mathcal{P}} = \varphi_{\mathcal{P}}(O^*_{\mathcal{P}}) \subseteq A_{n-1} \otimes \mathbb{R} \simeq \mathbb{R}^{n-1} \ .$$

Parameters. Let us estimate the parameters of $L_{\mathcal{P}}$.

Theorem 5.4.9. *Let* $L_{\mathcal{P}} = \varphi_{\mathcal{P}}(O^*_{\mathcal{P}})$. *Then*

a) $$d(L_{\mathcal{P}}) \geq \min_{f \in O^*_{\mathcal{P}} - \mathbb{F}^*_q} \sqrt{2 \cdot deg\ f} \geq \sqrt{\frac{2 \cdot |X(\mathbb{F}_q)|}{q+1}} \quad ,$$

b) $rk\ L_{\mathcal{P}} = n - 1$ *and*

$$det\ L_{\mathcal{P}} \leq \sqrt{n} \cdot |J_X(\mathbb{F}_q)| \leq \sqrt{n} \cdot \left(1 + q + \frac{|X(\mathbb{F}_q)| - q - 1}{g}\right)^g \quad .$$

Proof: a) Let $f \in O_{\mathcal{P}}^{*}$, $f \notin \mathbb{F}_{q}^{*}$,

$$\varphi_{\mathcal{P}}(f) = (x_1,\ldots,x_n) \in \mathbb{Z}^n .$$

Then

$$|\varphi_{\mathcal{P}}(f)| = \sqrt{\sum x_i^2} \geq \sqrt{\sum |x_i|} = \sqrt{2\cdot deg\ f} ,$$

since $x_i \in \mathbb{Z}$, $\sum x_i = 0$, $deg\ f = \sum_{x_i>0} x_i$. Any $f \in K$ maps $\mathbb{F}_q$-points to $\mathbb{F}_q$-points of $\mathbb{P}^1$. Therefore

$$|X(\mathbb{F}_q)| \leq (q+1)\cdot deg\ f$$

and we get the second inequality.

b) We know that $det\ A_{n-1} = \sqrt{n}$ and

$$det\ L_{\mathcal{P}} = [A_{n-1}:L_{\mathcal{P}}]\cdot det\ A_{n-1} .$$

Then

$$A_{n-1} \simeq Div_{\mathcal{P}}^{0}(X) \subset Div^{0}(X) ,$$

and

$$L_{\mathcal{P}} \simeq P_{\mathcal{P}}(X) = P(X) \cap Div_{\mathcal{P}}^{0}(X) .$$

Therefore

$$[A_{n-1}:L_{\mathcal{P}}] \leq [Div^{0}(X):P(X)] = |J_X(\mathbb{F}_q)| .$$

To prove the second inequality it is sufficient to establish the following bound for the number of points on the Jacobian:

$$|J_X(\mathbb{F}_q)| \leq \left(1 + q + \frac{|X(\mathbb{F}_q)| - q - 1}{g}\right)^g . \qquad (5.4.2)$$

Indeed, $|J_X(\mathbb{F}_q)| = \prod_{i=1}^{2g}(1 - \omega_i)$, ω_i being the Frobenius roots, $|\omega_i| = \sqrt{q}$, $\omega_{g+i} = \bar{\omega}_i$. The arithmetic mean geometric mean inequality yields

$$\prod_{i=1}^{2g}(1 - \omega_i) = \prod_{i=1}^{g}(q + 1 - \omega_i - \bar{\omega}_i) \le$$

$$\le \left(\frac{\sum_{i=1}^{g} (q + 1 - \omega_i - \bar{\omega}_i)}{g} \right)^g ,$$

and (5.4.2) follows from

$$-\sum_{i=1}^{g} (\omega_i + \bar{\omega}_i) = |X(\mathbb{F}_q)| - q - 1 \quad . \blacksquare$$

Asymptotic behaviour. As in Part 3 we consider families of curves of growing genus with

$$\frac{|X(\mathbb{F}_q)|}{g} \longrightarrow A ,$$

and set $\mathcal{P} = X(\mathbb{F}_q)$. We get

Theorem 5.4.10. *A family of curves X over $\mathbb{F}_q$ of growing genus g such that*

$$\frac{|X(\mathbb{F}_q)|}{g} \longrightarrow A > 0$$

yields an asymptotically good family of lattices $\{L_N \subset \mathbb{R}^N\}$ with

$$\lambda(\{L_N\}) \le -\log \sqrt{\pi e} + \log \sqrt{q + 1} + A^{-1} \cdot \log (1 + q + A) \quad . \blacksquare$$

We are again interested to take the largest possible A. Let $q = p^{2m}$, then we can consider curves with $A = \sqrt{q} - 1$. Recall (cf. Exercise 2.3.27) that for such curves we can do slightly better than (5.4.2):

Proposition 5.4.11. *For a family of curves* X *over* $\mathbb{F}_q$ *with*

$$\frac{|X(\mathbb{F}_q)|}{g} \longrightarrow \sqrt{q} - 1$$

there is an asymptotic equality

$$\frac{1}{g}\cdot \log |J_X(\mathbb{F}_q)| \sim \log q + (\sqrt{q} - 1)\cdot \log \frac{q}{q-1} \ . \blacksquare$$

Using this result we get

Theorem 5.4.12. *A family of curves* X *over* $\mathbb{F}_q$ *of growing genus* g *such that*

$$\frac{|X(\mathbb{F}_q)|}{g} \longrightarrow \sqrt{q} - 1$$

yields an asymptotically good family of lattices $\{L_N \subset \mathbb{R}^N\}$ *with*

$$\lambda(\{L_N\}) \le -\log \sqrt{\pi e} + \log \frac{\sqrt{q}+1}{q-1} + \frac{\sqrt{q}}{\sqrt{q}-1}\cdot \log q \ . \blacksquare$$

For $q = 9$ we get $\lambda \lesssim 1.8687\ldots$.

Remark 5.4.13. One can easily give an exponential construction algorithm for these lattices.

Problem 5.4.14. Find a polynomial algorithm of construction for these lattices.

Remark 5.4.15. For an algebraically closed basic field k of characteristic zero one can prove that $deg\, f \geq g/2$ for any "general" curve X and for any $f \in k(X)$. If it were the case for modular curves over $\mathbb{F}_q$ (which we do not know) we would get the estimate

$$\lambda(\{L_N\}) \leq -\log\sqrt{\frac{\pi e}{2}} + \log\frac{\sqrt{\sqrt{q}+1}}{q-1} + \frac{\sqrt{q}}{\sqrt{q}-1}\cdot\log q$$

that would give $\lambda(\{L_N\}) \leq 1.18$ for $q = 16$.

Exercise 5.4.16. Generalize the construction and the density estimates to the case when $\mathcal{P}$ does not lie in $X(\mathbb{F}_q)$, i.e. using points P_i of degree more than one.

5.4.3. Congruence constructions

Now we shall discuss some constructions depending on a divisor.

Multiplicative congruence sublattices (FMLD). The FML-construction of Section 5.4.2 can be slightly elaborated. We consider some specific sublattices of $L_{\mathcal{P}}$.

Let D be a positive divisor on X , $D = \sum a_i \cdot P_i$, $r_i = deg\, P_i$, $N(P_i) = q^{r_i}$,

$$a = deg\, D = \sum a_i \cdot r_i .$$

We write $f \equiv 1(mod\, D)$ iff $ord_{P_i}(f - 1) \geq a_i$ for any $P_i \in Supp\, D$. Suppose that $\mathcal{P} \cap Supp\, D = \emptyset$. Let

$$O^*_{\mathcal{P},D} = \{f \in O^*_{\mathcal{P}} \mid f \equiv 1(mod\, D)\} ,$$

and consider the lattice $L_{\mathcal{P},D} = \varphi_{\mathcal{P}}(O^*_{\mathcal{P},D}) \subseteq L_{\mathcal{P}}$.

Parameters. Here are the estimates.

Proposition 5.4.17. *Let* $L_{\mathcal{P},D} = \varphi_{\mathcal{P}}(O^*_{\mathcal{P},D})$. *Then*

a) $d(L_{\mathcal{P},D}) \geq \sqrt{2a}$,

b) $rk\ L_{\mathcal{P},D} = n - 1$ *and*

$$det\ L_{\mathcal{P},D} \leq \sqrt{n} \cdot |J_X(\mathbb{F}_q)| \cdot \frac{q^a}{q-1} \cdot \prod (1 - q^{-r_i}) \quad .$$

Proof: a) We use the first inequality of Theorem 5.4.9.a which in our case reads

$$d(L_{\mathcal{P},D}) \geq \min_{f \in O^*_{\mathcal{P},D} - \{1\}} \sqrt{2 \cdot deg\ f} \ ,$$

and notice that $deg\ f = deg(f - 1) \geq deg\ D = a$.

b) Theorem 5.4.9.b estimates $det\ L_{\mathcal{P}}$, and we have only to estimate $[L_{\mathcal{P}}:L_{\mathcal{P},D}]$. Look at the embedding $O^*_{\mathcal{P}} \hookrightarrow \prod \hat{O}^*_{P_i}$, where $\hat{O}^*_{P_i}$ is the group of units in the completion of the local ring at P_i . Let

$$\hat{O}^*_{P_i,a_i} = \{x \in \hat{O}^*_{P_i} \mid x \equiv 1 (mod\ a_i \cdot P_i)\} \ .$$

We have $O^*_{\mathcal{P},D} = O^*_{\mathcal{P}} \cap (\prod \hat{O}^*_{P_i,a_i})$ and

$$[O^*_{\mathcal{P}}:O^*_{\mathcal{P},D}] \leq \left[\prod \hat{O}^*_{P_i} : \prod \hat{O}^*_{P_i,a_i}\right] =$$

$$= \prod \left((q^{r_i} - 1) \cdot q^{r_i \cdot (a_i - 1)} \right) \quad .$$

Then $Ker\ \varphi_{\mathcal{P}} = \mathbb{F}_q^*$ and $O_{\mathcal{P},D}^* \cap Ker\ \varphi_{\mathcal{P}} = \{1\}$, therefore

$$[O_{\mathcal{P}}^*:O_{\mathcal{P},D}^*] = (q - 1)\cdot[L_{\mathcal{P}}:L_{\mathcal{P},D}] \ . \blacksquare$$

Asymptotic behaviour. Consider the same family of curves as in Section 5.4.2, let $\mathcal{P} = X(\mathbb{F}_q)$ and let D be such that

$$\lim \frac{\deg D}{|X(\mathbb{F}_q)|} = (2\cdot \ln q)^{-1}$$

(this choice appears to be optimal). We get

Theorem 5.4.18. *A family of curves* X *over* $\mathbb{F}_q$ *of growing genus* g *such that*

$$\frac{|X(\mathbb{F}_q)|}{g} \longrightarrow \sqrt{q} - 1$$

with the appropriate choice of divisors yields an asymptotically good family of lattices $\{L_N \subset \mathbb{R}^N\}$ *with*

$$\lambda(\{L_N\}) \leq -\log\sqrt{\frac{\pi}{2}} + \frac{1}{2}\cdot\log(\ln q) + \frac{\sqrt{q}}{\sqrt{q} - 1}\cdot\log q - \log(q - 1) \ . \blacksquare$$

For $q = 2209 = 47^2$ we get $\lambda \lesssim 1.3888\ldots$.

Number field congruence sublattices (NMLD). Now we return to the number field case and discuss an analogue of FMLD-lattices.

In the notation of Section 5.4.2 let $\mathfrak{a} \subset O_K$ be an ideal,

$$\mathfrak{a} = \prod \mathfrak{p}_i^{a_i} ,$$

such that $\mathfrak{p}_i \notin S_f$ (i.e. $S_f \cap Supp\, \mathfrak{a} = \emptyset$). Let

$$O^*_{S,\mathfrak{a}} = \{ f \in O^*_S \mid f \equiv 1 (mod\ \mathfrak{a}) \}$$

and consider the lattice $L_{S,\mathfrak{a}} = \varphi_S(O^*_{S,\mathfrak{a}}) \subseteq L_S$.

Parameters. Everything is quite similar to the function field case, though the estimates are slightly worse. Let $\mu_{\mathfrak{a}} = \mu_K \cap O^*_{S,\mathfrak{a}}$.

Proposition 5.4.19. *Let* $L_{S,\mathfrak{a}} = \varphi_S(O^*_{S,\mathfrak{a}})$. *Then*

a) $d(L_{S,\mathfrak{a}}) \geq \frac{2}{\sqrt{n}} \cdot (ln\ N_{K/\mathbb{Q}}(\mathfrak{a}) - (s + 2t) \cdot ln\ 2)$,

b) $rk\ L_{S,\mathfrak{a}} = n - 1$ *and*

$$det\ L_{S,\mathfrak{a}} \leq (det\ L_S) \cdot N_{K/\mathbb{Q}}(\mathfrak{a}) \cdot \prod_i \left(1 - \frac{1}{N_{K/\mathbb{Q}}(\mathfrak{p}_i)}\right) \cdot [\mu_K : \mu_{\mathfrak{a}}]^{-1} \leq$$

$$\leq \sqrt{(s + t) \cdot (n - s - t)} \cdot R \cdot h \cdot \left(\prod_{v \in S_f} ln\ N(v) \right) \cdot N_{K/\mathbb{Q}}(\mathfrak{a}) \cdot$$

$$\cdot \left(\prod_i \left(1 - \frac{1}{N_{K/\mathbb{Q}}(\mathfrak{p}_i)}\right) \right) \cdot [\mu_K : \mu_{\mathfrak{a}}]^{-1} .$$

Proof: a) Let

$$\varphi_S(f) = (x_1, x_2, \ldots, x_n) , \quad x_i = ln\ \|f\|_{v_i} .$$

Then we see that

$$|\varphi_S(f)| = \sqrt{\sum x_i^2} \geq \frac{1}{\sqrt{n}}\cdot\sum|x_i| = \frac{2}{\sqrt{n}}\cdot\sum_{x_i>0} x_i .$$

We have

$$\sum_{x_i>0} x_i = \sum_{\substack{v\in S\\ \|f\|_v>1}} \ln \|f\|_v =$$

$$= \sum_{\substack{v\in S_\infty\\ \|f\|_v>1}} \ln \|f\|_v + \sum_{\substack{v\in S_f\\ ord_v(f-1)<0}} (-ord_v(f-1))\cdot \ln N(v) =$$

$$= \sum_{\substack{v\in S_\infty\\ \|f\|_v>1}} \ln \|f\|_v - \sum_{\substack{v\in S_\infty\\ \|f-1\|_v>1}} \ln \|f-1\|_v +$$

$$+ \sum_{\substack{v\in S_\infty\\ \|f-1\|_v<1}} (-\ln \|f-1\|_v) + \sum_{\substack{v\notin S_\infty\\ \|f-1\|_v<1}} (ord_v(f-1))\cdot \ln N(v)$$

(we have used the equality $ord_v(f-1) = ord_v f$ for $ord_v f < 0$, the fact that $\|f-1\|_v \leq 1$ for $v \notin S_\infty \cup S_f$, and the product formula). We omit the third term (it is non-negative), the fourth term is at least $\ln N_{K/Q}(\mathfrak{a})$ since $f-1 \in \mathfrak{a}$, and the sum of the first two terms is at least $-\sum_{v\in S_\infty} \ln \|2\|_v$ since

$$\max\{0, \ln \|z\|\} - \max\{0, \ln \|1-z\|\}$$

is minimum for $z = -1$ (both for $z \in \mathbb{R}$ and $z \in \mathbb{C}$).

b) Knowing Proposition 5.4.7.a we have only to estimate $[L_S:L_{S,\mathfrak{a}}]$. Note that $Ker\ \varphi_S = \mu_K$, $Ker\ \varphi_S \cap O^*_{S,\mathfrak{a}} = \mu_{\mathfrak{a}}$, and proceed as in the proof of Proposition 5.4.17.b. ■

Asymptotic behaviour. The proof of Proposition 5.4.19.a shows also that if $\log N_{K/\mathbb{Q}}(\mathfrak{a}) > s + 2t$ then $\mu_{\mathfrak{a}} = \{1\}$ since $\varphi_S(f) = 0$ for $f \in \mu_{\mathfrak{a}}$. We choose $\mathfrak{a}$ in such a way that

$$\frac{1}{n}\cdot \log N_{K/\mathbb{Q}}(\mathfrak{a}) \sim \frac{s + 2t}{n} + \log e$$

(this is in fact optimal), K runs over an unramified tower. For $R\cdot h/|\mu_K|$ we use (as above) standard estimates for the residue of ζ-function of K. They give

$$\frac{R\cdot h}{|\mu_K|} \le \frac{2\cdot |D_K|}{\pi^{s+2t}} \quad .$$

Let us choose S in such a way that $\frac{s + 2t}{n}$ is constant in the tower and that S and $Supp\ \mathfrak{a}$ completely split in it (of course this restricts the choice of the tower). We get

Theorem 5.4.20. *If a number field* K_0 *of degree* N_0 *has an infinite unramified tower in which sets of its places* S_0 *and* $Supp\ \mathfrak{a}_0$ *split completely then it yields an asymptotically good family of lattices* $\{L_N \subset \mathbb{R}^N\}$ *with*

$$\lambda(\{L_N\}) \le -\log\sqrt{\frac{2\pi}{e}} + \frac{1}{n_0}\cdot \log|D_{K_0}| - \frac{N_0}{n_0}\cdot(\log \pi - 1) +$$

$$+ \frac{1}{n_0}\cdot \sum_{v\in S_{0f}} \log(\ln N(v)) + \frac{1}{n_0}\cdot \sum_{\mathfrak{p}_i|\mathfrak{a}} \log\left(1 - \frac{1}{N_{K_0/\mathbb{Q}}(\mathfrak{p}_i)}\right) \quad ,$$

where $n_0 = |S_0|$. ■

For the totally complex field $\mathbb{Q}(\cos\frac{2\pi}{11},\sqrt{-46})$ and $S_0 = S_\infty$ we get $\lambda \lesssim 11.1512...$. For the totally real field $\mathbb{Q}(\sqrt{2},\sqrt{3\cdot5\cdot7\cdot23\cdot29})$ and $S_0 = S_\infty$ we get $\lambda \lesssim 8.80$. This are not best choices but what we get is always much worse than for FMLD-case.

Congruence codes (FMCD). The construction we are now going to expose corresponds to function field congruence lattices in the same way as number field codes correspond to additive number field lattices.

In the notation of this section suppose that $D = p\cdot D'$ for some positive divisor D' (where p is the characteristic of $\mathbb{F}_q$, $q = p^m$).

Let

$$C = L_{\mathcal{P},D}/((p\mathbb{Z})^n \cap L_{\mathcal{P},D}) \subseteq (\mathbb{Z}/p)^n = \mathbb{F}_p^n .$$

It is a p-ary code of length n . To study C we have to use the Ω-construction. Let $C' = (X,\mathcal{P},D)_\Omega$ and thus $C' = Im(Res_{\mathcal{P}})$ where

$$Res_{\mathcal{P}} : \Omega(\Sigma P_i - D) \longrightarrow \mathbb{F}_q^n$$

$$Res_{\mathcal{P}} : \omega \longmapsto (Res_{P_1}(\omega), \ ..., \ Res_{P_n}(\omega)) .$$

There is the following obvious commutative diagram

$$\begin{array}{ccccccc} O^*_{\mathcal{P},D} & \xrightarrow{\varphi_{\mathcal{P}}} & L_{\mathcal{P},D} & \longrightarrow & C & \hookrightarrow & \mathbb{F}_p^n \\ {\scriptstyle d\ log}\downarrow & & & & & & \downarrow \\ \Omega(\Sigma P_i - D) & & & \xrightarrow{Res_{\mathcal{P}}} & & & \mathbb{F}_q^n \end{array}$$

where

$$d\ log : O^{*}_{\mathcal{P},D} \longrightarrow \Omega(\sum P_i - D)$$

$$f \longmapsto df/f\ ,$$

i.e. $C = C' \cap \mathbb{F}_p^n$.

Parameters. Now we are ready to estimate parameters.

Proposition 5.4.21. *Let* $D = p \cdot D' \geq 0$ *and let*

$$C = L_{\mathcal{P},D}/((p\mathbb{Z})^n \cap L_{\mathcal{P},D}) = C' \cap \mathbb{F}_p^n \subseteq \mathbb{F}_p^n \ .$$

Then

a) $$d \geq d' \geq a - 2g + 2\ ;$$

b) $$k \geq n - 1 - m \cdot \frac{p-1}{p} \cdot a \ .$$

Proof: a) The first inequality is obvious and the second is proved in Section 3.1.1.

b) Let $B = L_{\mathcal{P},D} \cap (p\mathbb{Z})^n$. Then

$$p^k = |C| = [L_{\mathcal{P},D} : B] =$$

$$= [A_{n-1} : A_{n-1} \cap (p\mathbb{Z})^n] \cdot [A_{n-1} \cap (p\mathbb{Z})^n : B] \cdot [A_{n-1} : L_{\mathcal{P},D}]^{-1} \ .$$

We have $[A_{n-1} : A_{n-1} \cap (p\mathbb{Z})^n] = p^{n-1}$. The multiplication by p maps isomorphically A_{n-1} onto $A_{n-1} \cap (p\mathbb{Z})^n$ and

$L_{\mathcal{P},D'}$ onto B . Therefore

$$[A_{n-1} \cap (p\mathbb{Z})^n : B]\cdot[A_{n-1}:L_{\mathcal{P},D}]^{-1} =$$

$$= [A_{n-1}:L_{\mathcal{P},D'}]\cdot[A_{n-1}:L_{\mathcal{P},D'}]^{-1} = [L_{\mathcal{P},D'}:L_{\mathcal{P},D}]^{-1} .$$

Proceeding as in the proof of Proposition 5.4.17.b we see that

$$[L_{\mathcal{P},D'}:L_{\mathcal{P},D'}] \le q^{deg\ D\ -\ deg\ D'} = p^{m\cdot\frac{p-1}{p}\cdot a} .$$

Summing up we get

$$k \ge n - 1 - m\cdot\frac{p-1}{p}\cdot a .$$

Asymptotic behaviour. As usual the best results are obtained for

$$\frac{|X(\mathbb{F}_q)|}{g} \longrightarrow A ,$$

A being as large as possible (recall that $A \le \sqrt{q} - 1$).

Theorem 5.4.22. *A family of curves* X *over* $\mathbb{F}_q$ *,* $q = p^m$ *, of growing genus* g *such that*

$$\frac{|X(\mathbb{F}_{p^m})|}{g} \longrightarrow A > \frac{2\cdot m\cdot(p-1)}{p}$$

yields a family of codes such that for any

$$\delta < \frac{p}{m\cdot(p-1)} - 2\cdot A^{-1}$$

there is an asymptotically good subfamily with parameters δ *and*

$$R \gtrsim 1 - m \cdot \frac{p-1}{p} \cdot (2A^{-1} + \delta) \quad . \blacksquare$$

One easily checks that this result is a particular case of Theorem 3.4.23, but we do not know whether the construction similar to that using $L_{\mathcal{P},D}$ can be given for all codes from Theorem 3.4.23, or not.

5.4.4. Remarks and open problems

Here we list some natural remarks and questions, without any particular order.

1. What is the best constant for NMLD-construction?

2. We have mostly restricted ourselves (in the function field case) to $\mathcal{P} \subseteq X(\mathbb{F}_q)$. All constructions in fact work for any set of places of K , though places of high degree usually spoil parameters. Can places of higher degree be of any use?

3. In the number field case we usually supposed that $S \supseteq S_\infty$. Can we get anything good without this condition?

4. Each construction considered above was encoded by three or four letters. Formally speaking there are 16 possibilities. What can we say about those we have not mentioned?

5. We obtained but estimates for asymptotic parameters

of our construction. What are their true values?

6. Can we use "explicit formulae" plus some other considerations to give *lower* bounds of the density exponent of what we are able to get by our constructions?

7. In the function field case the best families of curves (those with $|X(\mathbb{F}_q)|/g \longrightarrow \sqrt{q} - 1$) are provided by modular curves which form *ramified* towers (the ramification being rather "small"). What are their analogues in the number field case?

8. The results we have obtained concern packings either in $\mathbb{F}_q^N$ or in $\mathbb{R}^N$. Our constructions also lead to natural lattices in $\mathbb{F}_q((T))^N$ and in products of p-adic fields. What are the correct parameters (how to put the problem) in those cases?

9. Function field multiplicative lattices can be also constructed starting with curves over any field, provided that we know the finiteness of the subgroup in the Jacobian generated by $\mathcal{P}$. Consider modular curves (say, over $\mathbb{C}$) and $\mathcal{P}$ consisting of cusp points which are of finite order (the Manin-Drinfeld theorem). How to estimate parameters?

10. What can be done with varieties (over $\mathbb{F}_q$ and arithmetic) of dimension more than 1? For example what are the densities of Mordell-Weil lattices on abelian varieties?

11. $K^* = K_1(K)$ and the map we have used is the regulator map. What can be done with the help of higher regulators on $K_i(K)$?

Historical and bibliographic notes to Part 5

The sphere packing problem in Euclidian spaces is quite a classical one. Many generations of mathematicians appreciated its beauty. Packings in three dimensions were studied by I.Newton.

D.Hilbert included the sphere packing problem into his eighteenth problem (and it is the last unsolved question in it, see [Mil]).

There are several excelent books devoted to packings, for example [Co/Sl], [Rog], [FT], and we address the interested reader to them. It is worth mentioning that [Co/Sl] contains a bibliography of about 1500 titles.

The A_2 packing is known to humanity for long. Its being the densest in two dimensions was proved by A.Thue and L.Fejes Tóth. The fact that A_2 and A_3 are the densest lattice packings in their dimensions was proved by K.F.Gauss. The respective result for A_4 and D_5 is due to A.M.Korkine and E.I.Zolotareff, and for E_6, E_7, E_8 to H.F.Blichfeldt. The lattice Λ_{24} was discovered by J.Leech [Lee 1], [Lee 2] (who used codes).

The root lattices of types A_N, D_N, E_N appear in many different realms of mathematics (such as Lie groups and Lie algebras, singularities, rational surfaces, etc.). The

Leech lattice looks no less important.

The Kabatiansky-Levenshtein bound [Kab/Lev] is an amelioration of those by H.F.Blichfeldt ($\tilde{\lambda} \geq 0.5$) and V.M.Sidelnikov ($\tilde{\lambda} \geq 0.5096$). The bound of Proposition 5.1.16 was found by C.A.Rogers [Rog]. Theorem 5.1.19 was proved by J.A.Rush and N.J.A.Sloane [Ru/Sl], Proposition 5.1.23 and Remark 5.1.24 are taken from an elegant paper by J.A.Rush [Ru].

Analogies from Section 5.1.4 are quite natural and many of them are mentioned in different papers (see for example [Brou]).

The problem of explicit construction of high-dimensional packings is widely studied (see [Co/Sl]). The first explicit construction of an asymptotically good family of (non-lattice) packings is due to N.J.A.Sloane [Sl 2], he has proved that $\tilde{\lambda}^{pol} \leq 6$. The basic construction of Section 5.2.1 and Theorem 5.2.8 are due to A.Bos, J.H.Conway, and N.J.A.Sloane [Bos/Co/Sl], they have constructed a family of lattices that are not asymptotically good but whose density exponent tends to infinity quite slowly. The non-lattice version of the construction is due to A.Bos [Bos].

The complexity reduction trick (Theorem 5.2.19) is discovered by S.N.Litsyn [Li/Ts 3]. This trick and algebraic-geometric codes lead him and M.A.Tsfasman [Li/Ts 3] to the best known estimates $\tilde{\lambda}^{pol} \leq 1.31$ and $\tilde{\lambda}_{\ell}^{pol} \leq 2.30$. The rest part of Section 5.2.2 is taken from this paper.

The most part of Chapter 5.3 is classical. We refer to the books [Bor/Shf], [La 4], [We 4]. The existence of infinite class field towers (Theorem 5.3.19) was proved by E.S.Golod and I.R.Shafarevich [Gol/Shf]. Theorem 5.3.20 is due to J.Martinet [Mar], who has constructed a lot of examples of fields with infinite towers and rather small discriminants. The inverse estimate (the first formula of Theorem 5.3.21) was established by A.M.Odlyzko, its refinement subject to the Riemann hypothesis (the second formula of Theorem 5.3.21) is due to J.-P.Serre (see [Poi]). Serre was also the first to notice that the inequality of Theorem 2.3.22 (proved by V.G.Drinfeld and S.G.Vladuţ [Drf/Vl]) is an exact analogue of inequalities of Theorem 5.3.21 and can be obtained by the same method ("explicit formulae", cf. Remark 2.3.23).

The modern study of anologies between the number field case and the function field case was started by A.Weil.

The constructions of Chapter 5.4 are quite natural and most of them are well-known. The more surprising is that untill recently no one ever tried to look at these lattices from the point of their density. For the additive number field latices this was first done by M.A.Tsfasman (see [Li/Ts 3]. Proposition 5.4.1 and Theorem 5.4.2 are due to him. All results on additive number field codes (including Proposition 5.4.5 and Theorem 5.4.6) are due to H.W.Lenstra,Jr. [Len]. Sections 5.4.2, 5.4.3 (multiplicative lattices) are mostly based on the results of M.Yu.Rosenbloom and M.A.Tsfasman [Ro/Ts]. The FML-construction was independently discovered by H.-G.Quebbemann [Que 2], who used his formula (5.4.2) to prove Theorem 5.4.9; this formula gives a slightly worse constant ($\lambda \sim 1.906$). The FMCD-case is due to V.D.Goppa [Go 4]. The list of questions (Section 5.4.4) is borrowed from [Ts 5] which is also a general reference for Chapter 5.4.

There is some recent progress in the direction of Question 10 of Section 5.4.4. N.Elkies [El 1], [El 2] and T.Shioda [Sh 1], [Sh 2] have shown that Mordell-Weil lattices of certain elliptic curves over function fields are very dense. Some of these lattices can be also interpreted in terms of representation theory [Grs].

APPENDIX

SUMMARY OF RESULTS AND TABLES

In this appendix we give a brief summary of some results from our book. Section A.1 contains a list of simplest bounds for parameters of a code, a list of parameters of certain families of codes, a list of parameters of certain constructions, and a table of parameters of AG-codes obtained from curves of small genera. In Section A.2 we give a list of upper and lower asymptotic bounds, we show various connections between them, and present some tables of numerical values of these bounds. Section A.3 contains tables of asymptotic bounds for constant weight codes and for self-dual codes. In Section A.4 we give tables of density for some packings, both low-dimensional and asymptotical.

A.1. Codes of finite length

A.1.1. Bounds
A.1.2. Parameters of certain codes
A.1.3. Parameters of certain constructions
A.1.4. Binary codes from AG-codes

A.2. Asymptotic bounds

A.2.1. List of bounds
A.2.2. Diagrams of comparison
A.2.3. Behaviour at the ends
A.2.4. Numerical values

A.3. Additional bounds

A.3.1. Constant weight codes
A.3.2. Self-dual codes

A.4. Sphere packings

A.4.1. Small dimensions
A.4.2. Certain families
A.4.3. Asymptotic results

A.1. Codes of finite length

Here we give a brief summary of results on parameters of q-ary codes. We begin with bounds on the volume of a code, then we describe parameters of certain families of codes, and then we give parameters of codes obtained by various constructions. We conclude the section with a table of parameters of certain binary codes obtained from AG-codes corresponding to curves of small genera.

A.1.1. Bounds

A) If there exists an $[n,k,d]_q$-code C then:

1) $k + d \le n + 1$ (the Singleton bound);

2) $$log_q \sum_{i=1}^{\lceil (d-1)/2 \rceil} \binom{n}{i} \cdot (q-1)^i \le n - k$$ (the Hamming bound);

3) $$d \le n \cdot \frac{q^k}{q^k - 1} \cdot \frac{q-1}{q}$$ (the Plotkin bound);

4) for any integer w with $1 \le w \le n$ and

$$A = d - 2 \cdot w + \frac{q \cdot w^2}{(q-1) \cdot n} > 0$$

one has

$$n - k \ge log_q \binom{n}{w} + w \cdot log_q (q-1) - log_q d + log_q A$$

(the Bassalygo-Elias bound);

5) let

$$P_i(j) = \sum_{v=0}^{i} (-1)^v \cdot (q-1)^{i-v} \cdot \binom{j}{v} \cdot \binom{n-j}{i-v} \quad ,$$

then for any $a_1, \ldots, a_n \in \mathbb{R}$, $a_i \geq 0$ such that

$$1 + \sum_{i=1}^{n} a_i \cdot P_i(j) \leq 0$$

for any $j = d, d+1, \ldots, n$, one has

$$q^k \leq 1 + \sum_{i=1}^{n} a_i \cdot \binom{n}{i} \cdot (q-1)^i$$

(the linear programming bound);

6) if C is a linear code, then

$$n \geq \sum_{i=0}^{k-1} \left\lceil \frac{d}{q^i} \right\rceil \qquad \text{(the Griesmer bound);}$$

B) Conversely, if

$$q^{n-k} > \sum_{i=0}^{d-2} \binom{n-1}{i} \cdot (q-1)^i$$

(the Gilbert-Varshamov bound)

then there exists a linear $[n, k, d]_q$-code.

A.1.2. Parameters of certain codes

We give here parameters of certain families of codes. In fact, in some cases we can not give exact values of parameters since often we do not know them; it is true at least that exact values are not worse than those given below. If a family depends on some parameters then it means that they run over all possible values such that $1 \le d \le n$ and $1 \le k \le n$.

1) Trivial codes (n is arbitrary);

a) $[n,n,1]_q$-codes;

b) the even weight code: $[n,n-1,2]_q$;

c) the repetion code: $[n,1,n]_q$.

2) Hamming codes:

$$\left[\frac{q^m - 1}{q - 1} , \frac{q^m - 1}{q - 1} - m , 3\right]_q .$$

3) Reed-Solomon codes:

$$[q + 1, k, q + 2 - k]_q .$$

4) First order projective Reed-Muller codes:

$$\left[\frac{q^m - 1}{q - 1} , m , q^{m-1}\right]_q .$$

5) Affine Reed-Muller codes (of order r):

$$\left[q^m, \sum_{i=0}^{r} \sum_{j=0}^{\lceil i/q \rceil} (-1)^j \cdot \binom{m}{j} \cdot \binom{m-1+i-qj}{m-1} , (q - b) \cdot q^{m-a-1}\right]_q ,$$

where a and b are defined by

$$r = a \cdot (q - 1) + b , 1 \le b \le q - 1 .$$

In particular, for $q = 2$:

$$\left[2^m , \sum_{i=0}^{r} \binom{m}{i} , 2^{m-r}\right]_2 .$$

6) BCH-codes (primitive):

$$\left[q^m - 1 , q^m - 1 - m\cdot\left(d - 1 - \left\lceil\frac{d-1}{q}\right\rceil\right) , d\right]_q .$$

7) AG-codes: if there exists a curve $X/\mathbb{F}_q$ of genus g such that $|X(\mathbb{F}_q)| = n$ then:

$$[n , k , n + 1 - g - k]_q$$

(note that $|X(\mathbb{F}_q)| \le q + 1 + 2g\cdot\sqrt{q}$).

8) Golay codes:

$$[24,12,8]_2 , [12,6,6]_3 .$$

9) Goppa codes:

$$[q^m + 1, q^m + 1 - m\cdot(d - 1), d]_q ,$$

and for $q = 2$:

$$\left[2^m + 1, 2^m + 1 - m\cdot\left\lceil\frac{d}{2}\right\rceil , d\right]_q .$$

10) Justesen codes:

$$\left[2m\cdot(q^m + 1) , m\cdot r , \sum_{i=1}^{\ell} i\cdot\binom{2m}{i}\cdot(q - 1)^i\right]_q ,$$

where ℓ is the largest integer such that

$$\sum_{i=1}^{\ell} \binom{2m}{i} \cdot (q - 1)^i \le n - r + 1 .$$

11) Quadratic-residue codes ($q = p$ is a prime, ℓ is another prime and p is a quadratic residue modulo ℓ):

$$\left[\ell , \frac{\ell + 1}{2} , \rceil\sqrt{\ell}\lceil\right]_p .$$

A.1.3. Parameters of certain constructions

We give here the parameters of codes obtained by certain constructions from some codes whose parameters are known; in this section ℓ and m are positive integers such that for codes obtained by our constructions we have $0 \le k \le n$, $1 \le d \le n$. If we write the sign $\le$ between parameters of codes we mean that the corresponding codes are embedded.

Name of construction	Parameters of old codes	Parameters of new codes
Lengthening by zeroes	$[n,k,d]_q$	$[n + \ell,k,d]_q$
Omitting parity-checks	$[n,k,d]_q$	$[n - \ell,k - \ell,d]_q$
Projection	$[n,k,d]_q$	$[n - \ell,k,d - \ell]_q$
Direct sum	$[n_1,k_1,d_1]_q$ $[n_2,k_2,d_2]_q$	$[n_1+n_2,k_1+k_2,d]_q$ $d = min\{d_1,d_2\}$
Power	$[n,k,d]_q$	$[\ell \cdot n,\ell \cdot k,d]_q$

Name of construction	Parameters of old codes	Parameters of new codes
Tensor product	$[n_1,k_1,d_1]_q$ $[n_2,k_2,d_2]_q$	$[n_1n_2,k_1k_2,d_1d_2]_q$
Tensor power	$[n,k,d]_q$	$[n^\ell,k^\ell,d^\ell]_q$
Pasting	$[n_1,k,d_1]_q$ $[n_2,k,d_2]_q$	$[n_1+n_2,k,\ge d_1+d_2]_q$
Repetion	$[n,k,d]_q$	$[\ell n,k,\ell d]_q$
Code from embeded pair	$[n,k,d]_q \subseteq$ $\subseteq [n,k+1,d-1]_q$	$[n+1,k+1,d]_q$
$(u\|u+v)$-construction	$[n,k_1,d_1]_q$ $[n,k_2,d_2]_q$	$[2n,k_1+k_2,d]_q$ $d = min\{2d_1,d_2\}$
Suffix construction	$[n_1,k_1,d_1]_q \subseteq$ $\subseteq [n_1,k_2,d_2]_q$ $[n_3,k_2-k_1,d_3]_q$	$[n_1+n_3,k_2,d]_q$ $d \ge min\{d_1,d_2+d_3\}$
Combination of four codes	$[n_1,k_1,d_1]_q \subseteq$ $\subseteq [n_1,k_2,d_2]_q$ $[n_3,k_3,d_3]_q \subseteq$ $\subseteq [n_3,k_4,d_4]_q$ $d_3 \ge d_1, k_2-k_1=k_4-k_3$	$[n_1+n_3,k_2+k_3,d]_q$ $d \ge min\{d_1,d_2+d_4\}$
Shortening by the distance	$[n,k,d]_q$	$[n-d,k-1,\ge \rceil \frac{d}{q} \lceil]_q$
Shortening by the dual distance	$[n,k,d]_q$	$[n-d^\perp,k-d^\perp+1,\ge d]_q$
Parity-check	$[n,k,2t + 1]_2$	$[n+1,k,2t+2]_2$
Subfield restruction	$[n,k,d]_{q^m}$	$[n,\ge n-m\cdot(n-k),\ge d]_q$
Concatenation	$[n,k,d]_{q^m}$ $[N,m,D]_q$	$[N\cdot n,k\cdot m,d\cdot D]_q$
Subfield descent	$[n,k,d]_{q^m}$	$[m\cdot n,m\cdot k,d]_q$

Name of construction	Parameters of old codes	Parameters of new codes
Generalized concatenation	$[n,k_i,d_i]_{q^{m_i}}$	$[N\cdot n,k,d]_q$,
	$\left[N,\sum_{j=1}^{i} m_j,D_i\right]_q$	$k = \sum_{i=1}^{\ell} m_i \cdot k_i$,
	$i = 1,\ldots,\ell$	
	$\left[N,\sum_{j=1}^{i+1} m_j,D_{i+1}\right]_q \geq$	$d \geq \min_{1\leq i\leq \ell} \{d_i \cdot D_i\}$
	$\geq \left[N,\sum_{j=1}^{i} m_j,D_i\right]_q$	
Alphabet extension	$[n,k,d]_r$	$[n,k\cdot \log_q r,d]_q$
	$r \leq q$	
Restriction to subalphabet	$[n,k,d]_r$	$[n,k_r,\geq d]_q$
	$r \geq q$	$k_r \geq n-(n-k)\cdot \log_q r$

A.1.4. Binary codes from AG-codes

We give parameters of certain binary codes obtained from AG-codes of genus 1, 2 or 3 by concatenation, generalized concatenation and some other constructions. The table contains only those codes which are better than the best known (before discovery of AG-codes) codes.

A) $d = 25$:

n :	91	92	93	94	95	96
k :	24	24	25	26	27	27

97	98	100	101	102
27.585	27.585	29	29.585	30.585

103	104	128	129
31	32	50.856	51.856

;

B) $d = 27$:

n :	159
k :	69

;

C) $d = 29$:

n :	151	152	153	154	155	156
k :	58	59	60	61	61.087	62

157	158	159	160	161	162
63	64	65	65.087	66	67

163	164	165	166	168
68	69	69.087	69.585	70.459

169	170
71.044	72.044

.

A.2. Asymptotic bounds

AG-codes made it possible to obtain many new asymptotic lower bounds for the possible volume of a code. Here we give some diagrams which illustrate the relations between these bounds; we give also tables for asymptotic behaviour of bounds for $\delta \longrightarrow 0$ and $\delta \longrightarrow 1 - 1/q$ and tables of numerical values of these bounds (computed by A.M.Barg). For simplicity we assume that q is a power of a prime.

A.2.1. List of bounds

A) "True" bounds:

α_q is the "true" bound for q-ary codes (Theorem 1.3.1).

α_q^{lin} is the "true" bound for linear q-ary codes (Exercise 1.3.2).

α_q^{pol} is the "true" bound for polynomial families of q-ary codes (Theorem 1.3.25).

$\alpha_q^{pol,lin}$ is the "true" bound for polynomial families of linear q-ary codes (Theorem 1.3.25).

$\alpha_q^{pol\ dec}$ is the "true" bound for families of q-ary codes with polynomial complexity of construction and decoding (Exercise 3.4.33).

$\alpha_q^{pol\ dec,lin}$ is the "true" bound for families of lineary q-ary codes with polynomial complexity of construction and decoding (Exercise 3.4.33).

B) Upper bounds:

R_S is the Singleton bound:

$$R_S(\delta) = 1 - \delta \quad ;$$

R_P is the Plotkin bound (Theorem 1.3.6):

$$R_P(d) = 1 - \frac{q}{q-1}\cdot\delta \quad ;$$

R_H is the Hamming bound (Theorem 1.3.8):

$$R_H(\delta) = 1 - H_q(\delta/2) \quad ;$$

R_{BE} is the Bassalygo-Elias bound (Theorem 1.3.10):

$$R_{BE}(\delta) = 1 - H_q\left(\frac{q-1}{q} - \frac{q-1}{q}\cdot\sqrt{1 - \frac{q\cdot\delta}{q-1}}\right) \quad ;$$

R_4 is the bound of "four" (McElice-Rodemich-Ramsey-Welch bound, Theorem 1.3.11):

$$R_4(\delta) = H_q\left(\frac{q - 1 - \delta\cdot(q-2) - 2\sqrt{(q-1)\cdot\delta\cdot(1-\delta)}}{q}\right) \quad ;$$

$R_{4(2)}$ is the second bound of "four" for $q = 2$ (Theorem 1.3.12):

$$R_{4(2)}(\delta) = \min_{0<u\le 1-2\delta} \{1 + h(u^2) - h(u^2 + 2\cdot\delta\cdot u + 2\cdot\delta)\} \quad ,$$

where

$$h(x) = H_2\left(\frac{1 - \sqrt{1-x}}{2}\right) \quad ;$$

C) Lower bounds:

R_{GV} is the Gilbert-Varshamov bound (Theorem 1.3.15):

$$R_{GV}(\delta) = 1 - H_q(\delta) \quad ;$$

R_{BZ} is the Bloch-Zyablov bound (Remark 1.3.33):

$$R_{BZ}(\delta) = R_{GV}(\delta) - \delta \cdot \int_0^{R_{GV}(\delta)} \frac{dx}{R_{GV}^{-1}(x)} \quad ,$$

where R_{GV}^{-1} is the inverse function of R_{GV} ;

R_{TVZ} is the Tsfasman-Vladuţ-Zink bound (the basic AG-bound) for q which is an even power of a prime (Corollary 3.4.3):

$$R_{TVZ}(\delta) = 1 - (\sqrt{q} - 1)^{-1} - \delta \quad ;$$

R_V^0 is the Vladuţ bound (expurgatory bound, Theorem 3.4.11), q being a square, it is given by

$$(R_V^0(\delta) + (\sqrt{q} - 1)^{-1}) \cdot H_q\left(\frac{1 - \delta}{R_V^0(\delta) + (\sqrt{q} - 1)^{-1}}\right) + H_q(\delta) =$$

$$= 1 + (\sqrt{q} - 1)^{-1}$$

and is defined for $\delta \in [\delta_1', \delta_2'] \cup [\delta_3', \delta_4']$, where $\delta_1' < \delta_4'$ are the roots of the equation

$$H_q(\delta) + \frac{q}{q-1} \cdot (1 - \delta) = 1 + (\sqrt{q} - 1)^{-1} \quad ,$$

and $\delta'_2 < \delta'_3$ are the roots of the equation

$$H_q(\delta) + (1 - \delta)\cdot log_q(q - 1) = 1 + (\sqrt{q} - 1)^{-1}$$

$$R_V(\delta) = \begin{cases} R_V^0(\delta) & \text{for } \delta \in [\delta'_1,\delta'_2]\cup[\delta'_3,\delta'_4] \\ R_{GV}(\delta) & \text{for } \delta \in [0,\delta'_1]\cup[\delta'_4,1 - \frac{1}{q}] \\ R_{TVZ}(\delta) & \text{for } \delta \in [\delta'_2,\delta'_3] \ ; \end{cases}$$

R_{KTV} and $R_{KTV(lin)}$ are the Katsman-Tsfasman-Vladuţ bounds (concatenation bounds, Theorem 3.4.16)

$$R_{KTV}(\delta) = max \left\{(1 - (q^{k/2} - 1)^{-1})\cdot\frac{k}{n} - \frac{k}{d}\cdot\delta\right\} ,$$

where maximum is taken over all $[n,k,d]_q$-codes q^k being a square; $R_{KTV(lin)}(\delta)$ is defined similarly with maximum taken over linear codes;

R_{KT} is the Katsman-Tsfasman bound (restriction bound, Theorem 3.4.23)

$$R_{KT}(\delta) = \max_m \left\{1 - \frac{2m\cdot(q - 1)}{q\cdot(q^{m/2} - 1)} - \frac{m\cdot(q - 1)}{q}\cdot\delta\right\} ,$$

where maximum is taken over all positive integers m such that q^m is a square;

$$R_{KTV,KT}(\delta) = max\{R_{KTV}(\delta),R_{KT}(\delta)\} \ ;$$

$$R_{KTV(lin),KT}(\delta) = max\{R_{KTV(lin)}(\delta),R_{KT}(\delta)\} \ ;$$

$\tilde{R}_{SV}$ and $\tilde{R}_{SV(lin)}$ are the Skorobogatov-Vladuţ bounds (decoding bounds, Theorems 3.4.37 and 3.4.38)

$$\tilde{R}_{SV}(\delta) = max\{R'_{SV}(\delta), R''_{SV}(\delta)\} ,$$

$$\tilde{R}_{SV(lin)}(\delta) = max\{R'_{SV(lin)}(\delta), R''_{SV}(\delta)\} ,$$

where

$$R'_{SV}(\delta) = max \left\{(1 - 2\cdot(q^{k/2} - 1)^{-1})\cdot\frac{k}{n} - \frac{k}{d}\cdot\delta\right\} ,$$

and maximum is taken over all $[n,k,d]_q$-codes such that q^k is a square, $R'_{SV(lin)}(\delta)$ is defined in a similar way with maximum taken over linear codes,

$$R''_{SV}(\delta) = \max_m \left\{1 - \frac{3m\cdot(q - 1)}{q\cdot(q^{m/2} - 1)} - \frac{m\cdot(q - 1)}{q}\cdot\delta\right\} ,$$

where maximum is taken over all positive integers m such that q^m is a square;

R_{LT} is the Litsyn-Tsfasman bound (Theorem 3.4.31)

$$R_{LT}(\delta) = \max_m\{R'_{LT}(\delta), R''_{LT}(\delta)\} ;$$

where

$$R'_{LT}(\delta) = \max_{q'}\{1 - (1 - R_V^{(q')}(\delta))\cdot log_q(q')\} ,$$

here the maximum is taken over all even powers of primes $q' \geq q$, $R_V^{(q')}$ being the expurgatory bound for q'-ary codes,

$$R''_{LT}(\delta) = \max_{q'}\{R_V^{(q')}(\delta)\cdot \log_q(q')\} ,$$

the maximum is taken over all even powers of primes $q' \le q$.

A.2.2. Diagrams of comparison

In these diagrams we compare various bounds.

The notation $A \longrightarrow B$ means that $A(\delta) \ge B(\delta)$ for any $\delta \in \left[0, \frac{q-1}{q}\right]$ and we can prove this inequality.

The notation $A \dashrightarrow B$ means that $A(\delta) \ge B(\delta)$ for any $\delta \in \left[0, \frac{q-1}{q}\right]$ such that both $A(\delta)$ and $B(\delta)$ are known to us, but we have no proof of this inequality in general.

A) Comparison of bounds for $q = 2$

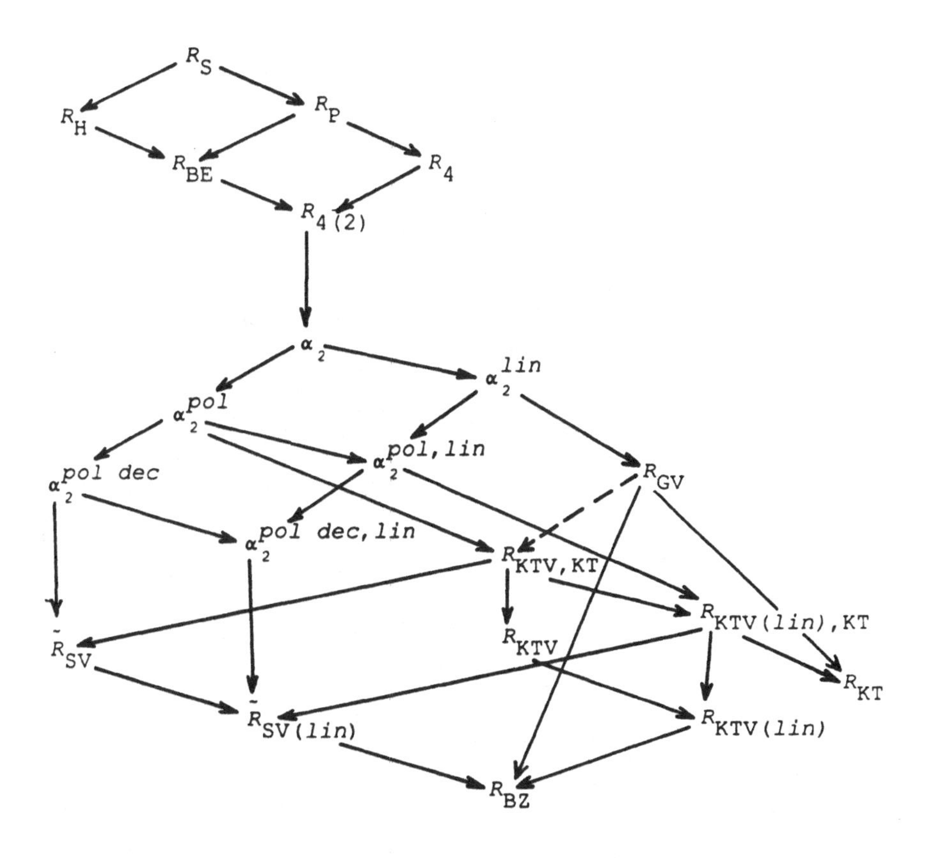

Comments: One can prove that $R_{GV}(\delta) > R_{KTV,KT}(\delta)$ for all δ except in a very small interval.

B) Comparison of bounds for $2 < q = p^{2m} < 49$

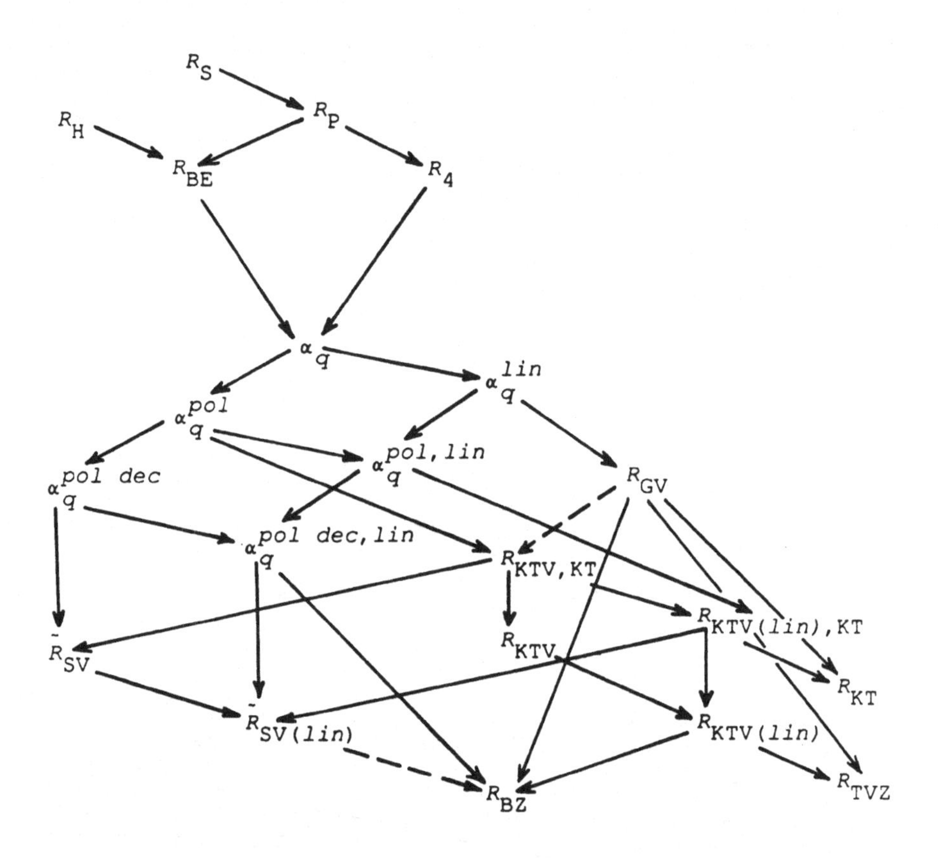

Comments: 1. If $q = 4$ (and also if $q = 3$) then we also have $R_S \longrightarrow R_H$.

2. If $2 < q = p^m < 49$ and m is odd, this diagram should be slightly changed. One should then set $R_{TVZ}(\delta) = 1 - \gamma_q - \delta$ where γ_q is not quite known. One should also consider R_{LT} which plays an important role, e.g. for $q = 47$ it intersects R_{GV} .

3. One can also consider the case of $2 < q < 49$, $q \neq p^m$. Then there is no proper notion of linear bounds, and non-linear constructions leading to R_{LT} play a crucial role.

C) Comparison of bounds for $q = p^{2m} \geq 49$

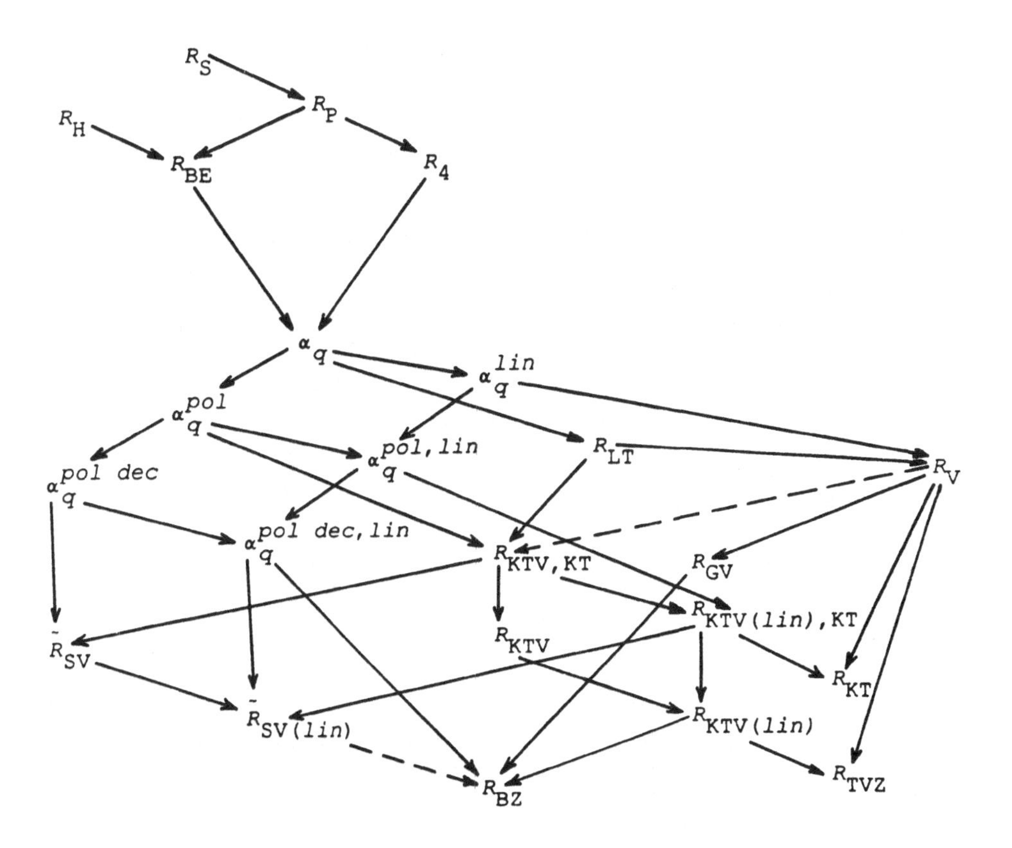

Comments: If $49 \leq q = p^{2m} \leq 289$ we also have $R_{GV} \longrightarrow R_{KT}$ and $R_{GV} \dashrightarrow \tilde{R}_{SV}$, which is wrong if $q \geq 361$.

A.2.3. Behaviour at the ends

Investigating bounds it is useful to know their behaviour for $\delta \longrightarrow 0$ and $\delta \longrightarrow 1 - 1/q$. The following tables display this behaviour for some bounds. For a bound $R(\delta)$ we give an asymptotic behaviour of $1 - R(\delta)$ for $\delta \longrightarrow 0$ and of $R\left(\frac{q-1}{q} - x\right)$ for $x \longrightarrow 0$.

For the bounds $\tilde{R}_{SV}$, $\tilde{R}_{SV(lin)}$, R_{KTV}, $R_{KTV(lin)}$ we do not know the "true" asymptotics, but we can bound them. Thus the "true" asymptotics are not worse than the asymptotics given in the tables.

A) Behaviour of bounds for binary codes at 0 and 1/2 .

Bound	$1 - R(\delta)$ for $\delta \longrightarrow 0$	$R\left(\frac{1}{2} - x\right)$ for $x \longrightarrow 0$
R_{BZ}	$\frac{\ln 2}{2}\cdot\delta\cdot(\log_2\delta)^2$	$\frac{4}{3\cdot\ln 2}\cdot x^3$
$\tilde{R}_{SV}$, $\tilde{R}_{SV(lin)}$	$-\delta\cdot\log_2\delta$	$-\frac{288}{343}\cdot x^3\cdot\log_2 x$
R_{KTV}, $R_{KTV(lin)}$	$-2\cdot\delta\cdot\log_2\delta$	$-\frac{1152}{343}\cdot x^3\cdot\log_2 x$
R_{KT}	$-\delta\cdot\log_2\delta$	< 0
R_{GV}	$-\delta\cdot\log_2\delta$	$\frac{2}{\ln 2}\cdot x^2$
$R_{4(2)}$	$-\frac{1}{2}\cdot\delta\cdot\log_2\delta$	$-2\cdot x^2\cdot\log_2 x$
R_4	$\frac{2}{\ln 2}\cdot\delta$	$-2\cdot x^2\cdot\log_2 x$
R_{BE}	$-\frac{1}{2}\cdot\delta\cdot\log_2\delta$	$\frac{1}{\ln 2}\cdot x$
R_H	$-\frac{1}{2}\cdot\delta\cdot\log_2\delta$	> 0
R_P	$2\cdot\delta$	$2\cdot x$
R_S	δ	> 0

B) Behaviour of bounds for q–ary codes at 0 and $\frac{q-1}{q}$.

Bound	$1 - R(\delta)$ for $\delta \longrightarrow 0$	$R\left(\frac{q-1}{q} - x\right)$ for $x \longrightarrow 0$
R_{BZ}	$\frac{\ln q}{2}\cdot\delta(\log_q\delta)^2$	$\frac{q^3}{6(q-1)^2\cdot\ln q}\cdot x^3$
$\tilde{R}_{SV}$, $\tilde{R}_{SV(lin)}$	$-2\cdot\frac{q-1}{q}\cdot\delta\cdot\log_q\delta$	$-\frac{(\sqrt{q}-1)\cdot q^4}{2(\sqrt{q}^3-1)^3}\cdot x^3\cdot\log_q x$ for $q = p^{2m}$, $-\frac{(q+1)^2\cdot q^6}{2\cdot(q^3-1)^3}\cdot x^3\cdot\log_q x$ for $q = p^{2m+1}$
R_{KTV}, $R_{KTV(lin)}$	$-2\cdot\delta\cdot\log_q\delta$	$-2\cdot\frac{(\sqrt{q}-1)\cdot q^4}{(\sqrt{q}^3-1)^3}\cdot x^3\cdot\log_q x$ for $q = p^{2m}$, $-2\cdot\frac{(q+1)^2\cdot q^6}{(q^3-1)^3}\cdot x^3\cdot\log_q x$ for $q = p^{2m+1}$
R_{KT}	$-2\cdot\frac{q-1}{q}\cdot\delta\cdot\log_q\delta$	< 0
R_{GV}	$-\delta\cdot\log_q\delta$	$\frac{q^2}{2\cdot(q-1)\cdot\ln q}\cdot x^2$
R_4	$\frac{2}{\ln q}\cdot\delta$	$-2\cdot x^2\cdot\log_2 x$
R_{BE}	$-\frac{1}{2}\cdot\delta\cdot\log_q\delta$	$\frac{q}{2\cdot\ln q}\cdot x$
R_H	$-\frac{1}{2}\cdot\delta\cdot\log_q\delta$	> 0
R_P	$\frac{q}{q-1}\cdot\delta$	$\frac{q}{q-1}\cdot x$
R_S	δ	> 0

A.2.4. Numerical values

In the following tables we give numerical values of certain bounds.

The values of three upper bounds (R_H , R_{BE} and R_4) are displayed (and for $q = 2$ the values of $R_{4(2)}$ are also displayed). For $q = 2, 4, 16$ we give values of five lower bounds (R_{BZ} , $\tilde{R}_{SV}$, R_{KTV} , R_{KT} and R_{GV}), and for $q = 64, 256$ of seven lower bounds (R_{BZ} , $\tilde{R}_{SV}$, R_{TVZ} , R_{KTV} , R_{KT} , R_V^0 , and R_{GV}). Note also that the given values for $\tilde{R}_{SV}$ and R_{KTV} are lower bounds rather than exact values, since their definitions involve maximization over all codes of certain type which we do not know.

The difference between the value of a bound and the displayed value is not greater than 10^{-4} . Note that we do not display R_{LT} in our tables since its deviation from R_V (if any) is always less than 10^{-4} . When the value is very small we write aE-b for $a \cdot 10^{-b}$.

$q = 2$

	Lower bounds					Upper bounds			
δ	R_{BZ}	$\tilde{R}_{SV}$	R_{KTV}	R_{KT}	R_{GV}	R_H	R_{BE}	R_4	$R_{4(2)}$
.0001	.9896	.9980	.9973	.9980	.9985	.9992	.9992	.9997	.9992
.001	.9546	.9836	.9801	.9842	.9886	.9938	.9938	.9971	.9938
.01	.7778	.8739	.8677	.8804	.9192	.9546	.9544	.9712	.9542
.02	.6609	.7738	.7884	.7847	.8586	.9192	.9185	.9427	.9180
.03	.5738	.7138	.7298	.6973	.8056	.8876	.8862	.9143	.8851
.04	.5039	.6538	.6798	.6173	.7577	.8586	.8562	.8862	.8544
.05	.4456	.6004	.6298	.5398	.7136	.8313	.8279	.8582	.8251
.06	.3959	.5504	.5896	.4698	.6726	.8056	.8008	.8305	.7971
.07	.3529	.5004	.5496	.3998	.6341	.7811	.7748	.8030	.7699
.08	.3152	.4504	.5096	.3298	.5978	.7577	.7498	.7758	.7436
.09	.2819	.4104	.4696	.2695	.5635	.7352	.7255	.7487	.7179
.10	.2524	.3704	.4347	.2095	.5310	.7136	.7019	.7219	.6927
.11	.2261	.3304	.4047	.1495	.5001	.6927	.6789	.6954	.6681
.12	.2025	.2904	.3747	.0895	.4706	.6726	.6565	.6691	.6439
.13	.1813	.2600	.3447	.0295	.4426	.6530	.6345	.6431	.6201
.14	.1622	.2467	.3147		.4158	.6341	.6130	.6173	.5966
.15	.1450	.2333	.2847		.3902	.6157	.5920	.5919	.5734
.16	.1294	.2200	.2547		.3657	.5978	.5713	.5667	.5506
.18	.1027	.1933	.2267		.3199	.5635	.5310	.5172	.5055
.20	.0808	.1667	.2000		.2781	.5310	.4920	.4690	.4614
.22	.0630	.1400	.1733		.2398	.5001	.4541	.4221	.4179
.24	.0485	.1133	.1467		.2050	.4706	.4172	.3767	.3750
.26	.0367	.0867	.1200		.1733	.4426	.3812	.3328	.3326
.28	.0273	.0667	.0933		.1445	.4158	.3461	.2906	.2906
.30	.0198	.0530	.0733		.1187	.3902	.3117	.2502	.2502
.32	.0140	.0444	.0533		.0956	.3657	.2781	.2118	.2118
.34	.0096	.0358	.0388		.0752	.3423	.2451	.1754	.1754
.36	.0062	.0273	.0302		.0573	.3199	.2127	.1414	.1414
.38	.0038	.0187	.0217		.0420	.2985	.1808	.1100	.1100
.39	.0029	.0144	.0174		.0352	.2882	.1651	.0954	.0954
.40	.0022	.0101	.0132		.0290	.2781	.1495	.0815	.0815
.41	.0015	.0058	.0097		.0235	.2682	.1340	.0684	.0684
.42	.0011	.0019	.0058		.0185	.2585	.1187	.0561	.0561
.43	.0007	.0015	.0046		.0142	.2491	.1035	.0448	.0448
.44	.0004	.0011	.0033		.0104	.2398	.0884	.0345	.0345
.45	.0003	.0007	.0021		.0072	.2308	.0734	.0253	.0253
.46	.0001	.0003	.0009		.0046	.2220	.0585	.0172	.0172
.47	5E-5	.0002	.0005		.0026	.2134	.0437	.0104	.0104
.48	2E-5	5E-5	.0002		.0012	.2050	.0290	.0051	.0051
.49	2E-6	7E-6	2E-5		.0003	.1967	.0145	.0015	.0015

$q = 4$

	Lower bounds						Upper bounds		
δ	R_{BZ}	$\tilde{R}_{SV}$	R_{KTV}	R_{KT}	R_{GV}		R_H	R_{BE}	R_4
.0001	.9920	.9985	.9986	.9985	.9992		.9996	.9996	.9999
.001	.9706	.9880	.9890	.9882	.9935		.9965	.9965	.9985
.01	.8481	.9143	.9241	.9103	.9517		.9733	.9733	.9851
.02	.7599	.8483	.8677	.8386	.9134		.9517	.9514	.9698
.04	.6330	.7355	.7733	.7129	.8472		.9134	.9124	.9388
.06	.5388	.6512	.6933	.6023	.7887		.8790	.8770	.9073
.08	.4637	.5899	.6171	.4973	.7355		.8472	.8437	.8756
.10	.4015	.5299	.5571	.4071	.6863		.8172	.8120	.8437
.12	.3489	.4796	.5067	.3171	.6402		.7887	.7815	.8117
.14	.3039	.4296	.4667	.2331	.5969		.7616	.7520	.7796
.16	.2648	.3796	.4267	.1581	.5560		.7355	.7234	.7475
.18	.2308	.3333	.3867	.0831	.5173		.7104	.6954	.7154
.20	.2009	.2933	.3467	.0081	.4805		.6863	.6681	.6834
.22	.1746	.2641	.3129		.4456		.6629	.6413	.6515
.24	.1515	.2441	.2829		.4123		.6402	.6149	.6196
.26	.1311	.2241	.2543		.3806		.6183	.5890	.5879
.28	.1130	.2041	.2343		.3504		.5969	.5635	.5564
.30	.0971	.1841	.2143		.3216		.5762	.5383	.5251
.32	.0830	.1641	.1943		.2942		.5560	.5133	.4940
.34	.0706	.1441	.1743		.2681		.5364	.4887	.4631
.36	.0597	.1241	.1543		.2434		.5173	.4643	.4325
.38	.0501	.1041	.1343		.2198		.4987	.4401	.4023
.40	.0418	.0841	.1143		.1975		.4805	.4161	.3725
.42	.0345	.0641	.0943		.1764		.4628	.3923	.3430
.44	.0282	.0573	.0743		.1565		.4456	.3686	.3140
.46	.0228	.0506	.0663		.1378		.4287	.3451	.2855
.48	.0182	.0439	.0596		.1202		.4123	.3216	.2576
.50	.0143	.0373	.0529		.1038		.3962	.2983	.2304
.52	.0110	.0306	.0463		.0885		.3806	.2750	.2038
.54	.0083	.0239	.0396		.0744		.3653	.2517	.1781
.56	.0061	.0173	.0329		.0614		.3504	.2285	.1532
.58	.0043	.0106	.0263		.0496		.3358	.2053	.1294
.60	.0030	.0042	.0196		.0390		.3216	.1820	.1067
.62	.0019	.0025	.0129		.0296		.3077	.1587	.0853
.64	.0012	.0016	.0063		.0215		.2942	.1353	.0654
.66	.0006	.0011	.0033		.0146		.2810	.1117	.0473
.68	.0003	.0006	.0017		.0089		.2681	.0878	.0313
.70	.0001	.0002	.0007		.0046		.2556	.0637	.0178
.72	2E-5	5E-5	.0002		.0017		.2434	.0390	.0074
.74	9E-7	3E-6	1E-5		.0002		.2314	.0135	.0010

$q = 16$

	Lower bounds					Upper bounds		
δ	R_{BZ}	$\tilde{R}_{SV}$	R_{KTV}	R_{KT}	R_{GV}	R_H	R_{BE}	R_4
.0001	.9932	.9992	.9992	.9991	.9995	.9998	.9998	.9999
.001	.9796	.9935	.9940	.9926	.9962	.9980	.9980	.9993
.01	.8922	.9522	.9561	.9440	.9700	.9838	.9837	.9920
.02	.8246	.9122	.9241	.8971	.9451	.9700	.9699	.9834
.04	.7220	.8483	.8641	.8033	.9004	.9451	.9446	.9654
.06	.6418	.7883	.8133	.7420	.8595	.9221	.9210	.9466
.08	.5751	.7283	.7733	.6857	.8213	.9004	.8985	.9271
.10	.5178	.6683	.7333	.6295	.7851	.8796	.8767	.9072
.12	.4676	.6267	.6933	.5732	.7505	.8595	.8555	.8868
.14	.4232	.5867	.6533	.5170	.7172	.8401	.8348	.8661
.16	.3834	.5467	.6133	.4607	.6851	.8213	.8144	.8451
.18	.3475	.5067	.5733	.4125	.6542	.8030	.7943	.8238
.20	.3151	.4667	.5333	.3750	.6242	.7851	.7745	.8022
.24	.2587	.3867	.4533	.3000	.5668	.7505	.7356	.7583
.28	.2117	.3067	.3867	.2250	.5127	.7172	.6972	.7136
.32	.1722	.2578	.3467	.1500	.4614	.6851	.6593	.6682
.36	.1390	.2178	.3067	.0750	.4127	.6542	.6217	.6223
.40	.1111	.1778	.2667		.3666	.6242	.5842	.5758
.44	.0877	.1378	.2267		.3228	.5951	.5468	.5288
.48	.0681	.1128	.1867		.2815	.5668	.5092	.4816
.50	.0596	.1028	.1667		.2616	.5530	.4904	.4578
.54	.0450	.0858	.1267		.2237	.5259	.4526	.4102
.58	.0330	.0705	.0867		.1881	.4996	.4144	.3625
.62	.0235	.0563	.0572		.1549	.4739	.3757	.3149
.66	.0160	.0449	.0458		.1242	.4489	.3363	.2674
.70	.0103	.0335	.0344		.0960	.4246	.2961	.2205
.72	.0080	.0277	.0297		.0829	.4127	.2755	.1973
.74	.0061	.0229	.0257		.0705	.4009	.2546	.1743
.75	.0052	.0209	.0237		.0646	.3951	.2441	.1629
.76	.0045	.0189	.0217		.0589	.3893	.2334	.1517
.78	.0032	.0149	.0177		.0481	.3779	.2117	.1294
.80	.0022	.0109	.0137		.0381	.3666	.1895	.1077
.82	.0014	.0069	.0097		.0291	.3554	.1666	.0867
.84	.0008	.0029	.0057		.0210	.3444	.1430	.0666
.86	.0004	.0006	.0023		.0140	.3336	.1183	.0477
.88	.0002	.0003	.0005		.0081	.3228	.0924	.0304
.90	5E-5	.0001	.0003		.0037	.3123	.0646	.0156
.92	6E-6	1E-5	3E-5		.0009	.3019	.0338	.0044
.93	5E-7	5E-7	4E-6		.0002	.2967	.0161	.0010

$q = 64$

	Lower bounds							Upper bounds		
δ	R_{BZ}	$\tilde{R}_{SV}$	R_{TVZ}	R_{KTV}	R_{KT}	R_V^0	R_{GV}	R_H	R_{BE}	R_4
.0001	.9936	.9994	.8570	.9995	.9994		.9997	.9998	.9998	.9999
.001	.9828	.9955	.8561	.9958	.9948		.9971	.9985	.9985	.9995
.01	.9087	.9661	.8471	.9680	.9589		.9766	.9874	.9874	.9941
.02	.8493	.9361	.8371	.9441	.9294		.9565	.9766	.9765	.9876
.04	.7569	.8883	.8171	.9041	.8703		.9198	.9565	.9561	.9734
.06	.6830	.8483	.7971	.8641	.8203		.8857	.9377	.9369	.9582
.08	.6203	.8083	.7771	.8241	.7809		.8533	.9198	.9183	.9423
.10	.5658	.7683	.7571	.7841	.7416		.8222	.9025	.9002	.9258
.12	.5173	.7283	.7371	.7441	.7022		.7922	.8857	.8824	.9089
.14	.4738	.6883	.7171	.7171	.6628		.7632	.8693	.8649	.8916
.16	.4344	.6483	.6971	.6971	.6234		.7349	.8533	.8477	.8738
.18	.3985	.6083	.6771	.6771	.5841		.7073	.8376	.8306	.8558
.20	.3657	.5683	.6571	.6571	.5447		.6804	.8222	.8136	.8374
.24	.3077	.4883	.6171	.6171	.4825	.6284	.6284	.7922	.7800	.7998
.28	.2582	.4343	.5771	.5771	.4431	.5797	.5785	.7632	.7465	.7611
.32	.2157	.3943	.5371	.5371	.4038		.5305	.7349	.7132	.7216
.36	.1792	.3543	.4971	.4971	.3644		.4842	.7073	.6799	.6811
.40	.1477	.3143	.4571	.4571	.3250		.4397	.6804	.6465	.6399
.44	.1206	.2743	.4171	.4171	.2856		.3967	.6541	.6128	.5980
.48	.0973	.2343	.3771	.3771	.2462		.3553	.6284	.5788	.5553
.50	.0870	.2143	.3571	.3571	.2266		.3352	.6157	.5616	.5337
.54	.0686	.1743	.3171	.3171	.1872		.2961	.5908	.5270	.4901
.58	.0531	.1343	.2771	.2771	.1478		.2586	.5663	.4917	.4458
.62	.0402	.0943	.2371	.2371	.1084		.2227	.5423	.4557	.4010
.66	.0295	.0543	.1971	.1971	.0691		.1884	.5188	.4188	.3556
.70	.0208	.0290	.1571	.1571	.0297		.1558	.4956	.3807	.3097
.72	.0172	.0253	.1371	.1371	.0100	.1402	.1402	.4842	.3612	.2865
.74	.0140	.0217	.1171	.1171			.1250	.4730	.3413	.2633
.76	.0112	.0181	.0971	.0971			.1104	.4618	.3210	.2400
.78	.0088	.0144	.0771	.0771			.0963	.4507	.3001	.2166
.80	.0067	.0124	.0571	.0571			.0827	.4397	.2787	.1931
.82	.0050	.0104	.0371	.0371			.0698	.4288	.2567	.1697
.84	.0035	.0083	.0171	.0171			.0575	.4180	.2338	.1462
.86	.0024	.0065		.0065			.0459	.4073	.2101	.1229
.88	.0015	.0048		.0048			.0351	.3967	.1852	.0997
.90	.0009	.0035		.0035			.0252	.3862	.1588	.0769
.92	.0004	.0022		.0023			.0165	.3758	.1305	.0546
.94	.0002	.0012		.0013			.0090	.3655	.0995	.0334
.96	3E-5	.0002		.0003			.0033	.3553	.0638	.0143
.98	2E-7	4E-7		2E-6			.0001	.3452	.0170	.0009

$q = 256$

	Lower bounds							Upper bounds		
δ	R_{BZ}	$\tilde{R}_{SV}$	R_{TVZ}	R_{KTV}	R_{KT}	R_V^0	R_{GV}	R_H	R_{BE}	R_4
.0001	.9938	.9996	.9932	.9996	.9995		.9997	.9999	.9999	.9999
.001	.9844	.9965	.9323	.9968	.9956		.9976	.9987	.9987	.9996
.01	.9173	.9722	.9233	.9761	.9687		.9799	.9893	.9893	.9950
.02	.8622	.9522	.9133	.9561	.9445		.9623	.9799	.9798	.9891
.04	.7751	.9122	.8933	.9161	.9047		.9297	.9623	.9620	.9761
.06	.7046	.8722	.8733	.8761	.8649		.8991	.9457	.9450	.9621
.08	.6443	.8322	.8533	.8533	.8250		.8698	.9297	.9284	.9474
.10	.5912	.7922	.8333	.8333	.7852		.8414	.9142	.9122	.9321
.12	.5438	.7522	.8133	.8133	.7477	.8142	.8139	.8991	.8962	.9162
.14	.5010	.7267	.7933	.7933	.7277		.7871	.8843	.8804	.9000
.16	.4619	.7067	.7733	.7733	.7078		.7608	.8698	.8648	.8833
.18	.4261	.6867	.7533	.7533	.6879		.7351	.8555	.8492	.8664
.20	.3931	.6667	.7333	.7333	.6680		.7099	.8414	.8337	.8491
.24	.3344	.6267	.6933	.6933	.6281		.6608	.8139	.8028	.8136
.28	.2838	.5867	.6533	.6533	.5883		.6133	.7871	.7719	.7771
.32	.2399	.5467	.6133	.6133	.5484		.5672	.7608	.7409	.7397
.36	.2016	.5067	.5733	.5733	.5086		.5224	.7351	.7098	.7014
.40	.1683	.4667	.5333	.5333	.4688		.4789	.7099	.6784	.6622
.44	.1393	.4267	.4933	.4933	.4289		.4366	.6851	.6466	.6223
.50	.1028	.3667	.4333	.4333	.3691		.3754	.6488	.5980	.5609
.54	.0826	.3267	.3933	.3933	.3293		.3360	.6250	.5648	.5191
.58	.0652	.2867	.3533	.3533	.2895		.2977	.6016	.5309	.4765
.62	.0504	.2467	.3133	.3133	.2496		.2607	.5786	.4961	.4333
.66	.0380	.2067	.2733	.2733	.2098		.2249	.5559	.4603	.3892
.70	.0277	.1667	.2333	.2333	.1699		.1903	.5335	.4231	.3445
.72	.0233	.1467	.2133	.2133	.1500		.1736	.5224	.4039	.3219
.74	.0193	.1267	.1933	.1933	.1301		.1572	.5114	.3843	.2990
.76	.0158	.1067	.1733	.1733	.1102		.1412	.5005	.3643	.2760
.78	.0127	.0867	.1533	.1533	.0902		.1255	.4897	.3436	.2527
.80	.0100	.0667	.1333	.1333	.0703		.1103	.4789	.3223	.2293
.82	.0077	.0467	.1133	.1133	.0504		.0956	.4682	.3003	.2057
.84	.0057	.0267	.0933	.0933	.0305		.0813	.4576	.2774	.1819
.86	.0041	.0067	.0733	.0733	.0105		.0676	.4471	.2535	.1579
.88	.0028	.0041	.0533	.0533			.0545	.4366	.2284	.1337
.90	.0017	.0028	.0333	.0333			.0420	.4262	.2017	.1094
.92	.0010	.0018	.0133	.0133			.0304	.4159	.1729	.0849
.94	.0005	.0010		.0010			.0197	.4057	.1414	.0604
.96	.0002	.0005		.0005			.0104	.3955	.1055	.0362
.98	2E-5	.0001		.0001			.0030	.3854	.0613	.0132
.99	1E-6	2E-5		2E-5			.0006	.3803	.0314	.0033

A.3. Additional bounds

Here we give some numerical values for constant weight codes and self-dual codes (cf. Section 3.4.4).

A.3.1. Constant weight codes

In this table (borrowed form [Er/Zi 1]) two bounds for constant weight codes are compared:

$R_G(\omega,\delta)$ - the "Gilbert bound", i.e. the random choice bound:

$$R_G(\omega,\delta) = H_2(\omega) - \omega\cdot H_2\left(\frac{\delta}{2\omega}\right) - (1-\omega)\cdot H_2\left(\frac{\delta}{2-2\omega}\right) \quad ;$$

and

$R_{EZ}(\omega,\delta)$ - the Ericson-Zinoviev bound obtained from R_{TVZ} for $\omega = p^{-2m}$

$$R_{EZ}(\omega,\delta) = -\left(\omega - \frac{\delta}{2} - \frac{\omega\cdot\sqrt{\omega}}{1-\sqrt{\omega}}\right)\cdot log_2\omega$$

For $\omega = p^{-2m} \le 1/81$ the values of $\kappa = 1 - \delta/2\omega$ are given such that for any $\delta \in (2\omega\cdot(1-\kappa_2), 2\omega\cdot(1-\kappa_1))$ one has

$$R_{EZ}(\omega,\delta) > R_G(\omega,\delta)$$

and the difference $\Delta(\omega,\delta) = R_{EZ}(\omega,\delta) - R_G(\omega,\delta)$ is maximal for $\delta = \delta_0 = 2\omega\cdot(1-\kappa_0)$; the values $R_{EZ}(\omega,\delta_0)$, $R_G(\omega,\delta_0)$, $\Delta(\omega,\delta_0)$ and $(\Delta(\omega,\delta_0)/R_G(\omega,\delta_0))\cdot 100\%$ are also included.

ω^{-1}	κ_1	κ_2	κ_0	$R_{EZ}(\omega,\delta_0)$	$R_G(\omega,\delta_0)$	$\Delta(\omega,\delta_0)$	%
81	.290	.482	.386	.02043	.01994	.00049	2.4
121	.250	.567	.385	.01630	.01512	.00117	7.7
169	.164	.609	.384	.01317	.01188	.00129	10.8
256	.127	.647	.383	.00989	.00870	.00118	13.6
289	.119	.657	.383	.00907	.00794	.00112	14.1
361	.105	.672	.383	.00771	.00670	.00101	15.0
529	.085	.694	.383	.00576	.00495	.00080	16.2
625	.078	.702	.382	.00506	.00434	.00072	16.6
729	.071	.709	.382	.00448	.00384	.00065	16.9
841	.066	.715	.382	.00400	.00342	.00058	17.0

A.3.2. Self-dual codes

Since for any self-dual or quasi-self-dual code $R = 0.5$, their asymptotical behaviour is determined by the value of δ. Here we give (for several different q) the numerical values of the Scharlau bound

$$\delta_{Sch} = \frac{1}{2} - \frac{1}{\sqrt{q} - 3}$$

and (for comparison) those of the basic AG-bound and of the Gilbert-Varshamov bound, i.e. δ_{TVZ} and δ_{GV} such that

$$R_{TVZ}(\delta_{TVZ}) = R_{GV}(\delta_{GV}) = \frac{1}{2}$$

Let us recall that δ_{Sch} is a lower bound for quasi-self-dual codes and for $q = 2^m$ it is a lower bound for self-dual codes.

q	25	49	64	81	121	169	256	65536
δ_{Sch}	0	.25	.3	.3333	.375	.4	.4231	.4960
δ_{GV}	.3113	.3375	.3462	.3532	.3639	.3718	.3805	.4382
δ_{TVZ}	.25	.3333	.3571	.375	.4	.4167	.4333	.4961

A.4. Sphere packings

We give here tables of densities for certain sphere packings in $\mathbb{R}^N$: for lattices with $N \le 8$ (cf. Section 5.1.1), for certain interesting families of lattices (cf. Section 5.1.2) and for the asymptotic behaviour (cf. Sections 5.1.3, 5.2.2 and Chapter 5.4).

A.4.1. Small dimensions

The densest lattice packings are known in dimensions up to 8. These are the root lattices

$$A_1 = \mathbb{Z}\ ,\ A_2\ ,\ A_3 = D_3\ ,\ D_4\ ,\ D_5\ ,\ E_6\ ,\ E_7\ ,\ E_8 = \Gamma_8\ .$$

We give here approximate values of various parameters which characterize the lattice density.

N	1	2	3	4	5	6	7	8
$\Delta_\ell(N)$	1	0.907	0.740	0.617	0.465	0.373	0.295	0.254
$\delta_\ell(N)$	0.5	0.229	0.177	0.125	0.088	0.072	0.063	0.063
$\lambda_\ell(N)$	0	0.070	0.144	0.174	0.221	0.237	0.251	0.247
$\nu_\ell(N)$	-1	-1.792	-2.5	-3	-3.5	3.792	-4	-4
$\gamma_\ell(N)$	1	1.155	1.260	1.414	1.516	1.665	1.811	2
$\delta_\ell(N)^{-2}$	4	12	32	64	128	192	256	256

A.4.2. Certain families

Let us give the values of $\delta(L_N)$ and $\gamma(L_N)$ for some interesting lattices $L_N \subset \mathbb{R}^N$:

L_N	$\delta(L_N)$	$\gamma(L_N)$
$\mathbb{Z}^N$	2^{-N}	1
A_N	$\frac{2^{-N/2}}{\sqrt{N+1}}$	$\frac{2}{\sqrt[N]{N+1}}$
D_N, $N \geq 3$	$\frac{2^{-N/2}}{2}$	$\frac{2}{\sqrt[N]{4}}$
E_N, $4 \leq N \leq 8$	$\frac{2^{-N/2}}{\sqrt{9-N}}$	$\frac{2}{\sqrt[N]{9-N}}$
Γ_N, $N \geq 8$, $N \equiv 0 (mod\ 4)$	$2^{-N/2}$	2
Λ_{24}	1	4

A.4.3. Asymptotic results

We give the best known to us asymptotic bounds for the density of sphere packings in $\mathbb{R}^N$ for $N \longrightarrow \infty$ (cf. Sections 5.1.3 and 5.2.2):

$$
\begin{array}{ccccc}
\tilde{\lambda}_T^{pol} \le 2.30 & \longrightarrow & \tilde{\lambda}_T^{exp} \le 1.30 & \longrightarrow & \tilde{\lambda}_T \le 1.30 \\
\downarrow & & \downarrow & & \downarrow \\
\tilde{\lambda}_\ell^{pol} \le 2.30 & \longrightarrow & \tilde{\lambda}_\ell^{exp} \le 1 & \longrightarrow & \tilde{\lambda}_\ell \le 1 \\
\downarrow & & \downarrow & & \downarrow \\
\tilde{\lambda}^{pol} \le 1.31 & \longrightarrow & \tilde{\lambda}^{exp} \le 1 & \longrightarrow & \tilde{\lambda} \le 1 \\
 & & & & \tilde{\lambda} \ge 0.599
\end{array}
$$

Here $\longrightarrow$ means $\ge$, and

$$\tilde{\lambda} = \liminf_{\{P_N\}} \lambda(\{P_N\}) = - \limsup_{\{P_N\}} \left(\frac{1}{N} \cdot \log_2 \Delta(\{P_N\})\right) ,$$

where the limit is taken over all families of packings $\{P_N\}$ with $N \longrightarrow \infty$;

$\tilde{\lambda}_\ell$ is defined in the same way for lattice packings;

$\tilde{\lambda}^{pol}$ is defined in the same way for families with polynomial construction complexity;

$\tilde{\lambda}_{\ell}^{pol}$ is the same for polynomial families of lattices;

$\tilde{\lambda}^{exp}$ and $\tilde{\lambda}_{\ell}^{exp}$ are the same for exponential families;

$\tilde{\lambda}_{T}$, $\tilde{\lambda}_{T}^{pol}$ and $\tilde{\lambda}_{T}^{exp}$ are the same for T-lattices.

The bound $\tilde{\lambda} \geq 0.599$ is the Kabatiansky-Levenstein bound; the bound $\tilde{\lambda} \leq \tilde{\lambda}_{\ell} \leq 1$ is the Minkowski bound; the bound $\tilde{\lambda}_{\ell}^{exp} \leq 1$ is due to Rush; all other bounds are due to Litsyn and Tsfasman.

Recall that for additive number field lattices we have $\lambda \lesssim 2.22$ (Tsfasman) and for multiplicative algebraic-geometric congruence lattices we have $\lambda \lesssim 1.39$ (Rosenbloom and Tsfasman).

Bibliography

It is quite impossible to give a comprehensive list of books and papers on all topics discussed in the book. We have taken the following way: here is the list of nearly everything we know on algebraic-geometric codes and a few titles on algebraic geometry, number theory, codes, and packings. Further references on codes are listed in [MW/Sl], on packings in [Co/Sl], on algebraic geometry in [Ha].

[Aa 1] **M.J.Aaltonen**, *Notes on the asymptotic behaviour of the information rate of block codes*, IEEE Trans.Info.Theory, IT-30 (1984), 84-85.

[Aa 2] **M.J.Aaltonen**, *A new upper bound on non-binary block codes*, in: Studies in honor of A.Tietäväinen (1987), 7-39.

[Ab] *Abrégé d'histoire des Mathématiques,1700-1900 (sous la direction de J.Dieudonné)*, v. 1-2, Hermann, Paris, 1979.

[Ah/Ho/Ul] **A.Aho, J.Hopcroft, J.Ullman**, *The design and analysis of computing algorithms*, Addison-Wesley, Reading, MA, 1974.

[At] **A.O.L.Atkin**, *Weierstrass points at cusps of* $\Gamma_0(N)$, Ann.Math., 85:1 (1967), 42-45.

[Ba/Ka/Ts] **A.M.Barg, G.L.Katsman, M.A.Tsfasman,** *Algebraic-geometric codes on curves of small genus*, Probl.Info.Trans., 23 (1987), 34-38.

[Ber 1] *Key papers in the development of coding theory (E.R.Berlekamp, editor)*, IEEE Press, NY, 1974.

[Ber 2] **E.R.Berlekamp**, *Algebraic coding theory*, Mc Graw-Hill, NY, 1968.

[Bet] **Th.Beth**, *Some aspects of coding theory between probability, algebra, combinatorics and complexity theory*, In: Combinatorial Theory, Lect.Notes Math., 969 (1982), 12-29.

[Bl/Z] **E.L.Blokh, V.V.Zyablov**, *Linear concatenated codes*, Nauka, Moscow, 1982 (in Russian).

[Bor/Shf] **Z.I.Borevich, I.R.Shafarevich**, *Number theory*, Acad. Press, NY, 1966.

[Bos] **A.Bos**, *Sphere packings in high dimensional space*, Preprint, 1980.

[Bos/Co/Sl] **A.Bos, J.H.Conway, N.J.A.Sloane**, *Further lattice packings in high dimensions*, Mathematika, 29:2 (1982), 171-180.

[Brou] **M.Broué**, *Codes correcteurs d'erreur auto-orthogonaux sur le corps à deux éléments et formes quadratiques entières définies positives à discriminant +1* , Discrete Math., 17 (1977), 247-269.

[Car] **P.Cartier**, *Une nouvelle opération sur les formes différentieles*, C.R. Acad. Sci. Paris, 244 (1957), 476-478.

[Cas] **J.W.S.Cassels**, *An introduction to the geometry of numbers*, Springer, Berlin e.a., 1959.

[Cheb] **N.G.Chebotarev**, *Introduction to the theory of algebraic functions of one variable*, Moscow, 1948 (in Russian).

[Chev] **C.Chévalley**, *Introduction to the theory of algebraic functions of one variable*, Math.Surv., n°VI, AMS, NY, 1951.

[Co/Sl] **J.H.Conway, N.J.A.Sloane**, *Sphere packings, lattices and groups*, Springer, NY, 1988.

[De/Hu] **P.Deligne, D.Husemoller**, *Drinfeld modular curves*, Contemp.Math., 67 (1987), 25-91.

[De/Ra] **P.Deligne, M.Rapoport**, *Schémas des modules de courbes elliptiques*, Lect. Notes Math., 349 (1973), 163-315.

[Deu] **M.Deuring**, *Die Typen der Multiplikatorenringe elliptischer Funktionenkörper*, Abh.Math.Sem.Univ.Hamburg, 14 (1941), 197-272.

[Die] **J.Dieudonné**, *Cours de Géométrie Algébrique, I, Aperçu Historique sur le Développement de la Géométrie Algébrique*, Press Univ. France, Collection Sup., 1974.

[Dr 1] **Y.Driencourt**, *Une exemple de codes géométriques: les codes elliptique*, Traitement du Signal, 4:2 (1987), 147-153.

[Dr 2] **Y.Driencourt**, *Some properties of elliptic codes over a field of characteristic* 2 , Lecture Notes in Comp.Sc., 229 (1985).

[Dr/Mi 1] **Y.Driencourt, J.F.Michon**, *Elliptic codes over fields of characteristic* 2, J. Pure Appl. Algebra, 45 (1987), 15-39.

[Dr/Mi 2] **Y.Driencourt, J.F.Michon**, *Remarques sur les codes géométriques*, C.R.Acad.Sci. Paris, 301 (1985), 15-17.

[Dr/Mi 3] **Y.Driencourt, J.F.Michon**, *Rapport sur les codes géométriques*, Preprint, 1986.

[Dr/St] **Y.Driencourt, H. Stichtenoth**, *A criterion for self-duality of geometric codes*, Commun. Alg., 17:4 (1989), 885-898.

[Drf 1] **V.G.Drinfeld,** *Elliptic modules, I,II.* Math. USSR Sbornik, 23:4 (1974), 561 - 592; 31:2 (1977), 159 - 170.

[Drf 2] **V.G.Drinfeld,** *Unpublished manuscript on the Langlands conjecture for* $GL(2)$, 1985 (in Russian).

[Drf/Vl] **S.G.Vladuţ, V.G.Drinfeld,** *Number of points of an algebraic curve,* Func.Anal. 17 (1983), 53-54.

[El 1] **N.D.Elkies,** *Letters to N.J.A.Sloane,* Aug.15, Sept.15, 1989.

[El 2] **N.D.Elkies,** *On Mordell-Weil lattices,* 29. Mathematische Arbeitstagung, Juni 1990.

[Er/Zi 1] **T.Ericson, V.A.Zinoviev,** *An improvement of the Gilbert bound for constant weight codes,* IEEE Trans. Info. Theory, IT-33:5 (1987), 721-722.

[Er/Zi 2] **V.A.Zinoviev, T.Ericson,** *On concatenated constant weight codes ameliorating the Varshamov-Gilbert bound,* Probl. Peredachi Inform., 23:1 (1987), 110-111 (in Russian).

[Fo] **G.D.Forney, Jr.,** *Concatenated codes,* Cambridge, MA, 1966.

[FT] **L.Fejes Tóth,** *Lägerungen in der Ebene, auf der Kugel und in Raum,* Springer-Verlag, 2nd ed., 1972.

[Fu] **W.Fulton,** *Algebraic curves,* Benjamin, N.Y., 1969.

[Gee/Lin] **J.H.van Lint, G.van der Geer,** *Introduction to coding theory and algebraic geometry,* Birkhäuser, 1988.

[Gee/Scf/Vlu] **G.van der Geer, R.Schoof, M.van der Vlugt,** *Weight formulas for ternary Melas codes,* Preprint, 1990.

[Gee/Vlu 1] **G.van der Geer, M.van der Vlugt,** *Artin-Schreier curves and codes,* J. Algebra, to appear.

[Gee/Vlu 2] **G.van der Geer, M.van der Vlugt,** *Reed-Muller codes and supersingular curves,* I, Preprint, 1990.

[Gol/Shf] **E.S.Golod, I.R.Shafarevich,** *On class field towers,* Amer. Math. Soc. Transl. (2) 48 (1965), 91-102.

[Go 1] **V.G.Goppa,** *Codes on algebraic curves,* Soviet Math. Dokl, 24 (1981), 170-172.

[Go 2] **V.G.Goppa,** *Algebraico-geometric codes,* Math. USSR Izvestiya, 21 (1983), 75-91.

[Go 3] **V.D.Goppa,** *Codes and information,* Russ. Math. Surveys, 39:1 (1984), 87-141.

[Go 4] **V.D.Goppa,** *Geometry and codes,* Kluwer Acad. Publ., 1988.

[Gos] **D.Goss,** *The algebraist's upper half-plane,* Bull Amer. Math. Soc., 2 (1980), 391-415.

[Gr/Ha] **P.A.Griffiths, J.Harris,** *Principles of algebraic geometry,* Wiley, NY, 1978.

[Grs] **B.Gross,** *Group representations and lattices,* Preprint, 1990.

[Gro] **A.Grothendieck** (with collaboration of **J.Dieudonné**), *Éléments de géométrie algébrique,* Publ.Math. IHES, 4 (1960), 8, 11 (1961), 17 (1963), 20 (1964), 24 (1965), 28 (1966), 32 (1967).

[Han 1] **J.P.Hansen,** *Codes on the Klein quartic, ideals and decoding,* IEEE Trans.Info.Theory, IT-33 (1987), 923-925.

[Han 2] **J.P.Hansen,** *Group codes on algebraic curves,* Mathematica Göttingensis, Heft 9 (Feb. 1987).

[Han/St] **J.P.Hansen, H.Stichtenoth,** *Group codes on algebraic curves associated to the Sylow-2-subgroups of the Suzuki groups,* Aarchus Univ., Preprint series, 1988/89, n^{0}7.

[Hart] **R.Hartshorne,** *Algebraic geometry,* Springer, NY, 1977.

[Hi] **J.W.P.Hirschfeld,** *Linear codes and algebraic curves,* in: Geometrical Combinatorics, Pitman, Boston, 1984, 35-53.

[Ih] **Y.Ihara**, *Some remarks on the number of rational points of algebraic curves over finite fields*, J. Fac. Sci. Tokyo, IA, 28 (1981), 721-724.

[Jan] **H.Janwa**, *Some optimal codes from algebraic geometry and their covering radii*, Europ.J.Combinatorics, to appear.

[Ju/al] **J.Justesen, K.J.Larsen, H.Elbrønd Jensen, A.Havemøse T.Høholdt**, *Construction and decoding of a class of algebraic geometry codes*, IEEE Trans. Info. Theory, IT-35 (1989), 811-821.

[Kab/Lev] **G.A.Kabatiansky, V.I.Levenshtein**, *Bounds for packings on a sphere and in space*, Probl. Info. Trans., 14:1 (1978), 1-17.

[Ka/Ts 1] **G.L.Katsman, M.A.Tsfasman**, *Spectra of algebraic-geometric codes*, Probl. Info. Trans., 23 (1987), 262-275.

[Ka/Ts 2] **G.L.Katsman, M.A.Tsfasman**, *A remark on algebraic-geometric codes*, Contemp Math., 93 (1989), 197-199.

[Ka/Ts/Vl 1] **G.L.Katsman, M.A.Tsfasman, S.G.Vladuţ**, *Modular curves and codes with a polynomial construction*, IEEE Trans. Info. Theory, IT-30 (1984), 353-355.

[Ka/Ts/Vl 2] **S.G.Vladuţ, G.L.Katsman, M.A.Tsfasman**, *Modular curves and codes with polynomial construction complexity*, Probl. Info. Trans. 20 (1984), 35-42.

[Kat/Liv] **N.Katz, R.Livné**, *Sommes de Kloosterman et courbes elliptiques universelles en caracteristiques 2 et 3*, C.R. Acad. Sci. Paris, sér.1, 309 (1989), 723-726.

[Kat/Maz] **N.Katz, B.Mazur**, *Arithmetic moduli of elliptic curves*, Princeton, 1985.

[Ko] **N.Koblitz**, *Introduction to elliptic curves and modular forms*, Springer, NY, 1984.

[Koch] **H.V.Koch**, *Unimodular lattices and self-dual codes*, Proc. Int. Congr. Math., Berkeley, Cal., 1986, 457-465.

[Kr] **V.Yu.Krachkovskii**, *Decoding of codes on algebraic curves*, Preprint, 1988 (in Russian).

[La 1] **G.Lachaud**, *Les codes géométriques de Goppa*, Sém.Bourbaki, n^{0}641 (Fev. 1985), 1-19.

[La 2] **G.Lachaud**, *Sommes d'Eisenstein et nombre de points de certaines courbes algébriques sur les corps finis*, C.R. Acad. Sci. Paris, 305 (1987), 729-732.

[La 3] **G.Lachaud**, *Exponential sums, algebraic curves, and linear codes*, J. Number Theory, to appear.

[La 4] **G.Lachaud**, *Distribution of the weights of the dual code of the Melas code*, Discr. Math., 79 (1990), 103-106.

[La 5] **G.Lachaud**, *Exponential sums and the Carlitz-Uchiyama bound*, Lect. Notes Comp. Sci., 388 (1989), 63-75.

[La 6] **G.Lachaud**, *The parameters of projective Reed-Muller codes*, Discr. Math., 81 (1990), 217-220.

[La/MD] **G.Lachaud, M.Martin-Deschamps**, *Nombres de points des jacobiennes sur un corps finis*, Acta Arithm., 1991, to appear.

[La/Wo 1] **G.Lachaud, J.Wolfmann**, *Sommes de Kloosterman, courbes elliptiques et codes cycliques en caracteristique 2* , C.R. Acad. Sci. Paris, sér. 1, 305 (1987), 881-883.

[La/Wo 2] **G.Lachaud, J.Wolfmann**, *The weights of the orthogonals of the extended quadratic binary Goppa codes*, IEEE Trans. Info. Theory, 1990, to appear.

[Lan 1] **S.Lang**, *Elliptic functions*, Addison-Wesley, Reading, 1973.

[Lan 2] **S.Lang**, *Algebra*, Addison-Wesley, Reading, 1984.

[Lan 3] **S.Lang**, *Introduction to modular forms*, Springer, NY, 1976.

[Lan 4] **S.Lang,** *Algebraic number theory*, Addison-Wesley, Reading, 1970.

[LB/Ri] **D.Le Brigand, J.-J.Risler,** *Algorithme de Brill-Noether et construction de codes de Goppa*, Bull. Soc. Math. France, 116 (1988), 231-253.

[Lee 1] **J.Leech,** *Some sphere packings in higher space*, Canad. J. Math., 16 (1964), 657-682.

[Lee 2] **J.Leech,** *Notes on sphere packings*, Canad. J. Math., 19 (1967), 251-267.

[Len] **H.W.Lenstra, Jr.,** *Codes from algebraic number fields*, In: Fundamental contributions in the Netherlands since 1945, North-Holand, Amsterdam, II (1986), 95-104.

[Lev] **V.I.Levenshtein,** *Bounds for packing in metric spaces and certain applications*, Probl. Kibernetiki, 40 (1983), 44-110.

[Lin] **J.H.van Lint,** *Introduction to coding theory*, Grad. Texts Math., 86, Springer, Berlin, 1982.

[Lin/Spr] **J.H.van Lint, T.A.Springer,** *Generalized Reed-Solomon codes from algebraic geometry*, IEEE Trans. Info. Theory, IT-33 (1987), 305-309.

[Li/Ts 1] **S.N.Litsyn, M.A.Tsfasman,** *A note on lower bounds*, IEEE Trans. Info. Theory, IT-32 (1986), 705-706.

[Li/Ts 2] **S.N.Litsyn, M.A.Tsfasman,** *Algebraic-geometric and number-theoretic packing of spheres in* $\mathbb{R}^N$, Russian Math. Surveys, 40 (1985), 219-220.

[Li/Ts 3] **S.N.Litsyn, M.A.Tsfasman,** *Constructive high-dimensional sphere packing*, Duke Math. J., 54 (1987), 147-161.

[Li/Zi] **V.A.Zinoviev, S.N.Litsyn,** *Codes that exceed the Gilbert bound*, Probl. Peredachi Inform., 21:1 (1985), 105-108 (in Russian).

[MW/Sl] **F.J.MacWilliams, N.J.A.Sloane**, *The theory of error-correcting codes*, North-Holland, Amsterdam, 1977.

[Ma] **Yu.I.Manin**, *What is the maximum number of points on a curve over* $\mathbb{F}_2$ *?*, J. Fac. Sci. Tokio, IA, 28 (1981), 715-720.

[Ma/Vl] **Yu.I.Manin, S.G.Vladuţ**, *Linear codes and modular curves*, J. Soviet. Math., 30 (1985), 2611-2643.

[Mar] **J.Martinet**, *Tours de corps de classes et estimation de discriminants*, Invent. Math., 44 (1978), 65-73.

[Mi 1] **J.F.Michon**, *Amélioration des paramèters des codes de Goppa*, Preprint, 1986.

[Mi 2] **J.F.Michon**, *Codes de Goppa*, Sém. Théorie Nombres Bordeaux (1983-1984), n^{0}7, 1-17.

[Mi 3] **J.F.Michon**, *Les codes BCH comme codes géométriques*, Preprint, 1985.

[Mil] **J.Milnor**, *Hilbert's problem 18: on crystallographic groups, fundamental domains, and on sphere packing*, Proc. Symp. Pure Math., AMS, Providence, RI, 28 (1976), 491-506.

[Mil/Hu] **J.Milnor, D.Husemoller**, *Symmetric bilinear forms*, Springer-Verlag, 1973.

[MF 1-6] *Modular functions of one variable*, I, II, III, IV, V, VI, Lect. Notes Math., 340, 349, 350 (1973), 476 (1975), 601, 626 (1977).

[Mor 1] **C.J.Moreno**, *Goppa codes and modular curves*, Preprint, 1985.

[Mor 2] **C.J.Moreno**, *Algebraic curves over finite fields with applications to coding theory*, Cambridge Univ. Press, 1990.

[Mor/Mor 1] **C.J.Moreno, O.Moreno**, *Exponential sums and Goppa codes*, I, II, III, IV, Preprints, 1988-1989.

[Mor/Mor 2] **C.J.Moreno, O.Moreno,** *An improved Carlitz-Uchiyama bound in characteristic two*, C.R. Acad. Sci. Paris, 1990, to appear.

[Mor/Mor 3] **C.J.Moreno, O.Moreno,** *An improved Bombieri-Weil bound in characteristic two and applications to coding theory*, J. Number Theory, to appear.

[Ogg] **A.Ogg,** *Hyperelliptic modular curves*, Bull. Soc. Math. France, 102 (1974), 449-462.

[Pel 1] **R.Pellikaan,** *On decoding linear codes by error correcting pairs*, Preprint, 1988.

[Pel 2] **R.Pellikaan,** *On a decoding algorithm for codes on maximal curves*, IEEE Trans. Info. Theory, to appear.

[Pel/She/Wee] **R.Pellikaan, B.Z.Shen, G.J.M.van Wee,** *On representing linear codes by curves*, Preprint, 1990.

[Per 1] **M.Perret,** *Sur le nombre de points d'une courbe sur un corps finis; application aux codes correcteur d'erreurs*, C.R. Acad. Sci. Paris, sér.1, 309 (1989), 177-182.

[Per 2] **M.Perret,** *Tours ramifiées infinies de corps de classes*, J. Number Theory, to appear.

[Pe/We] **W.W.Peterson, E.J.Weldon,** *Error-correcting codes*, MIT Press, Cambridge, Mass., 1972.

[Poi] **G.Poitou,** *Minorations de discriminants (d'apres A.M. Odlyzko)*, Sém.Bourbaki 1975/76, exp. 479, Lect. Notes Math., 567 (1977), 136-153.

[Por] **S.C.Porter,** *Decoding geometric Goppa codes*, IEEE Trans. Info. Theory, to appear.

[Que 1] **H.-G.Quebbemann,** *Cyclotomic Goppa codes*, IEEE Trans. Info. Theory, IT-34:5 (1988), 1317-1320.

[Que 2] **H.-G.Quebbemann,** *Lattices from curves over finite fields*, Preprint, 1989.

[Que 3] **H.-G.Quebbemann**, *On even codes*, Preprint, 1988.

[Que 4] **H.-G.Quebbemann**, *Estimates of regulators and class numbers in function fields*, Preprint, 1990.

[Rog] **C.A.Rogers**. *Packing and covering.*- Camb. Univ. Press, 1964.

[Rol] **R.Rolland**, *On hypersurfaces over finite fields and the parameters of the projective Reed-Muller codes*, Preprint, 1990.

[Roq] **P.Roquette**, *Abschätzung der Automorphismenzahl von Funktionenkörpern bei Primzahlcharacteristik*, Math. Zeit., 117 (1970), 157-163.

[Ro/Ts] **M.Yu.Rosenbloom, M.A.Tsfasman**, *Multiplicative lattices in global fields*, Invent. Math., 1990.

[Rüc] **H.-G.Rück**, *On Goppa codes defined by Kummer and Artin-Schreier extensions*, Preprint, 1989.

[Ru] **J.A.Rush**, *A lower bound on packing density*, Invent. Math., 98 (1989), 499-509.

[Ru/Sl] **J.A.Rush, N.J.A.Sloane**, *An improvement to the Minkowoki-Hlavka bound for packing superballs*, Mathematika, 34 (1987), 8-18.

[Ry/Bar] **S.S.Ryskov, E.P.Baranovskii**, *Classical methods in the theory of lattice packing*, Russian Math. Surv., 34:4 (1979), 1-68.

[Sch] **W.Scharlau**, *Selbstduale Goppa-codes*, Math. Nachr., 143 (1989), 119-122.

[Scho] **B.Schoeneberg**, *Uber die Weierstrasspunkte in den Korpern der elliptischen Modulfunktionen*, Abh. Math. Sem. Univ. Hamburg, 17 (1951), 104-111.

[Scf] **R.Schoof**, *Nonsingular plane curves over finite fields*, J. Combin. Theory, A, 46:2 (1987), 183-211.

[Scf/Vlu] **R.Schoof**, **M.van der Vlugt**, *Hecke operators and the weight distribution of certain codes*, J. Comb. Theory, A, to appear.

[Se 1] **J.-P.Serre**, *Sur le nombre des points rationnels d'une courbe algebrique sur un corps fini*, C. R. Acad. Sci. Paris, sér. 1, 296 (1983), 397-402. = Oeuvres, III, n^{0}128, 658-663.

[Se 2] **J.-P.Serre**, *Nombre de points de courbes algébriques sur* $\mathbb{F}_q$, Sém. Théorie Nombres Bordeaux (1982/1983), n^{0}22, 1-8. = Oeuvres, III, n^{0}129, 664-668.

[Se 3] **J.-P.Serre**, *Résumé des cours de* 1983-1984, Annuaire du Collége de France, Paris, (1985), 79-84. = Oeuvres,III, n^{0}132, 701-705.

[Se 4] **J.-P.Serre**, *The number of rational points on curves over finite fields*, Princeton lectures (Fall 1983), Notes by E. Bayer.

[Se 5] **J.-P.Serre**, *Groupes algébriques et corps de classes*, Hermann, Paris, 1959.

[Se 6] **J.-P.Serre**, *Lettre à M.Tsfasman*, Jullet 24, 1989.

[Shf] **I.R.Shafarevich**, *Basic algebraic geometry*, Springer, Berlin e.a., 1977.

[Shi] **G. Shimura**, *Introduction to the arithmetic theory of automorphic functions*, Iwanami Shoten, 1971.

[Sh 1] **T.Shioda**, *Mordell-Weil lattices and Galois representations*, I, II, III, Proc. Japan Acad., 65 (1989), 268-271, 296-299, 300-303.

[Sh 2] **T.Shioda**, *Mordell-Weil lattices and sphere packings*, Preprint, 1989.

[Sk] **A.N.Skorobogatov**, *Subfields subcodes of algebraic-geometric codes*, Probl. Peredachi Info., to appear (in Russian).

[Sk/Vl] **A.N.Skorobogatov, S.G.Vladuţ**, *On the decoding of algebraic-geometric codes*, IEEE Trans. Info. Theory, IT-36:5 (1990).

[Sl 1] **N.J.A.Sloane**, *The packing of spheres*, Scient. Amer., 250:1 (1984), 116-125.

[Sl 2] **N.J.A.Sloane**, *Sphere packings constructed from BCH and Justesen codes*, Mathematika, 19 (1972), 183-190.

[So 1] **A.B.Sørensen**, *Geometric Goppa-koder*, Master thesis, Aarhus, 1988.

[So 2] **A.B.Sørensen**, *Projective Reed-Muller codes*, Preprint, Aarhus, 1990.

[Spr] **G.Springer**, *Introduction to the theory of Riemann surfaces*, Addison-Wesley, Reading, 1957.

[St 1] **H.Stichtenoth**, *Uber die Automorphismengruppe eines algebraischen Funktionenkörpers von Primzahlcharakteris=tik*, Archiv der Math., I, II, 24 (1973), 527-544; 24 (1973), 615-631.

[St 2] **H.Stichtenoth**, *Self-dual Goppa codes*, J. Pure Appl. Algebra, 55 (1988), 199-211.

[St 3] **H.Stichtenoth**, *A note on Hermitian codes over $GF(q^2)$*, IEEE Trans. Info. Theory, IT-34:5 (1988), 1345-1348.

[St 4] **H.Stichtenoth**, *On the dimension of subfields subcodes*, IEEE Trans. Info. Theory, IT-36 (1990), 90-94.

[St 5] **H.Stichtenoth**, *On automorphisms of geometric Goppa codes*, J. Algebra, to appear.

[St 6] **H.Stichtenoth,** *Algebraic-geometric codes associated to Artin-Schreier extensions of* $\mathbb{F}_q[Z]$, Preprint, 1990.

[St/Vos] **C.Voss, H.Stichtenoth,** *Asymptotically good families of subfield subcodes of geometric Goppa codes,* Geometriae Dedicata, to appear.

[Sto/Vo] **K.O.Stöhr, J.F.Voloch,** *Weierstrass point and curves over finite fields,* Proc. London Math. Soc. (3), 52 (1986), 1-19.

[Ta] **J.Tate.** *The arithmetic of elliptic curves,* Invent. Math., 23 (1974), 179-206.

[Ti] **H.J.Tiersma,** *Remarks on codes from Hermitian curves,* IEEE Trans. Info. Theory, 33 (1987), 605-609.

[Ts 1] **M.A.Tsfasman,** *On Goppa codes which are better than the Varshamov-Gilbert bound,* Probl. Info. Trans., 18 (1982), 163-166.

[Ts 2] **M.A.Tsfasman,** *Group of points of an elliptic curve over a finite field,* In: Theory of numbers and its applications, Tbilissi (1985), 286-287.

[Ts 3] **M.A.Tsfasman,** *Group of points of an elliptic curve over a finite field,* Preprint, 1985.

[Ts 4] **M.A.Tsfasman,** *Algebraic-geometric codes and asymptotic problems,* AAECC meeting, Toulouse-1989, Discr. Appl. Math., 1990, to appear.

[Ts 5] **M.A.Tsfasman,** *Global fields, codes and sphere packings,* Journées Arithmétiques., Luminy-1989, "Asterisque", 1990, to appear.

[Ts/Vl/Z] **M.A.Tsfasman, S.G.Vladut, Th.Zink,** *Modular curves, Shimura curves, and Goppa codes, better that the Varshamov-Gilbert bound,* Math. Nachrichten, 109 (1982), 21-28.

[Vl 1] **S.G.Vladuţ**, *An exhaustion bound for algebraic-geometric modular codes*, Probl. Info. Trans., 23 (1987), 22-34.

[Vl 2] **S.G.Vladuţ**, *On the polynomiality of codes on classical modular curves*, Preprint, 1983.

[Vl 3] **S.G.Vladuţ**, *Kronecker's "Jugendtraum" and modular functions*, Gordon & Breach, 1990, to appear.

[Vl 4] **S.G.Vladuţ**, *Algebraic-geometric "modular" codes as group codes*, Preprint, 1989.

[Vl 5] **S.G.Vladuţ**, *On the decoding of algebraic-geometric codes over* $\mathbb{F}_q$ *for* $q \geq 16$, IEEE Trans. Info. Theory, to appear.

[Vo 1] **J.F.Voloch**, *Codes and curves*, Eureka, 43 (1983), 53-61.

[Vo 2] **J.F.Voloch**, *A note on elliptic curves over finite fields*, Bull. Soc. Math. France, 116 (1988), 455-458.

[Wat] **W.C.Waterhouse**, *Abelian vatieties over finite fields*, Ann. Sci. E.N.S. (4), 2 (1969), 521-560.

[We 1] **A.Weil**, *Number theory: an approach through history. From Hammurapi to Legendre*, Birkhäuser, 1981.

[We 2] **A.Weil**, *Sur les courbes algébriques et les variétés qui s'en déduisent*, Hermann, Paris, 1948.

[We 3] **A.Weil**, *Variétés abéliennes et courbes algébriques*, Hermann, Paris, 1948.

[We 4] **A.Weil**, *Basic number theory*, Springer, Berlin, 1967.

[We 5] **A.Weil**, *Sur les "formules explicites" de la théorie des nombres premiers*, Comm. Lund (1952), 252-265. = Oeuvres Sc., II, 48-61.

[Wi] **M.Wirtz**, *On the parameters of Goppa codes*, IEEE Trans. Info. Theory, IT-34:5 (1988), 1341-1343.

[Wo 1] **J.Wolfmann**, *Nombre de points rationnels de courbes algébriques sur des corps finis associées à des codes cycliques*, C. R. Acad. Sci. Paris, sér 1, 305 (1987), 345-348.

[Wo 2] **J.Wolfmann**, *The weights of the dual code to the Melas code over GF*(3) , Discr. Math., 74 (1989), 327-329.

[Wo 3] **J.Wolfmann**, *The number of points on certain algebraic curves over finite fields*, Commun. Algebra, 17 (1989), 2055-2060.

[Wo 4] **J.Wolfmann**, *New bounds on cyclic codes from algebraic curves*, Lect. Notes Comp. Sci., 388 (1989), 47-62.

[Z] **T.Zink**, *Degeneration of Shimura surfaces and a problem in coding theory*, Lect. Notes Comp. Sci. 199 (1986), 503-511.

Author Index

Index

List of symbols

The list of notation used in our book is divided into the following parts:

A. General notation.

B. Coding theory.

C. Algebraic geometry.

D. AG-codes.

E. Modular curves.

F. Sphere packings.

G. Number theory.

Section B corresponds to Part 1, Section C - to Part 2, Section D - to Part 3, Section E - to Part 4, Sections F and G - to Part 5. All notation except that of Section A is defined in the section where it appears for the first time.

A. General notation

$A \subset B$ proper subset ($A \neq B$);

$A \subseteq B$ subset (the case $A = B$ is not excluded);

$A \hookrightarrow B$ injective map;

$\emptyset$ empty set;

$\|M\|$	cardinality of a set;
$f \circ g$	composition of maps;
$Im\ \varphi$	image of a map;
$Ker\ \varphi$	kernel of a map;
$\mathbb{N}$	set of all non-negative integers;
$\mathbb{Z}$	ring of integers;
$\mathbb{Q}$	field of rational numbers;
$\mathbb{R}$	field of real numbers;
$\mathbb{C}$	field of complex numbers;
$\mathbb{Q}_p$	field of p-adic numbers;
$\mathbb{F}_q$	finite field of $q = p^a$ elements;
$\mathbb{Z}/n$	ring of residues modulo n ;
$\lceil a \rceil$	integer part;
$\rceil a \lceil$	upper integer part;
$Re\ z$	real part of $z \in \mathbb{C}$;
$Im\ z$	complex part of $z \in \mathbb{C}$;
$\binom{n}{m}$	binomial coefficient;
$\varphi(n)$	Euler function;
$\mu(n)$	Möbius function;
$\liminf$	lower limit of a real sequence;
$\limsup$	upper limit of a real sequence;
L/K	field extension;
$[L:K]$	degree of an extension;
$Gal(L/K)$	Galois group of an extension;
$L \cdot K$	composite of fields;
$\overline{k}$	algebraic closure of a field;

$char\ k$ characteristic of a field;

$N_{L/K}(a)$ norm of $a \in L$ in the extension L/K ;

$Tr_{L/K}(a)$ trace of $a \in L$ in L/K ;

$k[T_1,\ldots,T_n]$ ring of polynomials in n variables;

$k(T_1,\ldots,T_n)$ field of rational functions in n variables;

$k[[T_1,\ldots,T_n]]$ ring of power series in n variables;

$k((T_1,\ldots,T_n))$ field of Laurent series in n variables;

$deg\ f$ degree of a polynomial;

A^* multiplicative group of invertible elements in a ring;

$\mathbb{F}_q^*$ multiplicative group of a finite field;

$GL(n,A)$ general linear group of order n over a ring;

$M_n(A)$ algebra of $(n \times n)$-matrices over a ring;

V^* dual linear space;

$dim_k V$ dimension of a linear space over a field;

$V_1 \oplus V_2$ direct sum of spaces;

$V_1 \otimes_k V_2$ tensor product of spaces;

V^n n-th power of a space (direct sums of n copies);

$V^{\otimes n}$ n-th tensor power (tensor product of n copies);

$S^n V$ n-th symmetric power;

$det\ L$ determinant of a linear operator;

G/H factor-group;

$[G:H]$ index of a subgroup;

$\langle g \rangle$ cyclic group generated by g ;

G_x isotropy group (stabilizer);

$k[G]$ group ring;

ζ_n primitive n-th root of unity;

μ_n group of n-th roots of unity;

μ_k group of roots of unity in a field;

χ addive character.

B. Coding theory

Section 1.1.1: A ; $d(a,b)$; $q = |A|$; $C \subseteq A^n$; $M = |C|$; k ; d ; $[n,k,d]_q$; $R = k/n$; $\delta = d/n$; $\|a\|$; H .

Section 1.1.2. $\mathcal{P}$; $m_q(k)$.

Section 1.1.3. $A_r(C)$; $W_C(x:y)$; $W_C(x)$; B_i ; $C^\perp$; g ; $y \cdot C$.

Section 1.2.1. $P_i(x)$; $Ev_{\mathcal{P}}$; $Res_P F$; $Res_{\mathcal{P}}$; $s_j(v)$.

Section 1.2.2. C_{23} , C_{24} , C_{11} , C_{12} .

Section 1.2.3. $C_1 \otimes C_2$; C^ℓ ; $C_1 \circledast C_2$; $C^{\otimes \ell}$; $(C_1|C_2)$.

Section 1.3.1. V_q ; U_q ; α_q ; V_q^{lin} ; U_q^{lin} ; α_q^{lin} .

Section 1.3.2. R_P ; H_q ; R_H ; R_{BE} ; R_L ; R_4 ; $R_{4(2)}$; R_{GV} .

Section 1.3.4. U_q^{pol} ; $U_q^{pol,lin}$; α_q^{pol} ; $\alpha_q^{pol,lin}$; R_Z ; R_{BZ} .

Section 1.3.5. κ_q ; κ_q^{lin} ; δ_q ; δ_q^{lin} ; d_φ .

C. Algebraic geometry

Section 2.1.1. $\mathbb{A}^n$; $\mathbb{P}^n$; P , Q ; $\mathbb{P}(V)$; X ; U ; X_i ; $\dim X$; $\mathrm{codim}_X Y$; $k(X)$; $k[X]$; $f : X \longrightarrow Y$; U_f ; $f^*(g)$; $\deg f$; I_X ; $\mathcal{O}_P$; m_P ; Θ_P ; X_{smooth} .

Section 2.1.2. $X \times Y$; t_P ; $\frac{\partial F}{\partial x}(P)$; $\tau_P(f)$; $ord_P(f)$.

Section 2.1.3. D ; $\deg D$; $\mathrm{Supp}\, D$; $\mathrm{Div}(X)$; $\mathrm{Div}^0(X)$; $\mathrm{Div}^+(X)$; (f) ; $(f)_0$; $(f)_\infty$; (F) ; $P(X)$; $\mathrm{Cl}(X) = \mathrm{Pic}(X)$; $\mathrm{Cl}^0(X) = \mathrm{Pic}^0(X)$; $L(D)$; $\ell(D)$; $|M|$; φ_D ; L_P ; (L_P) ; $H^0(L)$; D_s ; $\mathcal{L}, \mathcal{M}, \mathcal{N}$; $\mathcal{O}(D)$; $\mathcal{L}\otimes\mathcal{M}$; $H^0(\mathcal{L})$; $h^0(\mathcal{L})$; $\overline{\mathcal{L}}_P$; $(\{U_i\},\{f_i\})$; $\varphi^*(D)$.

Section 2.1.4. $GL(n)$; $\mathbb{G}_m$; L_h ; J_X ; $g(X)$.

Section 2.1.5. X_{an} .

Section 2.2.1. df ; $\Omega[X]$; $\Omega(X)$; ω ; K_X ; $\Omega(D)$; $\varphi^*(\omega)$; $\mathrm{Aut}_k(X)$; X/H , $X/\langle g\rangle$, X^g .

Section 2.2.2. $\mathrm{Res}_P(\omega)$; φ_{mK} .

Section 2.2.3. B_f ; e_P ; F .

Section 2.2.4. $w(P)$.

Section 2.2.5. C ; C_q .

Section 2.3.1. X/k ; $X(k)$; $\mathrm{Hom}_k(X,Y)$.

Section 2.3.2. $N_r(X)$; $Z(X,t)$; $P(t)$; ω_i ; $M(X)$; $N_q(g)$; B_s .

Section 2.3.3. $A(q)$.

Section 2.4.1. λ .

Section 2.4.2. $j(E)$.

Section 2.4.3. f^t ; n_E ; V ; $\mathrm{Hom}(E,E')$; $\mathrm{End}(E)$; $\mathrm{End}^0(E)$; $H_{p,\infty}$.

Section 2.4.4. $\mathrm{End}_{\mathbb{F}_q}(E)$, Frob .

Section 2.4.5. Λ ; ω_1,ω_2 ; $\tau = \omega_2/\omega_1$; $\mathcal{P}(z)$; g_2 , g_3 .

Section 2.5.1. $\overline{A}_L$; x^ν .

Section 2.5.2. c_P ; n_P , δ_p ; D .

Section 2.5.3. $p_a(X)$.

Section 2.5.4. σ ; C' ; r_Q ; Γ_P .

Section 2.6.2. $\mathrm{Max}(A)$; $\mathrm{Spec}\, A$; $k(\mathfrak{p})$.

Section 2.6.3. $F(U)$; ρ_{UV} .

Section 2.6.4. $\mathcal{O}_X$; X/S ; $X\times_S Y$.

Section 2.6.5. (Sch); (Sch/S); $(Ens) = (Sets)$; $H_{X/S}$.

D. AG-codes

Section 3.1.1. $(X,\mathcal{P},D)_L$; $(X,\mathcal{P},D)_\Omega$; k_C; d_C; $(X,\mathcal{P},\mathcal{L})_H$.

Section 3.2.2. $M(n,\mathcal{P},h,a)$; $M(H,h,a)$; $M(H,a)$.

Section 3.3.2. $S(H)$; $s(H)$; $s_0(P)$.

Section 3.4.1. γ_q ; R_{TVZ} ; R_V ; R_{KTV} ; R_{KT} .

Section 3.4.4. R_{LT} ; $\tilde{R}_{SV}$; R'_{SV} ; R''_{SV} ; R_{EZ} .

E. Modular curves

Section 4.1.1. $SL(2,\mathbb{Z})$; $PSL(2,\mathbb{Z})$; $\Gamma(1)$; $\Gamma(N)$; $\Gamma_0(N)$; Γ ; $SL(2,\mathbb{Z}/N)$; $PSL(2,\mathbb{Z}/N)$; E_Z ; Y_Γ ; X_Γ ; $Y(N)$; $X(N)$; $Y_0(N)$; $X_0(N)$; w_m ; μ_N ; g_N ; $g_0(N)$; q ; c_m ; j_N ; $\Phi_N(x,j)$.

Section 4.1.2. $e(s,t)$; α_N ; E/S ; (Ell/S) ; $E(S)$; $Y(N)/\mathbb{Z}[1/N]$; $A_N = \mathbb{Z}[\zeta_N, 1/N]$; $Y_P(N)/\mathbb{Z}[1/N]$; $Y_0(N)/\mathbb{Z}[1/N]$; $B_N = K_N \cap A_N$; $X(N)/\mathbb{Z}[1/N]$; $X_P(N)/\mathbb{Z}[1/N]$; $X_0(N)/\mathbb{Z}[1/N]$; $X_N = X_0(N)/p$; $X'_N = X(N)/\mathfrak{p}$.

Section 4.1.3. $s(p)$.

Chapter 4.2. C ; ∞ ; A ; $|a|_\infty$; k ; k_∞ .

Section 4.2.1. "char" K ; $K\{\tau\}$; φ ; $Hom(\varphi,\psi)$; $End(\varphi)$; $Aut_K(\varphi)$; $j(\varphi)$; $G_\varphi(B)$; $T(\varphi)$.

Section 4.2.2. G_I ; λ ; $M(I)$; $\overline{M(I)}$; $Cusps(I)$; $\Gamma(I)$; $X(I)$; $\Gamma_0(I)$; $\Gamma_1(I)$; $H(I)$; $X_0(I)$, $X_1(I)$, $C(I)$; $Cusps_0(I)$; $Cusps_1(I)$; $F_I(x,y)$; $\overline{X_1(I)}$; $N(I)$; $g(I)$; $N_0(I)$; $g_0(I)$; $Cusps(I)$.

F. Packings

Section 5.1.1. P ; $\Delta(P)$; L ; $\det L$; $\operatorname{discr} L$; $d(L)$; V_N ; $\delta(P)$; $\lambda(P)$; $\nu(P)$; $\Delta(N)$; $\lambda(N)$; $\Delta_\ell(N)$; $\lambda_\ell(N)$; $L^\perp$.

Section 5.1.2. A_N ; D_N ; E_N ; Γ_N ; Λ_{24} ; Co_0 .

Section 5.1.3. $\tilde{\lambda}$; $\tilde{\lambda}_\ell$; $\tilde{\lambda}^{pol}$; $\tilde{\lambda}_\ell^{pol}$; $\tilde{\lambda}^{exp}$; $\tilde{\lambda}_\ell^{exp}$.

Section 5.2.1. $\tilde{\lambda}_T$, $\tilde{\lambda}_T^{pol}$, $\tilde{\lambda}_T^{exp}$.

G. Number theory

Section 5.3.1. k ; O_k ; O ; s, t; D_k ; R_k ; Cl_k ; h_k .

Section 5.3.2. $D_{K/k}$; $\mathfrak{D}_{K/k}$; K_1 .

Section 5.3.3. O_S ; O_F .

Section 5.4.1. S_f ; S_∞ ; Ev_S .

Section 5.4.3. $\gamma(q,r)$.